PROGRESS IN
Nucleic Acid Research and Molecular Biology

Volume 68

PROGRESS IN
Nucleic Acid Research and Molecular Biology

Base Excision Repair

edited by

KIVIE MOLDAVE

Department of Molecular Biology and Biochemistry
University of California, Irvine
Irvine, California

guest editors

SANKAR MITRA
Sealy Center for Molecular Science
University of Texas Medical Branch
Galveston, Texas

AMANDA McCULLOUGH
Sealy Center for Molecular Science
University of Texas Medical Branch
Galveston, Texas

R. STEPHEN LLOYD
Sealy Center for Molecular Science
University of Texas Medical Branch
Galveston, Texas

SAMUEL H. WILSON
Laboratory of Structural Biology
NIEHS, NIH
Research Triangle Park, North Carolina

Volume 68

ACADEMIC PRESS
A Harcourt Science and Technology Company

San Diego   San Francisco   New York   Boston
London   Sydney   Tokyo

This book is printed on acid-free paper.

Copyright © 2001 by ACADEMIC PRESS

All Rights Reserved.
No part of this publication may be reproduced or transmitted in any form or by any means, electronic or mechanical, including photocopy, recording, or any information storage and retrieval system, without permission in writing from the Publisher.

The appearance of the code at the bottom of the first page of a chapter in this book indicates the Publisher's consent that copies of the chapter may be made for personal or internal use of specific clients. This consent is given on the condition, however, that the copier pay the stated per copy fee through the Copyright Clearance Center, Inc. (222 Rosewood Drive, Danvers, Massachusetts 01923), for copying beyond that permitted by Sections 107 or 108 of the U.S. Copyright Law. This consent does not extend to other kinds of copying, such as copying for general distribution, for advertising or promotional purposes, for creating new collective works, or for resale. Copy fees for pre-2001 chapters are as shown on the title pages. If no fee code appears on the title page, the copy fee is the same as for current chapters.
0079-6603/01 $35.00

Explicit permission from Academic Press is not required to reproduce a maximum of two figures or tables from an Academic Press chapter in another scientific or research publication provided that the material has not been credited to another source and that full credit to the Academic Press chapter is given.

Academic Press
A Harcourt Science and Technology Company
525 B Street, Suite 1900, San Diego, California 92101-4495, USA
http://www.academicpress.com

Academic Press
Harcourt Place, 32 Jamestown Road, London NW1 7BY, UK
http://www.academicpress.com

International Standard Book Number: 0-12-540068-3

PRINTED IN THE UNITED STATES OF AMERICA
01 02 03 04 05 06 SB 9 8 7 6 5 4 3 2 1

Contents

PREFACE xv

KEYNOTE: PAST, PRESENT, AND FUTURE ASPECTS OF BASE
EXCISION REPAIR.. xvii

    Thomas Lindahl

    I. Discovery of Enzymes That Recognize Endogenously
       Generated DNA Lesions... xviii
    II. Reconstitution of Mammalian Base Excision Repair xxi
    III. Gene Knockout Mice Deficient in Base Excision Repair................ xxiii
    IV. Future Developments xxvi
        References xxix

Session 1   Multiple Pathways for DNA Base Excision Repair ... 1

    Eugenia Dogliotti

The Mechanism of Switching among Multiple BER Pathways ... 3

    Eugenia Dogliotti, Paola Fortini, Barbara Pascucci, and Eleonora Parlanti

    I. Introduction 4
    II. Factors That Control the Switch β among Multiple BER Pathways 9
    III. Conclusions and Future Research 24
        References 25

Yeast Base Excision Repair: Interconnections and Networks ... 29

    Paul W. Doetsch, Natalie J. Morey, Rebecca L. Swanson, and Sue Jinks-Robertson

    I. Yeast BER Mutants Demonstrate Wild-Type Sensitivity
       to Oxidizing Agents 31

II. Simultaneous Disruption of Multiple DNA Repair and Damage
    Tolerance Pathways Is Required to Sensitize Cells to Oxidizing Agents 33
III. Spontaneous Mutator and Hyperrecombination (Hyper-rec) Phenotypes
     in Single and Double Pathway Defect Mutants 33
IV. Other Characteristics of Repair Pathway Defect Mutants 36
V. Cytotoxicity, Mutagenicity, Recombination, and Growth Studies
   Suggest a Network of Interconnected Repair and Damage
   Processing Pathways... 36
   References 38

BER, MGMT, and MMR in Defense against Alkylation-Induced Genotoxicity and Apoptosis 41

Bernd Kaina, Kirsten Ochs, Sabine Grösch, Gerhard Fritz, Jochen Lips, Maja Tomicic, Torsten Dunkern, and Markus Christmann

I. Introduction 42
II. O^6-Methylguanine Is a Decisive Cytotoxic and Genotoxic Lesion 43
III. O^6-Methylguanine Is a Major Trigger of Apoptosis 44
IV. MMR Is Required for O^6-MeG-Triggered Apoptosis and Genotoxicity ... 44
V. When and Why Are N-Alkylation Lesions Genotoxic?................... 45
VI. Modulation of BER: Transfection and Knockout 46
VII. Regulation of BER, MGMT, and MMR............................... 47
VIII. Posttranscriptional Regulation of APE 49
IX. Induction of APE and Adaptive Response 50
X. An Alternative Pathway of Repair in Pol β-Deficient Cells? 50
XI. BER Intermediates Not Repaired in Pol β-Deficient Cells
    Trigger Apoptosis 51
XII. Overexpression of MGMT in Pol β-Deficient Cells 52
     References 52

Session 2   **Gene Targeting in the Mouse for Elucidating the Role of BER** 55

Samuel H. Wilson

Mammalian DNA β-Polymerase in Base Excision Repair of Alkylation Damage 57

Robert W. Sobol and Samuel H. Wilson

I. Introduction 58
II. Consequence of a β-Pol Null Genotype in Mammalian Cells............ 59

III. 5′-Deoxyribose Phosphate Mediates MMS-Induced Cytotoxicity	63
IV. Mutagenesis and Base Excision Repair	69
V. Summary	72
References	72

Regulation of Intracellular Localization of Human MTH1, OGG1, and MYH Proteins for Repair of Oxidative DNA Damage 75

Yusaku Nakabeppu

I. Introduction	76
II. hMTH1: An Oxidized Purine Nucleoside Triphosphatase	77
III. hOGG1 for Repair of 8-Oxoguanine Paired with Cytosine in DNA	83
IV. hMYH as a Bifunctional DNA Glycosylase for 2-Hydroxyadenine and Adenine Paired with 8-Oxoguanine	85
V. Discussion	89
References	92

Repair of 8-Oxoguanine and Ogg1-Incised Apurinic Sites in a CHO Cell Line 95

Serge Boiteux and Florence Le Page

I. Introduction	96
II. Materials and Methods	97
III. Results and Discussion	98
IV. Conclusions	103
References	103

Mammalian *Ogg1/Mmh* Gene Plays a Major Role in Repair of the 8-Hydroxyguanine Lesion in DNA 107

Susumu Nishimura

I. Background	108
II. Human OGG1/MMH Type 1a Protein Is a Major Enzyme for Repair of 8-OH-G Lesions	110
III. *Mmh* Knockout Results in Accumulation of 8-OH-G in DNA and Increases Mutation Frequency in Mice	113
IV. Discussion and Future Direction	119
References	121

Session 3 **Complexities of BER** **125**
   Michael Weinfeld

Molecular Mechanism of PCNA-Dependent Base Excision Repair 129
   Yoshihiro Matsumoto

   I. Introduction 129
   II. PCNA-Dependent AP Site Repair as an Alternative
       Pathway in Base Excision Repair 130
   III. Interaction of PCNA with Various Proteins 132
   IV. Replication-Coupled Base Excision Repair via Interaction
       between DNA-Glycosylases and PCNA 134
       References 137

Factors Influencing the Removal of Thymine Glycol from DNA in γ-Irradiated Human Cells 139
   Michael Weinfeld, James Z. Xing, Jane Lee,
   Steven A. Leadon, Priscilla K. Cooper,
   and X. Chris Le

   I. Assays for Oxidative DNA Damage 140
   II. Induction of Thymine Glycol: Comparison of CE-LIF
       to Other Assays 144
   III. Cellular Removal of Thymine Glycol from DNA 145
       References 148

Completion of Base Excision Repair by Mammalian DNA Ligases 151
   Alan E. Tomkinson, Ling Chen, Zhiwan Dong,
   John B. Leppard, David S. Levin, Zachary B. Mackey,
   and Teresa A. Motycka

   I. Introduction 152
   II. Mammalian *LIG* Genes and Their Protein Products 153
   III. DNA Ligase-Associated Proteins 155
   IV. Phenotype of DNA Ligase-Deficient Cell Lines 158
   V. Involvement of Mammalian DNA Ligases in BER 160

VI. Concluding Remarks	161
References	162

Uracil-Initiated Base Excision DNA Repair Synthesis Fidelity in Human Colon Adenocarcinoma LoVo and *Escherichia coli* Cell Extracts 165

Russell J. Sanderson, Samuel E. Bennett, Jung-Suk Sung, and Dale W. Mosbaugh

I. Introduction	166
II. Uracil-Initiated Base Excision DNA Repair Assay	167
III. Detection of Uracil-Initiated DNA Repair in Human Colon Adenocarcinoma LoVo Cell Extracts	171
IV. Uracil-DNA Repair Patch Size Produced by BER in LoVo Cell Extracts	175
V. Mutations Produced by Uracil-Initiated Base Excision DNA Repair Synthesis	181
VI. Concluding Remarks	186
References	187

Session 4   **DNA Glycosylases: Specificity and Mechanisms** **189**

Sankar Mitra

Multiple DNA Glycosylases for Repair of 8-Oxoguanine and Their Potential *in Vivo* Functions 193

Tapas K. Hazra, Jeff W. Hill, Tadahide Izumi, and Sankar Mitra

I. Repair of 8-Oxoguanine via the BER Pathway	194
II. Antimutagenic Processing of 8-OxoG: GO Model	194
III. Presence of Multiple OGGs in Bacteria and Eukaryotes	195
IV. Bipartite Antimutagenic Processing of 8-OxoG	197
V. Nei Is the OGG2 Paralog in *E. coli*	198
VI. Preferential Repair of 8-OxoG from the 8-OxoG·G Pair by OGG2 and Its Potential *in Vivo* Implication in Replication-Coupled Repair	199
VII. Mutual Interference of MutY and OGGs	200
VIII. Potential Role of OGG2 in Transcription-Coupled Repair	202
IX. Enzymatic Reaction of OGG1: The Handoff Process in Repair of 8-OxoG	203
References	204

DNA Substrates Containing Defined Oxidative Base Lesions and Their Application to Study Substrate Specificities of Base Excision Repair Enzymes 207

Hiroshi Ide

    I. DNA Containing Defined Oxidative Base Lesions 208
    II. Substrate Specificity of Endo III and Fpg Homologs 210
    III. Sequence Context Effects on Damage Recognition 216
    References 219

Mechanism of Action of *Escherichia coli* Formamidopyrimidine *N*-Glycosylase: Role of K155 in Substrate Binding and Product Release 223

Lois Rabow, Radhika Venkataraman, and Yoke W. Kow

    I. Introduction 224
    II. Materials and Methods 226
    III. Results 228
    IV. Conclusions 232
    References 233

Thymine DNA Glycosylase 235

Ulrike Hardeland, Marc Bentele, Teresa Lettieri, Roland Steinacher, Josef Jiricny, and Primo Schär

    I. Evidence for Short-Patch Mismatch Correction and Discovery of Thymine DNA Glycosylase 236
    II. The TDG Protein: Its Structure and Molecular Functions 237
    III. The Biochemistry of Thymine DNA Glycosylase 241
    IV. The Biology of Thymine DNA Glycosylase 246
    V. Conclusions 251
    References 252

Session 5   Mitochondrial DNA Repair 255

Vilhelm A. Bohr

Enzymology of Mitochondrial Base Excision Repair 257

Daniel F. Bogenhagen, Kevin G. Pinz, and Romina M. Perez-Jannotti

    I. mtDNA Damage 258
    II. Base Excision Repair of mtDNA 259

III. What Nuclear Genes Encode Mitochondrial DNA Repair Enzymes?	261
References	270

Base Excision Repair of Mitochondrial DNA Damage in Mammalian Cells ... 273

S. P. LeDoux and G. L. Wilson

I. Introduction	273
II. Evidence for BER in Mammalian Mitochondrial DNA	275
III. Repair of Oxidative Damage to mtDNA	276
IV. Repair of NO-Induced Damage in mtDNA	276
V. Evaluation of Damage and Repair to mtDNA at the Level of Individual Nucleotides	277
VI. Mechanisms Involved in the Repair of mtDNA	279
VII. Cell-Specific Differences in mtDNA Repair	279
VIII. Conclusions and Future Questions	282
References	283

Base Excision Repair in Nuclear and Mitochondrial DNA ... 285

Grigory L. Dianov, Nadja Souza-Pinto, Simon G. Nyaga, Tanja Thybo, Tinna Stevnsner, and Vilhelm A. Bohr

I. Oxidative DNA Damage Processing in Nuclear DNA	286
II. Mitochondrial DNA Repair in Mammalian Cells	291
III. Mitochondrial DNA Repair and Aging	295
References	296

Session 6 Structural Implications of BER Enzymes: Dragons Dancing—The Structural Biology of DNA Base Excision Repair ... 299

John A. Tainer

Crystallizing Thoughts about DNA Base Excision Repair .. 305

Thomas Hollis, Albert Lau, and Tom Ellenberger

I. Introduction	305
II. Structures of the Human Alkyladenine DNA Glycosylase	307
III. Structures of *E. coli* AlkA	309
IV. Recognition of Alkylated Bases	311
V. Catalysis of Glycosylic Bond Cleavage	312

| VI. Future Directions .... | 313 |
| References .... | 313 |

DNA Damage Recognition and Repair Pathway Coordination Revealed by the Structural Biochemistry of DNA Repair Enzymes 315

David J. Hosfield, Douglas S. Daniels, Clifford D. Mol, Christopher D. Putnam, Sudip S. Parikh, and John A. Tainer

I. DNA Damage Reversal and Avoidance	316
II. DNA Base Excision Repair Glycosylases	323
III. 5′ Apurinic/Apyrimidinic (AP) Endonucleases	328
IV. FEN-1 and Long-Patch Base Excision Repair	338
V. Perspectives and Implications	342
References	342

Potential Double-Flipping Mechanism by E. coli MutY 349

Paul G. House, David E. Volk, Varatharasa Thiviyanathan, Raymond C. Manuel, Bruce A. Luxon, David G. Gorenstein, and R. Stephen Llyod

I. DNA Glycosylases	350
II. MutY: An Adenine Glycosylase	350
III. Review of the Reaction Mechanism of DNA Glycosylases and Glycosylase/AP Lyases	352
IV. Structure of the Catalytic Domain of MutY	353
V. MutY: Glycosylase with an Opportunistic Lyase Activity	355
VI. Role for the C-Terminal Domain of MutY in Substrate Specificity	356
VII. Solution Structure of the C-Terminal Domain of MutY	357
VIII. Proposal for a Double-Flipping Mechanism by MutY	361
References	362

Properties and Functions of Human Uracil-DNA Glycosylase from the UNG Gene 365

Hans E. Krokan, Marit Otterlei, Hilde Nilsen, Bodil Kavli, Frank Skorpen, Sonja Andersen, Camilla Skjelbred, Mansour Akbari, Per Arne Aas, and Geir Slupphaug

| I. Introduction .... | 366 |
| II. Uracil-DNA Glycosylase from a Conserved Gene Family .... | 368 |

III. Recent Developments in Studies on Uracil-DNA Glycosylase: The Products of Human *UNG* and Murine *Ung*	370
IV. Regulation of Expression of the *UNG* Gene	376
V. *UNG* Mutants That Remove Normal Pyrimidines in DNA and Their Use in Studying the Biology of AP Sites	377
VI. Recent Information Indicates an Essential Role for Ung2 in Removal of Misincorporated Uracils, but Is This All?	378
VII. Is a Major Function of DNA Glycosylases a Long-Term Protection of the Mammalian Genome during Evolution?	381
References	384
INDEX	387

Preface

This special volume of *Progress in Nucleic Acid Research and Molecular Biology* represents research discussed during the DNA Base Excision Repair Workshop, which was held in Galveston, Texas, on March 10–13, 2000. In this preface, we provide the general reader with some background and rationale for the organization of the workshop.

DNA repair is a basic, universal process for maintaining the genetic integrity of organisms; it involves removal of DNA damage, which includes modified or abnormal bases, and repair of DNA strand breaks, which are produced both endogenously and exogenously after environmental stress. Furthermore, in spite of an exquisitely controlled enzymatic process, DNA replication is not error-free. Any error in replication would change the genomic sequence in the progeny, resulting in mutations which would be permanent. Even a single base-pair change in the mammalian genome, consisting of some 3 billion base pairs, could thus have devastating consequences.

Therefore, it is not surprising that multiple processes serve to repair the wide range of DNA damage. In addition, damage-tolerance processes have evolved to cope with the extreme situations in which the DNA damage cannot be removed. The base excision repair (BER) pathway has been shown to be selectively responsible for repairing reactive oxygen species–induced DNA damage and small base adducts, such as methylated bases.

It is perhaps to be expected that the repair processes are conserved from prokaryotes to mammals. This conservation is most striking for the BER pathway, which was heralded with the discovery of the first DNA glycosylase in 1979 by Thomas Lindahl, the keynote speaker at the workshop. Furthermore, his investigations revealed that, contrary to common belief, DNA is chemically unstable and abasic (AP) sites and uracil could be formed at a significant rate *in vivo*. Lindahl provides a brief history of the BER process and discusses future directions of research in BER in his review in this volume.

As is discussed in subsequent articles in this volume, several recent observations highlighting the complexity of the BER process rekindled interest in the field, including (a) degeneracy of the BER process observed in reconstituted systems *in vitro;* (b) coupling of repair of oxidized bases to transcription and the transcribed strand *in vivo*, which provides evidence for linkage between base excision repair and nucleotide excision repair; (c) linkage of different excision repair pathways *in vivo* mediated by transcription regulatory proteins, such as p53 and BRCA proteins; and (d) identification of multiple DNA glycosylases with overlapping, but not identical, substrate specificities *in vitro*, which have several potential implications *in vivo*, ranging from their backup or compensatory roles to their functions at distinct sites, cell cycle stages, or DNA strand specificity. Thus, DNA nascent strand-specific repair, either via the mismatch repair pathway or via the BER pathway under the newly labeled process of replication-associated repair, could involve many replication-related proteins yet to be identified.

In view of the newly found complexities of BER, coupled with recent successes in elucidating the tertiary structures of BER proteins, we believed that a workshop focused exclusively on the various facets of BER research would be timely and highly beneficial to investigators in the DNA repair field. Furthermore, the proceedings of the workshop would provide an opportunity to our scientific colleagues who are not directly involved in repair studies to obtain a comprehensive understanding of the current state of knowledge in the field.

Following the keynote lecture, the meeting was divided into several sessions. The articles of the speakers, based on their lectures, are grouped similarly, and topics of the lectures are prefaced by a short overview written by the chair of each session.

We acknowledge financial support by the University of Texas Medical Branch, Galveston, the National Institute of Environmental Health Sciences, the National Cancer Institute, the National Institute on Aging, and the U.S. Department of Energy, which made the meeting possible.

Finally, we gratefully recognize the invaluable administrative help of Ms. Wanda Smith and Mrs. Lisa Pipper Stephenson, who were crucial in all aspects of organizational and logistical support for the meeting and preparation of this volume.

Sankar Mitra
R. Stephen Lloyd
Priscilla K. Cooper
Samuel H. Wilson

Keynote: Past, Present, and Future Aspects of Base Excision Repair

THOMAS LINDAHL

ICRF Clare Hall Laboratories
South Mimms
Hertfordshire EN6 3LD, United Kingdom

I. Discovery of Enzymes That Recognize Endogenously
   Generated DNA Lesions xviii
II. Reconstitution of Mammalian Base Excision Repair xxi
III. Gene Knockout Mice Deficient in Base Excision Repair xxiii
IV. Future Developments xxvi
   A. BER of Chromatin xxvi
   B. Interactions of BER Proteins with Cellular Stress
      Response Factors xxvi
   C. BER Enzymes as Drug Targets xxvii
   D. Proteomics xxvii
   E. Accuracy of Base Excision Repair xxviii
   F. Malfunctioning BER Enzymes in Human Disease and Aging xxviii
   References xxix

Covalent alterations of DNA bases, which may have promutagenic or cytotoxic effects, are major consequences of endogenous DNA damage caused by hydrolysis, reactive oxygen species, and several metabolites and coenzymes. A common strategy for initiation of DNA base excision repair (BER) involves a DNA glycosylase that binds the altered deoxynucleoside in an extrahelical position and catalyzes cleavage of the base–sugar bond. Subsequently, an AP endonuclease or AP lyase activity incises the abasic site, followed by short-patch gap-filling, excision of the base-free sugar–phosphate residue, and ligation. The initial work that resulted in the discovery of DNA glycosylases and AP endonucleases is briefly reviewed. In recent years, it has been shown that the latter steps of the BER pathway differ greatly between mammalian cells and microorganisms such as yeast and bacteria. Three distinct subpathways of BER occur in mammalian cells, and these have been individually reconstituted with purified enzymes. Gene knockout mice are now revealing specific roles and backup mechanisms for repair functions in murine cells, and the results in general are also applicable to human cells. Future developments in the field of base excision repair include definition by proteomics of all factors involved in handling many different types of DNA lesions, clarification of mechanisms of repair of chromatin at a high level of accuracy, manifestation of repair proteins as drug targets for cellular sensitization to ionizing radiation and anticancer medicines, and elucidation of cross-talk between the base excision repair factors and other cellular proteins involved in a variety of stress responses. © 2001 Academic Press.

I. Discovery of Enzymes That Recognize Endogenously Generated DNA Lesions

Long-term incubation experiments with DNA radioactively labeled in specific base residues established that purine bases are lost from DNA by hydrolysis at significant rates under physiological conditions (1) and, similarly, that DNA cytosines are hydrolytically deaminated to premutagenic uracil residues at relevant rates (2). These studies and the subsequent enzymological work, which were carried out 25–30 years ago, have been reviewed relatively recently (3, 4) and have also been described in Errol Friedberg's historical account of the origins of DNA repair (5). An unavoidable conclusion from the chemical studies on the lability of DNA in neutral aqueous solution was that cells universally would be expected to possess correction factors, such as repair enzymes specific for endogenously generated DNA lesions, to counteract the continuously generated spontaneous damage.

A search for such repair factors was initiated. Since hydrolytic depurination of double-stranded DNA appeared to be the most frequent cause of damage, potential DNA substrates were generated by introduction of fixed small numbers of abasic sites by gentle heating in solution, guided by the previously determined rates of DNA depurination. An ~30-kDa enzyme that incised double-stranded DNA specifically at abasic sites was identified, it appeared to be widely distributed among different organisms. The active factor was purified from calf thymus (Fig. 1) and subjected to further characterization (6). The enzyme introduced single-strand breaks specifically at abasic sites, but did not recognize a variety of other DNA lesions. The first report on an endonuclease of this type, specific for apurinic sites in DNA, came from Verly and Paquette (7), who were investigating abasic sites generated as secondary lesions after treatment of DNA with alkylating agents. They concluded that such sites were susceptible to cleavage by an endonuclease, whereas the primary lesions, alkylated purine residues, were not recognized. The separate investigations of endogenous depurination rates of native, nonalkylated DNA clarified the cellular requirement for an enzyme of this specificity (1).

In an experimental approach similar to that used for investigating repair of abasic sites (6), a search was made for an enzyme that would specifically act on deaminated cytosine residues in DNA. Such an activity was identified (8) and surprisingly turned out to act by catalyzing hydrolysis of the uracil–deoxyribose glycosyl bond (Fig. 2) rather than an adjacent phosphodiester bond.

BASE EXCISION REPAIR   xix

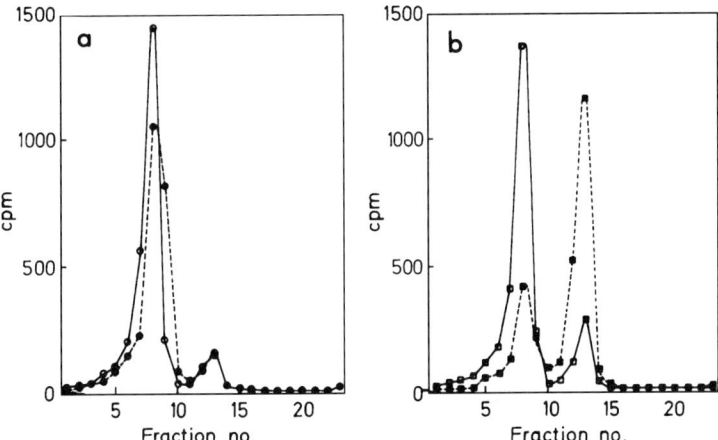

FIG. 1. Endonuclease for apurinic sites purified from calf thymus. The substrate was ^{32}P-labeled circular DNA, either intact (a) or containing an average of one apurinic site per molecule introduced by brief incubation at pH 5 and 70°C (b). The experiment shows neutral sucrose gradients, where the more rapidly sedimenting covalently closed circular DNA (fractions 7–9) is separated from nicked DNA (fractions 12–14). Solid symbols show DNA incubated with the AP endonuclease, and open symbols show DNA incubated without enzyme. For further details, see Ref. (6).

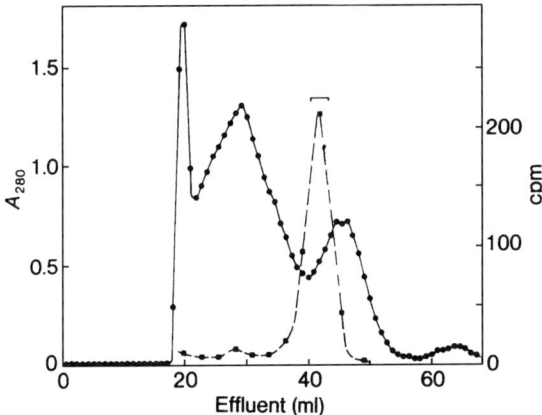

FIG. 2. Excision of free uracil from a polydeoxynucleotide containing scattered ^{3}H-labeled dUMP residues. An *E. coli* enzyme fraction was further purified by gel filtration, individual fractions were assayed, and the release of [^{3}H]uracil in free form was measured by chromatographic analysis. Solid line shows total protein; dashed line shows DNA glycosylase activity. For further details, see Ref. (8).

TABLE I
DNA GLYCOSYLASES IN HUMAN CELL NUCLEI

Enzyme	Size (amino acid residues)	Gene location at chromosome	Altered base removed from DNA
UNG	313	12q23–q24	U and 5-hydroxyuracil
TDG	410	12q24.1	U or T opposite G, ethenocytosine
hSMUG1	270	12q13.1–q14	U (preferentially from single-strand DNA)
MBD4	580	3q21	U or T opposite G at CpG sequences
hOGG1	345	3p25	8-oxoG opposite C, formamidopyrimidine
MYH	521	1p32.1–p34.3	A opposite 8-oxoG
hNTH1	312	16p13.2–p.13.3	Thymine glycol, cytosine glycol, dihydrouracil, formamidopyrimidine
MPG	293	16p (near telomere)	3-MeA, ethenoadenine, hypoxanthine

Reproduced with permission from Ref. (16).

Subsequently, several other DNA glycosylases with different substrate specificities were identified by our laboratory and by others, in most cases initially in *Escherichia coli*. The main types of DNA glycosylases have been retained during evolution, which is not surprising since endogenous DNA damage occurs in a similar way in all cells. The bacterial work therefore provided an excellent foundation for the detailed work carried out in recent years on mammalian enzymes. A current list of the DNA glycosylases known to be present in mammalian cell nuclei (Table I) shows a greater diversity than shown in bacteria. Four different DNA glycosylases appear to be involved in the correction of deaminated cytosine and 5-methylcytosine residues, whereas just two such functions have been found in *E. coli*. The only DNA glycosylase identified in several microbial systems but not found in higher eukaryotes is the pyrimidine dimer DNA glycosylase, which accounts for increased cellular resistance to ultraviolet light (for review, see Ref. 9). Thus, thermophilic bacteria have not been found to contain additional DNA glycosylases with substrate specificities different from those of the enzymes of mesophiles.

The discoveries of AP endonucleases and DNA glycosylases established the initial steps of the base excision repair (BER) pathway, in which enzymatic functions for excision of the base-free sugar–phosphate, gap-filling, and ligation are subsequently required (10–12). It is interesting to note that all the early work on BER used biochemical approaches. The determination of rates of endogenous hydrolytic damage to DNA predicted that repair factors specific for apurinic sites and deaminated cytosine residues in DNA must exist, and subsequent searches revealed enzymes with the exact specificities proposed. Genetic verification came later, with the isolation by mass screening of an *E. coli* strain lacking uracil-DNA glycosylase, and characterization of its mutation spectrum

(13). By contrast, bacterial mutator mutants deficient in DNA mismatch repair had been investigated long before the biochemistry of the mismatch correction process was understood, and mutant cells hypersensitive to exposure to ultraviolet light were of crucial importance for unraveling the nucleotide excision repair pathway, both in E. coli and human cells.

The existence of BER as a DNA repair pathway largely concerned with correction of endogenously produced DNA lesions provides a straightforward biochemical answer to a question that often caused concern and interest among physicists in the early days of molecular biology: How could genes be stable? In retrospect, no special laws of physics had to be predicted to overcome this problem. The double-stranded structure of DNA reiterates genetic information; thus, turnover of short single-strand stretches in DNA during correction of endogenous lesions by BER provides the desired stability. Developing tumors may relax these usually stringent requirements for minimizing spontaneous cellular mutation frequencies (14, 15).

II. Reconstitution of Mammalian Base Excision Repair

BER processes similar to those first established in E. coli occur in both nuclei and mitochondria of mammalian cells. DNA glycosylases and AP endonucleases show strong evolutionary conservation, but later steps in the BER pathway differ markedly between the microbial model systems and mammalian cells. Also, Saccharomyces cerevisiae and other lower eukaryotes do not appear to be good models for mammalian cells in that regard because they lack direct counterparts to mammalian Pol β, with its associated AP lyase activity, and the XRCC1 and DNA ligase III proteins.

Molecular cloning and overexpression of cDNA sequences encoding mammalian BER factors and extensive structural work on BER enzymes have contributed greatly to the detailed understanding of this mode of repair. A current scheme for the main reaction pathway is shown in Fig. 3. A set of consecutive interactions between pairs of monomeric enzymes occurs, and reaction intermediates are continuously protected (16–18). These specific protein interactions take place primarily at the damaged DNA site, and the XRCC1-DNA ligase III heterodimer is the only protein pair that is preformed (19). A similar strategy is employed by dimeric transcription factors, where assembly of the two relevant monomers by sequential binding to a DNA site allows more rapid and efficient detection of the site in preference to nonproductive alternative interactions (20).

Reconstitution of the major BER pathway (Fig. 3) with mammalian reagent enzymes has been demonstrated (21–23). Use of the thymine-DNA glycosylase (TDG), which shows an exceptionally low turnover number, necessitated

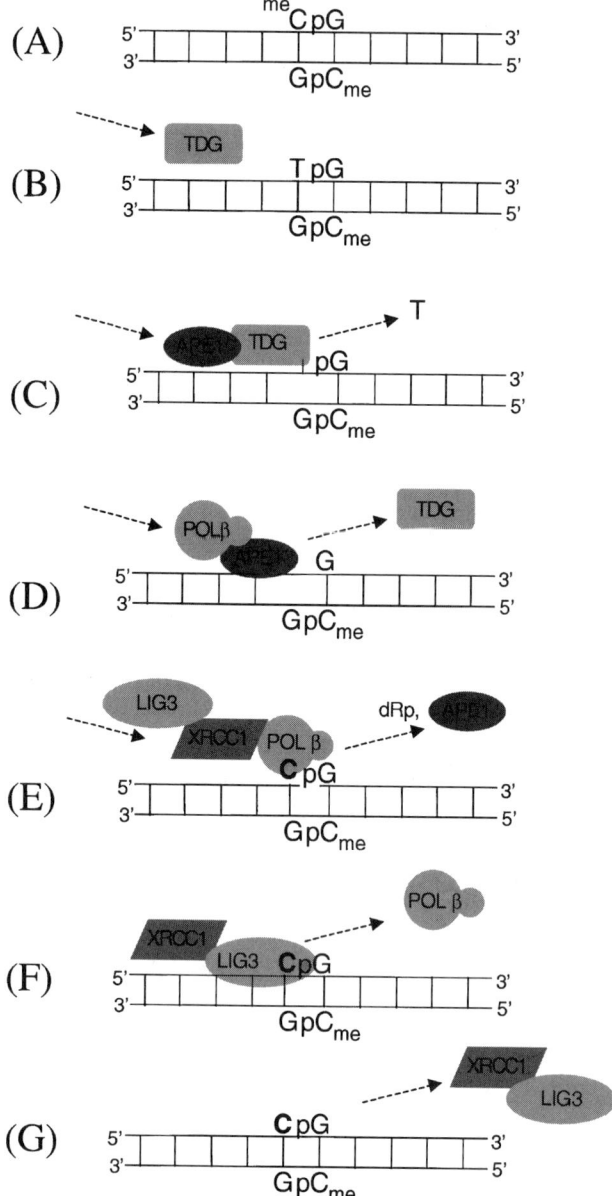

FIG. 3. The main pathway of base excision repair in mammalian cells. This scheme illustrates repair of a deaminated 5-methylcytosine residue. Pairwise interactions of monomeric enzymes are shown. The final product is a substrate for a maintenance methylase. For further details, see text. Reproduced with permission from Ref. (16).

very long incubation times in the first experiments (21), but employment of the human uracil-DNA glycosylase (UNG) and more detailed studies of the last reaction steps provided rapid and efficient reconstitution of human BER (22). Thus, the reconstituted *in vitro* reaction achieves 50–80% completion in less than 20 min, which is much more efficient than the more complex *in vitro* reactions of nucleotide excision repair or SV40 DNA replication with purified human factors.

DNA glycosylases acting on base damage generated by oxygen free radicals have an associated AP lyase activity, which cleaves the DNA chain on the 3′ side of the base lesion. This results in adherence to a single-nucleotide replacement mechanism (Fig. 4), and very little (if any) filling-in of large gaps occurs (24, 25). Since oxygen free radicals may produce clustered lesions, it may be of importance to minimize patch sizes during their repair in order to avoid the formation of double-strand breaks as an inadvertent consequence of BER processing of two closely located lesions in opposite strands (24).

Complex lesions with damage to both the base and deoxyribose residues can be repaired by an alternative BER pathway with slightly longer repair patches of 2–6 nucleotides rather than a single nucleotide (Fig. 4). This form of BER requires PCNA and other replication factors, so it is clearly a distinct subpathway (26, 27). This reaction pathway has also been reconstituted efficiently with purified human proteins (28–30). The availability of these reconstitution assays provides a direct means of characterizing potential stimulatory factors and inhibitors of the repair reaction. An example is the unexpected promoting effect of the human XPG protein on the excision of ring-saturated pyrimidines by the hNTH1 glycosylase (24). It is now important to extend this biochemical approach to the establishment of an experimental system to investigate putative transcription-coupled BER reactions (31). Many unsuccessful attempts have been made in different laboratories to reproduce the complex transcription-coupled NER pathway *in vitro*, but similar analysis of transcription-coupled BER may turn out to be less difficult since fewer factors apparently are required.

III. Gene Knockout Mice Deficient in Base Excision Repair

Many areas of mammalian DNA repair have been further elucidated by construction and analysis of relevant transgenic mice. In the BER field, attempts to obtain mice deficient in the APE1 (HAP1) endonuclease, DNA polymerase β, or the repair scaffold protein XRCC1 have yielded embryonic lethal phenotypes, indicative of the essential requirement for BER of endogenously produced DNA lesions. Knockout embryos deficient in APE1 or XRCC1 die very

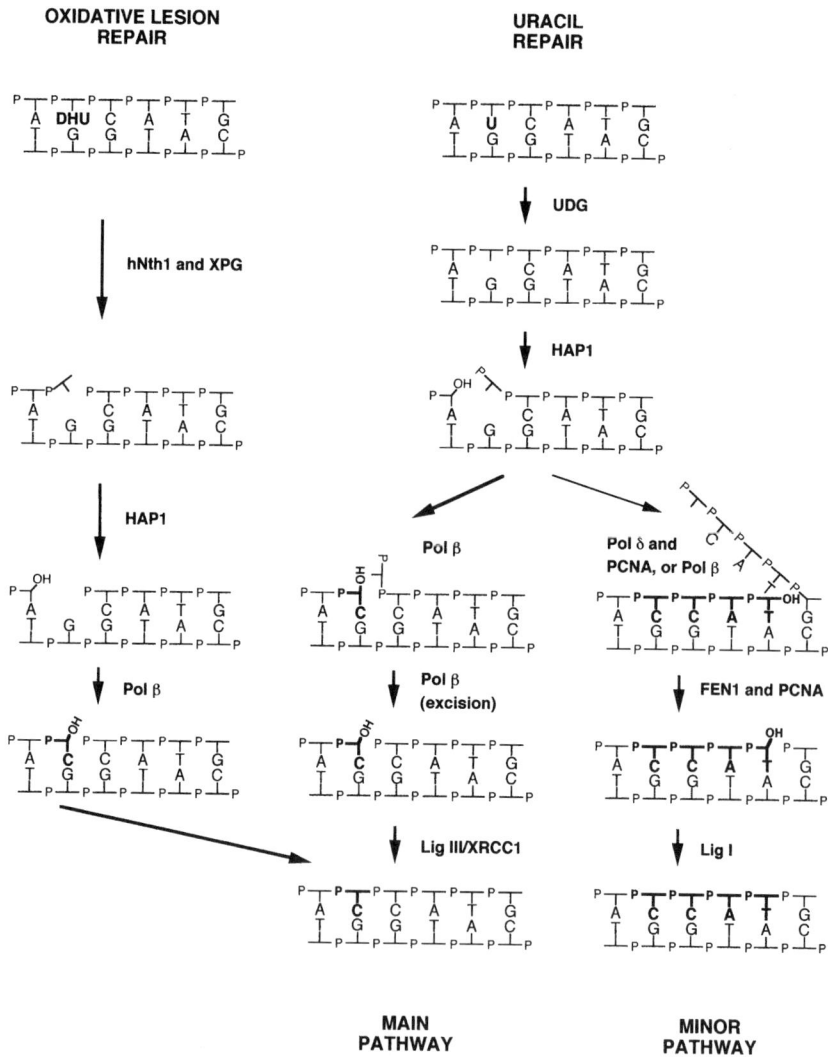

FIG. 4. Different subpathways of mammalian base excision repair. For further details, see text. DHU stands for dihydrouracil, a common radiation decomposition product of cytosine. Reproduced with permission from Ref. (24).

early, making cellular studies difficult, but DNA polymerase β–deficient mice survive through most of the gestation period and show an interesting phenotype associated with neuronal death (32). Repair studies *in vitro* have indicated that PCNA-dependent high-molecular-weight DNA polymerases can substitute to some extent for the gap-filling activity of Pol β during BER, and the AP lyase activity of the enzyme might be inefficiently substituted for by basic molecules such as polyamines, which would be consistent with Pol β-deficient embryos surviving longer than those deficient in APE1 or XRCC1.

In apparent contrast to these severe phenotypes, knockouts of specific DNA glycosylases in mice have shown unexpectedly mild phenotypes by comparison with bacterial or yeast mutants with corresponding defects. Several backup mechanisms seem to have evolved in higher eukaryotes to minimize mutagenic risks generated by endogenous damage, which might otherwise be substantial in large genomes.

In a collaborative project between our laboratory and that of Hans Krokan, mice have that are deficient in the major uracil-DNA glycosylase, UNG, been constructed. This mammalian enzyme shows strong sequence homology to the antimutagenic enzyme of S. cerevisiae and E. coli that counteracts the premutagenic effect of spontaneous cytosine deamination in DNA. However, these knockout mice show only a marginally increased mutation rate, are fertile, and do not exhibit gross pathological changes. Nuclear extracts from such mice exhibit low-level backup uracil-DNA glycosylase activity (33). This backup activity is indistinguishable from the recently detected hSMUG1 enzyme (34), which has no counterpart in yeast. The TDG DNA glycosylase (21) might also have a minor backup role in vivo. A major, different function specific to the abundant UNG enzyme in mammalian cells may be to avoid persistence of uracil in DNA that arises during replication as a consequence of the occasional use of dUTP instead of dTTP as a precursor. Recently, UNG was found to be present in replication factories together with PCNA and RPA (35), indicative of a role in editing newly synthesized DNA rather than dealing with deaminated cytosines that may arise throughout the genome.

Mice deficient in the hOGG1 DNA glycosylase (36, 37) are strongly impaired in their ability to excise the main mutagenic base lesion 8-hydroxyguanine (8-oxoG), generated by reactive oxygen species. These mice also show a surprisingly mild phenotype by comparison with E. coli mutM (also called fpg) mutants, which have a corresponding enzymatic defect. Substantial amounts of 8-oxoG residues gradually accumulate in cellular DNA of nonproliferating organs of these knockout mice, showing a strongly curtailed ability to remove the lesion. Recent work in the laboratory of Serge Boiteux (see Ref. 36) has shown, however, that our OGG1−/−cell line retained the capacity to remove 8-oxoG from actively expressed genes by a transcription-coupled repair process. This observation raises the interesting prospect that the massive amount of 8-oxoG (10,000–15,000 residues per genome) in hepatocytes of adult knockout animals might be less harmful than expected, because most of the persistent lesions may be in nontrascribed satellite and junk DNA sequences. Other backup possibilities for excision of 8-oxoG residues are discussed by Mitra elsewhere in this volume. In addition, removal of adenine residues from premutagenic 8-oxoG : A base pairs by the mammalian equivalent of the MutY DNA glycosylase (38) might serve to reduce the expected mutation rates owing to occurrence of oxidized G residues in the genome.

The first DNA glycosylase-deficient mice to be obtained lacked the 3-methyladenine-DNA glycosylase MPG (*39, 40*). *E. coli* mutants deficient in repair of 3-methyladenine, and also several minor DNA lesions, grow normally but are highly sensitive to the cytotoxic effect of simple alkylating agents such as methyl methanesulfonate (*41*). In this case, the mammalian mutants show a similar alkylation-sensitive phenotype.

IV. Future Developments

With the greatly increased interest in DNA repair processes in recent years, especially with regard to their crucial role in counteracting genomic instability, the removal of endogenously produced DNA damage in mammalian cells by base excision repair is being investigated in considerable detail. There are several aspects of BER where relatively little work has been performed to date, however, and some of these are briefly discussed below.

A. BER of Chromatin

The extensive studies in several laboratories on reconstitution of mammalian BER with purified mammalian proteins have, so far, been performed with naked linear or circular damaged DNA as substrates. An obvious extension of this protocol would be to use chromatin reconstituted from damaged DNA and histones, or minichromosomes exposed to DNA-damaging agents, as substrates. The high repair efficiency of the *in vitro* systems of BER should greatly aid such experimental approaches. BER must be able to function on tightly condensed heterochromatin in both proliferating and nonproliferating cells, because water and reactive oxygen species are ubiquitous DNA-damaging agents that cannot be excluded. However, it is not yet known if the relatively small monomeric DNA glycosylases and AP endonucleases find ready access to damaged nucleotide residues in chromatin, or if special protein factors exist to facilitate BER by local chromatin remodeling.

B. Interactions of BER Proteins with Cellular Stress Response Factors

The BER enzymes in general appear to be constitutively expressed in mammalian cells, although occasional reports have appeared in the literature on minor induction responses to very high, lethal levels of DNA damage. However, BER could be influenced and improved by interactions with a number of stress response signal proteins. A report (*42*) on promotion of the major, short-patch

BER pathway for correction of abasic sites by the p53 protein is of interest. Moderate amounts of DNA damage can cause stabilization and increased cellular levels of p53, but may not elevate p53 levels to such a high extent that the apoptotic machinery is activated. It would clearly be advantageous to promote BER by p53, after exposure to moderately increased levels of alkylating agents or reactive oxygen species. The rate-limiting step in the short-patch BER pathway of abasic sites may be the excision of a base-free deoxyribose phosphate from 5' termini of DNA strand breaks by the AP lyase function of DNA polymerase β (23), or under different experimental conditions the final joining step (22). But it is not known whether an increased level of p53 protein improves these steps specifically, or promotes BER by some other mechanism, such as facilitating protein–protein interactions between factors of the pathway. There are probably a number of strategies by which BER could be stimulated by signal transduction pathways triggered by cellular stress responses, but very little information is currently available.

C. BER Enzymes as Drug Targets

Next to surgery, radiotherapy remains the most important type of treatment of human cancer. Attempts to develop clinically useful radiosensitizers have usually focused on drugs that perturb cellular oxygen metabolism. A different approach would be to search for compounds that would temporarily inhibit cellular repair of DNA damage generated by reactive oxygen species. This damage could be either DNA double-strand breaks or base lesions. Cytotoxic DNA base lesions such as thymine glycols and formamidopyrimidines are major types of damage induced by ionizing radiation, and the human DNA glycosylases that excise these lesions have been cloned, overexpressed, and characterized. So far, there has been little effort to find or design specific inhibitors of these DNA repair enzymes.

D. Proteomics

Most of the nuclear DNA glycosylases found in mammalian cells (Table I) and in microorganisms have been known for 20 years. In spite of the high level of oxygen metabolism in mitochondria and consequent risks for oxidative damage to organelle DNA, no unique mitochondrial enzymes of this group have been found. Instead, alternatively spliced products of several of the nuclear enzymes occur in mitochondria. With regard to the multiplicity of DNA glycosylases, it should be noted that two of the four mammalian DNA glycosylases, hSMUG1 and MBD4, that act on G:U and G:T base pairs, respectively, were discovered only last year (34, 43). The hSMUG 1 enzyme (34) was revealed by *in vitro* expression cloning, initially in *Xenopus laevis*, and this general experimental approach could clearly be extended to search for additional DNA glycosylases acting on unknown substrates. Obviously, DNA glycosylase activity will not be revealed by standard biochemical assays unless the substrate contains the relevant form of DNA damage, and it remains a distinct possibility that some types of base lesions that have not yet been investigated or even detected are physiologically highly relevant and are excised by as yet undiscovered specific DNA glycosylases. It would be surprising, however, if there were more than a total of about 10 different DNA glycosylases in mammalian cell nuclei, with 8 of them (Table I) already having been investigated in considerable detail.

E. Accuracy of Base Excision Repair

DNA polymerase β has been firmly established as accounting for the short gap-filling step (44) in the major pathway of BER (Fig. 3). The absence of an editing 3'-exonuclease function of Pol β necessarily means that the enzyme has a relatively high error rate (45). A separate editing function seems to be required as an accessory factor to BER in mammalian cells. The major DNA 3' exonuclease of cell nuclei, called DNase III (46) or TREX1 (47), is an attractive candidate for this function because it efficiently excises mismatched residues from 3' termini *in vitro*, and shows sequence homology with microbial editing enzymes. Further experimental work to investigate this hypothesis is in progress. The strongly reduced ability of mammalian DNA ligase III to act at single-strand breaks with 3' mismatches would also be expected to contribute in a positive way to the high fidelity of BER (48).

F. Malfunctioning BER Enzymes in Human Disease and Aging

The essential requirement for BER, as shown by the early embryonic lethal phenotypes of knockout mice defective in one of the key components of the main pathway, means that a human disease resulting from a BER defect has not yet been identified. The severe developmental defects observed in XPG-deficient individuals unable to perform transcription-coupled repair of oxidative DNA damage (31) may reflect involvement of XPG as a cofactor in BER. This is also indicate by the ability of XPG to enhance activity of the hNTH1 DNA glycosylase that excises oxidized pyrimidines (24). There is currently much interest in the long-term effects of oxidative DNA damage with regard to aging and cancer, and since BER is the main repair pathway dealing with this damage, variations in BER capability could be relevant. Ongoing studies of genetic polymorphisms in genes encoding BER functions, such as the XRCC1 scaffold protein (49), and individual variations in repair capacity, when correlated with clinical observations, might serve to clarify the importance of BER in the cellular defense against pathological accumulation of endogenous DNA damage.

References

1. T. Lindahl and B. Nyberg, *Biochemistry* **11**, 3610–3618 (1972).
2. T. Lindahl and B. Nyberg, *Biochemistry* **13**, 3405–3410 (1974).
3. T. Lindahl, *Nature (London)* **362**, 709–715 (1993).
4. T. Lindahl, *Phil. Trans. R. Soc. London, Ser. B* **351**, 1529–1538 (1996).
5. E. C. Friedberg, "Correcting the Blueprint of Life: An Historical Account of the Discovery of DNA Repair Mechanisms." Cold Spring Harbor Laboratory Press, Plainview, NY, 1997.
6. S. Ljungquist and T. Lindahl, *J. Biol. Chem.* **249**, 1530–1540 (1974).
7. W. G. Verly and Y. Paquette, *Can. J. Biochem.* **50**, 217–224 (1972).
8. T. Lindahl, *Proc. Natl. Acad. Sci. U.S.A.* **71**, 3649–3653 (1974).
9. R. S. Lloyd, *Prog. Nucleic Acid Res. Mol. Biol.* **62**, 155–175 (1999).
10. T. Lindahl, *Nature (London)* **259**, 64–66 (1976).
11. J. Duncan, L. Hamilton, and E. C. Friedberg, *J. Virol.* **19**, 338–345 (1976).
12. J. Laval, *Nature (London)* **269**, 829–832 (1977).
13. B. K. Duncan, in "The Enzymes" (P. D. Boyer, ed.), 3rd ed., Vol. XIVA, pp. 565–586. Academic Press, New York, 1981.
14. A. L. Jackson and L. A. Loeb, *Genetics* **148**, 1483–1490 (1998).
15. J. H. Miller, *Cancer Surveys* **28**, 141–153 (1996).
16. T. Lindahl and R. D. Wood, *Science* **286**, 1897–1905 (1999).
17. C. D. Mol, T. Izumi, S. Mitra, and J. A. Tainer, *Nature (London)* **403**, 451–456 (2000).
18. S. H. Wilson and T. A. Kunkel, *Nature Struct. Biol.* **7**, 176–178 (2000).
19. K. W. Caldecott, J. D. Tucker, L. H. Stanker, and L. H. Thompson, *Nucleic Acids Res.* **23**, 4836–4843 (1995).
20. J. J. Kohler, S. J. Metallo, T. L. Schneider, and A. Schepartz, *Proc. Natl. Acad. Sci. U.S.A.* **96**, 11735–11739 (1999).
21. K. Wiebauer and J. Jiricny, *Proc. Natl. Acad. Sci. U.S.A.* **87**, 5842–5845 (1990).
22. Y. Kubota, R. A. Nash, A. Klungland, P. Schär, D. E. Barnes, and T. Lindahl, *EMBO J.* **15**, 6662–6670 (1996).
23. D. K. Srivastava, B. J. Vande Berg, R. Prasad, J. T. Molina, W. A. Beard, A. E. Tomkinson, and S. H. Wilson, *J. Biol. Chem.* **273**, 21203–21209 (1998).
24. A. Klungland, M. Höss, D. Gunz, A. Constantinou, S. G. Clarkson, P. W. Doetsch, P. H. Bolton, R. D. Wood, and T. Lindahl, *Mol. Cell* **3**, 33–42 (1999).
25. P. Fortini, E. Parlanti, O. M. Sidorkina, J. Laval, and E. Dogliotti, *J. Biol. Chem.* **274**, 15230–15236 (1999).
26. Y. Matsumoto, K. Kim, and D. F. Bogenhagen, *Mol. Cell. Biol.* **14**, 6187–6197 (1994).
27. G. Frosina, P. Fortini, O. Rossi, F. Carrozzino, G. Raspaglio, L. S. Cox, D. P. Lane, A. Abbondandolo, and E. Dogliotti, *J. Biol. Chem.* **271**, 9573–9578 (1996).
28. A. Klungland and T. Lindahl, *EMBO J.* **16**, 3341–3348 (1997).
29. B. Pascucci, M. Stucki, Z. O. Jónsson, E. Dogliotti, and U. Hübscher, *J. Biol. Chem.* **274**, 33696–33702 (1999).
30. Y. Matsumoto, K. Kim, J. Hurwitz, R. Gary, D. S. Levin, A. E. Tomkinson, and M.S. Park, *J. Biol. Chem.* **274**, 33703–33708 (1999).
31. F. Le Page, E. E. Kwoh, A. Avrutskaya, A. Gentil, S. A. Leadon, A. Sarasin, and P. K. Cooper, *Cell (Cambridge, Mass.)* **101**, 159–171 (2000).
32. N. Sugo, Y. Aratani, Y. Nagashima, Y. Kubota, and H. Koyama, *EMBO J.* **19**, 1397–1404 (2000).
33. H. Nilsen, I. Rosewell, P. Robins, C. F. Skjelbred, S. Andersen, G. Slupphaug, G. Daly, H. E. Krokan, T. Lindahl, and D. E Barnes, *Mol. Cell* **6**, 1059–1065 (2000).
34. K. A. Haushalter, P. T. Stukenberg, M. W. Kirschner, and G. L. Verdine, *Curr. Biol.* **9**, 174–185 (1999).

35. M. Otterlei, E. Warbrick, G. Slupphaug, T. A. Nagelhus, T. Haug, M. Akbari, P. A. Aas, K. Steinsbekk, O. Bakke, and H. E. Krokan, *EMBO J.* **18,** 3834–3844 (1999).
36. A. Klungland, I. Rosewell, S. Hollenbach, E. Larsen, G. Daly, B. Epe, E. Seeberg, T. Lindahl, and D. E. Barnes, *Proc. Natl. Acad. Sci. U.S.A.* **96,** 13300–13305 (1999).
37. O. Minowa, T. Arai, M. Hirano, Y. Monden, S. Nakai, M. Fukada, M. Itoh, H. Takano, Y. Hippou, H. Aburatani, K. Masumura, T. Nohmi, S. Nishimura, and T. Noda, *Proc. Natl. Acad. Sci. U.S.A.* **97,** 4156–4161 (2000).
38. M. M. Slupska, C. Baikolov, W. M. Luther, J.-H. Chiang, Y.-F. Wei, and J. H. Miller, *J. Bacteriol.* **178,** 3885–3892 (1996).
39. B. P. Engelward, G. Weeda, M. D. Wyatt, J. L. M. Broekhof, J. de Wit, I. Donker, J. M. Allan, B. Gold, J. H. J. Hoeijmakers, and L. D. Samson, *Proc. Natl. Acad. Sci. U.S.A.* **94,** 13087–13092 (1997).
40. R. H. Elder, J. G. Jansen, R. J. Weeks, M. A. Willington, B. Deans, A. J. Watson, K. J. Mynett, J. A. Bailey, D. P. Cooper, J. A. Rafferty, M. C. Heeran, S. W. P. Wijnhoven, A. A. van Zeeland, and G. P. Margison, *Mol. Cell. Biol.* **18,** 5828–5837 (1998).
41. P. Karran, T. Lindahl, I. Ofsteng, G. B. Evensen, and E. Seeberg, *J. Mol. Biol.* **140,** 101–127 (1980).
42. H. Offer, R. Wolkowicz, D. Matas, S. Blumenstein, Z. Livneh, and V. Rotter, *FEBS Lett.* **450,** 197–204 (1999).
43. B. Hendrich, U. Hardeland, H.-H. Ng, J. Jiricny, and A. Bird, *Nature (London)* **401,** 301–304 (1999).
44. R. W. Sobol, J. K. Horton, R. Kuhn, H. Gu, R. K. Singhal, R. Prasad, K. Rajewsky, and S. H. Wilson, *Nature (London)* **379,** 183–186 (1996).
45. W. P. Osheroff, H. K. Jung, W. A. Beard, S. H. Wilson, and T. A. Kunkel, *J. Biol. Chem.* **274,** 3642–3650 (1999).
46. M. Höss, P. Robins, T. J. P. Naven, D. J. C. Pappin, J. Sgouros, and T. Lindahl, *EMBO J.* **18,** 3868–3875 (1999).
47. D. J. Mazur and F. W. Perrino, *J. Biol. Chem.* **274,** 19655–19660 (1999).
48. A. S. Bhagwat, R. J. Sanderson, and T. Lindahl, *Nucleic Acids Res.* **27,** 4028–4033 (1999).
49. H. W. Mohrenweiser and I. M. Jones, *Mutat. Res.* **400,** 15 (1998).

Session 1
Multiple Pathways for DNA Base Excision Repair

EUGENIA DOGLIOTTI

Istituto Superiore di Sanità

Cellular DNA is under constant attack from various sources that jeopardize the integrity of genetic information. Among the most important sources of genetic damage are the spontaneous hydrolytic decay of DNA, the formation of oxidative bases and sugar lesions, and base and phosphate residue alkylations due to environmental carcinogens, endogenously formed metabolites, and various drugs used in cancer therapy. Many of the structural abnormalities that arise from these processes and treatments are corrected by the components of the base excision repair (BER) pathway. The persistence of various lesions represents a major threat to genomic stability, resulting in mutagenesis and cancer. Until recently, BER was thought to occur via a single pathway consisting of five sequential reactions, including the release of the base to leave an apurinic/apyrimidinic (AP) site, the incision of the AP site by a 5′ AP endonuclease, the excision of the resulting deoxyribose phosphate (dRP) residue at the incised AP site, the synthesis to fill in the one nucleotide gap, and, finally, the ligation to restore the intact DNA template. However, studies reported in the past few years have demonstrated the existence of alternative pathways for BER that involve new molecular partners. The following chapters discuss most of the key components of the BER pathways and provide the scientific background for speculations on the regulation of the cellular switch mechanism among these alternative pathways.

In addition to BER, cells have evolved other strategies to counteract the continuous attack on genomic integrity by reactive oxygen species, endogenous and exogenous alkylation pressure, or errors arising from normal DNA metabolism —namely, nucleotide excision repair, damage reversal by alkyltransferase, DNA mismatch repair, recombination, and translesion synthesis. Evidence is accumulating that these systems do not act in a single-track pathway, but rather in a network. The following chapters also show that the BER specificity, in fact, overlaps with that of the other cellular repair systems and provide a forum to discuss the mechanisms that control the participation of different repair/tolerance systems in DNA damage processing and its involvement in defense against or induction of genotoxicity.

The Mechanism of Switching among Multiple BER Pathways

EUGENIA DOGLIOTTI, PAOLA
FORTINI, BARBARA PASCUCCI,
AND ELEONORA PARLANTI

*Laboratory of Comparative Toxicology
and Ecotoxicology
Istituto Superiore di Sanità
Viale Regina Elena 299
00161 Rome, Italy*

I.	Introduction.	4
II.	Factors That Control the Switch among Multiple BER Pathways.	9
	A. Type of Damage.	9
	B. Influence of DNA Structure.	20
	C. Gene Expression of BER Factors.	21
	D. Cell Cycle.	23
III.	Conclusions and Future Research.	24
	References.	25

To preserve genomic β DNA from common endogenous and exogenous base and sugar damage, cells are provided with multiple base excision repair (BER) pathways: the DNA polymerase (Pol) β-dependent single nucleotide BER and the long-patch (2–10 nt) BER that requires PCNA. It is a challenge to identify the factors that govern the mechanism of switching among these pathways. One of these factors is the type of DNA damage induced in DNA. By using different model lesions we have shown that base damages (like hypoxanthine and 1,N^6-ethenoadenine) excised by monofunctional DNA glycosylases are repaired via both single-nucleotide and long-patch BER, while lesions repaired by a bifunctional DNA glycosylase (like 7,8-dihydro-8-oxoguanine) are repaired mainly by single-nucleotide BER. The presence of a genuine 5′ nucleotide, as in the case of cleavage by a bifunctional DNA glycosylase-β lyase, would then minimize the strand displacement events. Another key factor in the selection of the BER branch is the relative level of cellular polymerases. While wild-type embryonic mouse fibroblast cell lines repair abasic sites predominantly via single-nucleotide replacement reactions (80% of the repair events), cells homozygous for a deletion in the Pol β gene repair these lesions exclusively via long-patch BER. Following treatment with methylmethane sulfonate, these mutant cells accumulate DNA single-strand breaks in their genome in keeping with the fact that repair induced by monofunctional alkylating agents goes predominantly via single-nucleotide BER. Since the long-patch BER is strongly stimulated by PCNA, the cellular content of this cell-cycle regulated factor is also extremely effective in driving the repair reaction to either BER branch. These findings raise the interesting possibility that different BER pathways might be acting as a function of the cell cycle stage. © 2001 Academic Press.

I. Introduction

The genomes of mammalian cells are exposed to a variety of endogenous and environmental DNA-damaging agents. To counteract their potential mutagenic and cytotoxic effects, cells have evolved DNA repair mechanisms that excise and replace the damaged nucleotides. The most frequent lesions are efficiently repaired by base excision repair (BER). For many years, BER was thought to occur via a single pathway. This repair system was described as a multistep process including: (1) damage recognition by a DNA glycosylase that hydrolytically cleaves the N-glycosyl bond to generate an abasic (AP) site, simultaneously releasing the free base; (2) recognition of the AP site by an AP endonuclease that hydrolytically cleaves the phosphodiester bond 5′ of an AP site; (3) excision of the deoxyribose-phosphate group on the 5′ side of the strand break by a phosphodiesterase; (4) filling in of the resulting one-residue gap by a DNA polymerase (Pol); and (5) sealing of the nick by a DNA ligase. This pathway has been reconstituted *in vitro* with bacterial proteins (*1*) and later with purified human proteins (*2*). The combined action of human uracil-DNA glycosylase (UDG), the major mammalian AP endonuclease HAP1, Pol β, and either DNA ligase III or DNA ligase I has been shown to successfully repair a uracil residue present in a duplex oligonucleotide (*2*). In this pathway, Pol β first synthesizes the nucleotide to fill in the gap and then uses its deoxyribophosphodiesterase (dRPase) activity to cleave the DNA backbone 3′ of the AP site (*3*) (Fig. 1A). Today we know that, although the general scheme for BER depicted in Fig. 1A is commonly used by all organisms, it is by no means the only repair route for base damage. One alternative branch (Fig. 1B) is created by the possibility of excision of the modified base by a bifunctional DNA glycosylase that is endowed with an AP lyase activity. This AP lyase cleaves the phosphodiester backbone on the 3′ side of an AP site, leaving a fragmented sugar derivative at the 3′ side of the resulting strand break. This polymerase-blocking lesion requires removal by a phosphodiesterase in order for DNA polymerase to fill in the gap, while the 5′ terminus would not require further processing in this case. Human BER of thymine glycol, which is a substrate for the DNA-glycosylase/AP lyase hNTH1, has been reconstituted *in vitro* using HAP1 to remove the 3′ blocked terminus, Pol β to fill in the one-nucleotide gap, and DNA ligase III–XRCC1 heterodimer to seal the resulting nick (*4*). The involvement of Pol β in this pathway has also been confirmed by using a different target lesion, 7,8-dihydro-8-oxoguanine

THE SWITCH MECHANISM

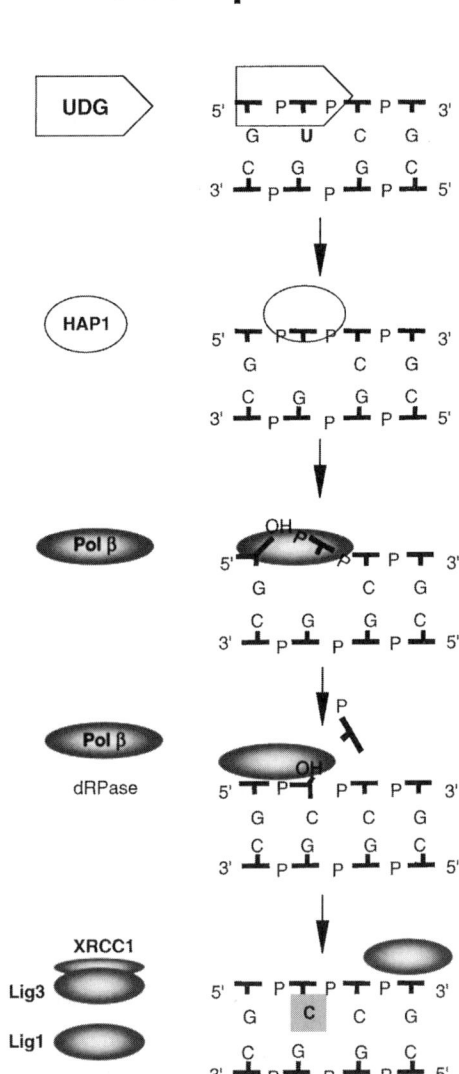

FIG. 1. Pathways for base excision repair in mammalian cells. (A and B) Short-patch BER. A damaged base—uracil (U) or 7,8-dihydro-8-oxoguanine (8-oxoG)—is excised by a DNA glycosylase to generate an AP site, which is then recognized by an AP endonuclease. In Panel 1A, following the base excision by UDG, an AP endonuclease hydrolyzes the phosphodiester bond immediately 5′ to the AP site, while in Panel 1B the AP lyase activity of OGG1 cleaves the phosphodiester bond 3′ to the AP site. The generated termini are processed by a phosphodiesterase that excises the 5′ deoxyribose phosphate in route A and the 3′ fragmented sugar derivative in route B. In both

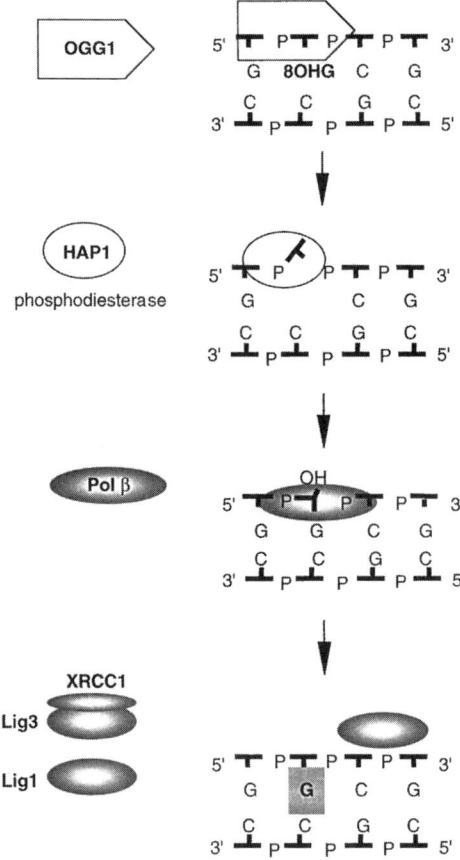

FIG. 1. (*continued*) pathways, a single-nucleotide gap is generated. Pol β fills this gap and the nick is sealed by a DNA ligase. (C) Long-patch BER. In the course of the removal of a damaged base (e.g., uracil), a gap greater than one nucleotide can also be generated. The additional players in this pathway are PCNA, RF-C, and FEN1, which are required for cleavage of the oligonucleotide containing the 5′ incised abasic site. The resynthesis step can be performed either by DNA Pol β or by Pol δ/ε and the nick is sealed by DNA ligase I.

THE SWITCH MECHANISM

C   Long-patch BER

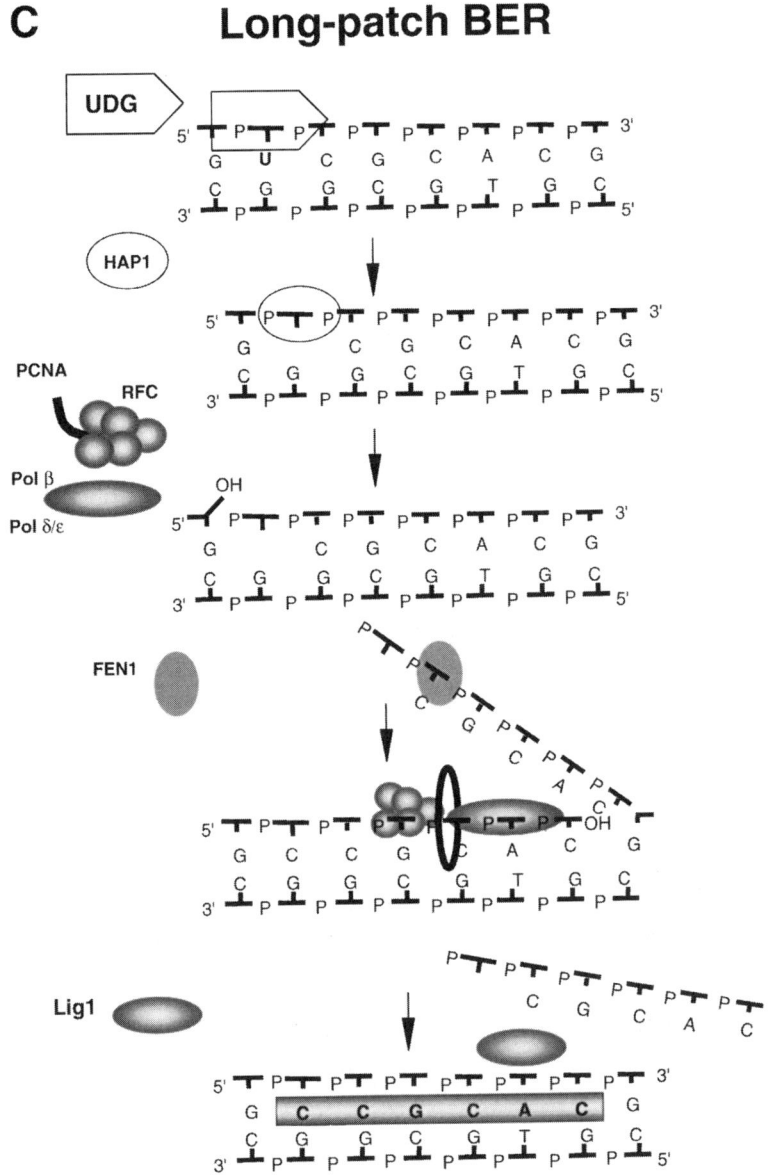

FIG. 1. (*continued*).

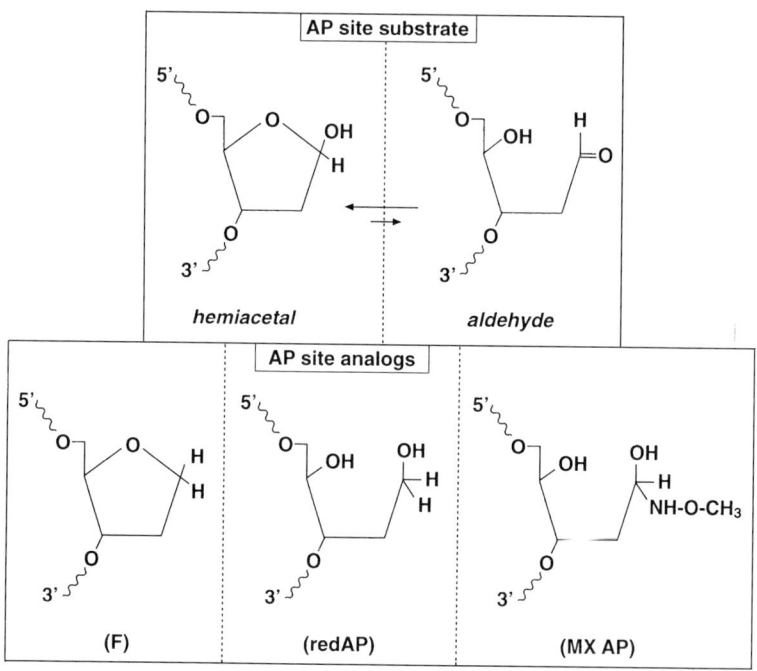

FIG. 2. The chemical structures of the regular AP site as present in DNA (the major cyclic hemiacetal form and the minor ring-opened aldehyde form) and the AP site analogs more frequently used in *in vitro* BER assays. F, tetrahydrofuran; redAP, reduced AP site; MXAP, methoxyamine-modified AP site.

(8-oxoG), which is excised by the mammalian bifunctional DNA glycosylase OGG1 (Fig. 1B) (5, 6). A third variation on the scheme (Fig. 1C) is the production of a gap that exceeds one nucleotide. This would be the consequence of a strand displacement reaction of the strand containing the dRP residue at the incised AP site. Replication factor C (RF-C), proliferating cell nuclear antigen (PCNA), and the structure-specific nuclease FEN1 are the additional players in this alternative branch. RF-C is required to load PCNA onto the DNA. PCNA functions as a DNA sliding clamp for Pol δ (reviewed in Ref. 7), and by interacting with FEN1, facilitates the excision of the oligonucleotide containing the 5′-incised AP site (8). This pathway has been reconstituted *in vitro* by using either Pol β or Pol δ/ε for the resynthesis step (9–11) in the presence of PCNA and FEN1. RF-C is required in the repair reaction when a circular plasmid is used as substrate (10, 11). *In vitro* data with mammalian cell extracts defective in the DNA ligase III-stabilizing factor XRCC1 show that the XRCC1/DNA ligase III complex is dispensable for long-patch BER, implying that the sealing step is performed by DNA ligase I (12).

Since this pathway requires PCNA, we will refer to it as PCNA-dependent BER or, due to the resynthesis of more than one nucleotide, as long-patch BER. Conversely, the main BER route that involves Pol β is named the Pol β-dependent BER pathway or single-nucleotide/short-patch BER. We have recently shown that Pol δ/ε cannot substitute for this Pol in one-nucleotide replacement reactions (*10, 13*).

If the cell has multiple choices in the priming of a specific BER pathway, the challenging question is identifying the factors that govern the switch mechanism among these alternative pathways. The current knowledge mainly relies upon *in vitro* studies performed with soluble cell extracts or purified proteins. This review summarizes the available experimental evidence with the objective of providing the basis for speculations on the regulation of these mechanisms *in vivo*.

II. Factors That Control the Switch among Multiple BER Pathways

A. Type of Damage

A decade ago, in their critical review on the use of permeabilized cell systems for analyzing mammalian DNA repair induced by a variety of BER-inducing agents, Keeney and Linn (*14*) concluded that these studies were consistent with a model in which "at least two different repair pathways exist in the cells" and that "the nature of the DNA lesion would determine which polymerase mediates repair synthesis." More recently, this hypothesis was confirmed and thoroughly analyzed using *in vitro* repair assays on single lesion-containing DNA substrates and BER mutant cell lines. These studies are presented in an historical perspective.

1. USE OF PERMEABILIZED CELL SYSTEMS

In 1970, Moses and Richardson (*15*) described the first permeable cell system for studying DNA repair in *Escherichia coli*. Following this report, several studies were performed using this experimental approach to investigate mammalian DNA repair enzymology. The great advantage of permeabilized cell systems over cell-free extracts is that the permeabilization procedure leaves intact the majority of the cellular DNA repair. The main disadvantage lies in the potential for artificial responses under nonbiological reaction conditions. Nevertheless, the repair processes observed in these systems are biologically relevant since they are damage- and repair-dependent. Moreover, the repair patches are similar to those measured *in vivo* (reviewed in Ref. *14*). There have been a number of reports of the effects of putatively selective DNA polymerase inhibitors

TABLE I
EFFECTS OF DNA POLYMERASE INHIBITORS ON DNA REPAIR INDUCED BY
BER-INDUCING AGENTS IN PERMEABLE MAMMALIAN CELL SYSTEMS

Inhibitor	DNA-damaging agent	Inhibition (%)			
		Repair	Pol α	Pol β	Pol δ
ddTTP	MNNG	29^a			
		34^b			
	MNU	54^b	$<5^a$	95^a	$<5^a$
	Bleomycin	79^b			
	Neocarzastatin	77^b			
	γ Rays	No effect^c			
Aphidicolin	MNNG	56^a			
		62^b			
	MNU	62^b	85^a	No effect^a	80^a
	Bleomycin	7^b			
	Neocarzastatin	22^b			
	γ Rays	Inhibition^c			

^{a}Data from Ref. 19. The concentrations of the inhibitors were: 100 μM ddTTP and 50 μg/ml aphidicolin.

^{b}Data from Ref. 17. The concentrations of the inhibitors were: 260 μM ddTTP and 10 μg/ml aphidicolin. The values reported refer to repair synthesis observed in the presence of 80 mM KCl.

^{c}Data from Ref. 18. The concentrations of the inhibitors were: 20 μM ddTTP and aphidicolin (30 μM) was used in combination with 1-β-D-arabinofuranosylcytosine (0.1 mM) to obtain maximal inhibition of repair synthesis.

on repair induced by a variety of chemical and physical agents. Many of these studies revolved around the different effects that aphidicolin and dideoxyTTP (ddTTP) have on these enzymes. Aphidicolin effectively inhibits DNA Pols $\alpha/\delta/\varepsilon$ but not Pol β, while ddTTP inhibits Pol β much more strongly than Pol $\alpha/\delta/\varepsilon$ (reviewed in Ref. 16). For the sake of comparison, some of the studies performed with BER-inducing agents are reported in Table I. With all the limitations intrinsic to the experimental system used, there is clear evidence that the effect of these inhibitors depends on the DNA-damaging agent used. Pol β seems to play a major role in the processing of lesions induced by bleomycin or neocarzastatin (17), while Pol α-like polymerases seem to be implicated in repair synthesis after exposure to ionizing radiation (18). In the case of monofunctional alkylating agents such as N-methyl-N'-nitro-N-nitrosoguanidine (MNNG) or N-methylnitrosourea (MNU), the data suggest that two repair pathways depending on different polymerases are involved in the cellular repair of DNA alkylation damage (17, 19).

These types of studies also suggested that an additional factor in determining which Pol would mediate repair synthesis is the dose of the damage. Dresler and Lieberman (20) reported that repair induced by low doses of MNU, bleomycin, N-acetoxy-2-acetylaminofluorene, and UV is more resistant to aphidicolin than is repair at higher doses. Although these data were not supported by other reports (21), it is tempting to speculate that alternative repair pathways might come into action at saturating damage doses. Interestingly, experiments performed in yeast POL3 (the yeast Pol δ) mutants showed that these cells were capable of repairing MMS-induced single-strand breaks (ssb) when only a small number of lesions was introduced into DNA (80% survival) (22). When the quantity of lesions increased (50% survival or less), the repair of ssb was blocked. This finding suggests that depending on the dose of DNA damage, different DNA Pols participate in the BER process.

2. Use of Cell-Free Extracts

a. DNA Substrates Containing AP Site Analogs The first *in vitro* BER system was developed by Matsumoto and co-workers (23) using crude extracts prepared from *Xenopus laevis* ovaries on a circular DNA substrate containing a synthetic AP site analog, tetrahydrofuran. In these studies, the choice of the AP site or its analogs was due to the fact that the AP site is the common repair intermediate during the BER process, and it is easy to construct in linear or circular DNA substrates. Figure 2 shows the structures of the regular AP site as present in DNA (i.e., the major cyclic hemiacetal form and the minor ring-opened aldehyde form) and of the AP site analogs more frequently used in BER assays. The *X. laevis* extracts were shown to be proficient in the repair of the synthetic abasic site tetrahydrofuran (23, 24). Later, the same results were obtained with a circular substrate containing a natural AP site and using mammalian cell extracts on either natural or methoxyamine (MX)-modified AP sites (25–27).

When the tetrahydrofuran was used as a model lesion, the PCNA-dependent BER was fully proficient (25), while the Pol β-dependent pathway, which utilizes β-elimination for excision, was unable to repair the tetrahydrofuran residue, as well as the reduced AP site (28). Although obtained with synthetic lesions, this is the first *in vitro* evidence that the type of damage was able to direct the selection of a specific BER branch. This was later confirmed when the repair of the reduced AP sites was reconstituted *in vitro* (9). HAP1, Pol β, and DNA ligase I were unable to complete repair when an oligonucleotide containing a single reduced AP site was used as substrate. The structure-specific nuclease FEN1 was essential for the repair of the reduced AP site, which occurred through the long-patch BER pathway.

Another modification of the AP site that has been extensively analyzed is the product of the interaction of the primary amine methoxyamine (MX) with the aldehyde of an AP site (29) which results in a stably adducted sugar molecule (30) (Fig. 2). In 1991, by using a synthetic oligonucleotide containing a single MX-modified AP site, we showed that this modification changed the substrate specificity of different AP endonucleases (31). While endonuclease III and endonuclease IV of *E. coli* were able to cleave the alkoxyamine-adducted site, a HeLa AP endonuclease and crude cell extracts were inhibited by this modification. This topic has recently been reanalyzed (32). Human HAP1/APE was able to cleave the MX-modified AP site, but as suggested by our previous observations, the turnover number of APE acting on the modified AP site was almost 330-fold lower than that for the normal AP site. In addition, the modification of the AP site with MX was shown to inhibit the β-elimination step involved in the dRP lyase activity of Pol β. Accordingly, a MX-adducted AP site, preincised with APE, was an active substrate for repair only when FEN1 and Pol β were added to the reaction. Similarly to what was observed with other AP sites with altered sugar moieties, when the incision step is not followed by the elimination of the dRP moiety, the long-patch BER takes over the single-nucleotide pathway. This phenomenon is clearly shown by the experiment displayed in Fig. 3. A circular duplex plasmid containing a single MX-modified AP site was used as substrate for an *in vitro* repair reaction in the presence of HeLa whole cell extracts. The results obtained with the reduced AP site are also shown for comparison. DNA repair replication was monitored as incorporation of labeled dTMP (lanes 1–3 and 6–8) or dCMP (lanes 4, 5 and 9, 10) within the *Xba*I–*Hin*dIII restriction fragment. Vigorous and comparable repair activity was observed with either dNTP on both lesions. The occurrence of long-patch BER is testified both by the strong incorporation of dCMP (C residues are located 3' to the lesion) and by the presence of repair intermediates exceeding the seven-nucleotide fragment that marks one-nucleotide incorporation at the lesion site (lanes 1, 2, 6, 7, and 9).

b. *DNA Substrates Containing BER Lesions*

i. OXIDIZED AP SITES   Reactive oxygen species (ROS) are generated continuously in the cells during either endogenous or exogenous processes such as exposure to ionizing radiation. The reaction of ROS with DNA determines the formation of a plethora of lesions, including oxidized purines and pyrimidines, DNA strand breaks, and oxidized AP sites (OAS). It has been proposed that OAS are among the major oxidative lesions in double-stranded DNA (32). ROS are able to abstract hydrogen atoms from all possible sugar positions, thus forming OAS. The chemical structure of AP sites oxidized at C1', C2', and C4' is shown in Fig. 4. These sites are thought to be refractory to β-elimination. In 1997, Lindahl and co-workers (9) showed that *in vitro* repair of plasmid treated with γ-irradiation or phenantroline/Cu^{2+} was increased 5-fold when FEN1 was

THE SWITCH MECHANISM

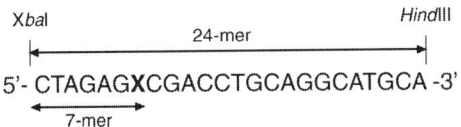

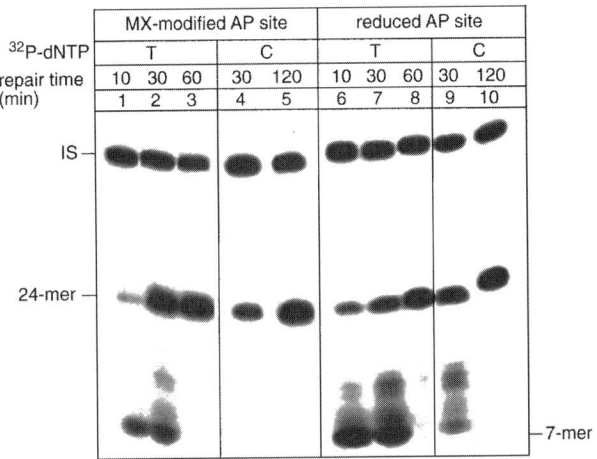

FIG. 3. Repair of MX-modified or reduced AP site by HeLa cell extracts. Autoradiograph of a denaturing polyacrylamide gel. Repair replication was performed for different periods of time on plasmid DNA containing either a single MX-modified AP site (lanes 1–5) or a single reduced AP site (lanes 6–10). DNA repair synthesis was monitored as incorporation of labeled dTMP (lanes 1–3 and 6–8) or dCMP (lanes 4, 5 and 9, 10) within the XbaI–HindIII restriction fragment of the plasmid (scheme on the top). The positions on the gel of the full length repair product (24-mer) and of the repair intermediate arising from single-nucleotide incorporation at the lesion site (7-mer) are indicated. IS, internal standard.

added to the reaction containing HAP1, Pol β, and DNA ligase I. No promotion was reported when the DNA substrate was a heat-depurinated plasmid, suggesting that OAS are specifically responsible for the involvement of FEN1, thus determining the occurrence of the long-patch BER. However, Demple and co-workers (34) have shown that the 5' nicked C4'-oxidized abasic sites induced by bleomycin, created by APE1, are excised by human Pol β. The excision efficiencies of Pol β for 5' nicked OAS directly induced by oxidative stress remain to be established. Recent data would suggest that these sites are indeed resistant to β-elimination [(35), see below].

ii. Lesions That Are Substrates for Different Types of DNA Glycosylases BER is characterized by an initial step, which is damage-specific and carried out by individual DNA glycosylases. The DNA glycosylases bind

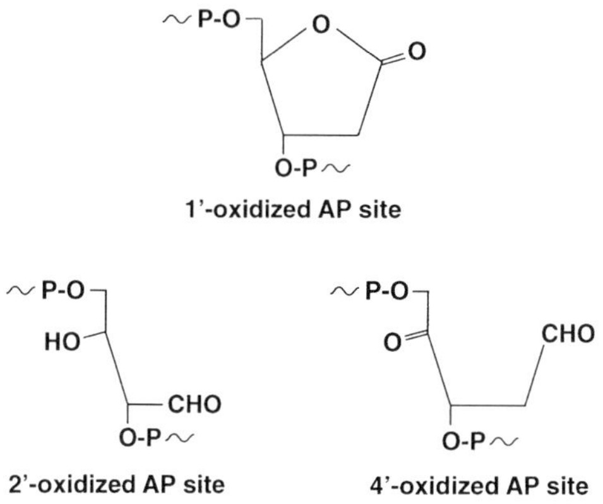

1'-oxidized AP site

2'-oxidized AP site

4'-oxidized AP site

FIG. 4. The chemical structure of AP sites oxidized at C1', C2', and C4'.

specifically to a damaged base and catalyze the cleavage of the N-glycosidic bond (monofunctional DNA glycosylases). In some cases, the DNA glycosylases are endowed with a concomitant AP lyase activity that cleaves the 3' site of the resulting AP site (bifunctional DNA glycosylase/β-lyases). In a recent study (6), we tested the hypothesis that the type of base damage present on DNA, by determining the specific DNA glycosylase in charge of its excision, drives the repair of the resulting AP site intermediate to either BER branch. The specific lesions for the two types of DNA glycosylases that we selected were hypoxanthine (HX) and 1,N^6-ethenoadenine (εA), which are the substrates for the monofunctional DNA-glycosylase 3-methyladenine DNA glycosylase (ANPG), and 8-oxoG, which is excised by a bifunctional DNA glycosylase/β-lyase 8-oxoG-DNA glycosylase (OGG1) (reviewed in Ref. 36). Circular plasmid molecules containing either a single HX, εA, or 8-oxoG were constructed and used in an *in vitro* repair assay with mammalian cell extracts. We showed that HX and εA were repaired via both short- and long-patch BER, while 8-oxoG was repaired mainly via the single-nucleotide pathway. The single-residue replacement pathway for the repair of oxidized bases excised by bifunctional DNA glycosylases has also been described by other groups (4, 5). These results are consistent with a model (Fig. 5) in which the intrinsic properties of the DNA glycosylase that recognizes the lesion select the branch of BER that will restore the intact DNA template. The formation of a 5' abasic terminus by the sequential action of a monofunctional glycosylase, such as ANPG, and of a 5'-AP endonuclease, such as HAP1/APE, and its slow processing by a dRPase activity will determine the long-patch repair events that occur in competition with the predominant one-gap-filling reactions (Fig. 5A). Conversely, the formation of a 3' blocked terminus by a glycosylase/AP lyase, such as OGG1, will be preferentially followed by single-nucleotide

THE SWITCH MECHANISM

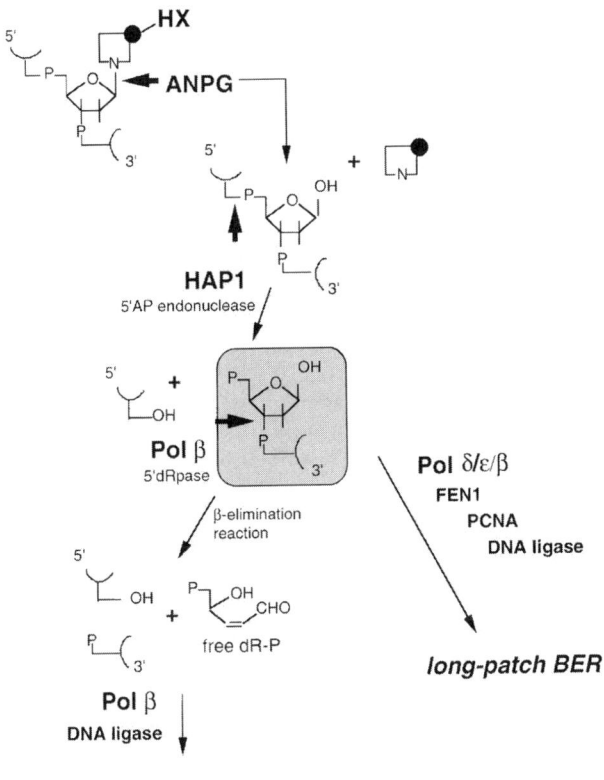

FIG. 5. Model showing how the type of DNA glycosylase that recognizes the lesion might select a specific BER pathway. (A) When the damaged base is recognized by a monofunctional DNA glycosylase followed by a 5′ AP endonuclease, the slow processing by dRPase activity of the resulting 5′ abasic terminus will determine strand displacement events. Long-patch BER will occur in competition with the predominant one-gap filling reactions. (B) The formation of a 3′ blocked terminus by a glycosylase/AP lyase, such as OGG1, will preferentially determine the occurrence of single-nucleotide replacement reactions owing to the presence of a genuine 5′ nucleotide residue produced by the AP lyase, which minimizes the strand displacement events.

replacement reactions due to the presence of a genuine 5′ nucleotide residue which minimizes the strand displacement events. In this case, the determinant of the repair synthesis step will be the production of 3′OH primers (likely by the 3′ phosphodiesterase activity of HAP1) since the 5′ terminus produced by the AP lyase is ready for the ligation step (Fig. 5B). As predicted by Keeney and

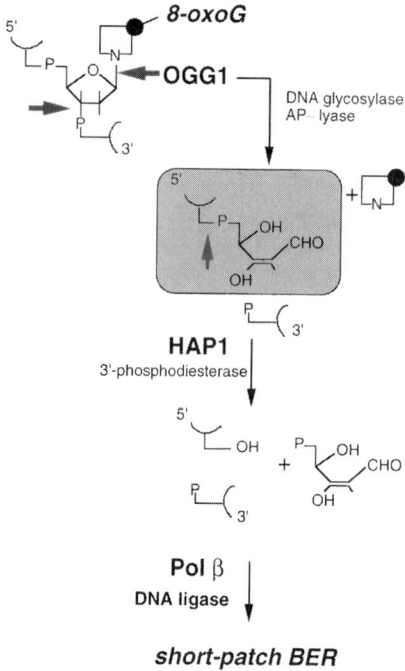

FIG. 5. (continued).

Linn (14) on the basis of the studies with DNA Pol inhibitors, a checkpoint for the selection of the repair route is likely to be "the nature of the gap or strand break to be resynthesized."

3. USE OF POL β-MUTANT MOUSE CELL LINES

Transgenic mice with a heterozygous germline deletion mutation in the Pol β gene (37) are viable up to day 18.5 postcoitum, but die perinatally (38). Pol β-null embryonic fibroblast cell lines were established from these embryos (39) and have been characterized to gain insight into the biological consequences of a defect in the Pol β-dependent BER. These cells are normal in viability and growth characteristics, but present a specific pattern of sensitivity to the cytotoxic effects of physical and chemical agents (Table II). Pol β-defective cells are hypersensitive to the cytotoxic effects of the alkylating agents methyl

TABLE II
CYTOTOXIC EFFECTS OF A VARIETY OF
DNA-DAMAGING AGENTS IN
POL β-DEFICIENT MOUSE CELLS AS
COMPARED WITH PROFICIENT CELLS

Agent	Hypersensitivity
MMS	Yes
EMS	Yes
MNNG	Yes
MNU	Yes
Mafosfamide	Yes
Mitomycin C	Yes
HeCNU	No
Cisplatin	No
Melphalan	No
BPDE	No
UV	No
γ Rays	No

Data from Refs. 39 and 40.

methanesulfonate (MMS), ethyl methanesulfonate (EMS), MNNG, and MNU and to the crosslinking antineoplastic drugs mafosfamide and mitomycin C, while they present the same sensitivity as normal cells to the killing effects of N-hydroxylethyl-N-chloroethylnitrosourea (HeCNU), cisplatin, melphalan, benzo(a)pyrene diol epoxide (BPDE), UV, and ionizing radiation [(39, 40); see also Sobol and Wilson, this volume]. The hypersensitivity to monofunctional alkylating agents that induce lesions processed via BER is an expected phenotype, as is the lack of hypersensitivity to NER-inducing agents such as UV and BPDE. Intriguing is the fact that the killing effect of ionizing radiation, which produces a plethora of BER lesions, is similar in Pol β-deficient and proficient cells. It is interesting to recall that in the old studies with DNA Pol inhibitors (reviewed in Ref. 14), the repair induced by UV or ionizing radiation was classified as ddTTP-insensitive while the MNNG- or MNU-induced repair was partially inhibited by the Pol β-selective inhibitor, ddTTP (see Table I).

More specific studies designed to investigate the molecular mechanisms of the damage-specific killing in Pol β-null cells were performed by using AP site reagents to inhibit the single-nucleotide BER, i.e., MX (32). In 1991, we reported that AP sites formed during the cell processing of alkylation-induced damage interact with MX, thus becoming resistant to alkali treatment (41). Accumulation in the cell genome of MX-modified AP sites resulting from the processing of S_N2-type alkylating agents, such as MMS, EMS, and diethyl sulfate, was shown to be associated with an increase in cytotoxicity (42). Sensitization by MX to the killing effects of MMS has recently been confirmed in Pol β-proficient mouse embryonic cells, and the interesting observation has been reported that Pol β-defective cells are instead refractory to MX sensitization (32). This suggests that the inhibition of the single-nucleotide BER by MX gives rise to the persistence of cytotoxic BER intermediates. Since Pol β-defective cells predominantly use the BER long-patch pathway [(10, 13)], MX cannot interfere with their cytotoxic response to alkylating agents. However, as remarked by Horton et al. (32), the hypersensitivity of the Pol β-null cells to alkylating agents cannot be explained solely by the inhibition of the single-nucleotide BER, since in the presence of MX they are still considerably more sensitive than wild-type cells. A possible explanation for this finding is that in vivo, Pol β is also involved in long-patch BER (43, 44) and cannot be efficiently replaced by Pol δ/ε. Alternatively, this might be due to incomplete modification by MX of the AP sites formed in the cells.

We have recently investigated the DNA ssb induction and repair by alkaline single-cell gel electrophoresis (SCGE) in Pol β-proficient and -deficient cells following exposure to MMS and hydrogen peroxide (H_2O_2) (13). Under these conditions, damaged cells present the shape of a comet with the tail representing supercoiled loops of DNA that, as a consequence of breaks, relax and migrate to form a tail (the tail moment is the integrated measure that is used to quantitate DNA damage) (45, 46). The alkaline SCGE allows one to monitor the balance between DNA ssb and alkali-labile sites (i.e., AP sites) arising from BER and breaks resealed by the same repair process. The results of a typical experiment are displayed in Fig. 6. A defect in Pol β was associated with a higher level of DNA strand breakage immediately after MMS treatment and with an accumulation of DNA ssb after 2-h repair in drug-free medium (although 70% of the initial breaks were repaired). In contrast, DNA ssb formation and repair by H_2O_2 was unaffected by the lack of Pol β. These data not only confirm that Pol β is the polymerase of choice for BER induced by alkylation damage, but also provide convincing evidence that in vivo an alternative, albeit kinetically slower, BER system is active on alkylated bases. In the case of oxidative damage, our data are compatible with the hypothesis that oxidation-induced DNA ssb are directly rejoined and/or do not require Pol β for their rejoining. Since the majority of ssb detected by alkaline SCGE are oxidized AP sites (frank strand breaks are a minor fraction of the total breaks) (33), their repair should be independent of a functional Pol β gene. The data are consistent with repair of OAS in the cells via long-patch BER, which is equally efficiently performed by Pol β or Pol δ/ε.

All these data were based on the use of Pol β-defective mouse cells and were obtained almost 10 years after the pioneering studies with permeable cell systems (reviewed in Ref. 14). The degree of agreement among these two sets

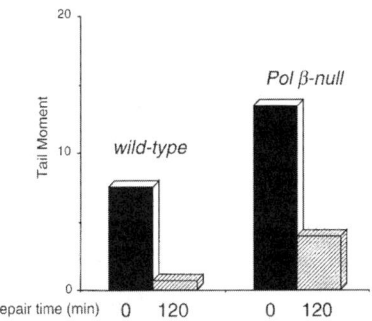

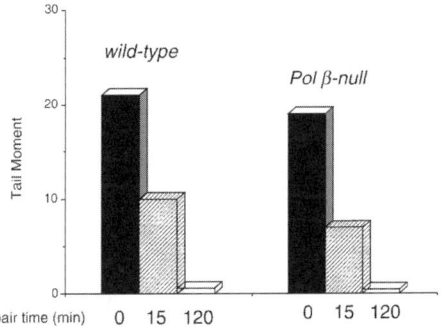

FIG. 6. DNA single-strand break induction and repair by MMS and H_2O_2 in Pol β-proficient and -deficient mouse cells. Cells were treated with 0.5 mM MMS or 50 μM H_2O_2 for 30 min and then allowed to repair in drug-free medium for different periods of time, as indicated. At different time intervals, samples were taken for SCGE analysis. The tail moment of 50 comets per experimental point was measured by computerized image analysis in two independent experiments. The results of a typical experiment are shown.

of data is striking: Damage induced by monofunctional alkylating agents is confirmed to be repaired via either BER pathway (either aphidicolin or ddTTP was able to partially inhibit alkylation damage-induced repair; see Table I), although the single-nucleotide pathway seems to be the favorite route, while ROS-induced DNA damage is predominantly repaired via long-patch BER (ionizing radiation-induced DNA damage was inhibited by aphidicolin but not by ddTTP; see Table I). This is further evidence that the type of damage induced in the cellular DNA drives the selection of the BER pathway.

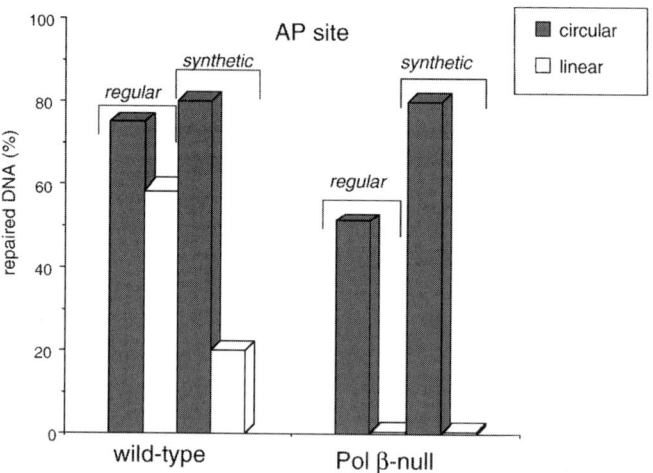

FIG. 7. Influence of DNA structure on the repair of AP sites. [Data derived from Biade et al. (47)]. Circular or linear DNA carrying either a natural or a synthetic AP site was incubated with Pol β-proficient or -deficient mouse cell extracts. The repair efficiency reported in the histograms refers to repair reactions carried out for 60 min.

B. Influence of DNA Structure

In 1994, we reported that repair replication by CHO cell extracts stimulated by AP sites was decreased when the substrate was linearized (26). Following the discovery that one of the BER pathways was PCNA-dependent, we hypothesized that the impairment of the repair reaction on the linear substrate could be due to unstable loading of the PCNA clamp onto linear DNA (27). More recently, Matsumoto and co-workers (47) showed that when the synthetic AP site was present on a circular plasmid, it was efficiently repaired by wild-type cells, while the repair efficiency dropped dramatically when the lesion was located on a linear DNA (Fig. 7). In contrast, the regular AP site was efficiently repaired independently of the DNA structure. When cell extracts from Pol β knockout cells were used, a significant reduction in repair efficiency due to DNA structure (circular versus linear DNA) was observed with both the regular and synthetic AP site. The lack of repair with Pol β-defective extracts when the target lesion is present in linear DNA indicates that only Pol β is functional on this DNA substrate. The circularity-dependent repair by the Pol β null cells was shown to be inhibited by PCNA antibody and fully restored by the addition of PCNA (47). Therefore, the PCNA-dependent pathway is defective on linear substrates, likely because of the unstable loading of PCNA on this structure.

These results have a more general, although indirect, implication for repair in the cells: The local presence of PCNA might be a key determinant in selecting the BER pathway (see also Section II,C).

C. Gene Expression of BER Factors

The yeast model gives the most striking evidence that the overall level of cellular polymerases determines the cellular repair capacity. The repair of methylation damage is partially defective in strains with mutations in the POL3 gene (*48*). This has been interpreted as an indication that Pol β is required for BER in yeast. In yeast, Pol β (Pol 4 gene) is not expressed in mitotically growing cells (*49*) but is induced after entry into meiosis, where it seems to play a nonessential role. Interestingly, Pol β is abundantly expressed in male germ cells and localizes at meiotic chromosomes during meiosis (*50–53*). However, if mammalian Pol β is expressed in POL3 mutants, it is able to suppress the defect in repair and complement the MMS sensitivity. The Pol β-null mouse fibroblasts behave like the yeast POL3 mutants. They are hypersensitive to MMS and repair methylation damage less efficiently than do wild-type cells (see Section II,A,3). This is due to the slower kinetics of the long-patch BER, the predominant BER route in the absence of Pol β, as compared with the single-nucleotide pathway (*54*). As shown in Fig. 8, when a circular duplex plasmid containing a single AP site was incubated with Pol β-deficient mouse cell extracts in the presence of labeled dTTP for short incubation times (15 and 30 min), a lower yield of repaired fragments as compared with the wild-type cell extracts was observed (compare lanes 1 and 2 with 3 and 4). This is expected on the basis of the kinetics of the two repair pathways. The addition of Pol β to the mutant extracts (lanes 5 and 6) was able to complement the defect by switching the repair pathway from the long-patch to the more efficient single-nucleotide pathway. The mapping of the repair patches confirmed that the majority of the repair events were in fact one nucleotide replacement reactions (B. Pascucci *et al.*, unpublished data). Therefore, the type of DNA polymerase involved in the BER reaction is associated with a specific BER branch: Pol β with the single-nucleotide BER, and Pol δ/ε with the long-patch BER. On the basis of these results, it is conceivable that the relative concentration in the cells of these Pols (or their availability at the lesion site) might affect not only the cell repair capacity, but also its repair mode.

Pol β is expressed at different levels depending on the cell type or differentiation stage (*55*) and is upregulated in response to DNA-damaging agents (*56*). Pol β level is significantly elevated in some types of adenocarcinomas and cell lines. Wilson and co-workers (*55*) have hypothesized that the increase in the expression of this protein might be due to an increased number of DNA lesions owing to intracellular production of genotoxic agents (endogenous oxidative and alkylation damage). What are the consequences of an overexpression

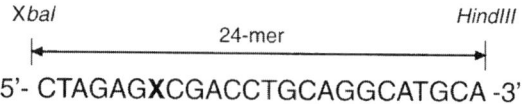

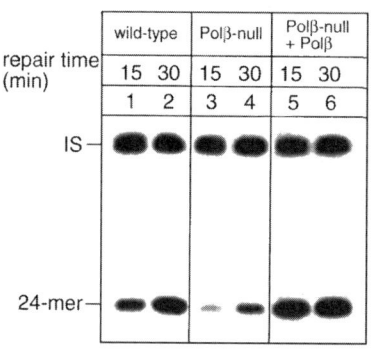

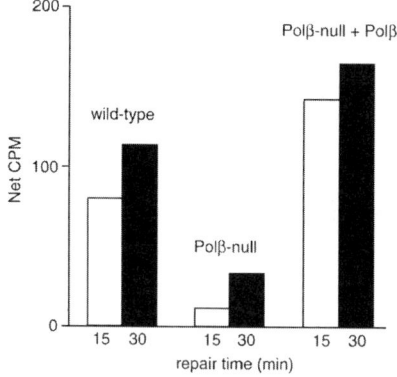

FIG. 8. Effect of additional Pol β on AP site repair by wild-type and Pol β-deleted mouse cell extracts. Repair replication was monitored as incorporation of labeled dTMP within the XbaI–HindIII restriction fragment of the plasmid (scheme on the top). The positions on the denaturing polyacrylamide gel of the full-length repair product (24-mer) and of the internal standard (IS) are indicated. Lanes 1 and 2: Repair synthesis by wild-type cell extracts as a function of the incubation time. Lanes 3 and 4: Repair synthesis by Pol β null cell extracts as a function of the incubation time. Lanes 5 and 6: After addition of 10 ng of purified rat Pol β. Bottom: The repair products were measured by electronic autoradiography, and relative incorporation, corrected for DNA recovery, is indicated on the ordinate (Net CPM).

of Pol β? It has been shown that hamster cells overexpressing Pol β acquire a spontaneous mutator phenotype (57). Interestingly, these cells displayed a decreased sensitivity to bifunctional cancer chemotherapeutic agents such as cisplatin and melphalan. No effect of these same agents was reported on the survival of Pol β-defective cells [(40), Table II], suggesting that Pol β imbalance "forces" the participation of this Pol into other cellular processes (translesion DNA synthesis? recombination?) that would normally rely upon other Pols with a higher fidelity.

Pol β is expressed constitutively in quiescent cells stimulated to proliferate and throughout the cell cycle, while Pol δ/ε, PCNA, and FEN1, as is typical of replication proteins, show a significant increase when quiescent cells are stimulated to proliferate. Moreover, in actively replicating cells, a trimeric form of PCNA is found in close association with the replicating forks; the same chromatin-bound PCNA complex is found following treatment of quiescent cells with UV or DNA-damaging agents such as alkylating agents or hydrogen peroxide (58, 59). In contrast, no change of FEN1 has been described either in protein levels or in tissue sections in response to UV treatment or γ irradiation (60). PCNA complex formation reflects NER activity as NER-deficient human XPA, XP-F, and XP-G cells fail to show any detectable PCNA complex after UV damage (61–64). Interestingly, recent data would suggest that the PCNA complex formation is also a marker of BER activity, since hydrogen peroxide-treated CS-B cells show a reduction in the ratio of PCNA relocated as compared to normal and XP-A cells (65). The presence of PCNA at the lesion site might apply a "selective pressure" on the occurrence of long-patch BER. This is suggested by our *in vitro* data showing that the addition of PCNA to PCNA-depleted wild-type mouse cell extracts determines the abrupt shift from the single-nucleotide pathway to the long-patch BER (B. Pascucci *et al.*, unpublished data).

D. Cell Cycle

The participation of replication proteins in the long-patch BER suggests that repair processes activated at or after the replication of chromosomal DNA might follow this route. Recent evidence has been provided that specific BER complexes are associated with different stages of the cell cycle. Otterlei *et al.* (66) have shown that a BER complex containing the major uracil-DNA glycosylase (UNG2), PCNA, and RPA is preferentially located in replication foci. Other BER factors have been reported to localize at replication foci such as Pol δ and DNA ligase I, while Pol β is not present in these multiprotein replication complexes. These findings strongly suggest that a Pol δ–PCNA-dependent long-patch BER is involved in the removal of misincorporated dUTP residues in a post-replicative stage. A specific cell cycle requirement for some BER factors was also recently reported during mammalian DNA strand break repair (67). XRCC1 protein is required for efficient DNA ssb repair, and similarly to what was reported for Pol β, this protein is essential for viability. Taylor and co-workers (67) observed that mutations within the XRCC1 BRCT domain, which interacts with DNA ligase III, abolished the XRCCI-dependent ssb repair occurring in G_1 phase of the cell cycle. In contrast, these same mutants had no effect in S phase, suggesting that the XRCC1 DNA ligase III complex would be not strictly required for DNA ssb repair in the S phase (67). It is important to recall that the XRCC1–DNA ligase III complex seems to be specifically involved in the short-patch BER, while DNA ligase I seems to be involved in the long-patch BER (12). It is tempting to speculate that the involvement of the two Pols in BER is controlled by the cell cycle stage. The overlap among the factors involved in DNA replication and those required for long-patch BER might favor Pol δ/ε in BER at post-replicative stages, while the single-nucleotide Pol β-mediated BER might be favored in pre-replicative stages.

III. Conclusions and Future Research

One interesting and specific feature of the BER process is its versatility and apparent redundancy. Certain BER enzymes, such as the DNA glycosylases, present a broad substrate specificity while others, such as Pol β or HAP1, are able to catalyze more than one BER step (e.g., Pol β has a dual polymerase and dRPase activity and HAP1 possess both AP endonuclease and 3' phosphodiesterase activities). On the other hand, there is an apparent redundancy at the level of the enzymes required for the resynthesis and ligation steps. Biochemical reconstitution has shown that either Pol β or δ/ε is capable of the resynthesis step, at least in the long-patch BER, and either DNA ligase I or III can seal the final DNA nick. In apparent contradiction with this notion, in mice, targeted deletions of Pol β (37, 38) and XRCC1 (68) lead to embryonic lethality. This is likely to reflect the fact that during specific stages of mouse development, impairment of the most efficient BER pathway (i.e., the single-nucleotide BER), resulting in accumulation of DNA ssb in specific tissues, is not tolerated by the embryo and results in death of the whole conceptus (69). This defect is compatible with life if it arises in later stages. Alternatively, we might hypothesize that the dramatic effects of Pol β or XRCC1 deletion in embryogenesis reflects a role of these proteins in important biological processes required for correct development other than BER (e.g., recombination, apoptosis, differentiation). Biochemical reconstitution studies, as well as genetic studies of damage repair, have allowed the identification of some of the potential "checkpoints" involved in the control of the multiple BER pathways. We have summarized the evidence that indicates how the type of DNA damage might exert its function via either the specificity of its recognition (e.g., via the DNA glycosylase responsible for its removal) or, after cleavage, by the constraints applied to the repair machinery via the structure of its 3′ and 5′ termini. Other important factors in the switch among multiple BER pathways seem to be the local structure of DNA and the expression level of BER-specific factors. Future research should investigate the control of the BER pathways in the cells in relation to the gene expression pattern and in connection with fundamental cell processes such as replication and transcription. The participation of replication proteins in one of the BER pathways opens the question of whether this subpathway might be specific for repair events occurring at replication foci. Pol δ, RFC, PCNA, and FEN1 recognize and process 3′ ends of gapped DNA and fill the gaps created during BER by the same mechanism that is used for joining Okazaki fragments. Moreover, PCNA might play an important role at the intersection among different excision repair strategies, namely BER, NER, and mismatch repair. PCNA interacts with multiple repair factors: It binds to damage recognition proteins, such as UDG (70) and MSH2/MLH1 (71), and to fragment processing and joining proteins, such as XP-G (72), FEN1 (73), and DNA ligase I (74, 75). It is tempting to speculate that this protein might play an important role in a more general "switch" mechanism that controls the participation of different excision repair pathways in DNA damage processing. The soluble, *in vitro* biochemically reconstituted repair reactions have facilitated the elucidation of which enzymes are the players in the different BER pathways; but at the current state of art, they have "exhausted" their role. In the intact cell, transactions are carried out by spatially organized arrays of enzymes on surfaces (e.g., the nuclear matrix) and in subcellular compartments. These features enhance the efficiency of repair by reducing the search for lesions and by concentrating particular repair events to those genomic domains in which the repair process is essential to normal cellular functions. Recently, evidence has been presented that specific BER pathways might be associated with specific cell cycle stages, and possible "DNA ssb factories" have been identified in association with replication (67). The challenge for the future is to develop tools to investigate the intracellular localization of BER repair events and to answer the key questions on the biological significance of the multiple BER pathways and on the "controllers" of the switch mechanism in less artificial experimental systems.

REFERENCES

1. G. Dianov and T. Lindahl, *Curr. Biol.* **4**, 1069–1076 (1994).
2. Y. Kubota, R. Nash, A. Klungland, P. Schar, D. Barnes, and T. Lindahl, *EMBO J.* **15**, 6662–6670 (1996).
3. D. K. Srivastava, B. J. Vande, R. Prasad, J. T. Molina, W. A. Beard, A. E. Tomkinson, and S. H. Wilson, *J. Biol. Chem.* **273**, 21203–21209 (1998).
4. A. Klungland, M. Hoss, D. Gunz, A. Constantinou, S. G. Clarkson, P. W. Doetsch, P. H. Boltin, R. D. Wood, and T. Lindahl, *Mol. Cell* **3**, 1–20 (1999).
5. G. Dianov, C. Bischoff, J. Piotrowski, and V. A. Bohr, *J. Biol. Chem.* **278**, 33511–33516 (1998).

6. P. Fortini, E. Parlanti, O. M. Sidorkina, J. Laval, and E. Dogliotti, *J. Biol. Chem.* **274**, 15230–15236 (1999).
7. Z. O. Jonsson and U. Hubscher, *BioEssays* **19**, 1–9 (1997).
8. R. Gary, K. Kim, H. L. Cornelius, M. S. Park, and Y. Matsumoto, *J. Biol. Chem.* **274**, 4354–4363 (1999).
9. A. Klungland and T. Lindahl, *EMBO J.* **16**, 3341–3348 (1997).
10. B. Pascucci, M. Stucki, Z. O. Jonsson, E. Dogliotti, and U. Hubscher, *J. Biol. Chem.* **274**, 33696–33702 (1999).
11. Y. Matsumoto, K. Kim, J. Hurwitz, R. Gary, D. S. Levin, A. E. Tomkinson, and M. S. Park, *J. Biol. Chem.* **274**, 33703–33708 (1999).
12. E. Cappelli, R. Taylor, M. Cevasco, A. Abbondandolo, K. Caldecott, and G. Frosina, *J. Biol. Chem.* **272**, 23970–23975 (1997).
13. P. Fortini, B. Pasevcci, F. Beusario, and E. Dogliotti, *Nucleic Acids Res.* **28**, 3040–3046 (2000).
14. S. Keeney and S. Linn, *Mutat. Res.* **236**, 239–252 (1990).
15. R. E. Moses and C. C. Richardson, *Proc. Natl. Acad. Sci. U.S.A.* **67**, 674–681 (1970).
16. M. Fry and L. A. Loeb, "Animal Cell DNA Polymerases." CRC Press, Boca Raton, FL, 1986.
17. M. R. Miller and D. N. Chinault, *J. Biol. Chem.* **257**, 46–49 (1982).
18. R. Mirzayans, M. R. Enns, S. Cubitt, K. Karimian, B. Radatus, and M. C. Paterson, *Biochim. Biophys. Acta* **1227**, 92–100 (1994).
19. R. A. Hammond, J. K. McClung, and M. R. Miller, *Biochemistry* **29**, 286–291 (1990).
20. S. L. Dresler and M. W. Lieberman, *J. Biol. Chem.* **258**, 9990–9994 (1983).
21. S. M. Keyse and R. M. Tyrrell, *Mutat. Res.* **146**, 109–119 (1985).
22. W. Suszek, H. Baranowska, J. Zak, and W. J. Jachymczyk, *Curr. Genet.* **24**, 200–204 (1993).
23. Y. Matsumoto and D. F. Bogenhagen, *Mol. Cell. Biol.* 3750–3757 (1989).
24. Y. Matsumoto and D. F. Bogenhagen, *Mol. Cell. Biol.* **11**, 4441–4447 (1991).
25. Y. Matsumoto, K. Kim, and D. F. Bogenhagen, *Mol. Cell. Biol.* **14**, 6187–6197 (1994).
26. G. Frosina, P. Fortini, O. Rossi, A. Abbondandolo, and E. Dogliotti, *Biochem. J.* **304**, 699–705 (1994).
27. G. Frosina, P. Fortini, O. Rossi, F. Carrozzino, G. Raspaglio, L. S. Cox, D. P. Lane, A. Abbondandolo, and E. Dogliotti, *J. Biol. Chem.* **271**, 9573–9578 (1996).
28. Y. Matsumoto and K. Kim, *Science* **269**, 699–702 (1995).
29. M. Liuzzi and M. Talpaert-Borlé, *J. Biol. Chem.* **260**, 5252–5258 (1985).
30. P. Fortini, A. Calcagnile, H. Vrieling, A. A. van Zeeland, M. Bignami, and E. Dogliotti, *Cancer Res.* **53**, 1149–1155 (1993).
31. S. Rosa, P. Fortini, P. Karran, M. Bignami, and E. Dogliotti, *Nucleic Acids Res.* **19**, 5569–5574 (1991).
32. J. K. Horton, R. Prasad, E. Hou, and S. H. Wilson, *J. Biol. Chem.* **275**, 2211–2218 (2000).
33. C. von Soontag, "The Chemical Basis of Radiation Biology" (Taylor and Francis, eds.), London, 1987.
34. Y.-J. Xu, E. Y. Kim, and B. Demple, *J. Biol. Chem.* **273**, 28837–28844 (1998).
35. J. Nakamura, D. K. La, and J. A. Swemberg, *J. Biol. Chem.* **275**, 5323–5328 (2000).
36. J. Laval, J. Jurado, M. Saparbaev, and O. Sidorkina, *Mutat. Res.* **402**, 93–102 (1998).
37. H. Gu, J. D. Marth, P. C. Orban, H. Mossmann, and K. Rajewsky, *Science* **265**, 103–106 (1994).
38. G. Esposito, G. Texido, U. A. K. Betz, H. Gu, W. Muller, U. Klein, and K. Rajewsky, *Proc. Natl. Acad. Sci. U.S.A.* **97**, 1166–1171 (2000).
39. R. W. Sobol, J. K. Horton, R. Kuhn, H. Gu, R. K. Singhal, R. Prasad, K. Rajewsky, and S. H. Wilson, *Nature (London)* **379**, 183–186 (1996).
40. K. Ochs, R. W. Sobol, S. H. Wilson, and B. Kaina, *Cancer Res.* **59**, 1544–1551 (1999).
41. P. Fortini, M. Bignami, and E. Dogliotti, *Mutat. Res.* **236**, 129–137 (1990).
42. P. Fortini, S. Rosa, A. Zijno, A. Calcagnile, M. Bignami, and E. Dogliotti, *Carcinogenesis* **13**, 87–93 (1992).

43. G. Dianov, R. Prasad, S. Wilson, and V. Bohr, *J. Biol. Chem.* **274,** 13741–13743 (1999).
44. R. Prasad, G. L. Dianov, V. A. Bohr, and S. H. Wilson, *J. Biol. Chem.* **275,** 4460–4466 (2000).
45. N. P. Singh, M. T. McCoy, R. R. Tice, and E. L. Schneider, *Exp. Cell. Res.* **175,** 184–191 (1988).
46. P. L. Olive, J. P. Banath, and R. E. Durand, *Radiat. Res.* **122,** 86–94 (1990).
47. S. Biade, R. W. Sobol, S. H. Wilson, and Y. Matsumoto, *J. Biol. Chem.* **273,** 898–902 (1998).
48. A. Blank, B. Kim, and L. A. Loeb, *Proc. Natl. Acad. Sci. U.S.A.* **91,** 9047–9051 (1994).
49. M. E. Budd and J. L. Campbell, *Methods Enzymol.* **262,** 108–130 (1995).
50. A. A. Alcivar, L. E. Hake, and N. B. Hecht, *Biol. Reprod.* **46,** 201–207 (1992).
51. R. Novak, M. Woszczynski, and J. A. Siedlecki, *Exp. Cell. Res.* **191,** 51–56 (1990).
52. R. Prasad, S. G. Widen, R. K. Singhal, J. Watkins, L. Prakash, and S. H. Wilson, *Nucleic Acids Res.* **21,** 5301–5307 (1993).
53. A. W. Plug, C. A. Clairmont, E. Sapi, T. Ashley, and J. B. Sweasy, *Proc. Natl. Acad. Sci. U.S.A.* **94,** 1327–1331 (1997).
54. P. Fortini, B. Pascucci, E. Parlanti, R. W. Sobol, S. H. Wilson, and E. Dogliotti, *Biochemistry* **37,** 3575–3580 (1998).
55. D. Srivastava, I. Husain, C. Arteage, and S. Wilson, *Carcinogenesis* **20,** 1049–1054 (1999).
56. K.-H. Chen, M. Yakes, D. K. Srivastava, R. K. Singhal, R. W. Sobol, J. K. Horton, B. Van Houten, and S. H. Wilson, *Nucleic Acids Res.* **26,** 2001–2007 (1998).
57. Y. Canitrot, C. Cazaux, M. Fréchet, K. Bouayadi, C. Lesca, B. Salles, and J. Hoffmann, *Proc. Natl. Acad. Sci. U.S.A.* **95,** 12586–12590 (1998).
58. R. Bravo and H. MacDonald-Bravo, *J. Cell. Biol.* **105,** 1549–1554 (1987).
59. M. Savio, L. A. Stivala, L. Bianchi, V. Vannini, and E. Prosperi, *Carcinogenesis* **19,** 591–596 (1998).
60. E. Warbrick, P. J. Coates, and P. A. Hall, *J. Pathol.* **186,** 319–324 (1998).
61. A. Aboussekhra and R. Wood, *Exp. Cell. Res.* **221,** 326–332 (1996).
62. A. S. Balajee, A. May, I. Dianova, and V. A. Bohr, *Mutat. Res.* **409,** 135–146 (1998).
63. M. Miura, S. Nakamura, T. Sasaki, T. Shiomi, and M. Yamaizumi, *Exp. Cell. Res.* **201,** 541–544 (1992).
64. M. Miura, S. Nakamura, T. Sasaki, Y. Takasaki, T. Shiomi, and M. Yamaizumi, *Exp. Cell. Res.* **226,** 126–132 (1996).
65. A. S. Balajee, I. Dianova, and V. A. Bohr, *Nucleic Acids Res.* **27,** 4476–4482 (1999).
66. M. Ottorlei, E. Warbrick, T. A. Nagelhus, T. Haug, G. Slupphaug, M. Akbart, P. A. Aas, K. {Steinsbekk}, O. Bakke, and H. E. Krokan, *EMBO J.* **18,** 3834–3844 (1999).
67. R. M. Taylor, D. J. Moore, J. Whitehouse, P. Johnson, and K. W. Caldecott, *Mol. Cell. Biol.* **20,** 735–740 (2000).
68. R. S. Tebbs, J. J. Meneses, R. A. Pedersen, L. H. Thompson, and J. E. Cleaver, *Environ. Mol. Mutagen.* **27,** 68 (1996).
69. R. S. Tebbs, M. L. Flannery, J. J. Meneses, A. Hartmann, J. D. Tucker, L. H. Thompson, J. E. Cleaver, and R. A. Pedersen, *Devel. Biol.* **208,** 513–529 (1999).
70. S. J. Muller and S. Cardonna, *Exp. Cell. Res.* **226,** 346–355 (1996).
71. A. Umar, A. B. Buermeyer, J. A. Simon, D. C. Thomas, A. B. Clark, M. Liskay, and T. A. Kunkel, *Cell (Cambridge, Mass.)* **87,** 65–73 (1996).
72. R. Gray, D. L. Ludwig, H. L. Cornelius, M. A. MacInnes, and M. S. Park, *J. Biol. Chem.* **272,** 24522–24529 (1997).
73. X. Wu, J. LI, X. LI, C.-L. Hsieh, P. M. Burgers, and M. R. Lieber, *Nucleic Acids Res.* **24,** 2036–2043 (1996).
74. D. S. Levin, W. Bai, N. Yao, M. O'Donnell, and A. E. Tomkinson, *Proc. Natl. Acad. Sci. U.S.A.* **94,** 12863–12868 (1997).
75. R. Mossi, E. Ferrari, and U. Hubscher, *J. Biol. Chem.* **273,** 14322–14330 (1998).

Yeast Base Excision Repair: Interconnections and Networks

> PAUL W. DOETSCH,[*,†] NATALIE J. MOREY,[*,‡] REBECCA L. SWANSON,[*] AND SUE JINKS-ROBERTSON[‡]
>
> Departments of *Biochemistry, ‡Biology, and †Radiation Oncology
> Emory University
> Atlanta, Georgia 30322

I. Yeast BER Mutants Demonstrate Wild-Type Sensitivity to Oxidizing Agents 31
II. Simultaneous Disruption of Multiple DNA Repair and Damage Tolerance Pathways Is Required to Sensitize Cells to Oxidizing Agents 33
III. Spontaneous Mutator and Hyperrecombination (Hyper-rec) Phenotypes in Single and Double Pathway Defect Mutants 33
IV. Other Characteristics of Repair Pathway Defect Mutants 36
V. Cytotoxicity, Mutagenicity, Recombination, and Growth Studies Suggest a Network of Interconnected Repair and Damage Processing Pathways 36
References 38

The removal of oxidative base damage from the genome of *Saccharomyces cerevisiae* is thought to occur primarily via the base excision repair (BER) pathway in a process initiated by several DNA N-glycosylase/AP lyases. We have found that yeast strains containing simultaneous multiple disruptions of BER genes are not hypersensitive to killing by oxidizing agents, but exhibit a spontaneous hyperrecombinogenic (hyper-rec) and mutator phenotype. The hyper-rec and mutator phenotypes are further enhanced by elimination of the nucleotide excision repair (NER) pathway. Furthermore, elimination of either the lesion bypass (*REV3*-dependent) or recombination (*RAD52*-dependent) pathway results in a further, specific enhancement of the hyper-rec or mutator phenotypes, respectively. Sensitivity (cell killing) to oxidizing agents is not observed unless multiple pathways are eliminated simultaneously. These data suggest that the BER, NER, recombination, and lesion bypass pathways have overlapping specificities in the removal of, or tolerance to, exogenous or spontaneous oxidative DNA damage in *S. cerevisiae*. Our results also suggest a physiological role for the AP lyase activity of certain BER N-glycosylases *in vivo*. © 2001 Academic Press.

Oxidative damage to cellular DNA, thought to be one of the most frequently occurring types of genetic injury, is mediated by a number of different endogenous and exogenous agents capable of generating reactive oxygen species. Reactive oxygen species cause a plethora of base modifications, and the repair of such lesions is generally thought to be primarily the responsibility of the base excision repair (BER) pathway as initiated by various DNA N-glycosylases (*1*). Several of these DNA N-glycosylases possess an associated apurinic/apyrimidinic (AP) lyase activity capable of causing strand breakage at AP sites, and it is currently unknown to what extent such lyase activity participates in BER and perhaps other DNA damage-processing pathways. Our studies of eukaryotic BER have utilized *Saccharomyces cerevisiae* as a model system owing to its biochemical and genetic tractability.

Saccharomyces cerevisiae possesses two homologs, Ntg1p and Ntg2p, of *Escherichia coli* endonuclease III (endo III). Ntg1p and Ntg2p are N-glycosylase/AP lyases that show significant sequence similarity to each other (41% identity, 63% similarity), as well as broad overlap with respect to the spectrum of oxidatively damaged bases (primarily pyrimidines) recognized (*2*–*6*). Although Ntg1p and Ntg2p share many similarities, they differ with respect to the presence of a C-terminal Fe-S cluster in Ntg2p, a hallmark feature of all known endo III homologs which is absent in Ntg1p, as well as the presence in Ntg1p of an N-terminal mitochondrial targeting sequence (*2*). In addition, the expression of *NTG1* appears to be inducible in response to oxidizing agent exposure, whereas *NTG2* is not induced (*3*, *5*). Ntg1p co-localizes to both the nucleus and mitochondria, and Ntg2p localizes exclusively to the nucleus (*6*, *7*). Biochemical studies on *ntg1* and *ntg2* single and double disruption mutants indicated that when both enzymes are eliminated, cell extracts lack N-glycosylase/AP lyase activities capable of processing pyrimidine-ring saturation products such as dihydrouracil (*3*). Wishing to confirm the notion that Ntg1p and Ntg2p might be critical for yeast cell repair of oxidative pyrimidine base damage *in vivo*, we assessed the sensitivities of single and double *ntg1* and *ntg2* mutants to several types of DNA damaging agents such as hydrogen peroxide, menadione, and methyl methane sulfonate (MMS). The results of these and related genetic studies led us to consider the possibility that other DNA repair and damage tolerance pathways play a role in the cellular response to DNA base damage in yeast.

In addition to BER, other DNA repair and damage tolerance pathways have evolved to reverse various types of genetic damage. The nucleotide excision repair (NER) pathway generally removes bulky DNA lesions, such as bipyrimidine adducts induced by ultraviolet light. However, recent reports have implicated NER in the repair of nonbulky lesions, such as those caused by alkylating and oxidizing agents (*8*). In yeast, recombination is mediated by proteins in the

RAD52 epistasis group and is involved in the repair of single and double strand breaks, and DNA lesions that block replication (e.g., UV-induced cyclobutane pyrimidine dimers) also appear to provoke recombination (9).

In addition, hydrogen peroxide exposure of yeast increases recombination rates, suggesting that some types of oxidative DNA damage are handled by this route (10). Recombinational processing (or recombinational repair, RR) of damaged DNA is probably best viewed as a DNA damage tolerance pathway since the recombined, damage-containing DNA segment, although allowing the cell to complete replication and mitosis, would have to be subsequently removed by a *bona fide* repair pathway. A second type of DNA damage tolerance pathway in yeast is translesion synthesis (TLS), which involves mutagenic bypass by DNA polymerase ζ (a complex of Rev3p and Rev7p) of several types of lesions including cyclobutane pyrimidine dimers and AP sites (11, 12). Such bypass allows cells to proceed through the cell cycle at the expense of an increased mutation rate.

I. Yeast BER Mutants Demonstrate Wild-Type Sensitivity to Oxidizing Agents

To determine the relative contributions of Ntg1p, Ntg2p, and BER in reversing the cytotoxic effects of agents that cause oxidative DNA damage, we performed an initial series of cell survival experiments with single (*ntg1* or *ntg2*) and double (*ntg1 ntg2*) mutants. These experiments were conducted with the expectation that such mutants might exhibit sensitivity to these agents. When such strains were examined for their sensitivities against either menadione (Fig. 1A) or hydrogen peroxide (not shown), neither the single mutants nor, surprisingly, the double mutant showed an increased sensitivity over the wild-type strain. One explanation for this result would be the existence of other BER *N*-glycosylases whose substrate specificities overlap with Ntg1p and Ntg2p and which could potentially repair the cytotoxic lesions induced by these agents. This situation would need to exist despite the biochemical evidence which suggests that Ntg1p and Ntg2p comprise the major cellular activities for the repair of oxidized pyrimidines in yeast (3). To address this possibility, we generated additional mutants in which the major AP endonuclease of yeast, Apn1p, was eliminated and assessed these strains for sensitivity to menadione and hydrogen peroxide. Surprisingly, the triple mutant, *ntg1 ntg2 apn1* also showed no increased sensitivity to these agents over wild-type strains (Fig. 1B and C). This finding led us to consider the role of additional DNA damage processing pathways in the cellular response to oxidative genotoxic stress.

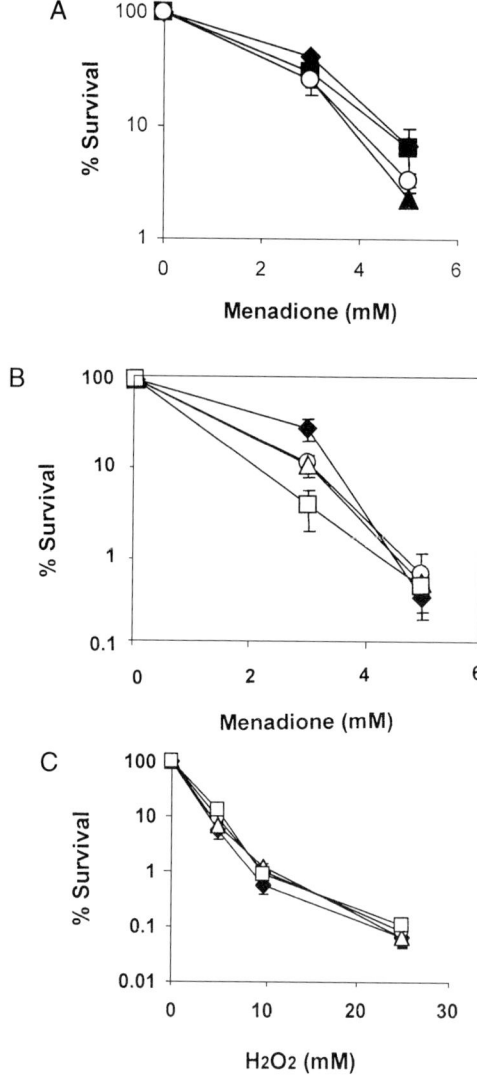

FIG. 1. Sensitivity of S. cerevisiae ntg1, ntg2, and apn1 combination mutant strains to menadione and hydrogen peroxide. Cells were exposed to increasing doses of menadione or hydrogen peroxide and assessed for survival by colony formation as described in Ref. (15). Descriptions of the strains used, which are isogenic derivatives of SJR751, and mutant construction (via gene disruption) can be found in Ref. (15). (A) Cytotoxicity of menadione on wild-type (closed diamonds), ntg1 (closed squares), ntg2 (closed triangles), and ntg1 ntg2 (open circles) mutants. Cytotoxicity of menadione (B) and hydrogen peroxide (C) on wild-type (closed diamonds), ntg1 ntg2 (open circles), apn1 (open triangles), and ntg1 ntg2 apn1 (open boxes) mutants. Error bars represent the standard deviation between three separate cultures.

II. Simultaneous Disruption of Multiple DNA Repair and Damage Tolerance Pathways Is Required to Sensitize Cells to Oxidizing Agents

The lack of increased sensitivity in the BER mutants to DNA-damaging agents prompted us to introduce additional mutations into these backgrounds, resulting in the simultaneous elimination of the major DNA excision repair (BER and NER) and/or damage-tolerance (TLS and RR) pathways. For NER, we targeted *RAD1*, which encodes one of the components (Rad1p) for the Rad1–Rad10 endonuclease complex responsible for the 5′ incision event in the early steps of damage removal (*13*). To eliminate the TLS damage-tolerance pathway, we disrupted *REV3*, which encodes one of two components (Rev3p and Rev7p) of the DNA polymerase ζ (*12*). Finally, the RR pathway for damage tolerance was eliminated via disruption of *RAD52*, an essential component of the yeast homologous recombination machinery (*14*). For simplicity, we will refer to the *ntg1 ntg2 apn1*, *rad1*, *rev3*, and *rad52* mutants as corresponding to defects in the BER, NER, TLS, and RR pathways, respectively. None of the single mutants defective in NER, TLS, or RR showed an increased sensitivity to either menadione (Fig. 2A) or hydrogen peroxide (Fig. 2B). A matrix of combination mutant strains was generated such that all possible combinations of two of these pathways were represented (BER plus NER, BER plus RR, etc.) and were also tested for sensitivity to menadione (Fig. 2A) and hydrogen peroxide (Fig. 2B).

The combination of defects in BER/NER, BER/RR, and RR/TLS resulted in a substantial increase in sensitivity (ranging from moderate to severe) to both oxidizing agents (*15*). In contrast, other double pathway defect combinations such as BER/TLS retained wild-type sensitivity. These results reveal the importance of the NER, TLS, and RR pathways in the tolerance of the cell to substantial, potentially cytotoxic levels of oxidative DNA damage and suggest that there is considerable overlap among these pathways for the repair of oxidative DNA lesions.

III. Spontaneous Mutator and Hyperrecombination (Hyper-rec) Phenotypes in Single and Double Pathway Defect Mutants

Because the aforementioned cytotoxicity experiments indicated that the BER, NER, TLS, and RR pathways can share a substantial part of the burden for processing oxidative DNA damage, it was of interest to determine whether an increase in shuttling the damage into alternative systems (such as TLS or RR) in the absence of BER or NER could be directly observed. For example, it is

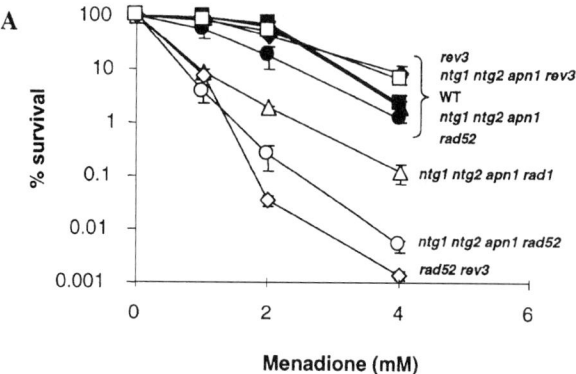

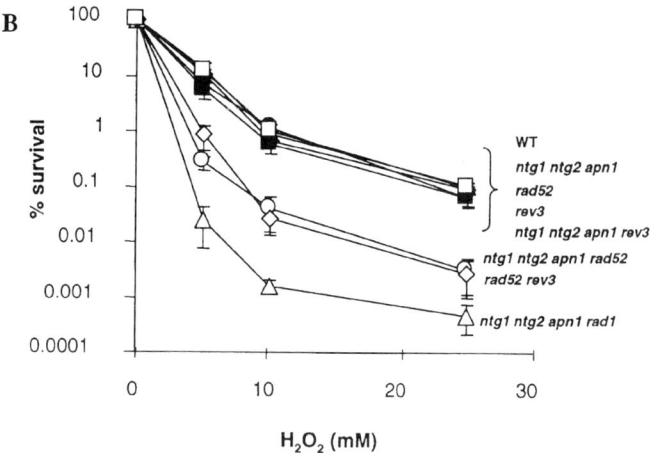

FIG. 2. Sensitivity of BER (*ntg1 ntg2 apn1*), NER (*rad1*), TLS (*rev3*), and RR (*rad52*) defect single and combination mutant strains to menadione (A) and hydrogen peroxide (B). Experiments were carried out as described in the Fig. 1 legend and in Ref. (*15*). Reprinted with permission from Ref. (*15*). Copyright 1999 American Society for Microbiology.

conceivable that the elimination of an error-free excision repair pathway such as BER or NER, singly or in combination, might result in an increase in mutation rates (due to shuttling into TLS) or recombination rates (due to shuttling into RR). Accordingly, elimination of TLS in conjunction with defects in excision repair might increase the recombination rate, and conversely, elimination of RR with defects in excision repair might result in increased mutation rates.

The rates of spontaneous frameshift and base substitution mutations were determined in various strains by measuring the reversion rates of the *lys2∆Bgl* frameshift allele and forward mutations rates of the *CAN1* locus, respectively (15). For spontaneous recombination rate measurements, a second *lys2* allele was introduced into the strains and recombination between the *lys2ΔBgl* allele (chromosome II) and the *lys2Δ3500* allele (chromosome V) was determined by measuring the rate of Lys$^+$ prototroph production (15). The results of these experiments are shown in Table I. Spontaneous mutation and recombination rates did not increase significantly over the wild-type strain for the *ntg1*, *ntg2*, *ntg1 ntg2*, *ntg1 apn1*, and *ntg2 apn1* strains (Table I and data not shown). However, the BER pathway defect strain (*ntg1 ntg2 apn1* triple mutant) showed significant increases in both mutation rates and recombination rates, indicating that both TLS and RR are involved in processing a lesion recognized in common by Ntg1p, Ntg2p, and Apn1p and repaired by the BER pathway. Based on the known substrate specificities of these three enzymes, we conclude that this lesion is likely to be an abasic site.

The elimination of both BER and NER had a dramatic, synergistic effect on increasing the rates for both spontaneous mutations (110-fold for frameshifts; 62-fold for base substitutions) and recombination (170-fold), suggesting that in a BER pathway defect mutant (*ntg1 ntg2 apn1* triple mutant) NER functions to repair abasic sites and in a NER pathway defect mutant (*rad1*) BER functions to repair abasic sites (Table I). In a situation where both BER and NER pathways are nonfunctional (*ntg1 ntg2 apn1 rad1* quadruple mutant) the TLS and RR damage-tolerance pathways must process the damage, suggesting that BER and

TABLE I
SPONTANEOUS RECOMBINATION AND MUTATION RATES[a]

Relevant genotype	Recombination rate × 10^{-8}	*lys2ΔBgl* reversion rate × 10^{-9}	Can^r Rate × 10^{-7}
WT	5.3 (1.0×)	3.2 (1.0×)	1.2 (1.0×)
ntg1 ntg2	5.5 (1.1×)	2.2 (0.8×)	1.5 (1.3×)
apn1	10 (1.9×)	3.8 (1.2×)	2.1 (1.8×)
ntg1 ntg2 apn1	93 (18×)	8.2 (2.6×)	13 (11×)
rad1	6.8 (1.3×)	5.9 (1.8×)	2.1 (1.8×)
ntg1 ntg2 rad1	8.2 (1.5×)	11 (3.4×)	5.5 (4.6×)
apn1 rad1	25 (4.7×)	23 (7.2×)	9.8 (8.2×)
ntg1 ntg2 apn1 rad1	890 (170×)	350 (110×)	74 (62×)
rad52	ND	22 (6.9×)	8.5 (7.1×)
ntg1 ntg2 apn1 rad52	ND	120 (38×)	180 (150×)
rev3	9.1 (1.7×)	1.5 (0.5×)	8.1 (0.7×)
ntg1 ntg2 apn1 rev3	310 (58×)	2.2 (0.7×)	2.4 (2.0×)

[a]Numbers in parentheses indicate the fold increase over wild-type. Rates are expressed as recombinants or mutants per cell per generation. Reprinted with permission from Ref. (15). Copyright 1999 American Society of Microbiology.

NER are competing pathways. The RR pathway defect mutant (*rad52*) showed a moderate mutator phenotype for both spontaneous frameshifts and base substitutions. When this RR defect background was combined with a BER defect (*ntg1 ntg2 apn1 rad52* quadruple mutant), a synergistic effect on both spontaneous frameshift (38-fold) and base substitution (150-fold) mutations was observed. In addition, when defects in BER and TLS were combined (*ntg1 ntg2 apn1 rev3* quadruple mutant), a synergistic effect on increasing the spontaneous recombination rate (58-fold) was observed. It should also be noted that the spontaneous mutation rates in BER/TLS defect strain were similar to wild-type, indicating that the bulk of mutations formed in the BER defect strain are generated by the TLS pathway. An important aspect of future studies with these strains will be to assess their mutator and recombination phenotypes in response to exogenous DNA damaging agents.

IV. Other Characteristics of Repair Pathway Defect Mutants

In addition to their spontaneous mutator and hyper-rec phenotypes, several of the pathway defect strains exhibited slow growth phenotypes on rich media [not shown and (15)]. These included the RR defect (*rad52*) mutant. The further elimination of either BER or TLS in the RR defect mutant background resulted in a synergistic decrease in growth rate, as did the simultaneous elimination of BER and NER [not shown and (15)]. Other pathway defect combinations did not exhibit obvious effects on growth rates.

V. Cytotoxicity, Mutagenicity, Recombination, and Growth Studies Suggest a Network of Interconnected Repair and Damage Processing Pathways

The BER pathway defect (*ntg1 ntg2 apn1* triple mutant) strain showed no increased sensitivity to oxidizing agents, yet exhibited a spontaneous hyper-rec and mutator phenotype. Because only the triple mutant and no double mutant combination of *ntg1*, *ntg2*, or *apn1* produced this phenotype, we conclude that the Ntg1p, Ntg2p, and Apn1p enzymes are competing for a common substrate, namely an abasic site. This finding also suggests that the AP lyase activity of Ntg1p and Ntg2p is sufficient for substituting for Apn1p for the initial processing of abasic sites and, as far as we are aware, provides the first evidence for the biological relevance of AP lyase activity in the removal of abasic sites in yeast. Thus the accumulation of abasic sites in the BER defect mutant drives the increase in mutation and recombination rates, conferring essentially a genetic instability phenotype. Along similar lines, recent studies by Samson and colleagues (16) reveal that overexpression of 3-methyladenine DNA glycosylase in yeast results in a mutator phenotype presumably due to an overproduction of AP sites which exceeds the capacities of the normal AP site processing machinery of the BER and NER pathways. An important implication of our studies is that if similar perturbations in the BER pathway occur in other eukaryotes, such as mammals, it might result in a genetic instability phenotype that may be important in certain scenarios of tumor development or the response of tumor cells to chemotherapeutic agents that damage DNA.

Our results suggest that the four pathways examined here—BER, NER, TLS, and RR—are all involved in processing of spontaneous DNA damage and compete for a common substrate, which we suggest is an abasic site. The repair of oxidative DNA damage has, for the most part, been believed to be a major function of the BER pathway, although some proposals for alternative processing have been discussed (17, 18). Our genetic analyses of the contribution of BER, NER, TLS, and RR in the response of yeast to oxidative DNA damage indicate that all four pathways have overlapping specificities and compete for the

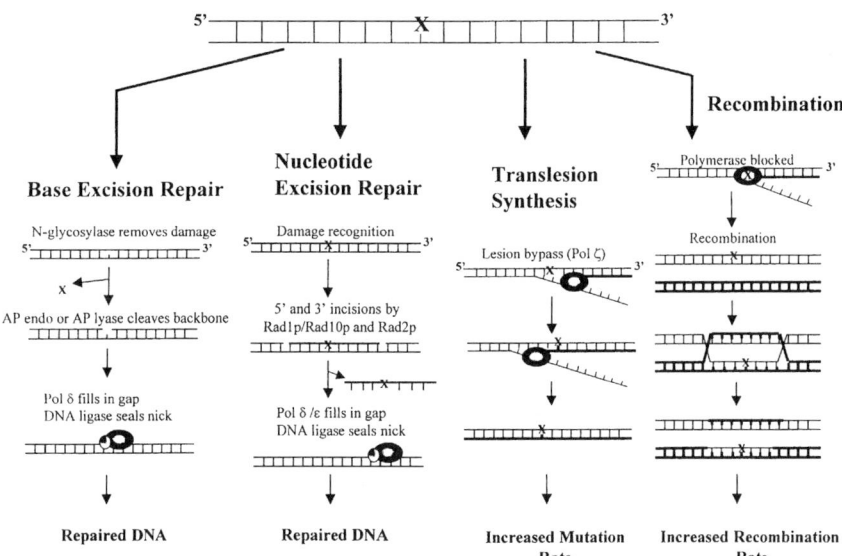

FIG. 3. Processing of oxidative and spontaneous DNA damage in S. cerevisiae. "X" represents either a base damage which can be recognized and removed by the BER or NER pathway, can be bypassed by TLS, or can block replication causing recombination to occur, or an abasic site which feeds into any one of the four pathways illustrated. The amount of damage that is processed by each pathway varies depending on the repair background genotype. Reprinted with permission from Ref. (15). Copyright 1999 American Society for Microbiology.

repair of base damage and/or abasic sites (Fig. 3). Multiple pathways can clearly process or tolerate certain DNA damages; however, the relative contribution of each pathway will probably vary, depending on the circumstances which may be influenced by cell cycle and lesion location within the genome, to name a few. A challenge for future work in this area will be to dissect the mechanisms of how these pathways interrelate to each other, as well as to ascertain the relative contributions of each pathway when the repertoire of systems for repair and damage tolerance is fully, simultaneously functional (i.e., is wild-type).

In conclusion, although we have acquired a great deal of knowledge concerning the linear (vertical) nature of the steps in individual pathways for dealing with genotoxic stress such as BER, NER, TLS, and RR, it is now apparent that attention must be focused on gaining an understanding of how these systems relate to each other horizontally (Fig. 3). The components that mediate the communication among various pathways for functioning in response to DNA damage are largely unknown. This aspect would be a major focus of our efforts in future work in this area.

Acknowledgments

We thank the members of our respective laboratories for their constructive discussions and suggestions during the course of this work. This work was supported by NIH grants CA73041, CA78622 (P.W.D.), and GM38464 (S.J.R.). N.J.M. and R.L.S. contributed equally to this work.

References

1. E. C. Friedberg, G. C. Walker, and W. Siede, "DNA Repair and Mutagenesis." ASM Press Washington,, D.C., 1995.
2. L. Augeri, Y.-M. Lee, A. B. Barton, and P. W. Doetsch, *Biochemistry* **36,** 721–729 (1997).
3. H. J. You, R. L. Swanson, and P. W. Doetsch, *Biochemistry* **37,** 6033–6040 (1998).
4. S. Senturker, P. Auffret van der Kemp, H. J. You, P. W. Doetsch, M. Dizdaroglu, and S. Boiteux, *Nucleic Acids Res.* **26,** 5270–5276 (1998).
5. L. Eide, M. Bjoras, M. Pirovano, I. Alseth, K. G. Berdal, and E. Seeberg, *Proc. Natl. Acad. Sci. U.S.A.* **93,** 10735–10740 (1996).
6. H. J. You, R. L. Swanson, C. Harrington, A. H. Corbett, S. Jinks-Robertson, S. Senturker, S. S. Wallace, S. Boiteux, M. Dizdaroglu, and P. W. Doetsch, *Biochemistry* **38,** 11298–11306 (1999).
7. I. Alseth, L. Eide, M. Pirovano, T. Rognes, E. Seeberg, and M. Bjoras, *Mol. Cell. Biol.* **19,** 3779–3787 (1999).
8. J. C. Huang, D. S. Hsu, A. Kazantsev, and A. Sancar, *Proc. Natl. Acad. Sci. U.S.A.* **91,** 12213–12217 (1994).
9. W. Y. Yap and K. N. Kreuzer, *Proc. Natl. Acad. Sci. U.S.A.* **88,** 6043–6047 (1991).
10. R. J. Brennan, B. E. Swoboda, and R. H. Schiestl, *Mutat. Res.* **308,** 159–167 (1994).
11. R. E. Johnson, C. A. Torres-Ramos, T. Izumi, S. Mitra, S. Prakash, and L. Prakash, *Genes. Dev.* **12,** 3137–3143 (1998).

12. J. R. Nelson, C. W. Lawrence, and D. C. Hinkle, *Science* **272,** 1646–1649 (1996).
13. A. A. Davies, E. C. Friedberg, A. E. Tomkinson, R. D. Wood, and S. C. West, *J. Biol. Chem.* **270,** 24638–24641 (1995).
14. T. D. Petes, R. E. Malone, and L. S. Symington, *in* "The Molecular and Cellular Biology of the Yeast Saccharomyces: Genome Dynamics, Protein Synthesis, and Energetics" (J. R. Broach, J. R. Pringle, and E. W. Jones, eds.), Vol. I, pp. 407–521. Cold Spring Harbor Laboratory Press, Plainview,, NY, 1991.
15. R. L. Swanson, N. J. Morey, P. W. Doetsch, and S. Jinks-Robertson, *Mol. Cell. Biol.* **19,** 2929–2935 (1999).
16. B. J. Glassner, L. J. Rasmussen, M. T. Najarian, L. M. Posnick, and L. D. Samson, *Proc. Natl. Acad. Sci. U.S.A.* **95,** 9997–10002 (1998).
17. P. Moller and H. Wallin, *Mutat. Res.* **410,** 271–290 (1998).
18. J. T. Reardon, T. Bessho, H. C. Kung, P. H. Bolton, and A. Sancar, *Proc. Natl. Acad. Sci. U.S.A.* **94,** 9463–9468 (1997).

BER, MGMT, and MMR in Defense against Alkylation-Induced Genotoxicity and Apoptosis

Bernd Kaina, Kirsten Ochs, Sabine Grösch, Gerhard Fritz, Jochen Lips, Maja Tomicic, Torsten Dunkern, and Markus Christmann

Division of Applied Toxicology
Institute of Toxicology
University of Mainz
Obere Zahlbacher Str. 67
D-55131 Mainz, Germany

I. Introduction 42
II. O^6-Methylguanine Is a Decisive Cytotoxic and Genotoxic Lesion 43
III. O^6-Methylguanine Is a Major Trigger of Apoptosis 44
IV. MMR Is Required for O^6-MeG-Triggered
    Apoptosis and Genotoxicity 44
V. When and Why Are N-Alkylation Lesions Genotoxic? 45
VI. Modulation of BER: Transfection and Knockout 46
    A. N-Methylpurine DNA Glycosylase (MPG)...................... 46
    B. Apurinic-apyrimidinic Endonuclease (APE)...................... 46
    C. DNA Polymerase β (Pol β)................................... 47
VII. Regulation of BER, MGMT, and MMR 47
VIII. Posttranscriptional Regulation of APE............................. 49
IX. Induction of APE and Adaptive Response.......................... 50
X. An Alternative Pathway of Repair in Pol β-Deficient Cells?............ 50
XI. BER Intermediates Not Repaired in Pol β-Deficient
    Cells Trigger Apoptosis 51
XII. Overexpression of MGMT in Pol β-Deficient Cells.................. 52
    References.. 52

Methylating carcinogens and cytostatic drugs induce different methylation products in DNA. In cells not expressing the repair protein MGMT or expressing it at a low level, O^6-methylguanine is the major genotoxic, recombinogenic, and apoptotic lesion. Genotoxicity and apoptosis triggered by O^6-methylguanine require mismatch repair (MMR). In cells expressing O^6-methylguanine-DNA methyl transferase (MGMT) at a high level or for agents producing low amounts of O^6-methylguanine, N-alkylations become the major genotoxic lesions.

N-Alkylations are repaired by base excision repair (BER). In mammalian cells, naturally occurring mutants of BER have not been detected, which points to the importance of BER for viability. In order to ascertain the role of BER in cellular defense, BER was modulated either by transfection or mutational inactivation. It has been shown that overexpression of N-methylpurine-DNA glycosylase (MPG) does not protect, but rather sensitizes cells to S_N2 agents. This has been interpreted in terms of an imbalance in BER. Regarding abasic site endonuclease (APE), transient but not stable overexpression of the enzyme was achieved upon transfection in CHO cells, which indicates that unphysiologic APE levels are not tolerated by the cell. Besides the repair function, APE (alias Ref-1) exerts redox capability by which the activity of various transcription factors is modulated. Therefore, it is possible that stable overexpression of mammalian APE impairs transcriptional regulation of genes, whereas transient overexpression may exert some protective effect. DNA polymerase β (Pol β) transfection was ineffective in conferring resistance to methylmethane sulfonate (MMS). On the other hand, Pol β-deficient cells proved to be highly sensitive to methylation-induced chromosomal aberrations and reproductive cell death. The dramatic hypersensitivity in the killing response is largely due to induction of apoptosis. Obviously, nonrepaired BER intermediates are clastogenic and act as a strong trigger of the apoptotic pathway. The elements of this pathway are currently under investigation. © 2001 Academic Press.

I. Introduction

A wide variety of agents, both endogenously produced and present in our environment, induce alkylation damage in DNA (1, 2). Also, many antineoplastic drugs used in tumor therapy (some of them are also used for immunosuppression in the treatment of autoimmune diseases) exert alkylating properties (3). In view of the permanent alkylation pressure imposed on cells and the wide utility of anticancer drugs, the mechanisms of cellular defense against alkylating agents deserve special attention. Development of alkylation resistance is a common obstacle in tumor therapy. Therefore, increasing our understanding of the mechanisms of alkylating drug-induced cytotoxicity and genotoxicity is of utmost practical importance and may, it is hoped, lead to improved health care strategies and tumor therapeutic procedures.

There is a general consensus that DNA is the main target of alkylating agents. This is true for endpoints such as mutagenicity, carcinogenicity and genotoxicity (formation of sister chromatid exchanges, DNA breakage, and chromosomal aberrations). For cytotoxicity, however, critical targets other than DNA must also be considered in view of the fact that apoptotic pathways can be triggered by membrane receptor activation, such as the TNF and CD95 receptors (4). Here we will show that repair of DNA alkylation damage protects against alkylation-induced apoptosis, providing evidence that DNA alkylation damage is a primary trigger of apoptotic cell death. In this brief review, we will focus on DNA methylation damage and its cellular consequences, as analyzed by modulation of various functions involved in the repair of DNA methylation lesions.

II. O^6-Methylguanine Is a Decisive Cytotoxic and Genotoxic Lesion

There are numerous reports proving O^6-methylguanine (O^6-MeG) to be a decisive genotoxic lesion (for review, see Ref. 5). Recent evidence suggests it to be majorly involved in triggering apoptosis as well. Evidence for the latter point is based mainly on a comparison of cell types deficient and proficient for the repair of O^6-MeG. Lack of O^6-methylguanine-DNA methyltransferase (MGMT) causes O^6-MeG to persist in DNA (6), indicating that O^6-MeG is repaired in mammalian cells exclusively by MGMT (7). Cells not expressing MGMT or expressing it at a low level and cells treated with a MGMT inhibitor (such as O^6-benzylguanine) are highly sensitive to all agents having the ability in common to generate O^6-MeG in DNA. These are notably S_N1 methylating agents, and the sensitivity pertains to all genotoxic endpoints, mutagenesis, and carcinogenesis (for review, see Ref. 8). It should be noted that nonrepaired O^6-MeG is an extremely highly recombinogenic lesion. Thus it has been calculated that 30–50 O^6-MeG molecules per cell give rise to 1 SCE, whereas >1000 O^6-MeG per cell are required to induce cell killing and even higher levels are needed to induce chromosomal aberrations (5, 9). Obviously, the major pathway of processing O^6-MeG in a replicating cell is homologous recombination, which has been proposed to be a process that circumvents potential replication blocking lesions (10). It is important to note that O^6-MeG itself does not block DNA replication and that it is completely unable to induce homologous recombination in the first DNA replication cycle. Recombination induced by O^6-MeG occurs only in the second post-exposure replication cycle (10), indicating that O^6-MeG-induced secondary DNA lesions interfere with DNA replication. This appears to trigger recombination, giving rise to SCEs and, at a lower probability, chromatid breaks and exchanges (for model, see Ref. 10).

Because the genotoxic effect of O^6-MeG is largely alleviated in cells deficient in DNA mismatch repair (MMR), it has been proposed that MMR is actively involved in O^6-MeG-induced genotoxicity (11). This model, which is supported by various lines of evidence, is based on the fact that O^6-MeG–thymine (O^6G–T) mispairs are substrates of MMR initiated by binding of MSH-2 and MSH-6 to the mismatch. Owing to the mispairing properties of O^6-MeG, this is thought to lead to repeated cycles of misincorporation of thymine, blockage of DNA replication, and finally DNA breakage. Thus, both MGMT deficiency and lack or downmodulation of MMR cause resistance of cells to O^6-MeG-generating agents. This has considerable implications for tumor formation, tumor cell resistance, and therapy. For instance, about 5% of tumors exhibit very low MGMT activity (*12*). This tumor group is expected to show a good response to chemotherapy, through utilization of CNU derivatives and methylating drugs inducing O^6-alkylguanine. Also, various tumors have been found to show alterations in MMR (*13*), which can be taken as a further marker of methylating drug resistance.

III. O^6-Methylguanine Is a Major Trigger of Apoptosis

O^6-Methylguanine is a powerful apoptotic DNA lesion. This has been inferred from studies in which the apoptotic response of MGMT-proficient versus MGMT-deficient cells was compared (*10, 14, 15*). With the same dose of an O^6-MeG-generating or chloroethylating agent, MGMT-deficient cells display a much higher frequency of apoptosis than do MGMT-proficient cells. One could argue that this difference is due to the cellular background, e.g., differences in growth factor receptor or apoptotic signaling elements. This, however, is not true since (1) the difference was observed in isogenic pairs of cells differing only in MGMT expression, and (2) the cells were equally responsive to DNA damaging agents not inducing O^6-MeG as they were to physiologic stimuli such as serum deprivation (*16*). It should be noted that a minor component of the cell-killing effect of O^6-MeG-induced in DNA is necrosis. For example, in CHO cells deficient for MGMT, a dose of N-methyl-N'-nitro-N-nitrosoguanidine (MNNG) reducing cell survival by about 70% (MTT assay) induces 40% apoptosis and 20% necrosis, as determined by annexing V flow cytometry (*16a*). Whereas the mechanism of necrosis triggered by O^6-MeG is completely unknown, the major pathway of cell killing, i.e., the apoptotic response, becomes more and more illuminated. The data currently available indicate that apoptosis triggered by O^6-MeG is related to the genotoxic pathway. It appears that O^6-MeG does not trigger the apoptotic pathway by itself. Rather, it requires cell proliferation and DNA replication in order to elicit an apoptotic response (T. Dunkern and B. Kaina, unpublished data). Also, apoptosis triggered by O^6-MeG occurs quite late, detectable not earlier than 3 days after pulse methylation with MNNG (*16a*). Although final proof is still lacking, it seems that cells undergo apoptosis in the second and third cell cycle after treatment, but not in the first cell cycle.

IV. MMR Is Required for O^6-MeG-Triggered Apoptosis and Genotoxicity

Apoptosis induced by O^6-MeG-generating agents is clearly suppressed in mismatch repair (MMR) compromised cells. This has been shown with a panel of isogenic CHO cells differing only in MGMT and MSH2 expression (*10*) and more recently with other cell types deficient in MutSα (*16*). Cells that do not express MGMT and exhibit very low MSH2 expression were highly resistant to MNNG (they exhibit the tolerance phenotype) and displayed a much lower frequency of apoptosis than did MGMT-deficient cells expressing MSH2. This clearly indicates that MMR is required for triggering the apoptotic pathway. O^6-MeG–T mispairs are recognized by MutSα (MSH2+MSH6) but not by MutSβ (MSH2+MSH3) complex (*17*). MutSα, but not MutSβ, is required for triggering the apoptotic pathway (*16*). p53 appears not to be an essential element in O^6-MeG-triggered apoptosis (*16, 16a*). Apoptosis induced by O^6-MeG is preceded by extensive nuclear DNA break formation and a decline in the Bcl-2 protein level. It is accompanied by cytochrome *c* release from mitochondria, caspase-9 and caspase-3 activation, and extensive DNA fragmentation (nucleosomal pattering). There was no significant Fas (CD95, Apo-1) and caspase-8 activation, indicating mitochondrial damage to be the major pathway involved (*16a*). On the basis of our data, we propose that O^6-MeG–T mispairs are subject to MMR, giving rise to DNA breaks which provide the ultimate signal for downmodulation of Bcl-2. This triggers the apoptotic cascade via cytochrome *c* release and caspase-9/caspase-3 activation. It would be interesting to determine whether the MutSα complex bound to O^6-MeG-T mismatches is somehow directly involved in triggering apoptotic signaling, e.g., by activating DNA break binding proteins or other signaling molecules targeting Bcl-2. Also, because extensive chromosomal aberration formation precedes apoptosis, loss of essential genes has to be taken into consideration as an apoptosis-triggering factor.

V. When and Why Are N-Alkylation Lesions Genotoxic?

Since S_N2 methylating agents (e.g., methyl methanesulfonate, dimethylsulfate) that induce only very low amounts of O^6-MeG in DNA (0.3% as compared to 8% for MMS and MNNG, respectively) are highly genotoxic, methylation lesions other than O^6-MeG must be taken into further consideration. One could argue that for S_N2 agents, a much higher overall DNA methylation level is required for eliciting a genotoxic effect; thus, the total level of O^6-MeG per cell would be similar to that induced by an S_N1 agent at equitoxic dose. However, this does not explain the S_N2 agent-induced cytotoxicity, since MGMT has only a very weak protective effect against S_N2 agents such as MMS (*18*). Also, MMR does not appear to be as important in mediating genotoxicity induced by S_N2 agents as those induced by S_N1 agents (*18, 19*). This supports the view that for S_N2 agents, other lesions and mechanisms are decisive in inducing cell killing and genotoxic effects. This paradigm, however, requires qualification, because the dose of the S_N1 agent applied and the cellular MGMT expression level must also be taken into account. Thus, in cells expressing MGMT at a high level, killing protection caused by MGMT (as measured by D_{37} level of the agent) reaches saturation (18). Therefore, killing resistance is not entirely a linear function of the number of MGMT molecules per cell, especially in the high-dose range. Obviously, with high doses of an S_N1 agent, alkylation lesions other than O^6-MeG become the preponderant pretoxic ones. These are very likely to be N-methylpurines, although phosphate methylations must also be considered since they can cause DNA single-strand breaks directly. That N-alkylations are highly cyto- and genotoxic lesions has been proved by modulating BER, which is the major pathway of removal of N-methylpurines from DNA.

VI. Modulation of BER: Transfection and Knockout

To address the question of the importance and rate limitation of the various enzymes involved in BER, we and other groups modulated the expression level of various BER enzymes. The results obtained can be summarized as follows (see also Ref. 8).

A. N-Methylpurine DNA Glycosylase (MPG)

Overexpression of human MPG in CHO cells did not provoke methylation resistance (20). Rather, the cells became more sensitive, which was most obvious for the endpoint aberrations induced by high MMS doses. This has been interpreted in terms of imbalanced BER: Apurinic sites in overlapping repair patches (in cases of long-patch repair) or interference of apurinic sites with DNA replication may cause DNA double-strand break formation, which is a critical clastogenic event (21). Interestingly, N7-methylguanine, which is the major DNA methylation product, is repaired more slowly than the minor product N3-methyladenine, although remarkable intragenomic heterogeneity does exist (22). This could be taken to indicate that fast repair of N7-methylguanine is disadvantageous for the cell. In accordance with this, overexpression of the *tag* gene of *E. coli*, which repairs N3-methyladenine but not N7-methylguanine, provoked some increase in methylation resistance (23). More recently, MPG-deficient cells and MPG knockout mice have been generated. Cells deficient in MPG are hypersensitive to S_N2-methylating agents (24); moreover, they are hypermutable (25). This clearly supports the view that N-methylpurines are not actively repaired in these cells, but are critical cytotoxic and genotoxic lesions.

B. Apurinic-apyrimidinic Endonuclease (APE)

In an attempt to overexpress APE, we transfected the human APE cDNA into CHO cells. Although APE mRNA and protein levels were enhanced, we could not find an increase of APE activity. On the other hand, transfection of yeast APE in CHO cells increased overall APE activity in cell extracts and conferred methylation resistance (26). This result has been interpreted in terms of posttranslational modification of human APE in the overexpressing cells, causing functional repression of the protein. It seems that long-term overexpression of APE is incompatible with the survival of cells. In line with this is the finding that transient overexpression of hAPE could be achieved by transfection with an inducible expression vector, causing transient increase in overall APE activity (27). Downmodulation of APE by transfection with an APE-antisense vector sensitized cells to MMS (28), a result which underlines the importance of APE for BER. APE homozygous knockout mice are not viable (29) and APE-deficient cells have not yet been generated. Thus it seems that APE is essential not only for BER, but also for viability and proliferation of nonmutagen exposed cells. Besides the apurinic endonuclease function, APE has two other activities. It acts as a redox factor for several transcription factors and as a repressor of its own and of other genes by binding to the nCARE element of the promoter (see below). These diverse functions may make APE absolutely essential for viability, thus also explaining the failure to overexpress APE stably. Assuming that yeast APE does not exert the same multiple regulatory functions as the human APE does, stable expression of it in CHO cells will be understandable.

C. DNA Polymerase β (Pol β)

Pol β exerts both DNA polymerase and dRPase activity (for review, see Ref. 30). Upon treatment of cells with MMS, transfection-mediated overexpression of Pol β did not significantly affect survival (J. Dosch and B. Kaina, unpublished data). On the other hand, Pol β overexpression rendered cells more resistant to some anticancer drugs such as cisplatin and melphalan (31). Thus, it appears that Pol β is rate-limiting in the repair of adducts induced by these agents, whereas it is not rate-limiting in the removal of MMS-induced lesions via BER. Whereas Pol β overexpression did not cause significant change in the cellular response to methylating agents, Pol β deficiency rendered cells highly sensitive to agents such as MNNG and MMS (32, 33). Interestingly, the cells displayed cross-sensitivity to various other agents such as mafosfamide and mitomycin C, but not to UV light and chloroethylnitrosourea (33). They are not significantly hypersensitive to ionizing radiation and are only slightly more sensitive to hydrogen peroxide (Fig. 1). This does not reject an involvement of Pol β in repair of oxidative bases, but rather indicates that oxidative bases (such as 8-oxoguanine) do not contribute much to the cell-killing effect of the agents.

VII. Regulation of BER, MGMT, and MMR

In a comparative study, we analyzed the expression of MPG, APE, Pol β, and MGMT on mRNA levels in rat H4IIE cells upon treatment with X-rays or

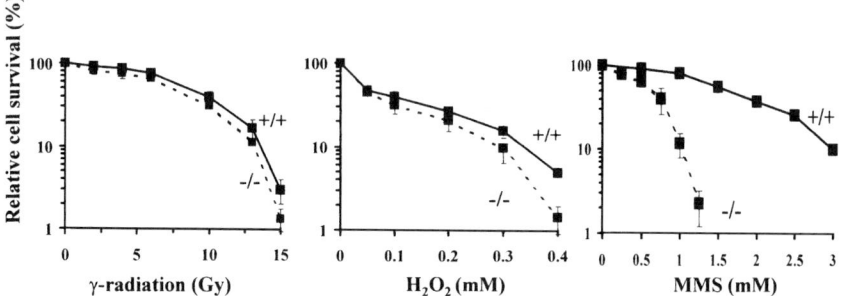

FIG. 1. Frequency of survival of Pol β-deficient and wild-type cells treated with ionizing radiation, hydrogen peroxide, and MMS. Cells were chemically treated for 1 h. They were seeded on 5-cm dishes 6 h prior to treatment.

MNNG. Significant induction (with moderate toxic doses) was found only for MGMT (34). Indeed, there is ample evidence that MGMT is a DNA damage-inducible gene. The transcription factors involved in induction have not yet been identified in detail, although, interestingly, p53 is required for induction (35, 36). As opposed to MGMT, the MPG promoter is not inducible by alkylation (37).

Although in the study noted above, which was performed on rat liver cells, BER genes were not found to have been induced, in other experimental systems transcriptional activation has been observed. Thus, APE is induced at the promoter, mRNA, and protein level upon treatment of CHO cells with ROS-generating agents, such as hydrogen peroxide and hypochlorite (27, 38). Induction of the cloned human APE promoter was found only at a borderline level after treatment with MMS, indicating that induction in these cells is most specific for ROS.

Another BER gene proven to be inducible by simple alkylating agents is Pol β (39). Promoter analysis revealed the CREB/ATF transcription factor-binding site to be required for activation of the promoter by MNNG (40). The Pol β promoter can be induced also by oxidative stress such as hydrogen peroxide or lipopolysaccharide treatment, and a transient increase in the Pol β mRNA level was accompanied by increased resistance of cells to MMS (41). This seems to contradict the finding that Pol β overexpression is ineffective in conferring methylation resistance (see above). It should be noted, however, that a similar case has been made for APE, pointing again to the difference between transient overexpression, which might be beneficial, and stable overexpression, which is not tolerated by or is even deleterious to the cell.

Little information is available on the regulation of MMR. The MSH2 promoter was found not to be inducible in CHO cells treated with various DNA-damaging agents. On the other hand, MutSα binding activity and the MSH2/MSH6 protein level were enhanced in cells upon treatment with MNNG due to nuclear translocation (*41a*). It will be an interesting topic of future research to elucidate the regulatory network of BER, MGMT, and MMR genes and proteins in cells exposed to various kinds of genotoxic stress.

VIII. Posttranscriptional Regulation of APE

As already noted, the APE/Ref-1 protein is a multifunctional protein that possesses several distinct physiological activities. Thus, besides the apurinic/apyrimidinic endonuclease activity, APE also acts as a 3′-diesterase removing phosphoglycolate residues from DNA damaged by ionizing radiation. Moreover, APE (alias Ref-1) is a redox factor involved in regulation of various transcription factors, such as AP-1, p53, and NF-kB (*42, 43*). For stimulation of transcriptional activity of AP-1, physical interaction of Ref-1 with thioredoxin is required (*44*). A further biological effect of APE/Ref-1 resides in its ability to act as a transcriptional repressor of its own and of other genes, such as the parathyroid hormone gene (*45*). Deletion analysis showed that the C-terminal part of APE is important for DNA binding and the AP-endonuclease activity of the protein (*46, 47*). Site-directed mutagenesis of Asp219 leads to loss of both DNA-binding and AP-endonuclease activity, indicating that this amino acid is crucial for the repair function of APE (*47*). Further amino acids that are functionally relevant for DNA repair are Asp90, Glu96, and Asp308; their replacement by alanine causes loss of endonuclease function without influencing DNA-binding activity. An ~6-kDa N-terminal domain is of particular relevance for the redox function of APE/Ref-1 (*48*). The amino acid that is essentially required for redox activation of transcription factors is Cys65 present in a reduced state (*49*). In an oxidized, redox-deficient state, Cys65 is thought to form a disulfide bridge with Cys93 (*49*).

Recently, it has been shown that APE can be phosphorylated *in vitro* by casein kinase II (CKII) (*50*). This phosphorylation was stimulated by MMS treatment. CKII-mediated APE phosphorylation did not impair the APE incision function, although a contrary report is available (*50a*); it rather affected redox regulation of AP-1 (*50*). Thus, the reduced state of AP-1 is maintained and AP-1 binding is promoted if APE/Ref-1 is present in its phosphorylated form. Furthermore, MMS-stimulated AP-1 binding is impaired by pharmacological inhibition of CKII activity (*50*). Based on this, one might speculate that CKII-mediated phosphorylation of APE and a concomitant increase in AP-1-binding activity contribute to the regulation of defense functions in cells exposed to genotoxic stress.

IX. Induction of APE and Adaptive Response

To address the question of whether APE is inducible by DNA-damaging agents, the expression of APE was analyzed in rodent cells, at the gene, protein, and activity level, upon exposure of cells to various forms of genotoxic stress. It was found that treatment with hydrogen peroxide and sodium hypochlorite increased the level of APE mRNA, protein, and APE incision activity about 3–9 h later. The effect was transient and declined 15 h after the beginning of treatment (27). Similar results have been obtained independently by another group (38). Pretreatment of cells with hypochlorite-inducing APE caused the cells to become more resistant to the cytotoxic and clastogenic effect of an oxidative agent, which points to the existence of an adaptive response pathway against oxidative stress (27, 38). In this study, there was no significant protection against the clastogenic effect of MMS. Studies with the cloned human APE promoter revealed that a CREB (CRE) binding site is required for induction of the gene (51). The CRE element binds c-Jun/ATF-2 and, under conditions of induction, the expression of c-Jun was enhanced. This indicates that induction of c-Jun is causally involved in APE promoter activation (51). The finding of induction of APE by oxidative agents raised the question of whether other agents can also elicit this response. We did not find clear induction of the APE promoter by alkylating agent treatment (S. Grösch and B. Kaina, unpublished data). Because there was also no clear cross-resistance to the clastogenic effect of alkylating compounds in pretreated cells, the response appears to be specific for protection against oxidative agents.

X. An Alternative Pathway of Repair in Pol β-Deficient Cells?

Lack of Pol β was shown to render cells highly sensitive to methylating agents, regarding the endpoints of reproductive cell death, apoptosis, and chromosomal aberrations (32, 33). One could argue that Pol β deficiency causes APE-mediated DNA single-strand breaks (ssb) not to be repaired. Therefore, strand breaks may be hypothesized to accumulate with time after mutagen exposure. In fact, the yield of DNA ssb was higher in Pol $\beta^{-/-}$ than in Pol $\beta^{+/+}$ cells (as detected both by alkaline single-cell gel electrophoresis and alkaline elution). However, this increase was only initially observed; a decline in ssb frequency occurred in both Pol $\beta^{-/-}$ and wild-type cells as a function of time after methylation (K. Ochs and B. Kaina, unpublished data). This indicates that ssb induced in Pol β-deficient cells by APE incision activity are repaired both by a fast- and slow-acting mechanism and that Pol β is not absolutely

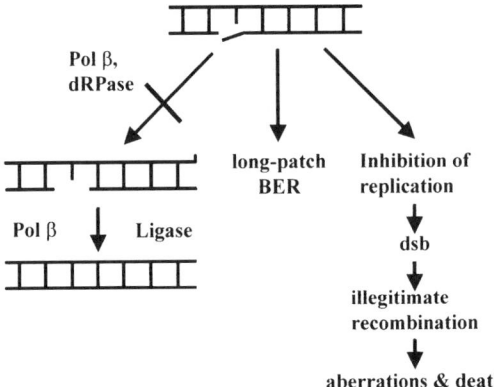

FIG. 2. Pathways of BER intermediates in Pol β-deficient cells. For details see text.

required for slow ssb repair. It is likely that the slow component of ssb repair in Pol β-deficient cells is due to long-patch BER involving exonuclease, Pol δ and Pol ε, proliferating cell nuclear antigen (PCNA), FEN I, and ligase I (52). We also speculate that the higher initial level of nonrepaired ssb in Pol β-deficient cells is critical, notably in replicating cells, because it may block DNA replication, thus causing chromosomal aberrations (53) and finally cell death (Fig. 2).

XI. BER Intermediates Not Repaired in Pol β-Deficient Cells Trigger Apoptosis

Increased cell killing in Pol β-deficient cells is by and large due to the induction of apoptosis (40% versus 15% for apoptosis and necrosis, respectively, 72 h after treatment with 1 mM MMS) (33). Apoptosis is, like in MGMT-deficient cells, a late response, becoming significantly detectable not earlier than 24 h after pulse treatment with MNNG. Apoptosis is accompanied by an increase in DNA breakage and chromosomal aberration formation, indicating that nonrepaired DNA breaks or chromosomal loss and rearrangements trigger the apoptotic pathway. Apoptotic downstream elements include Bcl-2, which declines in amount, and activation of caspase-3 and caspase-9. Fas (CD95) and caspase-8 appear not to be involved (K. Ochs and B. Kaina, unpublished data). Overall, the critical apoptotic factors in methylation-induced apoptosis in Pol β-deficient mouse cells appear to be the same as in MGMT-deficient Chinese hamster cells (16a).

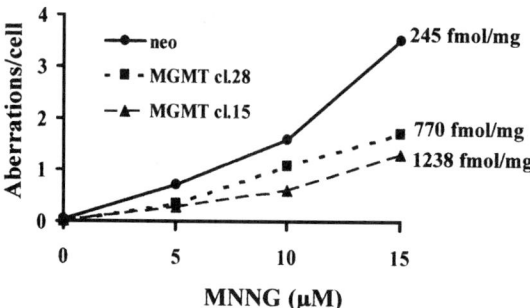

FIG. 3. Aberration frequencies in Pol β-deficient cells transfected with the neo gene only (neo) or with an MGMT expression vector (strains cl.28 and cl.15).

XII. Overexpression of MGMT in Pol β-Deficient Cells

The fact that both MGMT and Pol β-deficiency render cells highly sensitive to S_N1 agents such as MNNG raises the question of whether MGMT or BER is more important for cellular defense against the genotoxic effect of MNNG. We have addressed this question by transfecting MGMT cDNA into Pol β-deficient cells, thus elevating the level of MGMT expression. As shown in Fig. 3, an increase of MGMT expression in Pol β-deficient cells reduced the yield of chromosomal aberrations upon MNNG treatment. This is in line with the expectation based on the finding that nonrepaired O^6-MeG is a critical preclastogenic lesion. It should be noted that overexpression of MGMT did not fully complement the MNNG hypersensitivity of Pol β-deficient cells, indicating that both O^6-MeG and incompletely repaired N-alkylations are responsible for genotoxicity in Pol β knockout cells. Whether Pol β would be directly involved in the clastogenic pathway triggered by O^6-MeG is another interesting issue worthy of future study.

Acknowledgment

Studies were supported by DFG, SFB 519, B4. We gratefully acknowledge receipt of Pol β-deficient cells from Dr. R. Sobol.

References

1. H. Bartsch, H. Ohshima, D. E. G. Shuker, B. Pignetelli, and S. Calmels, *Mutat. Res.* **23**, 255–267 (1990).
2. R. Peto, R. Gray, P. Bantom, and P. Grosso, *IARC Sci. Publ.* **57**, 627–655 (1985).

3. K. Tew, M. Colvin, and B. A. Chabner, in "Cancer Chemotherapy: Principles and Practice" (B. A. Chabner and D. L. Longo, eds.), 2nd ed., pp. 297–332. J. B. Lippincott, Philadelphia, 1995.
4. S. Nagata, *Cell (Cambridge, Mass.)* **88,** 355–365 (1997).
5. B. Kaina, G. Fritz, and T. Coquerelle, *Environ. Mol. Mutagen* **22,** 283–292 (1993).
6. B. Kaina, A. A. van Zeeland, A. de Groot, and A. T. Natarajan, *Mutat. Res.* **243,** 219–224 (1990).
7. A. E. Pegg and T. L. Byers, *FASEB J.* **6,** 2302–2310 (1992).
8. B. Kaina, G. Fritz, K. Ochs, S. Haas, T. Grombacher, J. Dosch, M. Christmann, P. Lund, C. M. Gregel, and K. Becker, *Mutat. Res.* **405,** 179–191 (1998).
9. A. Rasouli-Nia, Sibghat-Ullah, R. Mirzayans, M. C. Paterson, and R. S. Day III, *Mutat. Res.* **314,** 99–113 (1994).
10. B. Kaina, A. Ziouta, K. Ochs, and T. Coquerelle, *Mutat. Res.* **381,** 227–241 (1997).
11. P. Karran and M. Bignami, *BioEssays* **16,** 833–839 (1994).
12. I. Preuss, S. Haas, U. Eichhorn, I. Eberhagen, M. Kaufmann, T. Beck, R. H. Eibl, P. Dall, T. Bauknecht, J. Hengstler, B. M. Wittig, W. Dippold, and B. Kaina, *Cancer Detect. Prev.* **20,** 130–136 (1996).
13. G. Aqiulina, P. Hess, P. Branch, C. MacGeoch, I. Casciano, P. Karran, and M. Bigami, *Proc. Natl. Acad. Sci. U.S.A.* **91,** 8905–8909 (1994).
14. W. Meikrantz, M. A. Bergom, A. Memisoglu, and L. Samson, *Carcinogenesis* **19,** 369–372 (1998).
15. Y. Tominaga, T. Tsuzuki, A. Shiraishi, H. Kawate, and M. Sekiguchi, *Carcinogenesis* **18,** 889–896 (1997).
16. M. J. Hickman and L. D. Samson, *Proc. Natl. Acad. Sci. U.S.A.* **96,** 10764–10769 (1999).
16a. K. Ochs and B. Kaina, *Cancer Res.* **60,** 5815–5824 (2000).
17. D. R. Duckett, J. T. Drummond, A. I. H. Hurchie, J. T. Reardon, A. Sancar, D. M. Lilley, and P. Modrich, *Proc. Natl. Acad. Sci. U.S.A.* **93,** 6443–6447 (1996).
18. B. Kaina, G. Fritz, S. Mitra, and T. Coquerelle, *Carcinogenesis* **12,** 1857–1867 (1991).
19. S. M. Galloway, S. K. Greenwood, R. B. Hill, C. I. Bradt, and C. L. Bean, *Mutat. Res.* **346,** 231–245 (1995).
20. G. Ibeanu, B. Hartenstein, W. C. Dunn, L. Y. Chang, E. Hofmann, T. Coquerelle, S. Mitra, and B. Kaina, *Carcinogenesis* **13,** 1989–1995 (1992).
21. T. Coquerelle, J. Dosch, and B. Kaina, *Mutat. Res.* **336,** 9–17 (1995).
22. N. Ye, G. P. Holmquist, and T. R. O'Connor, *J. Mol. Biol.* **284,** 269–285 (1998).
23. A. Klungland, L. Fairbairn, A. J. Watson, G. P. Margison, and E. Seeberg, *EMBO J.* **11,** 4439–4444 (1992).
24. B. P. Engelward, A. Dreslin, J. Christensen, D. Huszar, C. Kurahara, and L. Samson, *EMBO J.* **15,** 945–952 (1996).
25. R. H. Elder, J. G. Jansen, J. R. Weeks, M. A. Willington, B. Deans, A. J., Watson, K. J. Mynett, J. A. Bailey, D. P. Cooper, J. A. Rafferty, M. C. Heeran, S. W. Wijnhoven, A. A. van Zeeland, and G. P. Margison, *Mol. Cell. Biol.* **18,** 5828–5837 (1998).
26. M. Tomicic, E. Eschbach, and B. Kaina, *Mutat.* **383,** 155–165 (1997).
27. S. Grösch and B. Kaina, *Cancer Res.* **58,** 4410–4416 (1998).
28. L. J. Walker, R. B. Craig, A. L. Harris, and I. D. Hickson, *Nucleic Acids Res.* **22,** 4884–4889 (1994).
29. S. Xanthoudakis, R. J. Smeyne, J. D. Wallace, and T. Curran, *Proc. Natl. Acad. Sci. U.S.A.* **93,** 8919–8923 (1996).
30. S. H. Wilson, *Mutat. Res.* **407,** 203–215 (1998).
31. Y. Canitrot, C. Cazaux, M. Frechet, K. Bouayadi, C. Lesca, B. Salles, and J. S. Hoffmann, *Proc. Natl. Acad. Sci. U.S.A.* **95,** 12586–12590 (1998).

32. R. W. Sobol, J. K. Horton, R. Kühn, H. Gu, R. K. Singhai, R. Prasad, K. Rajewski, and S. H. Wilson, *Nature (London)* **379**, 183–186 (1996).
33. K. Ochs, R. W. Sobol, S. H. Wilson, and B. Kaina, *Cancer Res.* **59**, 1544–1551 (1999).
34. T. Grombacher, S. Mitra, and B. Kaina, *Carcinogenesis* **17**, 2329–2336 (1996).
35. J. A. Rafferty, A. R. Clarke, D. Sellappan, M. S. Koref, I. M. Frayling, and G. P. Margison, *Oncogene* **12**, 693–697 (1996).
36. T. Grombacher, U. Eichhorn, and B. Kaina, *Oncogene* **17**, 845–851 (1998).
37. T. Grombacher and B. Kaina, *DNA Cell Biol.* **15**, 581–588 (1996).
38. C. V. Ramana, I. Boldogh, T. Izumi, and S. Mitra, *Proc. Natl. Acad. Sci. U.S.A.* **95**, 5061–5066 (1998).
39. A. J. Fornace, Jr., B. Zmudzka, C. Hollander, and S. H. Wilson, *Mol. Cell. Biol.* **9**, 851–853, (1989).
40. P. S. Kedar, S. G. Widen, E. W. Englander, A. J. Fornace, and S. H. Wilson, *Proc. Natl. Acad. Sci. U.S.A.* **88**, 3729–3732, (1991).
41. K. H. Chen, F. M. Yakes, D. K. Srivastava, R. K. Singhai, R. W. Sobol, J. K. Horton, B. van Houten, and S. H. Wilson, *Nucleic Acids Res.* **8**, 2001–2007 (1998).
41a. M. Christmann and B. Kaina, *J. Biol. Chem.* **275**, 36256–36262 (2000).
42. L. Jayaraman, K. G. K. Murthy, C. Zhu, T. Curran, S. Xanthoudiakis, and C. Prives, *Genes Dev.* **11**, 558–570 (1997).
43. S. Xanthoudakis and T. Curran, *EMBO J.* **11**, 653–665 (1992).
44. K. Hirota, M. Matsui, S. Iwata, A. Nishiyama, K. Mori, and J. Yodoi, *Proc. Natl. Acad. Sci. U.S.A.* **94**, 3633–3638 (1997).
45. T. Izumi, W. D. Henner, and S. Mitra, *Biochemistry* **35**, 14679–14683 (1996).
46. T. Izumi and S. Mitra, *Carcinogenesis* **19**, 525–527 (1998).
47. G. Barzilay, L. J. Walker, C. N. Robson, and I. D. Hickson, *Nucleic Acids Res.* **23**, 1544–1550 (1995).
48. S. Xanthoudakis, G. G. Miao, and T. Curran, *Proc. Natl. Acad. Sci. U.S.A.* **91**, 23–27 (1994).
49. L. J. Walker, C. N. Robson, E. Black, D. Gillespie, and I. D. Hickson, *Mol. Cell. Biol.* **13**, 5370–5376 (1993).
50. G. Fritz and B. Kaina, *Oncogene* **18**, 1033–1040 (1999).
50a. A. Yacoub, M. R. Kelley, and W. A. Deutsch, *Cancer Res.* **57**, 5457–5459 (1997).
51. S. Grösch and B. Kaina, *Biochem. Biophys. Res. Commun.* **261**, 859–863 (1999).
52. A. Klungland and T. Lindahl, *EMBO J.* **16**, 3341–3348 (1997).
53. B. Kaina, *Mutat. Res.* **404**, 119–124 (1998).

Session 2
Gene Targeting in the Mouse for Elucidating the Role of BER

SAMUEL H. WILSON

National Institute of Environmental Health Sciences

In this section, the authors deal with further elucidation of pathways in human and rodent cells involved in repair of oxidative and alkylation agent-induced DNA lesions, with emphasis on the repair of the 8-hydroxyguanine (8-oxoG) lesion. Three of the presentations in the section contain information on cellular responses to oxidative stress or studies of DNA glycosylase enzymology and molecular biology. Key features emerging from the section are: (1) further information implicating DNA polymerase β in single-nucleotide BER; (2) the repair of plasmid containing a single 8-oxoG · C base pair; (3) the analysis of a single-nucleotide polymorphism in the human MTH1 gene which influences mRNA levels and probable intracellular distribution of the enzyme; and (4) further characterization of human and mouse 8-hydroxyguanine-DNA glycosylase (termed MMH/OGG), including studies of Mmh null mice demonstrating that MMH is the major glycosylase responsible for repair of the 8-oxoG lesion.

Mammalian DNA β-Polymerase in Base Excision Repair of Alkylation Damage

ROBERT W. SOBOL
AND SAMUEL H. WILSON

Laboratory of Structural Biology
National Institute of Environmental
Health Sciences
Research Triangle Park,
North Carolina 27709

I.	Introduction..	58
II.	Consequence of a β-Pol Null Genotype in Mammalian Cells	59
III.	5′-Deoxyribose Phosphate Mediates MMS-Induced Cytotoxicity	63
IV.	Mutagenesis and Base Excision Repair..............................	69
V.	Summary	72
	References..	72

DNA β-polymerase (β-pol) carries out two critical enzymatic reactions in mammalian single-nucleotide base excision repair (BER): DNA synthesis to fill the repair patch and lyase removal of the 5′-deoxyribose phosphate (dRP) group following cleavage of the abasic site by apurinic/apyrimidinic (AP) endonuclease (1). The requirement for β-pol in single-nucleotide BER is exemplified in mouse fibroblasts with a null mutation in the β-pol gene. These cells are hypersensitive to monofunctional DNA methylating agents such as methyl methanesulfonate (MMS) (2). This hypersensitivity is associated with an abundance of chromosomal damage and induction of apoptosis and necrotic cell death (3). We have found that β-pol null cells are defective in repair of MMS-induced DNA lesions, consistent with a cellular BER deficiency as a causative agent in the observed hypersensitivity. Further, the N-terminal 8-kDa domain of β-pol, which contains the dRP lyase activity in the wild-type enzyme, is sufficient to reverse the methylating agent hypersensitivity in β-pol null cells. These results indicate that lyase removal of the dRP group is a pivotal step in BER *in vivo*. Finally, we examined MMS-induced genomic DNA mutagenesis in two isogenic mouse cell lines designed for study of the role of BER. MMS exposure strongly increases mutant frequency in β-pol null cells, but not in wild-type cells. With MMS treatment, β-pol null cells have a higher frequency of all six base-pair substitutions, suggesting that BER plays a role in protecting the cell against methylation-induced mutations. © 2001 Academic Press.

I. Introduction

The predominant BER pathway in mammalian cells involves the formation of a single-nucleotide gap after AP-endonucleolytic hydrolysis of an abasic (AP) site (1, 4). This single-nucleotide gap is filled, in most cases, by DNA polymerase β (β-pol) (2, 5–7), and the 5′ deoxyribose phosphate (dRP) moiety in the gap is immediately removed by the dRP lyase activity in the N-terminal domain of β-pol (8). The nicked DNA product is then sealed either by DNA ligase I or by a complex of DNA ligase III and XRCC1 (8–11). *In vitro* studies have demonstrated an alternate BER pathway, termed long-patch BER (12–15). In long-patch BER, β-pol can play a gap-filling role, but other DNA polymerases are also active in gap filling (1, 12, 15–18). Specifically, long-patch BER differs from single-nucleotide BER in that the 3′-OH group formed by AP endonuclease activity is extended during gap filling by two to several nucleotides, involving strand displacement ahead of the growing strand. In one version of long-patch BER, strand displacement is mediated by a partnership between β-pol and FEN-1 and the displaced strand is then removed by FEN-1 in partnership with proliferating cell nuclear antigen (PCNA), replication protein A (RPA), or β-pol (19–23). Thus, research toward defining mammalian BER pathways has led to a working model of a complex and multicompensatory repair system. The paradigm of a four-enzyme pathway for complete BER (glycosylase + AP endonuclease + polymerase + ligase) has expanded to include many proteins and three coincident subpathways that complete DNA repair following AP endonuclease hydrolysis of an abasic residue (see Fig. 7). β-Pol plays a role in two of the subpathways for BER, as previously described (2–4, 15, 23).

β-Pol null mouse cells are hypersensitive to DNA monofunctional alkylating agents such as methylmethane sulfonate (MMS) (2, 3, 24), suggesting that β-pol-independent BER does not completely substitute for the β-pol-dependent BER pathways in removing MMS-induced lesions and that such DNA lesions may persist in exposed cells. However, the status of DNA lesion formation and DNA repair in β-pol null cells and the structure/function requirement of β-pol in methylating agent hypersensitivity have only recently been examined (24; Sobol *et al.*, manuscript in preparation). Further, any relationship between a deficiency in β-pol-dependent BER and mutagenesis has not been evaluated.

In the work presented here, isogenic cell lines (wild-type, β-pol null, and β-pol null cells modified to express mutant forms of β-pol) were characterized to define the importance of β-pol-dependent BER in the cellular response to MMS. Quantitative PCR (QPCR) analysis, AP site analysis, and the comet assay were used to assess MMS-induced DNA damage in the wt and β-pol null cell lines. DNA lesions were induced by MMS treatment in a dose-dependent manner, and the number of lesions initially formed per genome was similar in both cell lines. β-Pol null cells were deficient in their ability to repair DNA lesions, although repair clearly was seen after extended incubation times. Further, β-pol null cells expressing dRP-lyase-deficient or DNA polymerase-deficient forms of β-pol were used to address the requirement of each of these activities of β-pol with regard to methylating agent hypersensitivity. Finally, we examined mutagenesis in wt and β-pol null cells using a transgenic, chromosome-borne λ phage *cII* mutation assay (25). The β-pol null/BER deficient cells fail to exhibit a noteworthy spontaneous mutator effect, but have a high mutation frequency and a novel mutation spectrum after MMS treatment. These results are discussed in the context of a role for BER in mutation avoidance after MMS-induced DNA damage.

II. Consequence of a β-Pol Null Genotype in Mammalian Cells

β-Pol null cells were engineered to express epitope-tagged human β-Pol (FLAG-β-pol) using expression vector pRS1088 (Fig. 1A). During transient expression of this construct, FLAG-β-pol was highly expressed as a full-length 40-kDa protein that was recognized by both a polyclonal antibody (Ab) for β-Pol (Fig. 1B, lane 1) and a monoclonal Ab for the FLAG-epitope (Fig. 1B, lane 2). The level of expression was close to the endogenous level of expression after stable integration of pRS1088 (2). Expression of both endogenous and FLAG-β-pol was primarily in the nucleus (Fig. 1C). The BER capacity of cell extracts was measured with an *in vitro* assay using a 51-bp oligonucleotide with a unique G:U base pair (5). The results demonstrate that the β-pol null extract was deficient in BER as expected, but β-pol null cells expressing recombinant FLAG-β-pol were repair-proficient (Fig. 1D, lane 3), indicating that the recombinant FLAG-β-pol is functional as a repair enzyme. The BER activity of the FLAG-β-pol and wild-type cell extracts was similar.

The fact that cells deficient in β-pol are hypersensitive to MMS raises the question of the precise role of secondary BER pathways *in vivo*. The absence of β-pol results in a greater sensitivity to MMS-induced cytotoxicity at high and low dose, and after acute or chronic exposure to MMS. We examined dose- and time-dependent MMS-induced cytotoxicity in both wild-type and β-pol null cells. When exposed to 0.5 mM MMS for 1 h, 95% of wild-type and 40% of β-pol null cells survived; when exposed for 2 h, 75% of wild-type and only ~1% of β-pol null cells survived (Fig. 2). At lower doses (<0.5 mM), MMS exposure for 24 h also gave rise to a strong dose-dependent hypersensitivity. These results indicate that MMS-induced lethal lesions are formed in a dose- and time-dependent manner and have increased cytotoxicity in the absence of β-pol.

Since β-pol null cells are hypersensitive to MMS, more MMS-induced lesions may persist in the genomic DNA of β-pol null cells than wild-type cells.

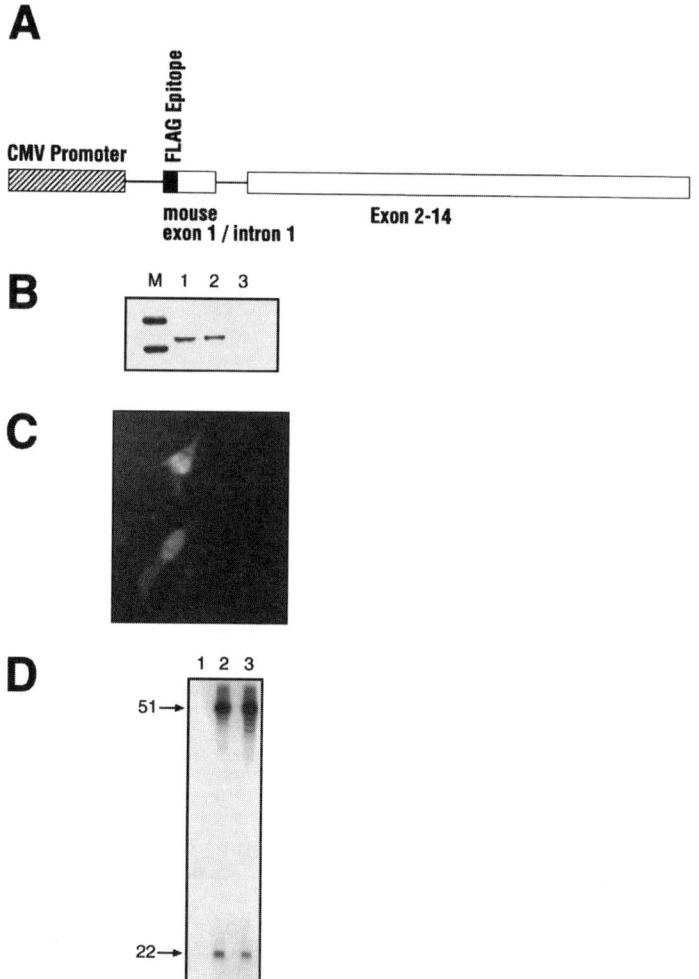

FIG. 1. Expression of recombinant β-pol corrects the β-pol null phenotype. (A) Diagrammatic representation of the β-pol minigene expression plasmid, which includes the CMV promoter, the mouse exon 1 fused with an N-terminal epitope tag (FLAG), mouse intron 1, and the remainder of the human β-pol cDNA. (B) Detection of β-pol by immunoblotting of whole-cell extracts from wild-type cells (lane 1), following FLAG-immunoprecipitation of FLAG-β-pol expressing cells (lane 2) and of whole-cell extracts from β-pol null cells (lane 3). Lane M shows molecular-weight markers of 35 kDa and 50 kDa (Perfect Protein Western blot markers, Novagen). (C) Immunofluorescent labeling of FLAG-β-pol expressing cells. FLAG-β-pol was detected by immunohistochemical labeling with an anti-FLAG bio-M2 antibody and FITC-strepavidin. A representative microphotograph is shown. (D) Base excision repair in extracts from fibroblast cell lines. Whole-cell extracts from β-pol null fibroblasts (lane 1) or from FLAG-β-pol expressing cells (lanes 2 and 3) were reacted with a 51-base pair synthetic DNA substrate containing a single G:U base pair. Base excision repair was determined by incorporation of $[\alpha^{32}P]dCMP$ exclusively to replace uracil. DNA was separated by electrophoresis on a 12% polyacrylamide gel containing 7 M urea. Gels were fixed, dried, and autoradiographed to visualize the products.

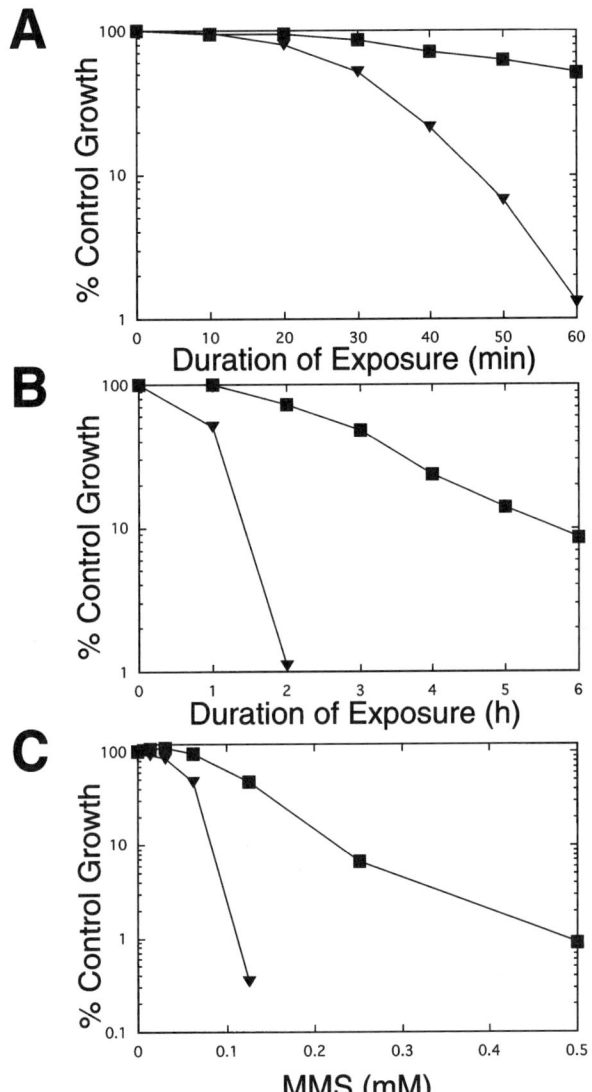

FIG. 2. Sensitivity of wild-type and β-pol null cells following acute or chronic exposure to MMS. Cytotoxicity was determined by growth inhibition assays. Wild-type cells (squares) and β-pol null cells (triangles) were exposed to MMS at varying doses 24 h after seeding (40,000 cells/well) in 6-well dishes. Cells (triplicate wells) were (A) treated with 1 mM MMS for 10–60 min, (B) treated with 0.5 mM MMS for 1–6 hours, or (C) treated with 0.1–0.5 mM MMS for 24 h at 34°C in a 10% CO_2 incubator and then allowed to grow for an additional 3 days. Cell numbers were determined by a nuclear lysis protocol. Results (average of at least two separate experiments) were determined as the number of treated cells relative to the cells in control wells (% control growth).

MMS-induced lesions include both abasic sites and alkylated bases. To differentiate between these lesions, abasic sites were quantified in genomic DNA using the aldehyde-reactive probe–slot–blot (ASB) method (26, 27). Surprisingly, abasic sites did not increase in the genomic DNA isolated from MMS-treated wild-type or β-pol null cells; however, MMS did cause an accumulation of lesions which gave rise to abasic residues following heat treatment of the purified DNA (not shown). These heat-labile lesions are likely N7-methylguanine and N3-methyladenine (83% and 8.3%, respectively), the predominant lesions formed by MMS (28). When assayed after a repair period (4 h after termination of MMS treatment), there was no significant difference in the number of abasic sites in genomic DNA from either cell type. This suggested that most of the MMS-induced base lesions remain intact without the accumulation of abasic sites or incised abasic sites in the genome. Further, we used QPCR to measure the number of MMS-induced DNA polymerase blocking lesions [e.g., N3-methyladenine and single-strand breaks (ssb)] in a segment of genomic DNA (29–38). Following MMS treatment, an equivalent amount of DNA damage was found in both cell lines immediately after exposure (not shown). Whereas wild-type cells repair all the polymerase blocking lesions within a 4-h period, cells deficient in β-pol repaired only a portion of these lesions in the same period. That only a fraction of MMS-induced DNA polymerase blocking lesions were repaired in the β-pol null cells indicates that they were BER-deficient and that β-pol-dependent and -independent repair may have a different substrate specificity and/or rate of lesion repair. Repair of the polymerase blocking base methylation lesion during a 4-h repair period may primarily reflect the removal of N3-methyladenine, which is known to be a polymerase blocking lesion and is repaired on a similar time frame in mammalian cells (39). This is consistent with full repair of DNA polymerase blocking lesions in wild-type cells; but these lesions are not expected to include N7-methylguanine, which is repaired on a much larger time frame and is not DNA polymerase blocking (28).

Finally, DNA damage was measured using the comet assay to detect genomic single-strand breaks (ssb) plus alkali-labile sites (40). This assay provides an analytical approach for repair of single-strand breaks and alkali-labile sites in genomic DNA (40–46). MMS induced a dose-dependent increase in comet assay tail moment, indicating an increase in DNA lesions in both cell lines (not shown). Following a repair period of 2 h, DNA damage was reduced in wild-type cells, but there was no reduction in DNA damage in β-pol null cells (not shown), indicating that β-pol null cells were inefficient in repair of MMS-induced lesions. These results are in accord with the conclusion that β-pol-dependent BER participates in repair of methylation damage and that β-pol-independent BER is not efficient enough to fully complement, even though it is able to fully repair model BER substrates *in vitro* after longer incubation times (12, 13).

III. 5′-Deoxyribose Phosphate Mediates MMS-Induced Cytotoxicity

The demonstration that DNA polymerases other than β-pol can conduct both single-nucleotide and long-patch BER DNA synthesis *in vitro* (12, 13) suggests that other polymerases might be able to functionally complement the β-pol deficit in β-pol null cells. However, as described previously (2, 3, 24) and as illustrated in Figs. 1 and 3a, β-pol null cells are hypersensitive to the cytotoxic effects of the simple methylating agent MMS. This hypersensitivity is, however, reversed by expression of wt β-pol (2, 24). β-pol null cells expressing Flag epitope tagged wild-type β-pol (β-pol null/WT β-pol cells) and wild-type cells were found to be resistant to MMS, whereas control β-pol null cells expressing the neomycin resistance gene (β-pol null/Neo cells) and untransfected β-pol null cells are hypersensitive to MMS (Fig. 3a,b). The polymerase active site of β-pol is in the 31-kDa C-terminal domain and the 5′-dRP lyase active site is in the 8-kDa N-terminal domain (1). We examined the functional contribution of these two domains and enzymatic activities toward resistance to methylating agents. The cDNA of β-pol was modified as shown in Fig. 3c; expression of the altered forms of β-pol in β-pol null cells yielded the expected domain polypeptides of 8 and 31 kDa, respectively (not shown). Other mutant β-pol cDNAs were also generated to express full-length β-pol protein with single or multiple enzymatic active site point mutations, as summarized in Fig. 3c.

Residue Arg283 of β-pol is critical to efficient nucleotide incorporation and discrimination (47, 48). In the β-pol/substrate crystal structures, Arg283 forms a hydrogen bond and makes van der Waals contact with the template strand in the DNA minor groove, and thus is proposed to stabilize the template base during recognition of the incoming dNTP (47, 49). Mutation of this residue to alanine results in a polymerase with a 5000-fold decreased catalytic efficiency in DNA synthesis. As shown in Fig. 3d (lane 2), the immunoaffinity-purified Flag-R283A-β-pol was found to be deficient in BER DNA synthesis activity, as expected (47). However, the purified Flag-R283A-β-pol enzyme had full dRP lyase activity (Fig. 3e, lane 2).

Surprisingly, β-pol null/R283A-β-pol cells showed a wild-type cell MMS sensitivity (Fig. 4a, closed circles), whereas the control β-pol/Neo cells (Fig. 4a, open circles) behaved like β-pol null cells. Therefore, there was no apparent requirement for DNA synthesis activity of β-pol in complementing the β-pol null phenotype. To further examine the requirement for β-pol DNA synthesis activity, we generated cell lines expressing a polymerase active-site deletion mutant of β-pol (Flag-Thumbless) and the polymerase active site point mutant, Flag-D256A-β-pol. For Flag-Thumbless, residues within the C-terminal deleted portion (residues 263–335) of β-pol are required for DNA synthesis

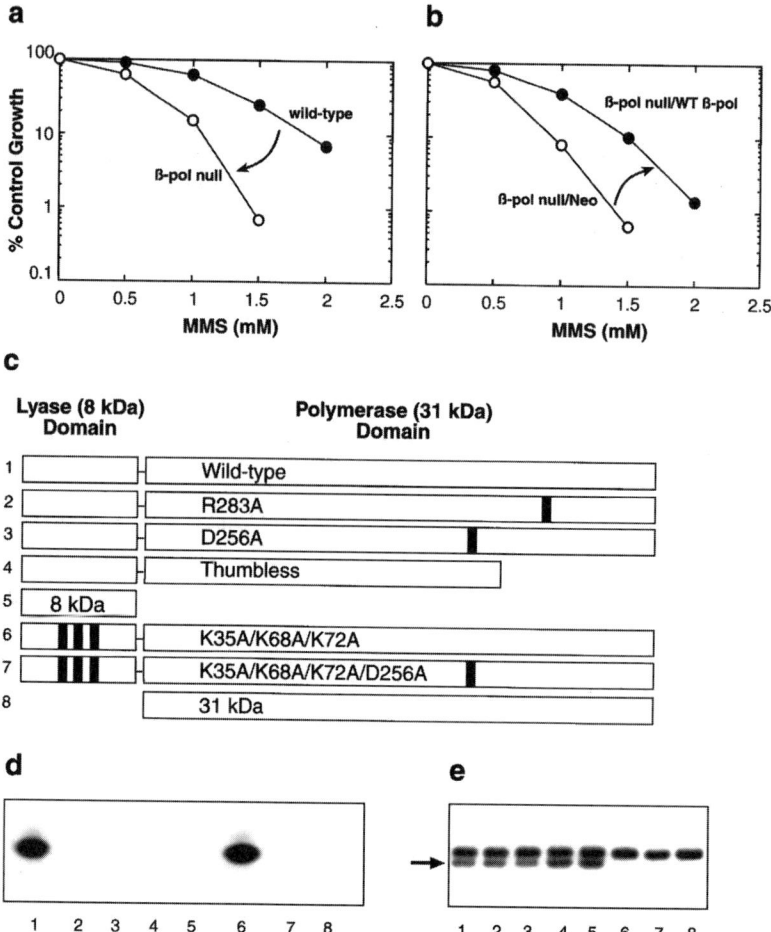

FIG. 3. Cytotoxic sensitivity of fibroblast cell lines to the alkylating agent MMS. (a) MMS sensitivity of wild-type cells (filled circles) as compared to β-pol null cells (open circles). (b) MMS sensitivity of β-pol null/WT β-pol cells (filled circles) as compared to β-pol null/Neo cells (open circles). The arrow in panels a and b represents the shift in MMS sensitivity due to the genetic loss of β-pol (panel a) or the gain of β-pol (panel b). (c) Diagrammatic representation of recombinant β-pol proteins expressed in β-pol null cells. Vertical bars represent the approximate location of specific point mutations. Numbers along the left side coincide with the lanes in panels d and e. (d) DNA synthesis activity. Flag-WT β-pol (lane 1) and Flag-β-pol mutant proteins (lanes 2-8) were immunoaffinity-purified from whole-cell extracts and were analyzed for DNA synthesis activity in a single-nucleotide gap-filling assay. A representative autoradiograph is shown. (e) dRP lyase activity. Immunoaffinity-purified Flag-WT β-pol (lane 1) and Flag-β-pol mutant proteins (lanes 2–8) were analyzed for dRP lyase activity on a pre-incised double-stranded oligonucleotide substrate. A representative autoradiograph is shown. The reaction product is indicated by the arrow. (d and e): Lane 1, Flag-WT β-pol; lane 2, Flag-R283A β-pol; lane 3, Flag-D256A β-pol; lane 4, Flag-Thumbless; lane 5, Flag-8 kDa; lane 6, Flag-K35A, K68A, K72A β-pol; lane 7, Flag-K35A, K68A, K72A, D256A β-pol; lane 8, Flag-31 kDa. Reprinted by permission from Nature (24), copyright 2000, Macmillan Magazines Ltd.

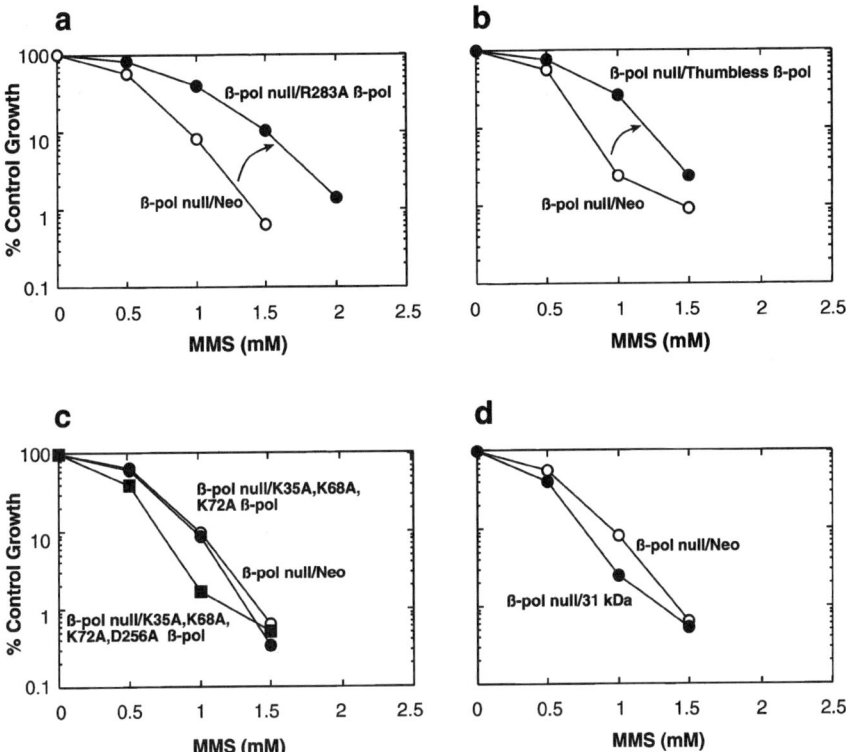

FIG. 4. DNA synthesis activity of β-pol is not required, but dRP lyase activity of β-pol is required to reverse the MMS hypersensitivity of β-pol null cells. (a) MMS sensitivity of β-pol null/R283A β-pol cells (filled circles) as compared to β-pol null/Neo cells (open circles). (b) MMS sensitivity of β-pol null/Thumbless β-pol cells (filled circles) as compared to β-pol null/Neo cells (open circles). (c) MMS sensitivity of β-pol null/K35A, K68A, K72A β-pol cells (filled circles) and β-pol null/K35A, K68A, K72A, D256A β-pol cells (filled squares) as compared to β-pol null/Neo cells (open circles). (d) MMS sensitivity of β-pol null/31 kDa β-pol cells (filled circles) as compared to β-pol null/Neo cells (open circles). Cytotoxicity was determined by growth inhibition assays. Results (mean of at least two separate experiments) were determined as the number of treated cells relative to the cells in control wells (% control growth). Reprinted by permission from *Nature* (24), copyright 2000, Macmillan Magazines Ltd.

activity (47, 50). Further, Asp256 is known to be critical to the nucleotidyltransferase mechanism, and substitution of alanine for aspartate (D256A) abolishes the polymerase activity (51). Both Flag-D256A-β-pol and Flag-Thumbless were found to be completely deficient in single-nucleotide BER gap-filling DNA synthesis (Fig. 3d, lanes 3 and 4), yet were active in dRP removal (Fig. 3e, lanes 3 and 4). Both β-pol null/Thumbless β-pol cells (Fig. 4b, closed circles) and β-pol null/D256A-β-pol cells (not shown) showed a sensitivity to MMS similar to wild-type cells. Based on these observations, it appears that DNA synthesis catalyzed by β-pol is not critical to cell survival following MMS treatment. This was reminiscent of observations that polymerase catalytic activity of yeast DNA polymerase ε, but not the protein, is dispensable for replication and repair (52, 53). The observations reported herein indicate that β-pol-dependent DNA synthesis is dispensable and suggest that the persistent dRP group may be lethal.

To investigate the idea that removal of the dRP group will alleviate the hypersensitivity to MMS, β-pol null cells were stably transfected with a full-length β-pol minigene with three alanine point mutations within the dRP lyase active site. Lys72 is the predominant Schiff base nucleophile in the dRP lyase β-elimination reaction mechanism, and mutating three active site lysine residues to alanine (K35A, K68A, K72A) completely eliminates dRP lyase activity (54–56). First, Flag-K35A, K68A, K72A-β-pol was expressed in β-pol null cells and immunoaffinity-purified to characterize its DNA synthesis and dRP lyase activities *in vitro*. As shown in Fig. 3, Flag-K35A, K68A, K72A-β-pol was proficient in DNA synthesis (panel d, lane 6), yet it was devoid of dRP lyase activity (panel e, lane 6). Next, β-pol null/K35A, K68A, K72A-β-pol cells were exposed to increasing doses of MMS. As shown in Fig. 4c, expression of this dRP lyase-deficient mutant (filled circles) failed to change the MMS hypersensitivity of β-pol null/Neo cells. In addition, a compound mutant (Flag-K35A, K68A, K72A, D256A β-pol; Fig. 3c), deficient in both dRP lyase and DNA synthesis activities, was expressed in β-pol null cells. The immunoaffinity-purified protein from these cells failed to show either gap-filling DNA synthesis activity (Fig. 3d) or dRP lyase activity (Fig. 3e) against BER substrates. β-pol null cells expressing Flag-K35A, K68A, K72A-D256A-β-pol (Fig. 4c, filled squares) were found to be similar to β-pol null/Neo cells in their sensitivity to MMS.

Complete removal of the N-terminal 8-kDa dRP lyase domain yields β-pol protein referred to as Flag-31 kDa (Fig. 3c). The C-terminal 31-kDa domain of β-pol contains the DNA synthesis active site, although the protein has little DNA-binding activity and is diminished in DNA synthesis activity (1). As shown in Fig. 3, purified Flag-31 kDa had neither gap-filling DNA synthesis nor dRP lyase activity (panels d and e, respectively). As with the other dRP lyase mutants described above, β-pol null/31 kDa β-pol cells were no more resistant to MMS than were β-pol null or β-pol null/Neo cells (Fig. 4d).

We next examined whether the N-terminal 8-kDa dRP lyase domain of β-pol alone is sufficient to complement the β-pol null MMS hypersensitivity. The N-terminal 8-kDa domain is responsible for single-stranded DNA binding activity, 5' phosphate recognition, and dRP lyase enzymatic activity (1). Flag-8 kDa was expressed in β-pol null cells and the protein was immunoaffinity-purified. This small β-pol domain had no gap-filling DNA synthesis activity when measured *in vitro* (Fig. 3d) but, as expected, exhibited robust dRP lyase activity (Fig. 3e) (55, 56). Since the entire C-terminal domain has been removed, there is no possibility for DNA synthesis activity from the Flag-8 kDa protein (see Fig. 3c). To determine if the 8-kDa domain is sufficient to restore MMS resistance, β-pol null/Neo, β-pol null/WT β-pol, and β-pol null/8 kDa β-pol cells were compared for MMS sensitivity. As shown in Fig. 5, β-pol null/8 kDa β-pol cells were wild-type–like in their resistance to MMS (Fig. 5, filled squares). Note the apparent increase in sensitivity of the β-pol null/8 kDa β-pol cells at higher MMS concentrations as compared to the β-pol null/WT β-pol cells. This suggests that at high concentrations of DNA damaging agent, direct dRP removal by the 8-kDa domain of β-pol is not entirely sufficient for cell survival. It is possible that the alternate long-patch BER pathway is more significant at high doses, requiring the DNA synthesis activity of β-pol to stimulate FEN-1 cleavage (23).

A correlation between failure to repair the 5'-dRP group and MMS hypersensitivity was unexpected since other proteins with 5'-dRP lyase activity have been observed in mammalian cells (57, 58). In addition, dRP lyase activity has been found in various *E. coli*, *Drosophila*, and yeast proteins, many of which have mammalian homologs (59–63). In an attempt to understand the apparent accumulation of the 5'-dRP-containing BER intermediate in the presence of an abundance of cellular dRP lyase activity, we prepared whole-cell extracts from wild-type and β-pol null cells, β-pol null/WT β-pol cells, β-pol null cells expressing the β-pol lyase-proficient point mutants Flag-R283A and Flag-D256A, and the β-pol null/8 kDa β-pol cells. Capacity to remove the 5'-dRP lesion from the BER intermediate was measured in a complete BER assay using a double-stranded oligonucleotide substrate with a single G:U base pair (the damaged site). As outlined in Fig. 6 (panel a), base removal and cleavage 5' to the abasic site is conducted by endogenous uracil DNA glycosylase and AP endonuclease,

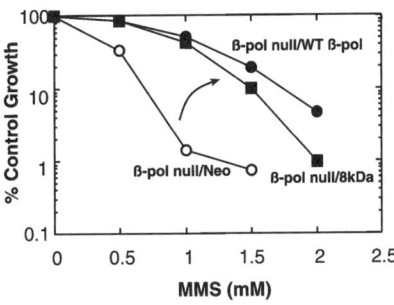

FIG. 5. dRP lyase activity of β-pol is sufficient to reverse the MMS hypersensitivity of β-pol null cells. MMS sensitivity of β-pol null/WT β-pol cells (filled circles) and β-pol null/8 kDa β-pol cells (filled squares) as compared to β-pol null/Neo cells (open circles). Cytotoxicity was determined by growth inhibition assays. Results (mean of at least two separate experiments) were determined as the number of treated cells relative to the cells in control wells (% control growth). Reprinted by permission from *Nature* (24), copyright 2000, Macmillan Magazines Ltd.

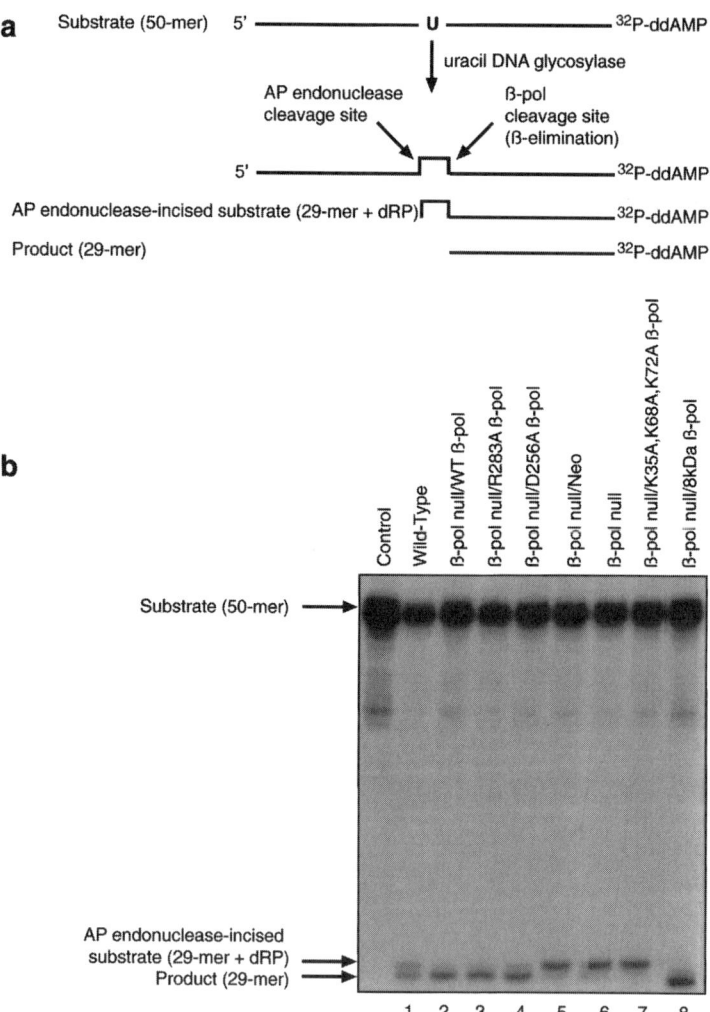

FIG. 6. dRP lyase activity in whole cell extracts from stable-transfected β-pol null cell lines. Dialyzed extracts from wild-type mouse fibroblasts (lane 1), β-pol null fibroblasts (lanes 5 and 6) or β-pol null fibroblasts stable-transfected to express wild-type or mutant Flag-β-pol proteins, as indicated (lanes 2–4, 7, and 8), were incubated with a ^{32}P-labeled (3') 50-base pair synthetic double-stranded DNA substrate containing a single G:U base pair, as outlined in panel a. Only the upper DNA strand is shown. The reaction products were separated by electrophoresis in a 20% polyacrylamide gel. Gels were fixed, dried, and autoradiographed to visualize the products. A representative autoradiograph is shown in panel b. The position of the substrate, the AP endonuclease-incised substrate, and the product are indicated. Reprinted by permission from Nature (24), copyright 2000, Macmillan Magazines Ltd.

respectively, giving rise to the 5′-dRP-containing BER intermediate. Extracts from all the cell lines expressing dRP lyase-proficient β-pol proteins (wild-type, Flag-WT β-pol, Flag-R283A β-pol, and Flag-D256A β-pol; Fig. 6, panel b, lanes 1–4) had a strong capacity to remove the dRP group from the AP endonuclease-incised substrate (upper band), thereby producing the 29-mer (lower band) product. However, cells deficient in β-pol (Fig. 6, panel b, lanes 5 and 6) or cells expressing a β-pol dRP lyase-deficient mutant (Flag-K35A, K68A, K72A β-pol; Fig. 6, panel b, lane 7) were devoid of dRP lyase activity. Finally, extract from the β-pol null/8 kDa β-pol cells was found to be highly active in BER-associated dRP lyase activity.

IV. Mutagenesis and Base Excision Repair

DNA lesions and/or the absence of DNA repair functions can result in specific mutations or unique mutation spectra. To evaluate this possibility, we characterized mutation spectra in wild-type and β-pol null cells as a function of MMS and MNNG treatment. Mutation spectra (either spontaneous or induced) were obtained using a recoverable λ-LIZ shuttle vector (64). To facilitate such an analysis, wild-type and BER-deficient cell lines, which were also transgenic for the λ-LIZ shuttle vector bearing the *cII* gene, were developed and analyzed for both spontaneous and MMS-induced mutations using a positive selection screen (25, 65–67). Independent mutants were generated for DNA sequence analysis using a modified fluctuation analysis (65). Wild-type cells and β-pol null cells had similar spontaneous mutant frequencies (not shown), demonstrating that a deficiency in β-pol dependent BER does not result in a large spontaneous hypermutable phenotype. The relative spontaneous mutation frequencies of −1 frameshift mutations and A:T ⇒ G:C transitions were comparable in wild-type and β-pol null cells, whereas the frequencies of all other mutations were elevated in β-pol null cells between 1.3- and 3.0-fold (not shown). There has been speculation that the known, relatively low fidelity of β-pol might result in more observed mutations *in vivo* (68, 69). However, there was no support for this hypothesis, since we failed to observe an "anti-mutator" effect in the absence of β-pol. Conversely, no significant spontaneous mutator effect was observed, suggesting that for spontaneous damage, the alternate BER pathways can sufficiently and accurately conduct the needed DNA repair.

Mutant frequencies and mutation spectra were determined in wild-type and β-pol null cells after treatment with MMS or MNNG (Sobol *et al.*, manuscript in preparation). MMS is generally a poor mutagen in repair-proficient mouse cells. Analysis of several different transgenic and endogenous genes in wild-type mice or cells showed that MMS did not increase mutant frequency (70–74). However, in some repair-deficient cells, MMS causes an accumulation of mutations (70,

71, 75, 76). The MMS hypersensitivity of cells deficient in β-pol-dependent BER, along with the DNA repair results indicated above, suggested the use of MMS as the DNA damaging agent for our mutation analysis. MMS treatment did not increase the mutant frequency in the wild-type cells. However, MMS was mutagenic in β-pol null cells, inducing a >4-fold increase in mutant frequency relative to MMS-treated wild-type cells. The mutation spectra also revealed that β-pol null cells are more readily mutated by MMS than are wild-type cells. MMS treatment increased the frequency of all types of base substitution mutations in β-pol null cells relative to wild-type cells. A:T ⇒ G:C transitions and G:C ⇒ C:G transversions were increased >8-fold, respectively. The most frequent mutation in β-pol null cells was G:C ⇒ T:A; yet, all six types of base substitutions were elevated and the amount of elevation was almost evenly balanced.

As in previous reports (70–74), the mutant frequency in the λ cII gene did not increase in MMS-treated repair-proficient mouse cells (Sobol et al., manuscript in preparation). However, the mutant frequency in MMS treated β-pol null cells was higher than the mutant frequency in untreated β-pol null cells and was ∼5-fold higher than in MMS-treated wild-type cells. These results indicate that these BER-deficient cells are hypermutable in the presence of alkylation DNA damage induced by MMS. A similar observation has been made for an XRCC1 defective cell line, in that the number of mutations at the *hprt* locus increased after MMS treatment, as compared to a wild-type CHO cell line (70). Interestingly, we found that the frequencies of all six base substitution mutations observed in the MMS-treatment spectra were elevated in β-pol null cells, indicating a general mutator effect of this BER deficiency in the presence of MMS-induced DNA methylation (Sobol et al., manuscript in preparation).

In the absence of β-pol-dependent BER, MMS is a general mutator in our mouse cell system. Several possible explanations exist for the observed increase in mutations in β-pol null cells following MMS treatment. This may be due to quantitatively minor levels of a range of base miscoding lesions, since the most abundant lesion (N7-methylguanine) fails to cause specific coding-based mutations, as in the case, for example, where O^6-methylguanine causes G:C ⇒ A:T base substitutions. Similarly, N3-methyladenine, the other relatively abundant MMS-induced lesion, is a DNA replication blocking lesion and would not be expected to directly lead to a coding-based mutation. One may consider that the fidelity of β-pol-independent BER is lower than that of β-pol-dependent BER, although this seems unlikely. Alternatively, the failure to repair MMS-induced lesions may saturate repair complexes to these sites and thereby contribute to a further general decrease in repair capacity and subsequent increase in mutations. Finally, it is interesting to consider the notion of coordination of repair and repair initiation. The absence of β-pol could somehow downregulate the initiation of glycosylase-mediated repair. In an uncoordinated pathway, the absence of β-pol would cause a buildup of incised abasic sites. On the contrary, the data presented herein are consistent with the idea of a coordinated pathway in which repair initiation is deficient in the absence of β-pol, leading to a persistence of polymerase blocking lesions such as N3-methyladenine. An increase in unrepaired N3-methyladenine (or other polymerase blocking lesions) in the β-pol null cells may induce an error-prone process. Such a concept has recently emerged for novel DNA polymerases that could play a role in lesion bypass (e.g., N3-methyladenine) in BER-deficient cells and confer a mutagenic phenotype to MMS treatment. Measurements of the fidelity and mutation spectra of novel translesion DNA polymerases are consistent with this hypothesis (77, 78). It is also noted that this hypothesis is similar to that already proposed by Klungland and coworkers, who observed MMS-induced base substitution mutagenesis in CHO cells as a function of persistence of N3-methyladenine in genomic DNA (79). Such hypotheses are consistent with general base substitution mutagenesis secondary to replication blockage in several other systems (79).

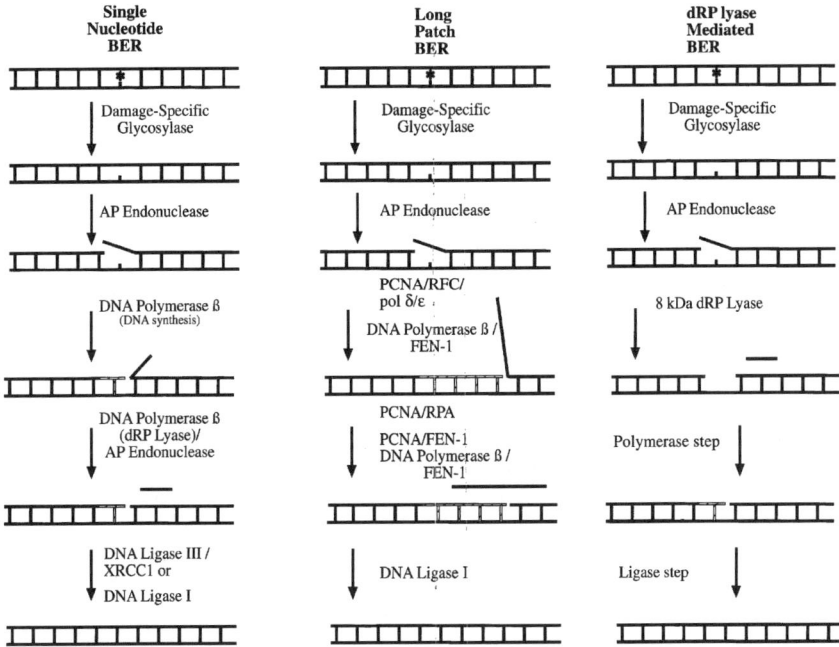

FIG. 7. Functional scheme for mammalian base excision repair. This model represents the mechanistic role of each of the proteins reported to be involved in BER and defines the involvement of each in single-nucleotide BER (left column), β-pol-dependent long-patch BER (center column) and dRP-lyase mediated BER (right column) (8–24).

V. Summary

The results presented herein document the involvement of DNA polymerase β in the repair of methylation induced DNA damage. In the absence of β-pol, a deficiency in the ability to initiate repair of MMS-induced lesions is observed. The failure to remove the 5'-deoxyribose phosphate lesion, and not a failure to conduct gap-filling DNA synthesis, results in the MMS-induced hypersensitivity in the absence of β-pol. Hence, the 5'-dRP group in the BER intermediate is a cytotoxic lesion produced after MMS exposure, and cellular proteins other than β-pol are not able to efficiently repair this 5'-dRP group (24). Since dRP removal is the rate-limiting step during single-nucleotide BER (8), the removal of the dRP group by the 8-kDa domain of β-pol offers other cellular polymerases the opportunity to incorporate a single-nucleotide residue without adversely affecting overall BER (see Fig. 7). Further, using isogenic wild-type and β-pol null mouse cells harboring a λ phage transgene (cII), we have found that BER is involved in avoidance of MMS-induced mutations. Interestingly, speculation regarding any mutator effect of β-pol itself was not observed in these studies, in that β-pol null cells do not have an anti-mutator phenotype. Instead, β-pol null mouse cells are more mutable for general base substitution than wild-type cells.

Acknowledgment

Portions of this manuscript were reprinted by permission from *Nature* (24) copyright 2000, Macmillan Magazines Ltd.

References

1. S. H. Wilson, *Mutat. Res.* **407**, 203–215 (1998).
2. R. W. Sobol, J. K. Horton, R. Kuhn, H. Gu, R. K. Singhal, R. Prasad, K. Rajewsky, and S. H. Wilson, *Nature (London)* **379**, 183–186 (1996).
3. K. Ochs, R. W. Sobol, S. H. Wilson, and B. Kaina, *Cancer Res.* **59**, 1544–1551 (1999).
4. J. K. Horton, R. Prasad, E. Hou, and S. H. Wilson, *J. Biol. Chem.* **275**, 2211–2218 (2000).
5. R. K. Singhal, R. Prasad, and S. H. Wilson, *J. Biol. Chem.* **270**, 949–957 (1995).
6. K. Nealon, I. D. Nicholl, and M. K. Kenny, *Nucleic Acids Res.* **24**, 3763–3770 (1996).
7. I. D. Nicholl, K. Nealon, and M. K. Kenny, *Biochemistry* **36**, 7557–7566 (1997).
8. D. K. Srivastava, B. J. Vande Berg, R. Prasad, J. T. Molina, W. A. Beard, A. E. Tomkinson, and S. H. Wilson, *J. Biol. Chem.* **273**, 21203–21209 (1998).
9. Y. Kubota, R. A. Nash, A. Klungland, P. Schar, D. E. Barnes, and T. Lindahl, *EMBO J.* **15**, 6662–6670 (1996).
10. E. Cappelli, R. Taylor, M. Cevasco, A. Abbondandolo, K. Caldecott, and G. Frosina, *J. Biol. Chem.* **272**, 23970–23975 (1997).
11. R. Prasad, R. K. Singhal, D. K. Srivastava, J. T. Molina, A. E. Tomkinson, and S. H. Wilson, *J. Biol. Chem.* **271**, 16000–16007 (1996).

12. P. Fortini, B. Pascucci, E. Parlanti, R. W. Sobol, S. H. Wilson, and E. Dogliotti, *Biochemistry* **37,** 3575–3580 (1998).
13. S. Biade, R. W. Sobol, S. H. Wilson, and Y. Matsumoto, *J. Biol. Chem.* **273,** 898–902 (1998).
14. Y. Matsumoto, K. Kim, and D. F. Bogenhagen, *Mol. Cell. Biol.* **14,** 6187–6197 (1994).
15. G. L. Dianov, R. Prasad, S. H. Wilson, and V. A. Bohr, *J. Biol. Chem.* **274,** 13741–13743 (1999).
16. M. Stucki, B. Pascucci, E. Parlanti, P. Fortini, S. H. Wilson, U. Hubscher, and E. Dogliotti, *Oncogene* **17,** 835–843 (1998).
17. R. Sanderson and D. Mosbaugh, *J. Biol. Chem.* **273,** 24822–24831 (1998).
18. Y. Matsumoto, K. Kim, J. Hurwitz, R. Gary, D. S. Levin, A. E. Tomkinson, and M. S. Park, *J. Biol. Chem.* **274,** 33703–33708 (1999).
19. G. L. Dianov, B. R. Jensen, M. K. Kenny, and V. A. Bohr, *Biochemistry* **38,** 11021–11025 (1999).
20. R. Gary, K. Kim, H. L. Cornelius, M. S. Park, and Y. Matsumoto, *J. Biol. Chem.* **274,** 4354–4363 (1999).
21. K. Kim, S. Biade, and Y. Matsumoto, *J. Biol. Chem.* **273,** 8842–8848 (1998).
22. M. S. DeMott, S. Zigman, and R. A. Bambara, *J. Biol. Chem.* **273,** 27492–27498 (1998).
23. R. Prasad, G. L. Dianov, V. A. Bohr, and S. H. Wilson, *J. Biol. Chem.* **275,** 4460–4466 (2000).
24. R. W. Sobol, R. Prasad, A. Evenski, A. Baker, X. P. Yang, J. K. Horton, and S. H. Wilson, *Nature (London)* **405,** 807–810 (2000).
25. J. L. Jakubczak, G. Merlino, J. E. French, W. J. Muller, B. Paul, S. Adhya, and S. Garges, *Proc. Natl. Acad. Sci. U.S.A.* **93,** 9073–9078 (1996).
26. J. Nakamura, V. E. Walker, P. B. Upton, S. Y. Chiang, Y. W. Kow, and J. Swenberg, *Cancer Res.* **58,** 222–225 (1998).
27. J. Nakamura and J. A. Swenberg, *Cancer Res.* **59,** 2522–2526 (1999).
28. D. T. Beranek, *Mutat. Res.* **231,** 11–30 (1990).
29. D. P. Kalinowski, S. Illenye, and B. Van Houten, *Nucleic Acids Res.* **20,** 3485–3494 (1992).
30. G. Deng, J. Su, K. J. Ivins, B. Van Houten, and C. W. Cotman, *Exp. Neurol.* **159,** 309–318 (1999).
31. S. W. Ballinger, B. Van Houten, G. F. Jin, C. A. Conklin, and B. F. Godley, *Exp. Eye Res.* **68,** 765–772 (1999).
32. B. Van Houten, Y. Chen, J. A. Nicklas, I. R. Rainville, and J. P. O'Neill, *Mutat. Res.* **403,** 171–175 (1998).
33. M. J. McCarthy, J. I. Rosenblatt, and R. S. Lloyd, *Photochem. Photobiol.* **66,** 356–362 (1997).
34. F. M. Yakes and B. Van Houten, *Proc. Natl. Acad. Sci. U.S.A.* **94,** 514–519 (1997).
35. M. J. McCarthy, J. I. Rosenblatt, and R. S. Lloyd, *Mutat. Res.* **363,** 57–66 (1996).
36. S. Cheng, C. Fockler, W. M. Barnes, and R. Higuchi, *Proc. Natl. Acad. Sci. U.S.A.* **91,** 5695–5699 (1994).
37. H. Yang, G. P. Kotturi, J. G. de Boer, and B. W. Glickman, *Environ. Mol. Mutagen.* **33,** 21–27 (1999).
38. K. H. Chen, F. M. Yakes, D. K. Srivastava, R. K. Singhal, R. W. Sobol, J. K. Horton, B. Van Houten, and S. H. Wilson, *Nucleic Acids Res.* **26,** 2001–2007 (1998).
39. B. P. Engelward, J. M. Allan, A. J. Dreslin, J. D. Kelly, M. M. Wu, B. Gold, and L. D. Samson, *J. Biol. Chem.* **273,** 5412–5418 (1998).
40. Y. Miyamae, K. Zaizen, K. Ohara, Y. Mine, and Y. F. Sasaki, *Mutat. Res.* **393,** 99–106 (1997).
41. R. Helbig and G. Speit, *Mutat. Res.* **377,** 279–286 (1997).
42. P. Fortini, G. Raspaglio, M. Falchi, and E. Dogliotti, *Mutagenesis* **11,** 169–175 (1996).
43. S. Nocentini, *Radiat. Res.* **144,** 170–180 (1995).
44. Y. F. Sasaki, E. Nishidate, F. Izumiyama, N. Matsusaka, and S. Tsuda, *Mutat. Res.* **391,** 215–231 (1997).
45. L. Henderson, A. Wolfreys, J. Fedyk, C. Bourner, and S. Windebank, *Mutagenesis* **13,** 89–94 (1998).

46. D. Anderson and M. J. Plewa, *Mutagenesis* **13,** 67–73 (1998).
47. W. A. Beard, W. P. Osheroff, R. Prasad, M. R. Sawaya, M. Jaju, T. G. Wood, J. Kraut, T. A. Kunkel, and S. H. Wilson, *J. Biol. Chem.* **271,** 12141–12144 (1996).
48. J. Ahn, B. G. Werneburg, and M. D. Tsai, *Biochemistry* **36,** 1100–1107 (1997).
49. W. A. Beard and S. H. Wilson, *Chem. Biol.* **5,** 7–13 (1998).
50. H. Pelletier, M. R. Sawaya, A. Kumar, S. H. Wilson, and J. Kraut, *Science* **264,** 1891–1903 (1994).
51. K. L. Menge, Z. Hostomsky, B. R. Nodes, G. O. Hudson, S. Rahmati, E. W. Moomaw, R. J. Almassy, and Z. Hostomska, *Biochemistry* **34,** 15934–15942 (1995).
52. T. Kesti, K. Flick, S. Keranen, J. E. Syvaoja, and C. Wittenberg, *Mol. Cell* **3,** 679–685 (1999).
53. R. Dua, D. L. Levy, and J. L. Campbell, *J. Biol. Chem.* **274,** 22283–22288 (1999).
54. Y. Matsumoto, K. Kim, D. S. Katz, and J. A. Feng, *Biochemistry* **37,** 6456–6464 (1998).
55. R. Prasad, W. A. Beard, P. R. Strauss, and S. H. Wilson, *J. Biol. Chem.* **273,** 15263–15270 (1998).
56. R. Prasad, W. A. Beard, J. Y. Chyan, M. W. Maciejewski, G. P. Mullen, and S. H. Wilson, *J. Biol. Chem.* **273,** 11121–11126 (1998).
57. A. Price and T. Lindahl, *Biochemistry* **30,** 8631–8637 (1991).
58. M. J. Longley, R. Prasad, D. K. Srivastava, S. H. Wilson, and W. C. Copeland, *Proc. Natl. Acad. Sci. U.S.A.* **95,** 12244–12248 (1998).
59. M. Sandigursky, A. Yacoub, M. R. Kelley, W. A. Deutsch, and W. A. Franklin, *J. Biol. Chem.* **272,** 17480–17484 (1997).
60. M. Sandigursky, A. Yacoub, M. R. Kelley, Y. Xu, W. A. Franklin, and W. A. Deutsch, *Nucleic Acids Res.* **25,** 4557–4561 (1997).
61. M. Sandigursky and W. A. Franklin, *Nucleic Acids Res.* **22,** 247–250 (1994).
62. R. J. Graves, I. Felzenszwalb, J. Laval, and T. R. O'Connor, *J. Biol. Chem.* **267,** 14429–14435 (1992).
63. G. Dianov, B. Sedgwick, G. Daly, M. Olsson, S. Lovett, and T. Lindahl, *Nucleic Acids Res.* **22,** 993–998 (1994).
64. M. J. Dycaico, G. S. Provost, P. L. Kretz, S. L. Ransom, J. C. Moores, and J. M. Short, *Mutat. Res.* **307,** 461–478 (1994).
65. D. E. Watson, M. L. Cunningham, and K. R. Tindall, *Mutagenesis* **13,** 487–497 (1998).
66. G. L. Erexson, M. L. Cunningham, and K. R. Tindall, *Mutagenesis* **13,** 649–653 (1998).
67. P. R. Harbach, D. M. Zimmer, A. L. Filipunas, W. B. Mattes, and C. S. Aaron, *Environ. Mol. Mutagen.* **33,** 132–143 (1999).
68. Y. Canitrot, C. Cazaux, M. Frechet, K. Bouayadi, C. Lesca, B. Salles, and J. S. Hoffmann, *Proc. Natl. Acad. Sci. U.S.A.* **95,** 12586–12590 (1998).
69. Y. Canitrot, M. Frechet, L. Servant, C. Cazaux, and J. S. Hoffmann, *FASEB J.* **13,** 1107–1111 (1999).
70. C. W. Op het Veld, J. Jansen, M. Z. Zdzienicka, H. Vrieling, and A. A. van Zeeland, *Mutat. Res.* **398,** 83–92 (1998).
71. R. Helbig, E. Gerland, M. Z. Zdzienicka, and G. Speit, *Mutat. Res.* **336,** 307–316 (1995).
72. H. Tinwell, P. A. Lefevre, and J. Ashby, *Environ. Mol. Mutagen.* **32,** 163–172 (1998).
73. T. Suzuki, S. Itoh, N. Takemoto, N. Yajima, M. Miura, M. Hayashi, H. Shimada, and T. Sofuni, *Mutat. Res.* **388,** 155–163 (1997).
74. T. M. Brooks and S. W. Dean, *Mutat. Res.* **388,** 219–222 (1997).
75. W. E. Glaab, J. I. Risinger, A. Umar, J. C. Barrett, T. A. Kunkel, and K. R. Tindall, *Carcinogen* **19,** 1931–1937 (1998).
76. W. E. Glaab, K. R. Tindall, and T. R. Skopek, *Mutat. Res.* **427,** 67–78 (1999).
77. B. A. Bridges, *Curr. Biol.* **9,** 475–477 (1999).
78. E. C. Friedberg, W. J. Feaver, and V. L. Gerlach, *Proc. Natl. Acad. Sci. U.S.A.* **97,** 5681–5683 (2000).
79. A. Klungland, K. Laake, E. Hoff, and E. Seeberg, *Carcinogenesis* **16,** 1281–1285 (1995).

// Regulation of Intracellular Localization of Human MTH1, OGG1, and MYH Proteins for Repair of Oxidative DNA Damage

Yusaku Nakabeppu

Division of Neurofunctional Genomics
Medical Institute of Bioregulation
Kyushu University
and
CREST
Japan Science and Technology Corporation
Fukuoka, 812-8582, Japan

I. Introduction 76
II. hMTH1: An Oxidized Purine Nucleoside Triphosphatase 77
III. hOGG1 for Repair of 8-Oxoguanine Paired with Cytosine in DNA 83
IV. hMYH as a Bifunctional DNA Glycosylase for 2-Hydroxyadenine
    and Adenine Paired with 8-Oxoguanine 85
V. Discussion 89
   References 92

In mammalian cells, more than one genome has to be maintained throughout the entire life of the cell, one in the nucleus and the other in mitochondria. It seems likely that the genomes in mitochondria are highly exposed to reactive oxygen species (ROS) as a result of their respiratory function. Human MTH1 (hMTH1) protein hydrolyzes oxidized purine nucleoside triphosphates, such as 8-oxo-dGTP, 8-oxo-dATP, and 2-hydroxy (OH)-dATP, thus suggesting that these oxidized nucleotides are deleterious for cells. Here, we report that a single-nucleotide polymorphism (SNP) in the human *MTH1* gene alters splicing patterns of *hMTH1* transcripts, and that a novel hMTH1 polypeptide with an additional mitochondrial targeting signal is produced from the altered *hMTH1* mRNAs; thus, intracellular location of hMTH1 is likely to be affected by a SNP. These observations strongly suggest that errors caused by oxidized nucleotides in mitochondria have to be avoided in order to maintain the mitochondrial genome, as well as the nuclear genome, in human cells. Based on these observations, we further characterized expression and intracellular localization of 8-oxoG DNA glycosylase (hOGG1) and 2-OH-A/adenine DNA glycosylase (hMYH) in human cells. These two enzymes initiate base excision repair reactions for oxidized bases in DNA generated by direct oxidation of DNA or by incorporation of oxidized nucleotides. We describe the detection of the authentic hOGG1 and hMYH proteins in mitochondria, as well as nuclei in human cells, and how their intracellular localization is regulated by alternative splicing of each transcript. © 2001 Academic Press.

I. Introduction

Oxidative phosphorylation in mitochondria makes it feasible for eukaryotic organisms to produce energy to maintain life. By electrons leaked from the respiratory chain, about 1–3% of consumed oxygen molecules are partially reduced, thus generating reactive oxygen species (ROS) such as superoxide, hydrogen peroxide, and hydroxyl radicals (1). ROS are highly reactive and can readily oxidize macromolecules in living cells, including lipids, proteins, and nucleic acids, leading to various cellular damages including cell death and mutagenesis (2, 3).

Organisms are equipped with defense mechanisms to minimize accumulation of ROS. For example, superoxide dismutase converts superoxide to oxygen and hydrogen peroxide, and the latter is further decomposed by catalase into water and oxygen. Mice lacking the *SOD2* gene that encodes mitochondrial superoxide dismutase have severe abnormalities in development and growth, including cardiomyopathy and neurodegeneration (4, 5). Once excessive ROS accumulates in the cells, these cells can no longer avoid severe oxidative damage. Even in the presence of functional superoxide dismutases, accumulation of oxidized macromolecules in human tissues gradually occurs during normal aging; hence oxidative damage has been implicated in aging and degenerative diseases.

Among various oxidative damages in cellular macromolecules, damage to nucleic acid is particularly hazardous because genetic information present in genomic DNAs, nuclear and mitochondrial, can be altered. Damage to genomic DNAs often leads to cell death, and degenerative diseases occur or related mutations result in neoplasia and hereditary diseases (6). ROS lead to various base or sugar modifications in DNA and free nucleotides, and strand breaks in DNA are introduced (7). 8-Oxoguanine (8-oxoG), one oxidized form of guanine, is produced by ROS in a fairly large amount in both DNA and nucleotide pools; and, being fairly stable, it is likely to accumulate in genomic DNAs in nuclei and mitochondria during normal aging of humans (1, 8, 9). Accumulation of 8-oxoG in DNA, as a result of incorporation of 8-oxo-dGTP from nucleotide pools or direct oxidation of DNA, increases the occurrence of A:T to C:G or G:C to T:A transversion mutations, respectively. This is because 8-oxoG forms a stable base pair with adenine as well as with cytosine (10–12).

Studies on mutator mutants revealed that *Escherichia coli* has several error-avoiding mechanisms which minimize the deleterious effects of 8-oxoG and in which MutT, FPG (MutM), and MutY proteins play important roles (13, 14). MutT protein hydrolyzes 8-oxo-dGTP to 8-oxo-dGMP and pyrophosphate (12), thereby avoiding spontaneous occurrence of A:T to C:G transversion mutation during DNA synthesis, the rate of which in a *mutT*-deficient strain increases hundreds to thousand-fold compared to the wild type (*15*). FPG (MutM) protein, originally identified as a formamidopyrimidine DNA glycosylase, removes the 8-oxoG paired with cytosine and introduces a single strand gap as a result of the accompanying apurinic/apyrimidinic (AP) lyase activity (*16, 17*). MutY protein with its DNA glycosylase activity excises adenine paired with guanine or 8-oxoG (*18*). The rate of spontaneous occurrence of G:C to T:A transversion mutation in *fpg* (*mutM*) or *mutY* deficient strains is 10–50 times higher than that in the wild-type strain (*19, 20*). Double mutants of *mutM* and *mutY* exhibit a spontaneous mutation frequency equivalent to that of the *mutT* mutant.

Mammalian cells also have similar error-avoidance mechanisms. cDNAs encoding MutT homolog (MTH1) proteins with 8-oxo-dGTPase activity have been cloned from human, mouse, and rat (*21–23*), and genomic organizations and regulation of expression have been well characterized (*24–27*). An activity that excises adenine paired with guanine or 8-oxoG in DNA has been identified in human cells (*28*), and a human gene encoding a MutY homolog (MYH) protein has been cloned (*29*). In contrast to the *mutT* and *mutY* gene families, there are at least two divergent genes in eukaryotes that encode 8-oxoG DNA glycosylase, one is the *fpg* or *mutM* homolog, *AtMMH*, identified in *Arabidopsis thaliana* (*30*), and the other is a novel gene found in yeast, *ogg1* (*31–32*). Thereafter, human and mouse genes encoding proteins homologous to the yeast Ogg1 protein have been identified (*33*). Thus, mammalian cells have developed error-avoiding mechanisms against 8-oxoG, combining those which evolved from *E. coli* and yeast.

In mammalian cells, more than one genome in a single cell has to be maintained throughout the entire life of the cell, one in the nucleus and the other in mitochondria. The genome in the mitochondria is likely to be more susceptible to ROS-induced oxidative damage as oxygen metabolism is high. We reported that all three gene products—hMTH1, hMYH, and hOGG1—which are considered to minimize oxidative DNA damage in human cells locate both in nuclei and mitochondria of human cells (*34–36*).

II. hMTH1: An Oxidized Purine Nucleoside Triphosphatase

The human *MTH1* gene located on chromosome 7p22 consists of 5 major exons. There are two alternative exon 1 sequences, namely exon 1a and 1b, and three contiguous segments (exon 2a, 2b, and 2c) in exon 2 that are alternatively spliced (*24, 26, 27*). Thus, the *hMTH1* gene produces seven types (type 1, 2A, 2B, 3A, 3B, 4A, and 4B) of mRNAs, as shown in Fig. 1. The B-type mRNAs with exon 2b–2c segments direct synthesis of three forms of hMTH1

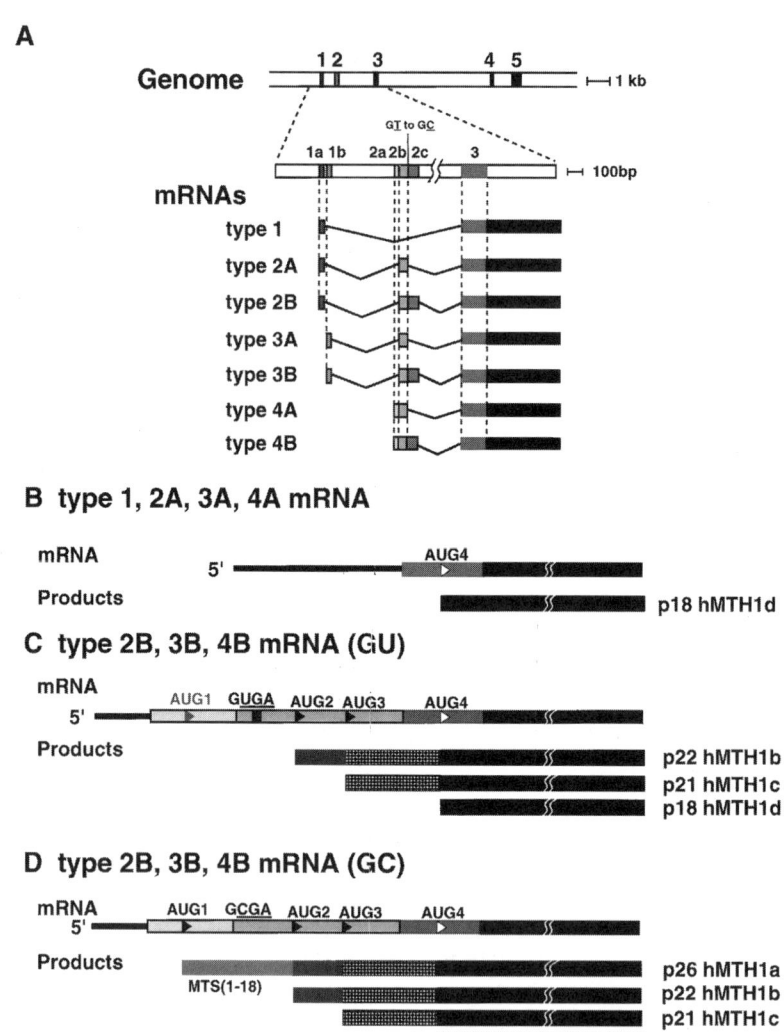

FIG. 1. Schematic representation of human *MTH1* genome structure, mRNAs, and translation products from each transcript. (A) Genomic structure and alternative splicing of the *hMTH1* gene. The upper part shows the overall structure of the *hMTH1* gene located on chromosome 7p22. Each hatched box represents an exon. In the lower part, seven types of *hMTH1* mRNA produced by alternative transcription initiation and splicing are shown together with part of the genomic structure. Alternative splicing is not observed in exons 3–5. Polymorphic alteration (G<u>T</u> to G<u>C</u> located at the beginning of exon 2c segment) is shown. (B) Types 1, 2A, 3A, and 4A *hMTH1* mRNAs and their translation products. AUG4 is a unique initiation codon in these *hMTH1* transcripts. (C) B-type *hMTH1* mRNAs with the GU polymorphism at the beginning of exon 2c and their translation products. In these transcripts, AUG1 encounters a stop codon UGA soon after initiation

hMTH1d   hOGG1-2a   hMYH

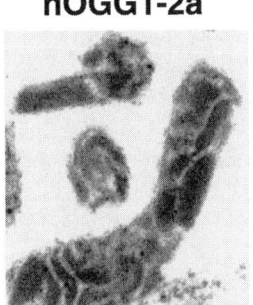
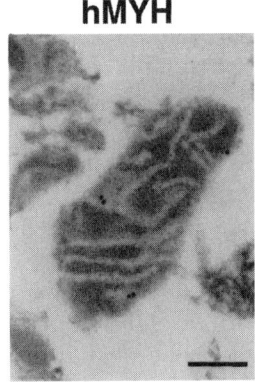

FIG. 2. Submitochondrial localization of human MTH1d, OGG1-2a, and MYH proteins, determined by electron microscopic immunocytochemistry. After each mitochondrion was isolated from HeLa MR cells, thin sections (about 0.1 μm) were prepared for electron microscopic immunocytochemistry with anti-hMTH1, anti-2a-CT, and anti-hMYH in combination with protein A-gold. Bars indicate 0.2 μm. [Adapted from Kang et al. (34), Nishioka et al. (35), and Ohtsubo et al. (36) with permission]

polypeptides—hMTH1b (p22), hMTH1c (p21), and hMTH1d (p18)—by alternative initiation of translation, while the others encode only hMTH1d (Fig. 1B,C). In human cells, hMTH1d, the major form is mostly localized in the cytoplasm with about 5% in the mitochondrial matrix (Fig. 2) (34). A single cell of Jurkat and HeLa lines contains about 4×10^5 or 2×10^5 molecules, respectively, of the hMTH1d; thus, 1 to 2×10^4 molecules of hMTH1d are present in mitochondria of each cell (27). The other two forms are localized in the cytoplasm, with amounts being about 10% of those for hMTH1d.

We found a single-nucleotide polymorphism (SNP), which affects the splicing pattern of *hMTH1* mRNAs, GT to GC base substitution at the beginning of exon 2c segment (Fig. 1A,D) (26, 27). This site serves as the 5′ splice site during maturation of types 1, 2A, 3A, and 4A mRNA; thus, the base change abolishes

---

of translation. (D) B-type *hMTH1* mRNAs with the GC polymorphism at the beginning of exon 2c and their translation products. This SNP abolishes the stop codon UGA; thus AUG1 can be utilized as an initiation codon. Each box in *hMTH1* mRNAs represents an exon segment shown in panel A. Boxes below each mRNA represent translation products. Closed boxes show translation products in frame to 18-kDa hMTH1d protein. Apparent molecular mass of the translation products is shown at the right end of the line. Hatched boxes show amino-terminal leader sequences for each hMTH1 polypeptide (27). MTS, mitochondrial targeting sequence (see Fig. 3). [Adapted from Oda et al. (27) with permission]

proper splicing at this site, leaving segments 2b and 2c connected. Cells with the homozygous GC allele produce only types 1, 2B, 3B, and 4B mature transcripts of *hMTH1* (27).

In B-type mRNAs, the polymorphic alteration (GU to GC) at the beginning of exon 2c converts the in-frame UGA stop codon, present upstream of the AUG for the hMTH1b, to a sense codon, CGA, hence yielding another in-frame AUG further upstream. Thus, B-type mRNAs from the GC allele produces an additional polypeptide hMTH1a (p26) (Fig. 1D).

We analyzed the probability of each hMTH1 polypeptide for mitochondrial targeting, using the MitoProt II Program (Fig. 3) (27, 37). Our results suggested that hMTH1a and hMTH1d can be imported into mitochondria and that the former is likely to have a much better mitochondrial targeting signal than does the latter (34). Our evidence shows that the 18-amino acid leader sequence of hMTH1a functions as the mitochondrial targeting signal (MTS) when fused to the green fluorescent protein (GFP) (Y. Sakai and Y. Nakabeppu, unpublished).

Thus, in human cells synthesis of multi-forms of hMTH1 polypeptides is regulated by both alternative splicing of its transcripts and alternative initiation of their translation, both of which are further altered by SNP, resulting in production of an additional mitochondrial form of hMTH1.

We have shown that recombinant human MTH1d, but not *E. coli* MutT, efficiently hydrolyzes two forms of oxidized dATP, 2-OH-dATP and 8-oxo-dATP, as well as 8-oxo-dGTP (38–40), and that both proteins hydrolyze 8-oxo-GTP, to which hMTH1 has a much lower affinity compared to findings with MutT protein (12, 41). Thus, we concluded that MTH1 has a much wider substrate specificity than does MutT. hMTH1a, hMTH1b, and hMTH1c also have 8-oxo-dGTPase and 8-oxo-GTPase activities, and whether or not these hydrolyze oxidized forms of dATP and ATP is now under investigation.

Thirty amino acid residues are identical between hMTH1 and *E. coli* MutT, and there is a highly conserved region consisting of 23 residues (MTH1: Gly36 to Gly58), with 14 identical residues (Fig. 4A). A chimeric protein, hMTH1-Ec, in which the 23-residue sequence of hMTH1 was replaced with that of MutT (Fig. 4B), retains the potential to hydrolyze 8-oxo-dGTP, thus indicating that the 23-residue sequences of hMTH1 and MutT are not only functionally and structurally equivalent, but actually constitute a functional phosphohydrolase module (42). Saturation mutagenesis of the module in hMTH1 indicated that an amphipathic property of α-helix I consisting of 14 residues of the module (Thr44 to Gly58) is essential to maintain the stable catalytic surface for 8-oxo-dGTPase (42, 43). Human MTH1 with a unique C-terminal region (aa 130–156), which is missing in MutT, is likely to determine its unique substrate specificity (Fig. 4), as discussed later (36).

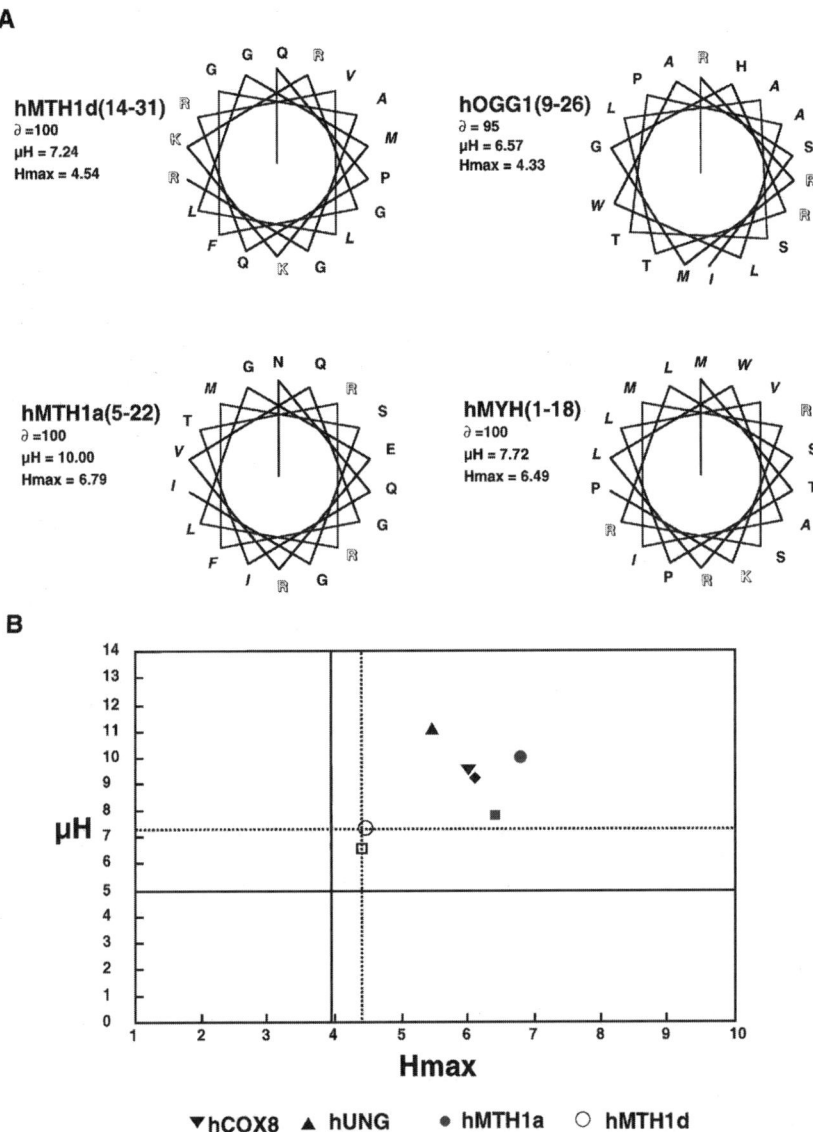

FIG. 3. Mitochondrial targeting sequences in human MTH1d, MTH1a, OGG1, and MYH proteins. (A) Each putative MTS predicted by MitoProt II (37) is shown in a helical structure. Numbers of amino acid residues constituting each MTS are shown in parentheses. Basic residues are outlined, and hydrophobic ones are italicized. The ∂, μH, and H_{max} values are shown. (B) Amphiphilic plot of each putative MTS. μH versus H_{max} are plotted. The points of $\mu H > 5.0$ and $H_{max} > 3.9$ have sufficient amphiphilicity to be considered MTS. Those of $\mu H > 7.3$ and $H_{max} > 4.4$ are more likely MTS (37). Values for MTS of human cytochrome c oxidase polypeptide VIII (hCox8), human DNA ligase III (hLigIII), and human uracil DNA glycosylase 1 (hUNG) are also plotted.

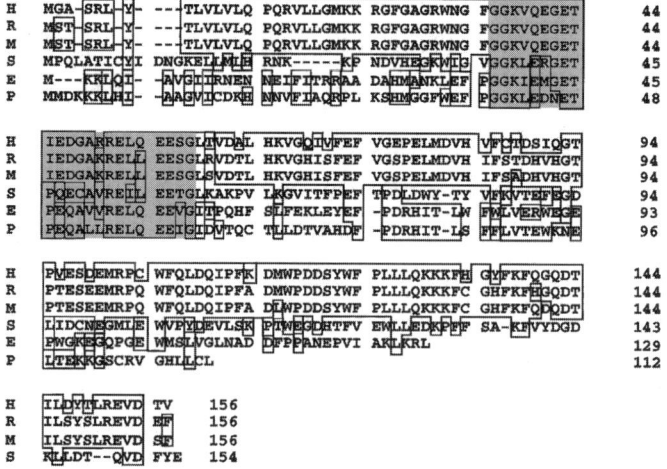
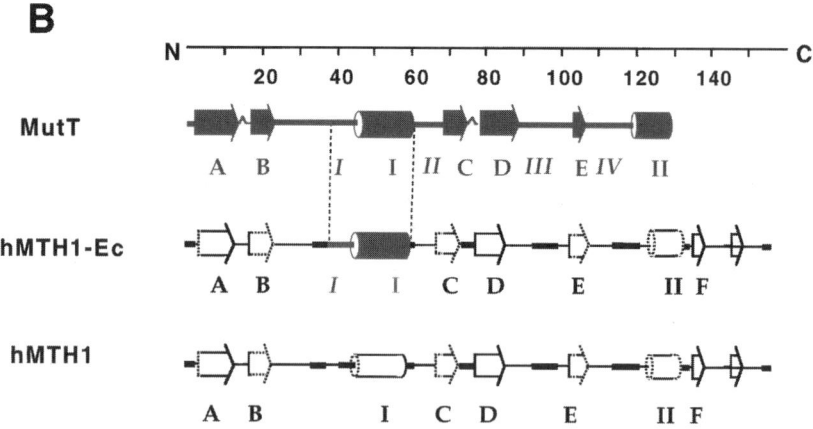

FIG. 4. Primary structures of MutT homolog proteins and secondary structures of *E. coli* MutT and human MTH1. (A) Comparison of primary structures of bacterial MutT homologs and mammalian MTH1 proteins. The 23-residue sequence, namely the phosphohydrolase module, is the sole conserved sequence among the MutT homologs, and is shown in a hatched box. Identical amino acid residues among them are boxed. The 23-residue sequences are shaded. (B) Comparison of the secondary structure of MutT, hMTH, and a chimeric protein MTH1-Ec. Secondary structures of MTH1 predicted previously (38, 42). In the chimeric protein, MTH1-Ec, the 23-residue sequence, which consists of Loop-I and α-Helix I in hMTH1 protein, is replaced with that of *E. coli* MutT protein (shaded), as described in Ref. (42). The numbers represent numbers of amino acid residues from the N-terminal end. Arrowed shapes indicate β-strand; cylindrical shapes indicate α-helix; thick lines indicate loop. Thin lines in hMTH1 and MTH1-Ec indicate no prediction made for the residues. [Adapted from Fujii *et al.* (42) with permission]

III. hOGG1 for Repair of 8-Oxoguanine Paired with Cytosine in DNA

The human *OGG1* gene located on chromosome 3p25, has 8 major exons (*33*). We identified seven alternatively spliced forms of human 8-oxoG DNA glycosylase (hOGG1) mRNAs, classified into two types based on their last exons (type 1 with exon 7: 1a and 1b; type 2 with exon 8: 2a to 2e) (Fig. 5) (*35*). hOGG1 polypeptides encoded by these mRNAs share the first 190 aa in common, which are encoded by exons 1–3, and each has a unique C-terminal region, with the exception that polypeptides encoded by type 2a and 2b mRNAs share the common C-terminal 108 aa. Five polypeptides (hOGG1-1a, 1b, 2a, 2d, and 2e) carry the helix–hairpin–helix PVD (HhH-PVD) motif which seems to be essential for 8-oxoG DNA glycosylase activity (*32, 44*). Types 1a and 2a mRNAs are major in various human tissues. Using MitoProt II (*37*), we predicted that all forms of hOGG1 polypeptides carry a relatively poor MTS (Fig. 3), which consists of residues 9–26 at the common N-terminal region and is likely to be processed at residue 23 (W) after being translocated into mitochondria as shown in Fig. 5. Among all the polypeptides, only hOGG1-1a has a nuclear localization signal (NLS) at the C-terminal end.

We transfected HeLa MR cells with expression plasmids encoding HA epitope-tagged hOGG1-1a or hOGG1-2a, and cells were labeled for mitochondria with MitoTracker and stained for the HA epitope. Cells transiently expressing hOGG1-1a:HA showed strong immunoreactivity exclusively in their nuclei and did not colocalize with signals for MitoTracker. On the other hand, hOGG1-2a:HA was present in the cytosol on punctuate structures distributed around the nucleus, and the structures were also labeled with MitoTracker. These results suggest that alternative splicing determines differential intracellular localization of human 8-oxoG DNA glycosylase (*35*).

A 36-kDa polypeptide, corresponding to hOGG1-1a and recognized only by antibodies against the region containing the HhH-PVD motif, was copurified from the nuclear extract prepared from Jurkat cells, with activity which introduced a nick at 8-oxoG paired with cytosine in double-stranded oligonucleotides (*35*). A 40-kDa polypeptide corresponding to a processed form of hOGG1-2a was detected in their mitochondria when we used antibodies against its C-terminus. Electron microscopic immunocytochemistry and subfractionation of the mitochondria revealed that hOGG1-2a locates on the innermembrane of mitochondria (Fig. 2), in contrast to hMTH1d present in mitochondrial matrix (*35*). Deletion mutant analyses revealed that the unique C-terminus of hOGG1-2a and its MTS are essential for mitochondrial localization and that nuclear localization of hOGG1-1a depends on NLS at its C-terminus (*35*).

The unique C-terminal region of hOGG1-2a consists of two distinct regions: One is the N-terminal-sided acidic region (aa from Ile^{345} to Asp^{381}) and the other

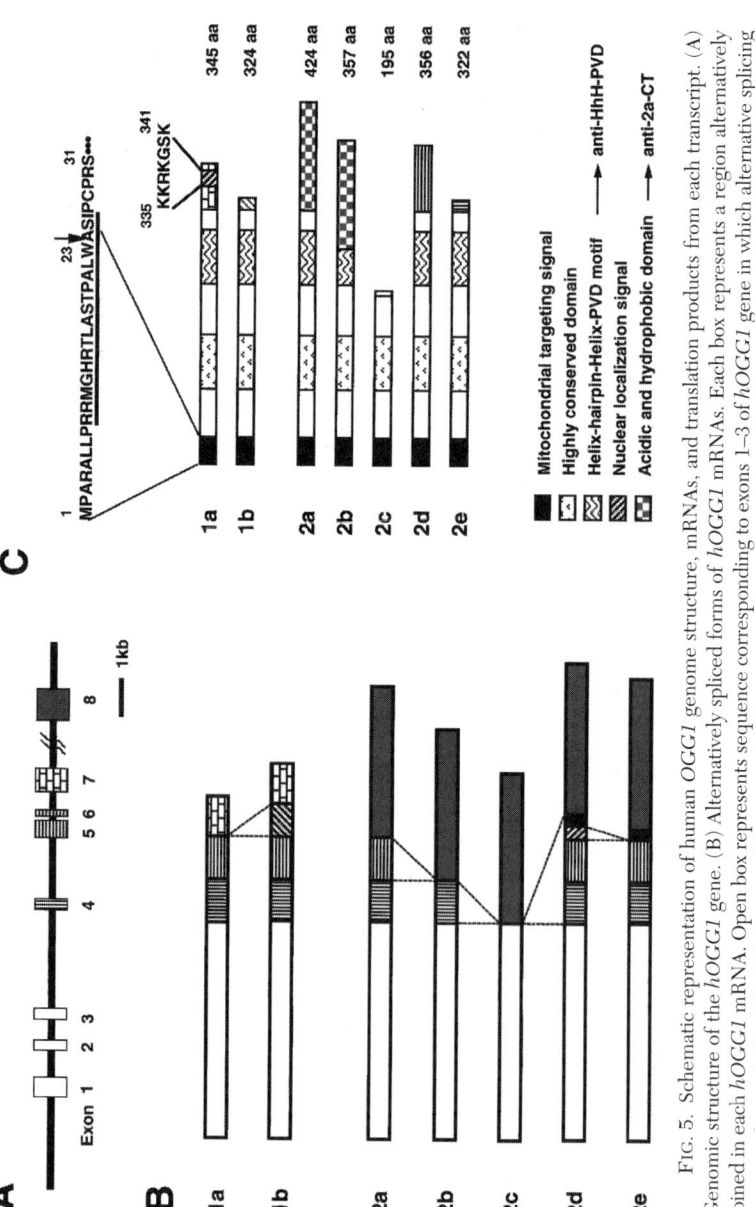

FIG. 5. Schematic representation of human *OGG1* genome structure, mRNAs, and translation products from each transcript. (A) Genomic structure of the *hOGG1* gene. (B) Alternatively spliced forms of *hOGG1* mRNAs. Each box represents a region alternatively joined in each *hOGG1* mRNA. Open box represents sequence corresponding to exons 1–3 of *hOGG1* gene in which alternative splicing was not observed (33, 35). (C) Structural features of different hOGG1 proteins. All forms of the hOGG1 proteins contain a putative MTS at the common N-terminal region, and the MTS may be processed at residue 23 (W) after being translocated into the mitochondria (arrow) (35). There is a NLS only in the C-terminal end of hOGG1-1a (residues 335–341). The region exhibiting the highest homology to the corresponding region of yeast Ogg1 protein (aa 128–156) is shown as the highly conserved region. Anti-HhH-PVD recognizes the region containing HhH-PVD motif (aa 221–290), and anti-2a-CT recognizes the C-terminal regions of both hOGG1-2a (aa 322–424) and 2b (aa 255–357). [Adapted from Nishioka *et al.* (35) with permission]

is the C-terminal hydrophobic one (the last 20 residues). We therefore speculate that the latter mediates a hydrophobic interaction between hOGG1-2a and the inner membrane of the mitochondria. This affinity may be comparable to that in the association of Bcl2 with the mitochondrial membrane (35, 45). Recombinant hOGG1-2a, but not hOGG1-1a, expressed in *E. coli,* was mostly insoluble, which means that this C-terminal region contributes to insolubility as a result of its hydrophobicity. On the other hand, the acidic region with a value of 3.63 for local isoelectric pH may possibly provide a protein–protein interacting surface for other component(s) and, if so, may be involved in its stabilization and/or in the base excision repair (BER) pathway in mitochondria, as discussed below.

IV. hMYH as a Bifunctional DNA Glycosylase for 2-Hydroxyadenine and Adenine Paired with 8-Oxoguanine

We examined the enzyme activity that introduces an alkali-labile site into 2-OH-A containing oligonucleotides in human cells. We obtained evidence for DNA repair activity that is likely to function as a DNA glycosylase (36). This activity co-eluted with activities toward adenine paired with guanine or 8-oxoG as a single peak and corresponding to a 55-kDa molecular mass determined by gel filtration chromatography. When further copurification was done, Western blotting revealed that these activities also copurified with a 52-kDa polypeptide, which reacted with antibodies against human MYH (anti-hMYH). Recombinant hMYH has activities essentially similar to those of the partially purified enzyme. Thus, hMYH is likely to have both adenine and 2-OH-A DNA glycosylase activities.

Our data indicate that hMYH recognizes 2-OH-A paired with 8-oxoG or purines as well as adenine paired with guanine or 8-oxoG. hMYH in partially purified fractions preferentially acts on adenine paired with guanine in DNA rather than with 8-oxoG (Fig. 6); however, it binds more efficiently to the substrate containing the latter (36). These characteristics of authentic hMYH are the same as those of calf MYH (46). However, recombinant hMYH in crude extracts showed the strongest activity for oligonucleotides containing adenine paired with 8-oxoG (47, 48), but substantially less nicking activity for oligonucleotides containing adenine or 2-OH-A paired with guanine. Thus, partially purified fractions of hMYH from Jurkat cells may contain other factors that modify its substrate specificity.

In crude extracts from *E. coli* cells, we found no activity on substrates containing 2-OH-A, and purified MutY does not act on 2-OH-A (T. Ohtsubo,

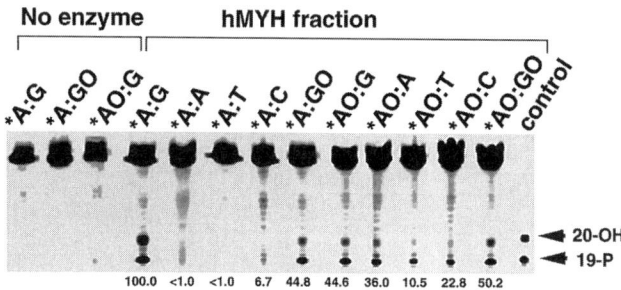

FIG. 6. Substrate specificity of human MYH protein purified from nuclear extracts of Jurkat cells. Partially purified fractions of hMYH were incubated with various double-stranded oligonucleotides (36). Reaction products were fractionated on a 6% LongRanger gel containing 7 M urea after mild alkaline treatment. Relative amounts of cleaved product of each substrate to that of A:G substrate are shown. 19-P, FAM-labeled control oligonucleotide corresponding to the cleaved product at the 5′ side of adenine or 2-OH-A, with 3′-phosphate. 20-OH, FAM-labeled control oligonucleotide corresponding to the cleaved product at the 3′ side of adenine or 2-OH-A, without 3′-phosphate. [Adapted from Ohtsubo et al. (36) with permission]

H. Kamiya, H. Kasai, and Y. Nakabeppu, unpublished). Thus, hMYH, but not MutY, has novel repair activity for 2-OH-A in DNA, as does hMTH1 for 2-OH-dATP. There is a weak homology between MutT protein and the C-terminal region of MutY protein which binds to oligonucleotides containing 8-oxoG paired with adenine or opposite abasic site, thus indicating that the conserved residues between the two may be involved in the recognition of 8-oxoG (49; also see House et al., this volume). In MutT protein, the conserved residues are scattered outside of the phosphohydrolase module.

We also found that there are significantly conserved residues between human MYH (369–499 aa in MYHα3) and the entire human MTH1d molecule, as shown in Fig. 7 (36). Thirty-nine residues in hMTH1d (156 aa) are identical to those in the human MYH C-terminal region, and 18 are also conserved in E. coli MutT or MutY. hMYH protein has novel DNA glycosylase activity excising 2-OH-A in DNA as well as adenine DNA glycosylase; therefore, the 21 residues conserved only between the hMTH1 and hMYH are likely to be involved in the recognition of 2-OH-A, and eight of the conserved residues between hMTH1 and hMYH are located in the C-terminal region of hMTH1 which is missing in MutT (Fig. 4).

In human cells, Western blots showed authentic hMYH in nuclei and in mitochondria and that the molecular masses differed, p52/53 and p57, respectively, indicating that there are multi-forms of hMYH (Fig. 8) (36). Based on results obtained from 5′-RACE and RT-PCR of *hMYH* transcripts, we concluded that there are three major *hMYH* transcripts, namely hMYHα, β, γ, with a different 5′ sequence or first exons, and further that each transcript is alternatively

```
MutT    -----MKK-- ---------- ---------L -QIAVGIIRN ENNEIFITRR  23
hMTH1   -----MGA-S R--------- ---------L -YTLVLVLQP QRVLLGMKKR  25
hMYH    VVNFPRKA-S RKPPREESSA TCVLEQPGAL GAQIILVQRP NSGLLA-GLW 388
MutY    AANNSWALYP GKKPKQTLPE RTGYFLLLQH EDEVLLAQRP PSGLWG-GLY 259

MutT    AADAHMANKL EFPGGKIEMG ETPEQAVVRE LQEEVG-ITP QHFSLFEKLE  72
hMTH1   GFGAGRWN-- GF-GGKVQEG ETIEDGARRE LQEESG-LTV DALHKVGQIV  71
hMYH    EFPSVTW--- E-P----SEQ LQRKALLQ-E LQRWAGPLPA THLRHLGEVV 429
MutY    CFPQFAD--- ---------- --EESLRQWL AQRQ-L-IAA DNLTQLTAFR 291

MutT    YEFPDRHITL ---WFWLVER WEGEPWGKEG QPGEWMSLVG LNADDFPPAN 119
hMTH1   FEFV-GEPEL MDVHVFCTDS IQGTPVESDE MRPCWFQLDQ IHFKDMWPDD 120
hMYH    HTF--SHIKL TYQ-VYGLAL EGGTPVTTVP PGARW--LTQ EEF--HTAAV 472
MutY    HTF--SHFHL DIVPMWLPVS SF-TGCMDEG N-ALWYNLAQ PPS-VGLAAP 336

MutT    EP----VIAK LKRL                                         129
hMTH1   SYWFPLLLQK KKFHGYFKFQ GQDTILDYT- LREVDTV                156
hMYH    STAM-----K KVFRVY---Q GQQPGTCMGS KRSQVSS                501
MutY    V---ERLLQQ LRTGAPV                                      350
```

FIG. 7. Sequence alignment of the *E. coli* MutT, hMTH1, hMYH C-terminal region (341–501 aa), and *E. coli* MutY C-terminal region (211–350 aa). Alignment for MutT, hMYH, and MutY is based on data of Noll *et al.* (49), and that for hMTH1 and MutT is based on data of Fujii *et al.* (42). Residues conserved among more than three proteins are shown in a box with a solid line, and residues conserved only between MutT and hMTH1 or hMYH and MutY, respectively, are shown in a box with a gray line. Residues conserved only between hMTH1 and hMYH and which may be involved in 2-OH-A recognition are shown in a gray box with a solid line. [Adapted from Ohtsubo *et al.* (36) with permission]

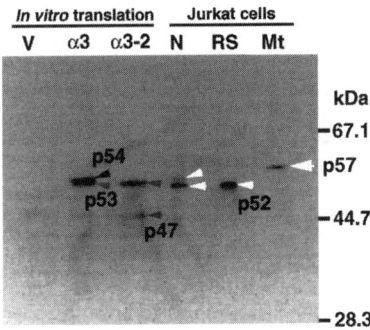

FIG. 8. Immunological detection of authentic hMYH in human cells. *In vitro* translation products of pT 7Blue vector itself (V), pT 7Blue:hMYHα3 (α3), and pT 7Blue:hMYHα3-2 (α 3-2), and isolated nuclei (N), mitochondria (Mt) (equivalent to 50 μg of protein) from Jurkat cells, and partially purified hMYH in RESOURCE S fraction (RS) were subjected to Western blot analysis, using anti-hMYH (36). [Adapted from Ohtsubo *et al.* (36) with permission]

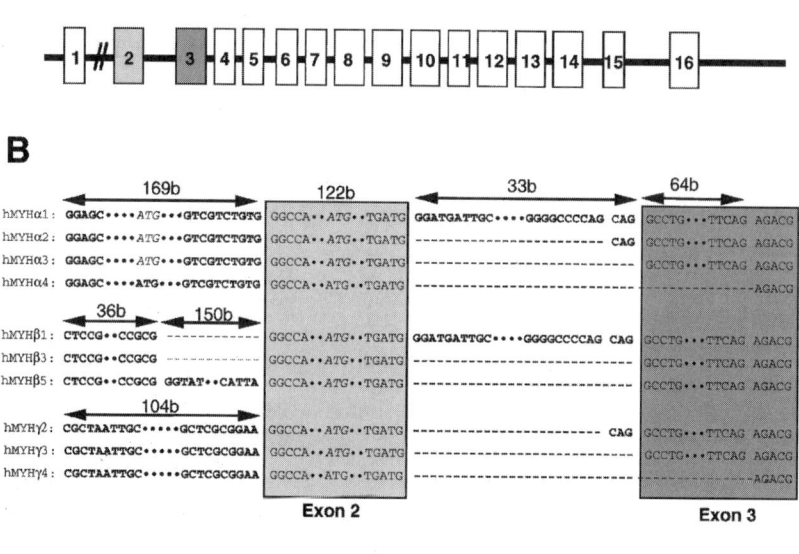
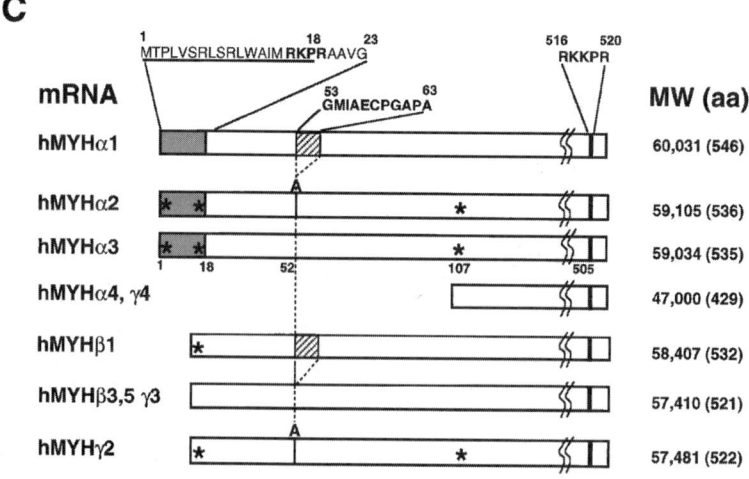

FIG. 9. Schematic representation of human *MYH* genome structure, mRNAs, and translation products from each transcript. (A) Genomic structure of the *hMYH* gene (29). (B) Alternatively spliced forms of *hMYH* mRNAs. Sequences for 5' regions (exons 1–3) of ten different forms of *hMYH* cDNA are shown. The reported exon 1 sequence (29) corresponds to the 5' sequence of α type mRNAs. Two putative initiation codons, ATG in exon 1 and 2 are shown in italics. (C) Predicted polypeptides encoded by the various *hMYH* transcripts. Asterisks indicate a methionine residue. Gray box indicates the mitochondrial targeting signal (See Fig. 3) (29, 48), and hatched box indicates 11-amino acid insertion. Putative NLS are shown in bold letters over the hMYH polypeptides. [Adapted from Ohtsubo *et al.* (36) with permission]

spliced, thus forming over 10 mature transcripts. A major transcript, hMYHα3 essentially corresponds to the *hMYH* cDNA originally reported (29) and encodes two polypeptides, p54 and p53, the former translated from the first AUG and the latter from the second AUG (Fig. 9).

It has been reported that the translation product from hMYHα3 transcript is exclusively localized in the mitochondria and that the amino terminal sequence of the hMYH translated from the first AUG functions as a mitochondrial targeting signal (Fig. 3) (48, 50); however, the authentic hMYH polypeptide detected in mitochondria is p57 (Fig. 8). The hMYHα1 transcript has a 33-nucleotide insertion into hMYHα3, and thus encodes a polypeptide with an expected molecular mass of 60,031; this may be the p57 detected in mitochondria (Fig. 9). Moreover, the nuclear form of hMYH, p52 partially purified from Jurkat cells, is likely to correspond to p53 translated from the second AUG of hMYHα3 transcript; and hMYHβ1, β3, and γ2 transcripts which are missing the first AUG may encode the nuclear form of p52 hMYH. It is also possible that the hMYHα3 transcript produces p53 and p54 by alternative translation initiation, and one may correspond to the p52 detected in nuclei of Jurkat cells. Determination of primary sequences of purified hMYH will show which polypeptide is encoded by each *hMYH* transcript.

Electron microscopic immunocytochemistry revealed that hMYH in mitochondria associates with the inner-membrane structure (Fig. 2), as does hOGG1-2a (35, 36), suggesting that the machinery required for base excision repair in mitochondria generally associates with the inner membrane to achieve efficient repair of mitochondrial DNA.

V. Discussion

Among many oxidized bases, 8-oxoG has been intensively studied as it is implicated in various diseases, such as cancer, neurodegeneration, and teratogenicity. 8-oxoG has mutagenic potential and various enzymes specifically act on the lesion (6, 13, 14, 33). Here, we propose that 2-OH-A, an oxidized form of adenine, must also be involved in biological responses, as hMTH1 efficiently hydrolyzes 2-OH-dATP, and 2-OH-A in DNA can be efficiently excised by a DNA glycosylase encoded by the *hMYH* gene. Furthermore, our studies on intracellular localization of hMTH1, hOGG1, and hMYH indicate that such oxidized nucleotides and oxidative DNA damage are likely to be highly deleterious to human mitochondrial genomes; thus mitochondria as well as nuclei come equipped with such a defense system.

hOGG1-2a and hMYH reside on the inner membrane of mitochondria, while hMTH1 locates in the matrix. Since mitochondrial DNAs exist in association with the inner membrane (51), it seems likely that some hOGG1-2a and hMYH molecules in mitochondria colocalize with DNA. Indeed, the number of authentic hOGG1-2a molecules in mitochondria is comparable to findings on mitochondrial DNA, a few thousand molecules per cell (35).

Activities of an AP endonuclease, DNA ligase III, and DNA polymerase γ were fractionated from mitochondria of *Xenopus* oocytes (52). It is likely that there is a repair complex containing the entire machinery essential for base excision repair, and which associates with the inner-membrane and/or mitochondrial DNAs as expected for hOGG1-2a and hMYH. The efficient repair would contribute to maintaining functions and the integrity of mitochondrial DNAs, even if threatened by ROS attacks produced during oxidative phosphorylation. hMTH1 in the matrix hydrolyzes oxidized purine nucleoside triphosphates generated in ribo- and deoxyribonucleotide pools in mitochondria; thus, incorporation of such oxidized nucleotides into mitochondrial genomes and their transcripts is eliminated (Fig. 10).

2-OH-A (also known as isoguanine) has been detected in DNA isolated from human tissues or from experimental animals. The amounts of 2-OH-A increased following exposure to various sources of reactive oxygen species, both *in vitro* and *in vivo* (53–55). In certain human cancerous tissues, 2-OH-A levels are increased severalfold compared to findings in noncancerous tissues (56). Concerning the origin of 2-OH-A, the extent of oxidation of the second position of the adenine base by hydroxyl radical is higher in its free nucleotide form than that in DNA (53), indicating that incorporation of 2-OH-dATP from the nucleotide pool is

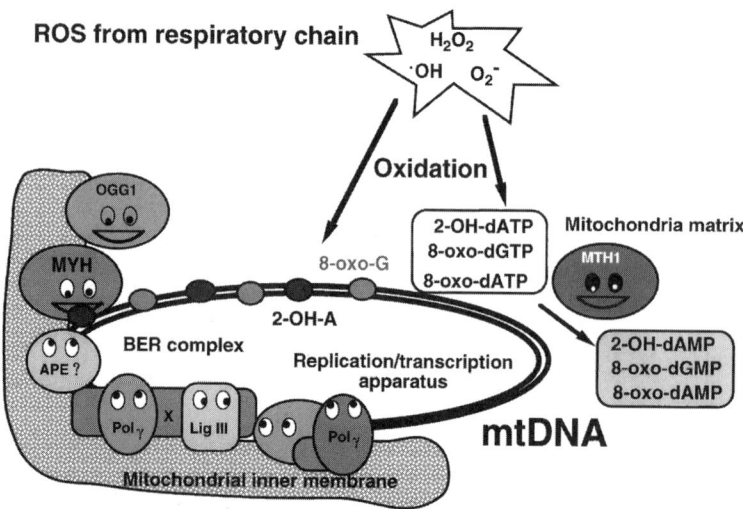

FIG. 10. Hypothetical representation of human MTH1, OGG1, and MYH proteins located in mitochondria. See text for details.

the major source of 2-OH-A in DNA. 2-OH-A in DNA thus incorporated forms a relatively stable base pair with thymine and cytosine (57, 58), and may also pair with the *syn* forms of guanine and adenine (53). It has been shown that 2-OH-A in plasmid DNA causes various base substitutions and deletion mutations in *E. coli* and in mammalian cells (59, 60). Introduction of 2-OH-dATP into *E. coli* cells induces mostly a G:C to T:A transversion mutation (61).

hMTH1 has the lowest K_m value with 2-OH-dATP among oxidized purine nucleoside triphosphates, while *E. coli* MutT, a prototype of 8-oxo-dGTPase, has little activity to hydrolyze 2-OH-dATP. hMYH, but not *E. coli* MutY, has repair activity to excise 2-OH-A in DNA. Thus, one can argue that human or mammalian cells are exposed to a higher risk of incorporation of 2-OH-dATP into their genome than are *E. coli* cells; thus, they eliminate the oxidized precursors from nucleotide pools. 2-OH-A incorporated into DNA causes mutations such as G:C to T:A, A:T to G:C, or frameshift mutations, which can be minimized by base excision repair initiated by hMYH. Thus, hMTH1 and hMYH with their coordinated actions protect genomic integrity in human cells.

The accumulation of 8-oxoG in DNA is minimized by coordinated actions of hMTH1 and hOGG1, and hMYH which has adenine DNA glycosylase to excise adenine paired with 8-oxoG. hMYH excises adenine paired with 8-oxoG and binds tightly to DNA containing the 8-oxoG opposite abasic site. Binding of hMYH may protect 8-oxoG from the action of OGG1 during the first round of BER; otherwise, the DNA may be incised in both strands by the actions of 8-oxoG DNA glycosylase and AP-lyase activities of hOGG1. Once cytosine is inserted opposite 8-oxoG by DNA polymerase β or proliferating cell nuclear antigen (PCNA)-dependent DNA polymerase δ and the ends are joined by DNA ligase, hMYH dissociates from the DNA and OGG1 may initiate the next round of BER of 8-oxoG. It has been suggested that hMYH has a consensus sequence for PCNA-binding motif, and may interact with PCNA (Matsumoto, this volume); thus, coordinated action between MYH and PCNA may support the sequential BER discussed above. Both adenine misinserted against 8-oxoG in template DNA and 2-OH-A incorporated from nucleotide pools are present on a newly replicated strand of DNA, and both must be repaired before the next round of replication to avoid mutation fixation. The interaction of hMYH with PCNA may support strand recognition.

Recognition or binding domains in human MYH for 2-OH-A and 8-oxoG are likely to be independent and both are conserved in human MTH1 protein (Fig. 7). It is noteworthy that *E. coli* MutT and MutY recognize 8-oxoG but not 2-OH-A, and have only the 8-oxoG recognition domain. Our working hypothesis is that hMTH1 and hMYH have co-evolved to capture the recognition domain for 2-OH-A during the evolution of eukaryotic cells; there is no such conserved sequence in the MutY homolog of *Schizosaccharomyces pombe* (SpMYH) (62).

ACKNOWLEDGMENTS

I extend special thanks to all members of my laboratory and to Drs. H. Kasai, H. Kamiya, M. Takahashi, M. Shirakawa, D. Kang, and M. Sekiguchi for helpful discussions. M. Ohara provided language assistance.

REFERENCES

1. D. Kang, K. Takeshige, M. Sekiguchi, and K. K. Singh, in "Mitochondrial DNA Mutations in Aging, Disease and Cancer" (K. K. Singh, ed.) pp. 1–15 Springer/Verlag Berlin, (1998).
2. B. N. Ames, M. K. Shigenaga, and T. M. Hagen, Proc. Natl. Acad. Sci. U.S.A. **90,** 7915–7922 (1993).
3. M. K. Shigenaga, T. M. Hagen, and B. N. Ames, Proc. Natl. Acad. Sci. U.S.A. **91,** 10771–10778 (1994).
4. Y. Li, T. T. Huang, E. J. Carlson, S. Melov, P. C. Ursell, J. L. Olson, L. J. Noble, M. P. Yoshimura, C. Berger, P. H. Chan, D. C. Wallace, and C. J. Epstein, Nature Genet. **11,** 376–381 (1995).
5. R. M. Lebovitz, H. Zhang, H. Vogel, J. Cartwright, Jr., L. Dionne, N. Lu, S. Huang, and M. M. Matzuk, Proc. Natl. Acad. Sci. U.S.A **93,** 9782–9787 (1996).
6. B. N. Ames and L. S. Gold, Mutat. Res. **250,** 3–16 (1991).
7. B. Demple and L. Harrison, Annu. Rev. Biochem. **63,** 915–948 (1994).
8. H. Kasai and S. Nishimura, Nucleic Acids Res. **12,** 2137–2145 (1984).
9. M. Hayakawa, S. Sugiyama, K. Hattori, M. Takasawa, and T. Ozawa, Mol. Cell. Biochem. **119,** 95–103 (1993).
10. S. Shibutani, M. Takeshita, and A. P. Grollman, Nature (London) **349,** 431–434 (1991).
11. K. C. Cheng, D. S. Cahill, H. Kasai, S. Nishimura, and L. A. Loeb, J. Biol. Chem. **267,** 166–172 (1992).
12. H. Maki and M. Sekiguchi, Nature (London) **355,** 273–275 (1992).
13. M. L. Michaels and J. H. Miller, J. Bacteriol. **174,** 6321–6325 (1992).
14. M. Sekiguchi, Genes Cells **1,** 139–145 (1996).
15. T. Tajiri, H. Maki, and M. Sekiguchi, Mutat. Res. **336,** 257–267 (1995).
16. V. Bailly, W. G. Verly, T. O'Connor, and J. Laval, Biochem. J. **262,** 581–589 (1989).
17. M. L. Michaels, C. Cruz, A. P. Grollman, and J. H. Miller, Proc. Natl. Acad. Sci. U.S.A. **89,** 7022–7025 (1992).
18. K. G. Au, S. Clark, J. H. Miller, and P. Modrich, Proc. Natl. Acad. Sci. U.S.A. **86,** 8877–8881 (1989).
19. M. Cabrera, Y. Nghiem, and J. H. Miller, J. Bacteriol. **170,** 5405–5407 (1988).
20. Y. Nghiem, M. Cabrera, C. G. Cupples, and J. H. Miller, Proc. Natl. Acad. Sci. U.S.A. **85,** 2709–2713 (1988).
21. K. Sakumi, M. Furuichi, T. Tsuzuki, T. Kakuma, S. Kawabata, H. Maki, and M. Sekiguchi, J. Biol. Chem. **268,** 23524–23530 (1993).
22. T. Kakuma, J. Nishida, T. Tsuzuki, and M. Sekiguchi, J. Biol. Chem. **270,** 25942–25948 (1995).
23. J. P. Cai, T. Kakuma, T. Tsuzuki, and M. Sekiguchi, Carcinogenesis **16,** 2343–2350 (1995).
24. M. Furuichi, M. C. Yoshida, H. Oda, T. Tajiri, Y. Nakabeppu, T. Tsuzuki, and M. Sekiguchi, Genomics **24,** 485–490 (1994).
25. H. Igarashi, T. Tsuzuki, T. Kakuma, Y. Tominaga, and M. Sekiguchi, J. Biol. Chem. **272,** 3766–3772 (1997).
26. H. Oda, Y. Nakabeppu, M. Furuichi, and M. Sekiguchi, J. Biol. Chem. **272,** 17843–17850 (1997).

27. H. Oda, A. Taketomi, R. Maruyama, R. Itoh, K. Nishioka, H. Yakushiji, T. Suzuki, M. Sekiguchi, and Y. Nakabeppu, *Nucleic Acids Res.* **27**(22), 4335–4343 (1999).
28. Y. C. Yeh, D. Y. Chang, J. Masin, and A. L. Lu, *J. Biol. Chem.* **266**, 6480–6484 (1991).
29. M. M. Slupska, C. Baikalov, W. M. Luther, J. H. Chiang, Y. F. Wei, and J. H. Miller, *J. Bateriol.* **178**, 3885–3892 (1996).
30. T. Ohtsubo, O. Matsuda, K. Iba, I. Terashima, M. Sekiguchi, and Y. Nakabeppu, *Mol. Gen. Genet.* **259**, 577–590 (1998).
31. P. A. van der Kemp, D. Thomas, R. Barbey, R. de Oliveira, and S. Boiteux, *Proc. Natl. Acad. Sci. U.S.A.* **93**, 5197–5202 (1996).
32. H. M. Nash, S. D. Bruner, O. D. Schärer, T. Kawate, T. A. Addona, E. Spooner, W. S. Lane, and G. L. Verdine, *Curr. Biol.* **6**, 968–980 (1996).
33. S. Boiteux and J. P. Radicella, *Biochimie* **81**, 56–67 (1999).
34. D. Kang, J. Nishida, A. Iyama, Y. Nakabeppu, M. Furuichi, T. Fujiwara, M. Sekiguchi, and K. Takeshige, *J. Biol. Chem.* **270**, 14659–14665 (1995).
35. K. Nishioka, T. Ohtsubo, H. Oda, T. Fujiwara, D. Kang, K. Sugimachi, and Y. Nakabeppu, *Mol. Biol. Cell* **10**, 1637–1652 (1999).
36. T. Ohtsubo, K. Nishioka, Y. Imaiso, S. Iwai, H. Shimokawa, H. Oda, T. Fujiwara, and Y. Nakabeppu, *Nucleic Acids Res.* **28**, 1355–1364 (2000).
37. M. G. Claros and P. Vincens, *Eur. J. Biochem.* **241**, 779–786 (1996).
38. H. Yakushiji, F. Maraboeuf, M. Takahashi, Z. S. Deng, S. Kawabata, Y. Nakabeppu, and M. Sekiguchi, *Mutat. Res.* **384**, 181–194 (1997).
39. H. Hayakawa, A. Hofer, L. Thelander, S. Kitajima, Y. Cai, S. Oshiro, H. Yakushiji, Y. Nakabeppu, M. Kuwano, and M. Sekiguchi, *Biochemistry* **38**, 3610–3614 (1999).
40. K. Fujikawa, H. Kamiya, H. Yakushiji, Y. Fujii, Y. Nakabeppu, and H. Kasai, *J. Biol. Chem.* **274**, 18201–18205 (1999).
41. F. Taddei, H. Hayakawa, M. Bouton, A. Cirinesi, I. Matic, M. Sekiguchi, and M. Radman, *Science* **278**, 128–130 (1997).
42. Y. Fujii, H. Shimokawa, M. Sekiguchi, and Y. Nakabeppu, *J. Biol. Chem.* **274**, 38251–38259 (1999).
43. J. P. Cai, H. Kawate, K. Ihara, H. Yakushiji, Y. Nakabeppu, T. Tsuzuki, and M. Sekiguchi, *Nucleic Acids Res.* **25**, 1170–1176 (1997).
44. P. M. Girard, N. Guibourt, and S. Boiteux, *Nucleic Acids Res.* **25**, 3204–3211 (1997).
45. M. Nguyen, D. G. Millar, V. W. Yong, S. J. Korsmeyer, and G. C. Shore, *J. Biol. Chem.* **268**, 25265–25268 (1993).
46. J. P. McGoldrick, Y. C. Yeh, M. Solomon, J. M. Essigmann, and A. L. Lu, *Mol. Cell. Biol.* **15**, 989–996 (1995).
47. M. M. Slupska, W. M. Luther, J.-H. Chiang, H. Yang, and J. H. Miller, *J. Bacteriol.* **181**, 6210–6213 (1999).
48. M. Takao, Q.-M. Zhang, S. Yonei, and A. Yasui, *Nucleic Acids Res.* **27**, 3638–3644 (1999).
49. D. M. Noll, A. Gogos, J. A. Granek, and N. D. Clarke, *Biochemistry* **38**, 6374–6379 (1999).
50. M. Takao, H. Aburatani, K. Kobayashi, and A. Yasui, *Nucleic Acids Res.* **26**, 2917–2922 (1998).
51. M. Albring, J. Griffith, and G. Attardi, *Proc. Natl. Acad. Sci. U.S.A.* **74**, 1348–1352 (1977).
52. K. G. Pinz and D. F. Bogenhagen, *Mol. Cell. Biol.* **18**, 1257–1265 (1998).
53. H. Kamiya and H. Kasai, *J. Biol. Chem.* **270**, 19446–19450 (1995).
54. Z. Nackerdien, K. S. Karsprzak, G. Rao, B. Halliwell, and M. Dizdaroglu, *Cancer Res.* **51**, 5837–5842 (1991).
55. R. Olinski, T. Zastawny, J. Budzbon, J. Skokowski, W. Zegarski, and M. Dizdaroglu, *FEBS Lett.* **309**, 193–198 (1992).
56. S. Toyokuni, T. Mori, and M. Dizdaroglu, *Int. J. Cancer* **57**, 123–128 (1994).

57. H. Robinson, Y.-G. Gao, C. Bauer, C. Roberts, C. Switzer, and A. H.-J. Wang, *Biochemistry* **37,** 10897–10905 (1998).
58. X.-L. Yang, H. Sugiyama, S. Ikeda, I. Saito, and A. H.-J. Wang, *Biophys. J.* **75,** 1163–1171 (1998).
59. H. Kamiya and H. Kasai, *Biochemistry* **36,** 11125–11130 (1997).
60. H. Kamiya and H. Kasai, *Nucleic Acids Res.* **25,** 304–311 (1997).
61. M. Inoue, H. Kamiya, K. Fujikawa, Y. Ootsuyama, N. Murata-Kamiya, T. Osaki, K. Yasumoto, and H. Kasai, *J. Biol. Chem.* **273,** 11069–11074 (1998).
62. A.-L. Lu and W. P. Fawcett, *J. Biol. Chem.* **273,** 25098–25105 (1998).

Repair of 8-Oxoguanine and Ogg1-Incised Apurinic Sites in a CHO Cell Line

SERGE BOITEUX AND
FLORENCE LE PAGE

CEA, DSV
Département de Radiobiologie
et Radiopathologie
UMR217 CNRS-CEA "Radiobiologie
Moléculaire et Cellulaire"
BP6
92265-Fontenay aux Roses, France

I. Introduction	96
II. Materials and Methods	97
A. Cell Line and Culture Conditions	97
B. Preparation of Closed Circular Plasmids Carrying a Unique 8-oxoG:C Base Pair	97
C. Assay for the Removal of 8-oxoG from Plasmid DNA in AA8 Cells	98
D. Assay for the Repair of Ogg1-Incised Apurinic Sites	98
III. Results and Discussion	98
A. A Novel Assay for the Removal of 8-oxoG in AA8 CHO Cells	98
B. Transcription-Coupled Repair of 8-oxoG in AA8 CHO Cells	100
C. Repair of Ogg1-Incised Apurinic Sites in AA8 CHO Cells	102
IV. Conclusions	103
References	103

The repair mechanisms involved in the removal of 8-oxo-7,8-dihydroguanine (8-oxoG) in damaged DNA have been investigated using cell-free extracts or purified proteins. However, *in vivo* repair assays are required to further dissect mechanisms involved in the repair of 8-oxoG in the cellular context. In this study, we analyzed the removal of 8-oxoG from plasmids that contain a single 8-oxoG·C base pair in a sequence that can be transcribed (TS) or nontranscribed (NTS) in a chinese hamster ovary (CHO) cell line. The results show that 8-oxoG located in a TS is removed faster than in a NTS, indicating transcription-coupled repair (TCR) of 8-oxoG in rodent cells. The results also show that CHO cells efficiently repair DNA molecules that contain an Ogg1-incised AP site, which is the first intermediate in the course of base excision repair of 8-oxoG. © 2001 Academic Press.

I. Introduction

Oxidative DNA damage induced by reactive oxygen species (ROS) is involved in the process of carcinogenesis and may play a role in the pathogenesis of aging (1–4). In the case of cancer, oxidative damage to DNA is thought to cause mutations that activate oncogenes or inactivate tumor suppressor genes. As many as fifty different modifications have been identified in DNA exposed to γ-irradiation in air (5, 6). Most of these lesions are substrates for DNA repair systems in prokaryotes and eukaryotes (7–10). A particularly abundant lesion, 8-oxo-7,8-dihydroguanine (8-oxoG), can be generated not only by oxidative stresses from the environment but also as a by-product of normal cellular metabolism (1–6). 8-OxoG is highly mutagenic, yielding GC to TA transversions upon its replication by DNA polymerases (11–13). *Escherichia coli* possesses two DNA glycosylases that prevent mutagenesis by 8-oxoG: the Fpg protein, excises 8-oxoG in damaged DNA, and the MutY protein, which excises the adenine residues incorporated by DNA polymerases opposite 8-oxoG (11–13). Inactivation of both the *fpg* and *mutY* genes of *E. coli* results in a strong GC to TA mutator phenotype (11–13). In *Saccharomyces cerevisiae*, the *OGG1* gene encodes a DNA glycosylase activity that catalyzes the removal of 8-oxoG from damaged DNA (14–16). Furthermore, Ogg1-deficient strains of *S. cerevisiae* exhibit a spontaneous mutator phenotype and specifically accumulate GC to TA transversions (17). These results strongly suggest that base excision repair (BER) of 8-oxoG by the Fpg or Ogg1 proteins protects genomes from the mutagenic action of endogenous ROS in prokaryotes or in the simple eukaryote *S. cerevisiae* (reviewed in Ref. 13). Recent studies in yeast also show that other excision repair pathways such as nucleotide excision repair (NER) (18) or mismatch repair (MMR) (19) are involved in the repair of 8-oxoG in DNA.

In human cells, MutY and Ogg1 homologs have been identified (20–27). The human *OGG1* gene encodes two isoforms, α-hOgg1 and β-hOgg1, resulting from an alternative splicing after transcription (reviewed in Ref. 13). α-hOgg1 is a 37-kDa protein localized in the nucleus whereas the β-Ogg1 is a 44-kDa protein targeted to the mitochondria (28–29). The predicted mouse Ogg1 is a 37-kDa protein which is 84% identical to the human nuclear α-hOgg1 (27, 30, 31). Both, human and mouse Ogg1 are DNA glycosylases/AP lyases that excise 2,6-diamino-4-hydroxy-5-*N*-methylformamidopyrimidine (Me-FapyGua), 2,6-diamino-4-hydroxy-5-formamidopyrimidine (FapyGua), and 8-oxoG and incise DNA at apurinic/apyrimidinic (AP) sites (reviewed in Ref. 13). Repair studies using mammalian cell-free extracts show that 8-oxoG is preferentially removed by the short-patch pathway of BER (32, 33). However, the elimination of 8-oxoG can also be performed through a long-patch pathway of BER, albeit with a lower efficiency (32, 33). Although not demonstrated, removal of 8-oxoG in mammalian cells is probably initiated by the Ogg1 protein. Thus, homozygous $ogg1^{-/-}$ null mice have been shown to accumulate 8-oxoG in their genomes and exhibit a significantly higher spontaneous mutation rate in nonproliferative tissues compared to the wild-type animals (34). Furthermore, cell-free protein extracts from $ogg1^{-/-}$ cell lines do not exhibit detectable activity for cleavage of short DNA fragments containing a unique 8-oxoG·C base pair (34). However, in vivo repair studies using the same $ogg1^{-/-}$ cell lines indicated a slow but substantial repair of Fpg-sensitive sites, suggesting the presence of alternative repair for oxidative DNA damage in the absence of Ogg1 (34). Indeed, human cells also possess another repair activity, Ogg2, able to cleave DNA duplexes containing 8-oxoG·A or 8-oxoG·G base pairs (35). In addition, purified mammalian NER systems have been shown to cleave at 8-oxoG·C in vitro (36) and the contribution of MMR cannot be excluded. Finally, recent studies have demonstrated the existence of a transcription-coupled repair mechanism (TCR) for the removal of 8-oxoG in human cells that requires proteins such as XPG, TFIIH, and CSB (37).

These results reveal an unexpected complexity for the repair of 8-oxoG in mammalian cells and consequently the need for specific assays to dissect repair mechanisms in the cellular context. In this study, we describe a novel repair assay using plasmids that contain a single 8-oxoG·C base pair located in a plasmid DNA sequence that can be transcribed (TS) or not (NTS) in the AA8 chinese hamster ovary (CHO) cell line. We also investigate the repair of an Ogg1-incised AP site which is the first intermediate in the course of BER of 8-oxoG in the same cell line.

II. Materials and Methods

A. Cell Line and Culture Conditions

AA8 chinese hamster ovary (CHO) cells were cultured in Dulbecco's modified Eagle's medium supplemented with 10% (v/v) fetal calf serum, fungizone, penicillin (100 U/ml), and streptomycin (100 μg/ml).

B. Preparation of Closed Circular Plasmids Carrying a Unique 8-oxoG:C Base Pair

Plasmids derived from pS189 were a generous gift of Dr. M. Seidman. Plasmids pSΔoriSV and pSΔ(ori-p)SV were a generous gift of Dr. A. Sarasin (CNRS, Villejuif). Plasmid pSΔoriSV harbors a deletion of the SV40 origin of replication whereas plasmid pSΔ(ori-p)SV harbors a deletion at both the SV40 origin of replication and the promoter region. Plasmids pSΔoriSV-[8-oxoG·C] and pSΔ(ori-p)SV-[8-oxoG·C] containing a unique 8-oxoG·C base pair at the same site were prepared as described (37, 38). The 19-mer oligodeoxyribonucleotide carrying a unique 8-oxoG (5'-GATCGGCGCCG[**8-oxoG**]CGGTGTG-3') was a kind gift of Dr. Jean Cadet (CEA, Grenoble).

C. Assay for Removal of 8-oxoG from Plasmid DNA in AA8 Cells

Eight hundred nanograms of closed circular double-stranded plasmids pSΔoriSV-[8-oxoG·C] (TS) or pSΔ(ori-p)SV-[8-oxoG·C] (NTS) were transfected into AA8 cells (semiconfluent cell culture in 10-cm^2 petri dishes) using the cationic liposome Dotap procedure (Boehringer). Cells were then incubated for 2–12 hours and harvested. Elimination of any contaminating extracellular input DNA was performed by treatment of cell cultures with DNase I prior to extraction. Extrachromosomal plasmid DNA was recovered by a small-scale alkaline lysis method (38). Recovered plasmid DNA was treated (+) or not (−) with 5 ng of homogeneous *E. coli* Fpg protein (39) and directly analyzed on a 0.8% agarose gel containing ethidium bromide (EtBr) to separate covalently closed (CC) and nicked plasmid molecules (OC). Plasmid DNA was detected by Southern blotting using the bacterial *Amp* sequences present in all constructs as a fluorescent-labeled probe (Amersham-Pharmacia). Visualization and quantification were done with a PhosphorImager (Molecular Dynamics). The repair of 8-oxoG in AA8 cells was calculated as a ratio of covalently closed molecules to the total amount of recovered plasmid DNA.

D. Assay for the Repair of Ogg1-Incised Apurinic Sites

Eight hundred nanograms of closed circular double-stranded plasmid pSΔ(ori-p)SV-[8-oxoG·C] (NTS) were incubated in the presence of 10 ng of homogeneous yeast Ogg1 protein for 15 min at 37°C. Complete cleavage was monitored by 0.8% agarose gel containing ethidium bromide (EtBr). Ogg1-incised DNA molecules were transfected, recovered, and analyzed as described for 8-oxoG-containing plasmids, except they were not incubated with Fpg after recovery.

III. Results and Discussion

A. A Novel Assay for Removal of 8-oxoG in AA8 CHO Cells

Biochemical studies of the BER of 8-oxoG in mammalian cells have unveiled the presence of two distinct pathways according to the proteins required as well as to their intermediate products (32, 33). The short-patch repair requires, after the initial steps performed by DNA glycosylase and AP lyase activities, the release of the 3'-dRP by the Hap1 endonuclease, the resynthesis by DNA polymerase β, followed by ligation by the ligaseIII/XRCC1 complex (Fig. 1). On the other hand, the long-patch repair involves different polymerases (possibly Pol δ or Pol ε), PCNA, the specific nuclease Fen1 and Ligase I (40). In

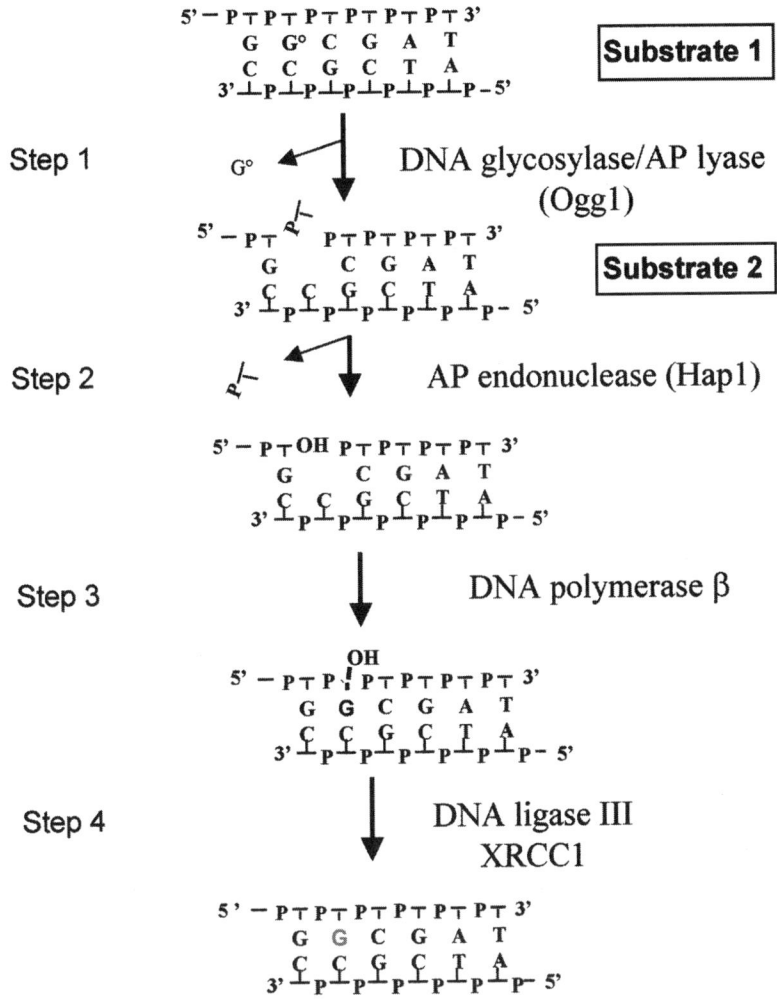

FIG. 1. Short-patch BER of 8-oxoG in mammalian cells. The short-patch BER pathway is the primary pathway for the repair of 8-oxoG; however, long-patch BER can substitute for it when components of the short patch are saturated or missing. Substrates 1 and 2 indicate BER intermediates used in the present study.

mammalian cell free extracts, 8-oxoG is eliminated preferentially by the short-patch pathway of BER (32, 33). However, cell-free extracts may only partially reflect the repair pathways used in the nucleus of living cells. Therefore, *in vivo* assays are necessary to understand DNA repair of unusual or altered DNA bases in the cellular context. For these purposes, we have developed an assay using two nonreplicative plasmids that contain a unique 8-oxoG·C base pair (37, 38)(Fig. 2). The sequence containing the 8-oxoG·C base pair was included in the 3' untranslated SV40 TAg unit in the strand that may be transcribed (TS) from the early promoter of SV40 in pSΔoriSV-[8-oxoG·C] or nontranscribed (NTS) when the promoter region has also been deleted in pSΔ(ori-p)SV-[8-oxoG·C] (Fig. 2). Therefore, these two constructs allow the analysis of the removal of an 8-oxoG lesion in the same sequence context but different transcription status (37). After transfection and incubation in AA8 cells for 2–12 hours, plasmid DNA was recovered and analyzed for removal of 8-oxoG using the Fpg protein that specifically nicks DNA at 8-oxoG·C (Fig. 2). After treatment with Fpg, covalently closed (CC) molecules correspond to repaired molecules, where the 8-oxoG·C base pair has been replaced by a G·C base pair. On the other hand, open circles (OC) correspond to unrepaired molecules that kept the 8-oxoG·C base pair and consequently are sensitive to Fpg. Therefore, removal of 8-oxoG is assessed by the increasing fraction of CC molecules after Fpg treatment (Fig. 2A). Control experiments show that the plasmid DNA preparations used migrate as CC molecules which are fully relaxed after treatment by 5 ng of Fpg protein (Fig. 2B). This assay also requires that the vast majority of plasmid DNA molecules be CC after extraction from mammalian cells. Figure 2B shows that plasmid DNA recovered from AA8 cells transfected and incubated for 2–12 hours entirely migrates as CC molecules. Thus, formation of OC is dependent upon Fpg treatment, indicating the persistence of the 8-oxoG·C base pair. The novelty of the present assay resides in the direct analysis of recovered DNA without amplification in bacteria (37, 38).

B. Transcription-Coupled Repair of 8-oxoG in AA8 CHO Cells

To study the repair of 8-oxoG, nonreplicative plasmids pSΔoriSV-[8-oxoG·C] (TS) and pSΔ(ori-p)SV-[8-oxoG·C] (NTS) were transfected into AA8 cells and incubated for 2–12 hours before recovery. It should be noted that cell-free extracts of CHO cell lines AS52 and AA8 possess an Ogg1-like activity catalyzing the cleavage of duplexes containing an 8-oxoG·C base pair (41, and data not shown). Figure 3 (left panel) illustrates the repair kinetics of 8-oxoG either on NTS or TS in AA8 cells. Southern blots show that after transfection plus 2 hours of incubation, recovered plasmid DNA essentially migrates as OC after treatment with Fpg, indicating that 8-oxoG has not yet been repaired. This delay

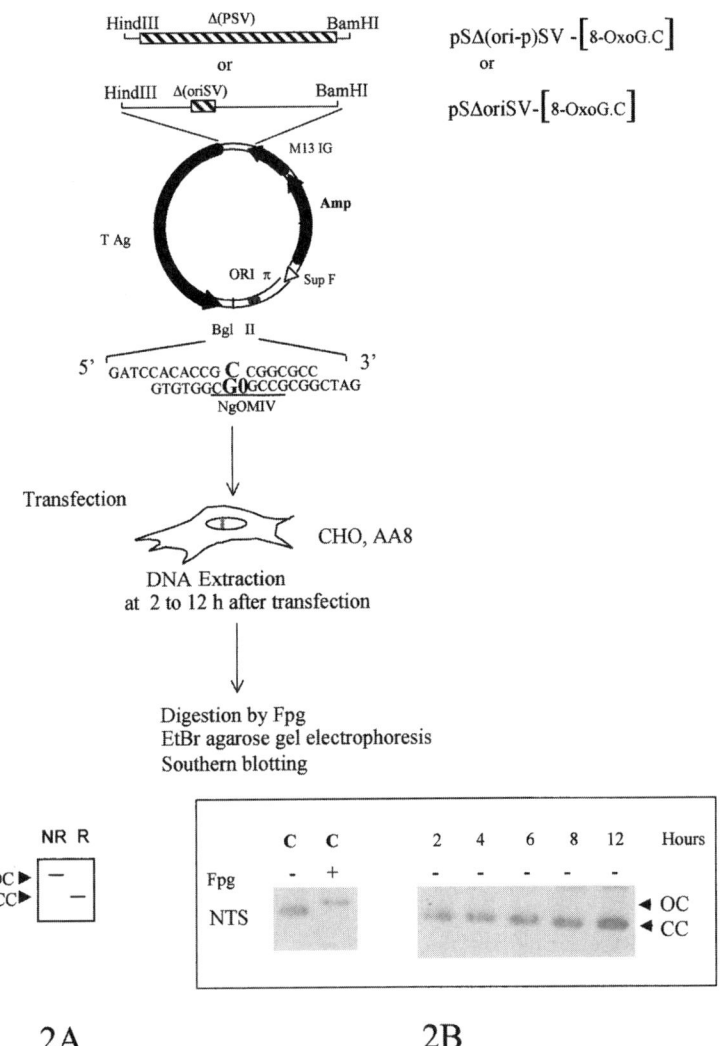

FIG. 2. Repair of 8-oxoG·C in AA8 CHO cell line. (A) Experimental scheme: Plasmid pSΔoriSV-[8-oxoG·C] has a deletion of the SV40 origin of replication (hatched) whereas pS(ori-p)SV-(8-oxoG·C) has a deletion of both the SV40 origin of replication and the SV40 promoter from −10 to 273 that eliminates transcription of TAg and of the 8-oxoG (hatched). After transfection and incubation in CHO cells, plasmid DNA was recovered and incubated with 5 ng of Fpg protein before EtBr–agarose gel electrophoresis and Southern blot analysis. In this assay, removal of 8-oxoG (R) is assessed by the presence of covalently closed (CC) plasmid molecules after digestion by Fpg, whereas unrepaired (NR) plasmid molecules migrate as open circle (OC). (B) Controls. Left panel: DNA substrate (pSΔoriSV-[8-oxoG·C]) analyzed before transfection: incubated (C+) or not (C−) with Fpg. Right panel: DNA extracted from wild-type AA8 CHO cell line, not incubated with Fpg. The same result was obtained with pS(ori-p)SV-[8-oxoG·C].

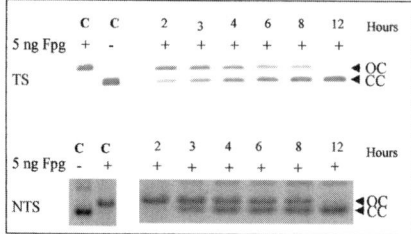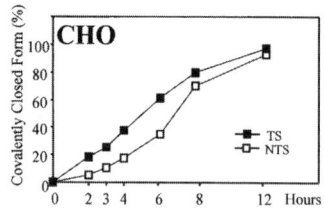

FIG. 3. Kinetics of 8-oxoG removal in AA8 CHO cell line. The nonreplicating, monomodified shuttle vectors, pSΔoriSV-[8-oxoG·C] (TS) and pSΔ(ori-p)SV-[8-oxoG·C] (NTS), were transfected into AA8 CHO cell line, incubated for 2–12 hours and recovered. Extracted DNA was analyzed for repair of 8-oxoG. Control experiments show constructs incubated (C+) or not (C−) with Fpg. Left panel: Southern blots after transfection of plasmids in the AA8 CHO cell line and Fpg-cleavage. Right panel: Quantitative analysis of Southern blots. Experimental values are the average of two independent experiments.

probably reflects the time required by plasmid DNA to reach the nucleus of the cell. From 3 to 12 hours, an increasing fraction of plasmid molecules migrates as CC molecules after Fpg treatment, indicating removal of 8-oxoG in NTS and TS (Fig. 3). However, 8-oxoG removal occurs at a faster rate when the lesion is located on the TS compared to NTS indicating transcription-coupled repair (TCR) of 8-oxoG in the AA8 CHO cell line (Fig. 3, right panel).

C. Repair of Ogg1-Incised Apurinic Sites in AA8 CHO Cells

The first step in the course of BER of 8-oxoG in mammalian cells is the release of the lesion and the subsequent incision of DNA at the 3′ side of the resulting AP site by the Ogg1 protein (Fig. 1). Such a repair intermediate can be generated after cleavage of pSΔ(ori-p)SV-[8-oxoG·C] (NTS) by the yeast Ogg1 protein *in vitro* (Fig.1; substrate 2). In fact, this repair intermediate is common to any lesion whose repair is initiated by a DNA glycosylase/AP lyase such as Ogg1 or Nth1 (40). Thus, Ogg1-treated plasmid pSΔ(ori-p)SV-[8-oxoG·C] (NTS) was transfected into CHO cell line. After incubation for 2–12 hours, the plasmid DNA molecules were recovered and analyzed by agarose gel electrophoresis, without Fpg treatment. The recovery of CC molecules indicates DNA repair of the Ogg1-incised DNA, presumably after action of a phosphodiesterase, a DNA polymerase, and a DNA ligase. Figure 4 illustrates repair of Ogg1-incised AP site in the AA8 CHO cell line. The results show significant repair after 2 hours, which was not observed for the same plasmid harboring an 8-oxoG·C base pair (Fig. 4). The presence of CC DNA at 2 hours is clearly due to DNA repair because the input DNA was quantitatively incised OC after Ogg1-treatment

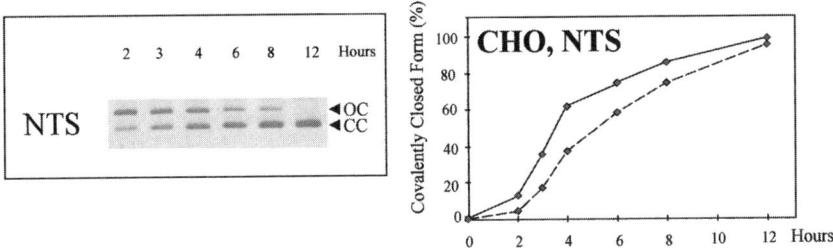

FIG. 4. Kinetics of repair of Ogg1-incised apurinic site in AA8 CHO cell line. The nonreplicating shuttle vector, pSΔ(ori-p)SV-[8-oxoG·C] (NTS), was quantitatively pre-incised by the yeast Ogg1 protein and transfected into AA8 CHO cell line. After incubation for 2–12 hours, plasmid DNA was recovered and analyzed for repair of the Ogg1-incised AP site. Left panel: Southern blots after transfection and incubation in the AA8 CHO cell line. Right panel: Quantitative analysis of Southern blots. Solid line: Repair of Ogg1-incised AP site in NTS. Dotted line: Removal of 8-oxoG from NTS (Fig. 3). Experimental values are the average of two independent experiments.

(data not shown). The fast repair of the Ogg1-incised AP site compared to the 8-oxoG·C base pair is confirmed at 4, 6, and 8 hours incubation times (Fig. 4).

IV. Conclusions

In this study, we have investigated the repair of two DNA substrates occurring in the course of the BER pathway of 8-oxoG. The same assay can also be used to study the repair of other BER intermediates generated *in vitro* with purified enzymes (Fig. 1). The comparison of the repair kinetics of these various substrates in rodent cell lines specifically disrupted for DNA repair genes will allow us to dissect mechanisms of BER of oxidized DNA bases *in vivo*.

Acknowledgments

This work was supported by the Centre National de la Recherche Scientifique (CNRS) and the Commissariat à l'Energie Atomique (CEA). The authors thank Dr. Alain Sarasin (CNRS-Villejuif, France) for the kind gift of the plasmids used in this study and Dr. Jean Cadet (CEA-Grenoble, France) for the kind gift of the 8-oxoG containing oligodeoxyribonucleotides.

References

1. D. I. Feig, T. M. Reid, and L. A. Loeb, *Cancer Res. Suppl.* **54**, 1890–1894 (1994).
2. H. Wiseman, H. Kaur, and B. Halliwell, *Cancer Lett.* **93**, 113–120 (1995).

3. K. B. Beckman and B. N. Ames, *J. Biol. Chem.* **272,** 19633–19636 (1997).
4. S. Boiteux, *in* "Oxidative Stress in Cancer, AIDS and Neurodegenerative diseases" (L. Montagnier, R. Olivier, and C. Pasquier, eds.), pp. 351–358. Marcel Dekker, New York, 1998.
5. M. Dizdaroglu, *Free Radical Biol. Med.* **10,** 225–242 (1991).
6. J. Cadet, M. Berger, T. Douki, and J. L. Ravanat, *Rev. Physiol. Biochem. Pharmacol.* **131,** 1–87 (1997).
7. S. Boiteux and J. Laval, *in* "Base Excision Repair of DNA Damage" (I. D. Hickson, ed.), pp. 31–44. Landes-Springer, Austin TX, 1997.
8. H. E. Krokan, R. Standal, and G. Slupphaug, *Biochem. J.* **325,** 1–16 (1997).
9. S. Boiteux and J. P. Radicella, *in* "Advances in DNA Damage and Repair" (M. Dizdaroglu, ed.), pp. 35–45. Kluwer Academic, New York, 1999.
10. T. Lindahl and R. D. Wood, *Science* **286,** 1897–1905 (1999).
11. M. L. Michaels and J. H. Miller, *J. Bacteriol.* **174,** 6321–6325 (1992).
12. A. P. Grollman and M. Moriya, *Trends Genet.* **9,** 246–249 (1993).
13. S. Boiteux and J. P. Radicella, *Biochimie* **81,** 59–67 (1999).
14. P. Auffret van der Kemp, D. Thomas, R. Barbey, R. De Oliveira, and S. Boiteux, *Proc. Natl. Acad. Sci. U.S.A.* **93,** 5197–5202 (1996).
15. H. M. Nash, S. D. Bruner, O. D. Scharer, T. Kawate, T. A. Addona, E. Spooner, W. S. Lane, and G. L. Verdine, *Curr. Biol.* **6,** 968–980 (1996).
16. P. M. Girard, N. Guibourt, and S. Boiteux, *Nucleic Acids Res.* **25,** 3404–3411 (1997).
17. D. Thomas, A. D. Scott, R. Barbey, M. Padula, and S. Boiteux, *Mol. Gen. Genet.* **254,** 171–178 (1997).
18. A. D. Scott, M. Neishabury, D. H. Jones, S. H. Reed, S. Boiteux, and R. Waters, *Yeast* **15,** 205–218 (1999).
19. T. T. Ni, G. T. Marsichky, and R. D. Kolodner, *Mol. Cell* **4,** 439–444 (1999).
20. M. M. Slupska, C. Baikalov, W. M. Luther, J. H. Chiang, W. F. Wei, and J. H. Miller, *J. Bacteriol.* **178,** 3885–3892 (1996).
21. H. Aburatani, Y. Hippo, T. Ishida, R. Takashima, C. Matsuba, T. Kodama, M. Takao, A. Yasui, K. Yamamoto, and M. Asano, *Cancer Res.* **57,** 2151–2156 (1997).
22. K. Arai, K. Morishita, K. Shinmura, T. Kohno, S. R. Kim, T. Nohmi, M. Taniwaki, S. Ohwada, and J. Yokota, *Oncogene* **14,** 2857–2861 (1997).
23. M. Bjoras, L. Luna, B. Johnsen, B. Hoff, T. Haug, T. Rognes, and E. Seeberg, *EMBO J.* **16,** 6314–6322 (1997).
24. R. Lu, H. M. Nash, and G. L. Verdine, *Curr. Biol.* **7,** 397–407 (1997).
25. J. P. Radicella, C. Dhérin, C. Desmaze, M. S. Fox, and S. Boiteux, *Proc. Natl. Acad. Sci. U.S.A.* **94,** 8010–8015 (1997).
26. T. Roldan-Arjona, Y. F. Wei, K. C. Carter, A. Klungland, C. Anselmino, R. P. Wang, M. Augustus, and T. Lindahl, *Proc. Natl. Acad. Sci. U.S.A.* **94,** 8016–8020 (1997).
27. T. A. Rosenquist, D. O. Zharkov, and A. P. Grollman, *Proc. Natl. Acad. Sci. U.S.A.* **94,** 7429–7434 (1997).
28. M. Takao, H. Aburatani, K. Kobayashi, and A. Yasui, *Nucleic Acids Res.* **26,** 2917–2922 (1998).
29. K. Nishioka, T. Ohtsubo, H. Oda, T. Fujiwara, D. Kang, K. Sugimachi, and Y. Nakabeppu, *Mol. Biol. Cell* **10,** 1637–1652 (1999).
30. S. Boiteux, C. Dhérin, F. Reille, F. Apiou, B. Dutrillaux, and J. P. Radicella, *Free Radical Res.* **29,** 487–497 (1998).
31. M. Tani, K. Shinmura, T. Kohno, T. Shiroishi, S. Wakana, S. R. Kim, T. Nohmi, H. Kasai, S. Takenoshita, Y. Nagamuchi, and J. Yokota, *Mammalian Genome* **9,** 32–37 (1998).
32. G. Dianov, C. Bischoff, J. Piotrowski, and V. A. Bohr, *J. Biol. Chem.* **273,** 33811–33816 (1998).
33. P. Fortini, E. Parlanti, O. M. Sidorkina, J. Laval, and E. Dogliotti, *J. Biol. Chem.* **274,** 15230–15236 (1999).

34. A. Klungland, I. Rosewell, S. Hollenbach, E. Larsen, G. Daly, B. Epe, E. Seeberg, T. Lindahl, and D. E. Barnes, *Proc. Natl. Acad. Sci. U.S.A.* **96,** 13300–13305 (1999).
35. T. K. Hazra, T. Izumi, L. Maidt, R. A. Floyd, and S. Mitra, *Nucleic Acids Res.* **26,** 5116–5122 (1998).
36. J. T. Reardon, T. Bessho, H. C. Kung, P. H. Bolton, and A. Sancar, *Proc. Natl. Acad. Sci. U.S.A.* **94,** 9463–9468 (1997).
37. F. Le Page, E. E. Kwoh, A. Avrutskaya, A. Gentil, S. A. Leadon, A. Sarasin, and P. K. Cooper, *Cell (Cambridge, Mass.)* **101,** 159–171 (2000).
38. F. Le Page, A. Guy, J. Cadet, A. Sarasin, and A. Gentil, *Nucleic Acids Res.* **26,** 1276–1281 (1998).
39. S. Boiteux, T. R. O'Connor, F. Lederer, A. Gouyette, and J. Laval, *J. Biol. Chem.* **265,** 3916–3922 (1990).
40. A. Klungland and T. Lindahl, *EMBO J.* **16,** 3341–3348 (1997).
41. S. Hollenbach, A. Dhénaut, I. Eckert, J. P. Radicella, and B. Epe, *Carcinogenesis* **20,** 1863–1868 (1999).
42. E. C. Friedberg and L. B. Meira, *Mutat. Res.* **433,** 69–87 (1999).

Mammalian *Ogg1/Mmh* Gene Plays a Major Role in Repair of the 8-Hydroxyguanine Lesion in DNA

SUSUMU NISHIMURA

Banyu Tsukuba Research Institute in Collaboration with Merck Research Laboratories Okubo 3, Tsukuba, Ibaraki 300-2611, Japan

I. Background 108
II. Human OGG1/MMH Type 1a Protein Is a Major Enzyme for Repair of 8-OH-G Lesions.. 110
   A. Expression of hMMH Type 1a Protein in Human Cell Lines 110
   B. Depletion of hMMH Type 1a in a Whole HeLa S3 Cell Extract 110
   C. The Presence of hMMH Type 1a in Various Human Cell Lines...... 112
III. *Mmh* Knockout Results in Accumulation of 8-OH-G in DNA and Increases Mutation Frequency in Mice 113
   A. Generation of *Mmh* Knockout Mice............................ 113
   B. Loss of AP Lyase Activity in Liver Extracts of *Mmh* Mutant 113
   C. Accumulation of 8-OH-G in DNA of *Mmh* Mutant Mice 115
   D. Mutation Frequency in *Mmh* Mutants........................... 115
IV. Discussion and Future Direction 119
   References... 121

8-Hydroxyguanine (7,8-dihydro-8-oxoguanine, abbreviated as 8-OH-G or 8-oxoG) is the site of a frequent mutagenic DNA lesion produced by oxidative damage. MutM of *E. coli* and OGG1 of *Saccharomyces cerevisiae* are known to possess 8-OH-G glycosylase activity and apurinic (AP) site lyase activity to repair 8-OH-G lesions. Recently, cDNA clones of four isoforms (types 1a, 1b, 1c, and type 2) of human *OGG1* homologs (*hMMH*) were isolated. However, it is unknown whether expression of endogenous hMMH proteins actually occurs in mammalian cells.

We have chosen two approaches to clarify this issue. First, using hMMH type 1a-specific antibody and cells overexpressing tag-fused hMMH type 1a, we found that hMMH type 1a protein is in fact expressed in many types of human cells, showing that endogenous hMMH type 1a protein has 8-OH-G glycosylase/AP lyase activity. Furthermore, we have shown that upon antibody-mediated depletion of hMMH type 1a protein in a whole-cell extract, most of the AP lyase activity is lost, indicating that hMMH type 1a protein is a major enzyme for repair of 8-OH-G lesion in human cells. In our second approach we have generated a mouse line carrying a mutant *Mmh* allele by targeted gene disruption. *Mmh* homozygous mutant mice were found to be physically normal in appearance, but to have lost the nicking activity for substrate DNA containing 8-OH-G in liver extracts. In addition, the amount of endogenous 8-OH-G in liver DNA of the homozygous mutant mice at 8 weeks of age was 3-fold higher compared with wild-type or heterozygous mice. A further increase of 8-OH-G up to 7-fold was observed in 14-week-old animals. These results indicate that exposure of DNA to internal oxidative species constantly produces the mutagenic DNA adduct 8-OH-G in mice, and that Mmh plays an essential role in the repair of this type of oxidative DNA damage. © 2001 Academic Press.

I. Background

In 1983, we discovered 8-hydroxyguanine (7,8-dihydro-8-oxoguanine, abbreviated as 8-OH-G or 8-oxoG)[1] as an oxidized product of guanine (*3, 4*). 8-OH-G was produced in DNA by various oxygen-radical–forming agents in *in vitro* reactions and also in *in vivo* systems (*3, 4*). In 1986, R. Floyd and his colleagues showed that electrochemical detection coupled with HPLC was able to detect 8-OH-dG with >1000 times greater sensitivity than measured by conventional UV detection (*5*). By using their method, we were able to detect an increase of 8-OH-G in liver DNA, proportional to the dose of γ-radiation to the whole body of mice (*6*). Another important observation at that time was a rapid decrease of 8-OH-G in liver DNA following γ-radiation, indicating that there was a repair system for 8-OH-G (*6*). The presence of the repair system for 8-OH-G was, and continues to be, a strong indication for the importance of repairing the 8-OH-G lesion in living organisms.

8-OH-G is also formed *in vivo* by chemical carcinogens which are known to generate oxygen radicals. In 1985, we first showed that administration of the renal carcinogen potassium bromate to rats increased the amount of 8-OH-G in kidney DNA, but not in liver DNA, suggesting that 8-OH-G was a good marker for target organ identification of potential chemical carcinogens (*7*). Subsequently, and up to now, many investigators have reported an increase of 8-OH-G in target tissues of rats and mice by oxygen-radical–forming carcinogens.

In 1987, we reported that 8-OH-G was misread during DNA replication in *in vitro* reactions (*8*). Subsequently, A. P. Grollman and his colleagues clearly showed that 8-OH-G is misread to incorporate adenine in the opposite strand by either the Klenow fragment or DNA polymerase α in *in vitro* reactions (*9*).

---

[1]8-Hydroxyguanine is the same as 7,8-dihydro-8-oxoguanine, abbreviated as 8-OH-G or 8-oxoG. 8-Oxoguanine is not a formal name; it is an abbreviation of 7,8-dihydro-8-oxoguanine. We previously showed that a favorable conformation of 8-OH-G is the tautomeric 8-oxo form, rather than the 8-hydroxy form (*1, 2*). Thus, 8-OH-G can be named either 8-hydroxyguanine or 7,8-dihydro-8-oxoguanine.

Misincorporation of adenine opposite 8-OH-G in *Escherichia coli* was also shown in *in vivo* systems by several investigators (*10–12*).

If the presence of 8-OH-G is deleterious to living organisms, it follows that there must be a mechanism by which to repair it. Therefore, we tried to isolate an enzyme from *E. coli* that cleaves DNA specifically at the site of 8-OH-G residues. Double-stranded oligonucleotides containing 8-OH-G in a specific position were used as substrate for identification of specific glycosylases. The enzyme that was purified was found to be specific for double-stranded DNA containing 8-OH-G, but was not active against mispaired DNA with normal bases (*13*). It was later found that the enzyme is the same as the FPG protein (*14*). Since then, much work has been done to clarify mechanisms of repair of 8-OH-G in DNA in *E. coli* by many investigators. Specifically, three genes were found to be involved in repair of 8-OH-G: FPG (*MutM* product) is the specific glycosylase/AP lyase for 8-OH-G, and with its inactivation, G to T transversions are enhanced (*15*); *MutY* is the gene for the specific mismatch repair enzyme for the removal of an adenine base paired with 8-OH-G (*16*); and *MutT* codes for a specific phosphatase cleaving 8-OH-dGTP to 8-OH-dGMP (*17*). Oxygen radicals produce not only 8-OH-G in DNA but also 8-OH-dGTP from dGTP in the nucleotide pool. 8-OH-dGTP is incorporated into DNA opposite A instead of C, and thus, if not repaired prior to a subsequent round of replication, induces a mutation. Therefore, inactivation of *MutT* induces A to C transversions. The existence of these three gene products for repair of 8-OH-G in *E. coli* DNA suggests that they function coordinately to minimize mutagenesis through the pathway called the GO system (*16*). This is an indication of the importance of 8-OH-G in living organisms.

An important question arose as to whether or not a similar mechanism for repair of 8-OH-G exists in mammalian cells. It has been shown that homologs of *MutY* and *MutT* exist in mammalian cells (*18, 19*). On the other hand, the gene for a *MutM* homolog was not identified in mammalian cells until 1997. Meanwhile, a gene encoding a functional *MutM* homolog called *OGG1* was isolated from the yeast *Saccharomyces cerevisiae* in 1996 by S. Boiteux and his colleagues (*20*) and independently by G. L. Verdine and his colleagues (*21*).

In 1997, several groups independently obtained a human or mouse homolog of *OGG1* by a similarity search of the human EST database with the yeast *OGG1* sequence (*22–28*). Among them, H. Aburatani and his group, in collaboration with us and with others, isolated four isoforms of cDNA produced by alternative splicing, as well as genomic DNA from human and mouse cells (*22*). This gene is called *hOGG1/hMMH* or *mOGG1/mMMH*.

One of the human isoforms, type 1a, was expressed in *E. coli* and its gene product was purified to homogeneity. It showed both glycosylase and lyase activity specific for the 8-OH-G residue. Transfection of each for the four isoforms of the cDNA into $MutM^-$ and $MutY^-$ *E. coli* reduced mutation frequency of

G to T transversions. The isoform type 1a contains a nuclear localization signal at its 3' end, but the others do not, suggesting that type 1a is involved in repair of 8-OH-G in nuclear DNA, and the others in mitochondrial DNA repair. Later, this speculation was confirmed by A. Yasui and his colleagues (29).

II. Human OGG1/MMH Type 1a Protein Is a Major Enzyme for Repair of 8-OH-G Lesions

Although expression of human and mouse *Mmh/Ogg1* (hereafter abbreviated as *Mmh*) was confirmed at the mRNA level, it has not yet been shown at the protein level. In addition, it is not clear whether hMMH type 1a, type 1b, type 1c, type 2, or some other enzyme is most important for repair or 8-OH-G lesions. To answer this question, we prepared hMMH type 1a specific antibody; by using this antibody, we showed that hMMH type 1a protein is actually expressed in many human cell lines and that endogenous hMMH type 1a protein has repair activity for 8-OH-G lesions. In addition, by adopting a hMMH type 1a depletion assay, it was shown that it is a major protein for repair of 8-OH-G lesions (30).

A. Expression of hMMH Type 1a Protein in Human Cell Lines

First, HeLa S3 cell extract was analyzed by Western blotting using Ab1a311 (hMMH type 1a specific antibody αM). The antibody detected four major proteins (60, 40, 38, and 28 kDa). The 40-kDa protein is likely to be endogenous hMMH type 1a protein (Fig. 1A, arrow). To examine this possibility, HeLa S3 cell extracts were incubated with Ab1a311 (αM), and the precipitate was analyzed by Western blotting. The 40-kDa protein was found as a major protein immunoprecipitated with the antibody. The much larger band found at 50 kDa is a heavy chain of the antibody, since the same band was observed in the buffer control. Examination of DNA nicking activity showed that the immunoprecipitate has AP lyase activity (Fig. 1B) similar to the purified recombinant hMMH type 1a protein. These results indicated that the 40-kDa endogenous hMMH type 1a protein is expressed in human cells and has MutM-like activity.

B. Depletion of hMMH Type 1a in a Whole HeLa S3 Cell Extract

To examine whether hMMH type 1a is a major enzyme for repair of 8-OH-G in human cells, we examined the total AP lyase activity against 8-OH-G containing oligonucleotide in whole-cell extracts. For this purpose, HeLa S3 cell extracts were incubated with Ab1a311 or control rabbit IgG, and the immunoprecipitates were separated by protein A–Sepharose CL-4B. As shown in Fig. 2A, almost

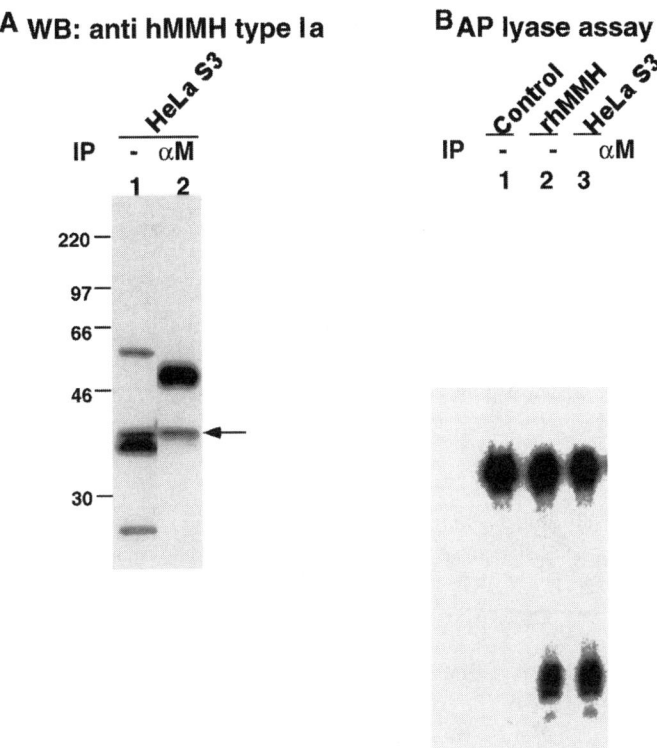

FIG. 1. Detection of endogenous hMMH type 1a protein in HeLa S3 cells. (A) Western blotting analysis of hMMH type 1a protein in HeLa S3 cells. Cell extract (lane 1) and immunoprecipitate (lane 2) were fractionated by PAGE in a 12.5% gel and analyzed by Western blotting. The antibody used for immunoprecipitation (IP) or Western blotting (WB) is indicated (αM, anti-hMMH type 1a, Ab1a331). The arrow shows the position of endogenous hMMH type 1a protein. Positions of marker proteins with molecular mass (kDa) are indicated on the left. (B) AP lyase activity of hMMH type 1a protein in HeLa S3 cells. DNA containing an 8-OH-G/C base pair labeled at the 5' end of the 8-OH-G containing strand with [γ-^{32}P]ATP was incubated with the immunoprecipitate and the nicked products were analyzed by PAGE in 20% gel. The antibody used for immunoprecipitation (IP) is indicated (αM, anti-hMMH type 1a). A sample of 1 ng of recombinant hMMH type 1a protein (rhMMH type 1a) was used as a positive control (lane 2).

all of the 40-kDa protein, corresponding to hMMH type 1a protein (Fig. 2A, arrow), was depleted by the treatment with Ab1a311, but not by control rabbit IgG treatment. As indicated in Fig. 2B, nicking activity of the depleted extract was about 10% of that of control extract treated with control rabbit IgG (lane 4 versus lane 5). This finding indicated that the AP lyase activity of the whole-cell extract was mainly due to the hMMH type 1a enzyme. The size of the

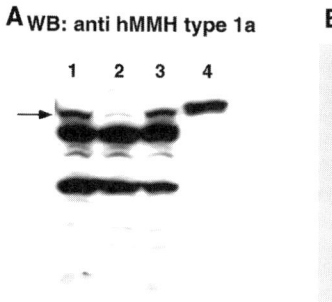
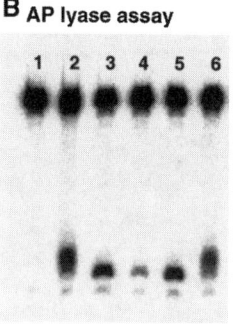

FIG. 2. Depletion analysis of hMMH type 1a protein in a whole HeLa S3 cell extract. HeLa S3 cell extracts were incubated with anti-hMMH type 1a antibody (Ab1a331) or control rabbit IgG and the immunoprecipitates were separated by treatment with protein A–Sepharose CL-4B. The depleted extracts and control extract were analyzed by Western blotting using Ab1a331 antibody (A) or AP lyase assay (B). (A) Lane 1, whole-cell extract; lane 2, hMMH type 1a-depleted extract; lane 3, control rabbit IgG-treated extract; lane 4, immunoprecipitate of a whole-cell extract with Ab1a331. The arrow shows the position of hMMH type 1a protein. (B) Lane 1, control; lane 2, rhMMH type 1a (1 ng); lane 3, whole-cell extract; lane 4, hMMH type 1a-depleted extract; lane 5, control rabbit IgG-treated extract; lane 6, immunoprecipitate of a whole-cell extract with Ab1a331.

nicked product in the whole-cell extract was slightly smaller than that of purified recombinant enzyme or the immunoprecipitate by Ab1a311 (Fig. 2B; lane 3 versus lane 2 and 6). Because the hMMH type 1a protein yields only β-elimination, the δ-elimination in the cell extract could be catalyzed by other enzyme(s).

C. The Presence of hMMH Type 1a in Various Human Cell Lines

We examined the expression of hMMH type 1a protein in various human cell lines. As shown in Fig. 3, Western blotting analysis showed that all cell lines examined expressed hMMH type 1a protein, although expression levels differed.

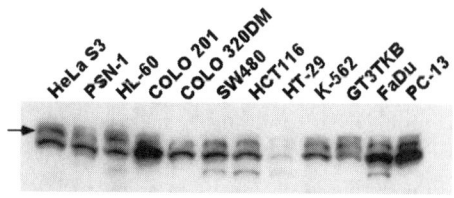

FIG. 3. Expression of endogenous hMMH type 1a protein in various human cell lines. Cell extracts were analyzed by Western blotting with Ab1a331 antibody. The arrow shows the position of hMMH type 1a protein.

III. *Mmh* Knockout Results in Accumulation of 8-OH-G in DNA and Increases Mutation Frequency in Mice

A. Generation of *Mmh* Knockout Mice

As described in the preceding section, it was shown that hMMH type 1a is a major enzyme present in human cells to remove 8-OH-G in DNA in *in vitro* reactions. An important question asked is whether or not MMH is in fact involved in repair of 8-OH-G lesions *in vivo*. To answer this question, we generated *Mmh* knockout mice (*31*). The work was in collaboration with T. Noda of the Cancer Institute (Tokyo) and H. Aburatani of the School of Medicine, University of Tokyo. Figure 4A shows a schematic diagram of how the targeted disruption of the mouse *Mmh* gene was made. Two independent ES clones carrying mutant alleles of *Mmh* were obtained. Chimeric mice were generated and germ-line transmission of mutant *Mmh* alleles to F_1 offspring was confirmed (Fig. 4B). Subsequently, F_2 progeny were produced by interbreeding of F_1 heterozygotes for the study. Genotype analysis of F_2 offspring showed that the ratio of the number of mice with wild-type, heterozygous, and homozygous alleles were 25%, 48%, and 23%, indicating that the *Mmh* gene product does not play an essential role in embryonic development. Both homozygous and heterozygous mutants appeared healthy, with no obvious difference from wild-type littermates in survival and gross appearance until 50 weeks of age. It should be mentioned that no tumor has been found in homozygous mutants thus far. Both male and female homozygous mice were found to be fertile.

B. Loss of AP Lyase Activity in Liver Extracts of *Mmh* Mutant

AP lyase activity in liver extracts of *Mmh* mutant mice was analyzed using substrate DNA containing 8-OH-G in a defined position. As shown in Fig. 5, a clear decrease of AP lyase activity was observed, depending upon the gene dosage. It is noteworthy that slight activity is observed even in the case of homozygous mice. This residual activity of unknown origin was at most <5% of that of wild-type mice, indicating that the results are consistent with the previous study using hMMH type 1a antibody. The substrate specificity of nicking activity in liver extracts was in the following order: 8-OH-G/C > 8-OH-G/T > 8-OH-G/G > 8-OH-G/A ≫ G/C, reminiscent of the case of recombinant human MMH type 1a (*31*) (Fig. 6). Although reduced to half of the wild type, the extracts obtained from heterozygous mice showed similar specificity. Activity against all substrates almost disappeared in liver extracts of homozygous mice. Thus, it can be concluded that the major repair of 8-OH-G in mouse liver cells is carried out by MMH.

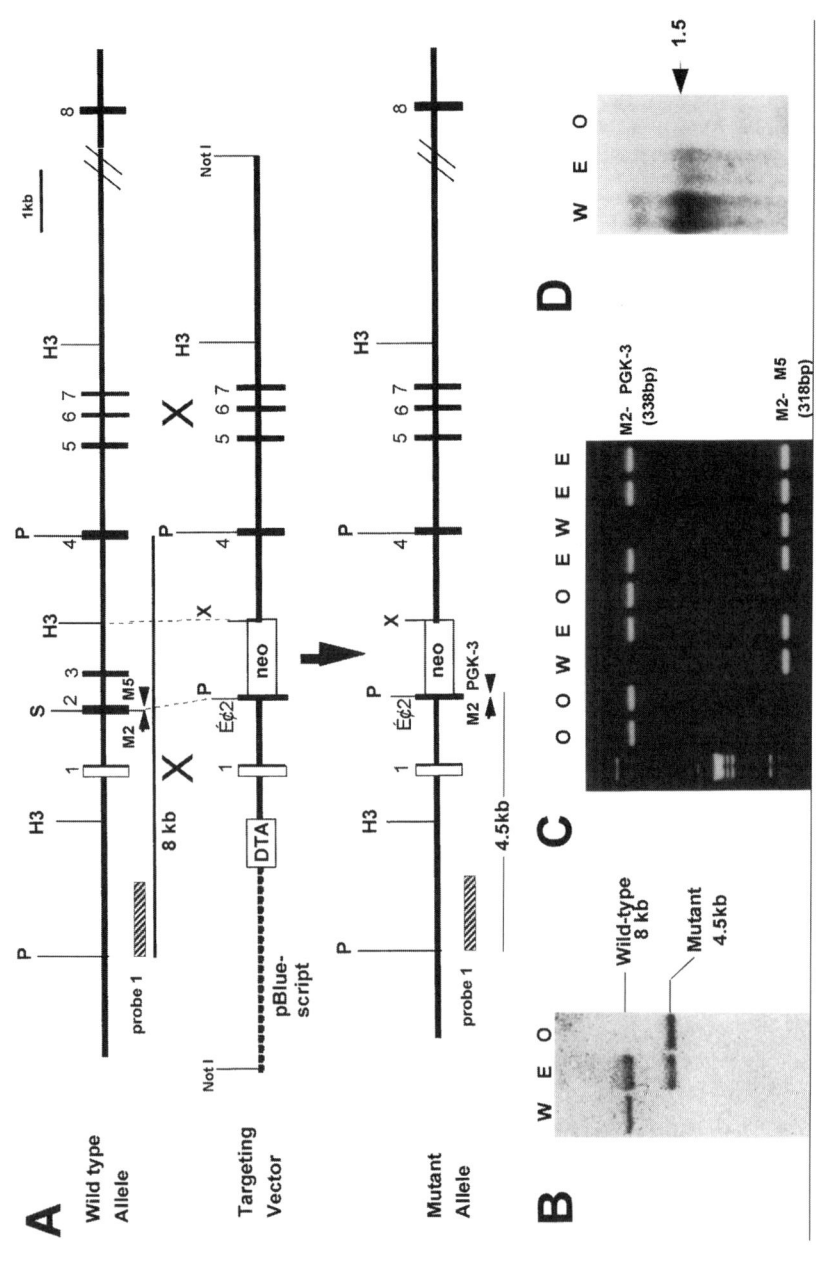

C. Accumulation of 8-OH-G in DNA of *Mmh* Mutant Mice

To analyze the effect of *Mmh* deficiency on oxidative DNA lesions, the amount of 8-OH-G in genomic DNA of the mutant mice was analyzed by HPLC coupled with the electrochemical detection method (32). As shown in Fig. 7A and B, the amounts of 8-OH-G in DNA of wild-type mice were found to be 1.0–2.0 per 10^6 dG, in good agreement with our previous results (6). A significant increase of 8-OH-G was observed in the liver DNA of *Mmh* homozygous mice. At 9 weeks of age, the level of 8-OH-G in liver DNA was approximately 4.2-fold higher, as compared to that of wild type. Further increases in 8-OH-G levels were observed in 14-week-old mice, reaching 7-fold that of wild-type of the same age. Although heterozygous mice possessed only half of the AP lyase activity of wild-type mice, this amount of enzyme seemed to be sufficient to process the usual 8-OH-G load, since no statistically significant elevation of 8-OH-G was observed.

D. Mutation Frequency in *Mmh* Mutants

Mmh homozygous (−/−) and wild-type (+/+) mice hemizygous for the *gpt* transgene were established in order to assess spontaneous mutation frequency in the mutant mice. As shown in Table I, the mutation frequency for *gpt/Mmh*$^{+/+}$ transgenic mice was under 1×10^{-5}, although the value fluctuated among individual mice. On the other hand, in liver DNA obtained from *gpt/Mmh*$^{-/-}$ mutants, the mutation frequency tended to increase, and on average, a 2.3-fold elevation was observed. This increase is statistically significant (Fisher's test, $P < 0.05$). To analyze the spectrum of mutations, plasmids were recovered from *gpt* mutant colonies and the sequences of ORFs of the *gpt* gene were analyzed.

FIG. 4. Targeted disruption of mouse *Mmh*. (A) Restriction map of wild-type *Mmh* and structures of the targeting vector and the mutant allele generated by homologous recombination. Coding exons are represented by filled boxes and the noncoding exon by a white box. The genomic fragment used as a probe (probe 1) for Southern-blot analysis is shown by a hatched box. neo, neomycin phosphotransferase gene flanked by the phosphoglycerate kinase (Pgk) promoter and a polyadenylation signal; DTA, diphtheria toxin A gene driven by the MC1 promoter; P, *Pst*I; H3, *Hind*III; S, *Sca*I; X, *Xba*I. (B) Southern-blot analysis of liver genomic DNA extracted from F_2 mice obtained by F_1 heterozygous intercrosses. *Pst*I digestion generated hybridized bands of the expected sizes for wild-type and mutant *Mmh* of 8 kb and 4.5 kb respectively. W, wild-type; E, heterozygote; O, homozygous mutant. (C) Genotyping of F_2 offspring by PCR amplification. Tail lysates were used as the templates. Amplification of fragments from the mutant and wild-type allele was achieved by using M2/PGK-3 and M2/M5 primer pairs, respectively. The positions of the primers are indicated in (A). (D) Northern-blot analysis of *Mmh* expression in liver RNA extracted from F_2 mice. Each lane contained 2μg of poly(A)-RNA. (Reprinted with permission from Reference 31. Copyright 2000 National Academy of Sciences, U.S.A.)

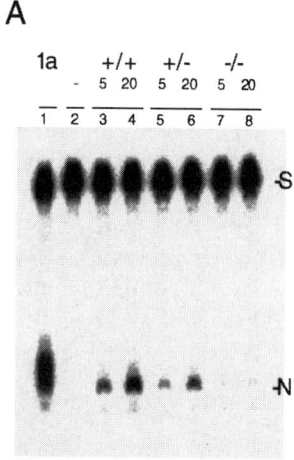

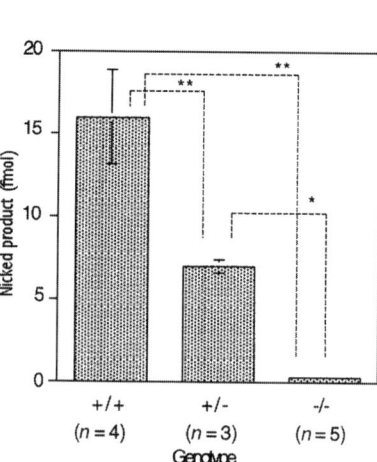

FIG. 5. AP lyase activity in mouse liver crude extracts. A 5′-end-labeled oligonucleotide containing an 8-OH-G hybridized with the complementary oligonucleotide was incubated with the crude extracts, and then nicked products produced were analyzed by 20% PAGE. (A) Representative AP lyase assay: lane 1, 1 ng of purified recombinant human MMH type 1a protein as a positive control; lane 2, the lysis buffer as a negative control; lanes 3 and 4, 5 and 20 μg of crude extract of liver from an Mmh wild-type mouse; lanes 5 and 6, 5 and 20 μg of crude extract of liver from an Mmh heterozygous mutant mouse; lanes 7 and 8, 5 and 20 μg of crude extract of liver from an Mmh homozygous mutant mouse. S, substrate oligonucleotide; N, nicked product. (B) Quantification of the nicked products by analyzing radioactivity with a bioimaging analyzer BAS2000. Numbers of mice used are indicated under the genotype. Results are mean and SD. *, $P < 0.0003$; **, $P < 0.0001$ (Fisher statistical analysis). (Reprinted with permission from Reference 31. Copyright 2000 National Academy of Sciences, U.S.A.)

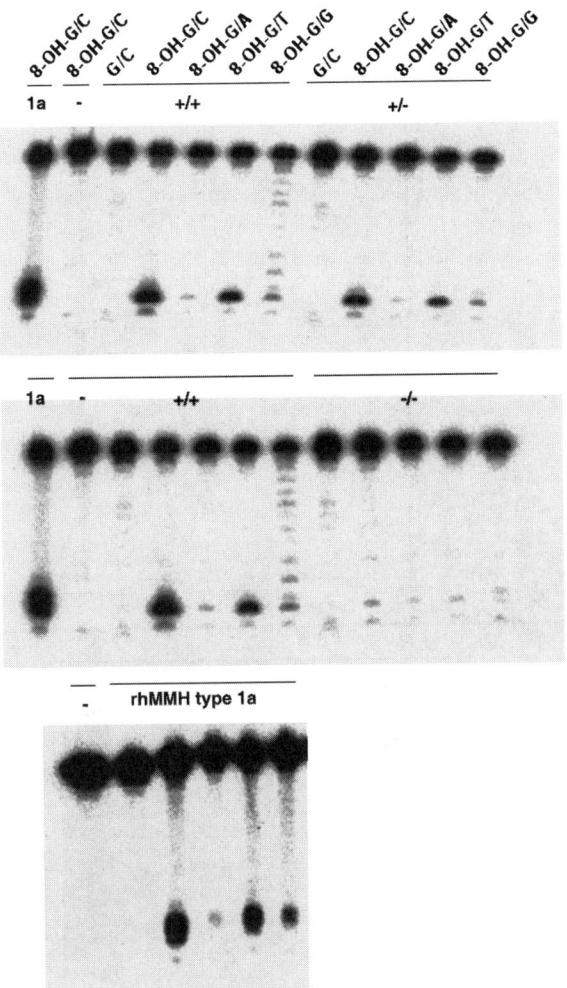

FIG. 6. Substrate specificity of AP lyase activity in crude liver extracts of *Mmh* wild-type, *Mmh* heterozygous mutant, and *Mmh* homozygous mutant mice. The 8-OH-G- or G-containing oligonucleotide was labeled and annealed with one of the four complementary oligonucleotides having C, A, T, or G opposite the 8-OH-G or G. The crude liver extracts, purified recombinant human MMH type 1a protein (rhMMH type 1a or 1a) or lysis buffer (−) were incubated with the substrates listed above and the nicked products were analyzed by 20% PAGE. (Reprinted with permission from Reference 31. Copyright 2000 National Academy of Sciences, U.S.A.)

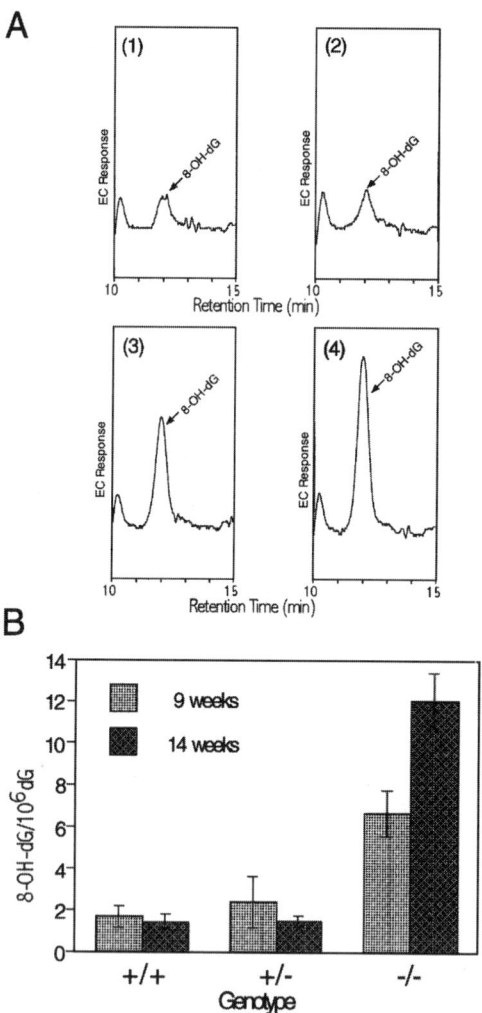

FIG. 7. Accumulation of 8-OH-G in DNA of *Mmh*-deficient mice. (A) Electrochemical chromatograms of 8-OH-dG in *Mmh* mutant mouse liver genomic DNAs. The HPLC patterns were traced by an electrochemical (EC) detector. Patterns are for 8-OH-dG level, in: 1, a wild-type mouse; 2, an *Mmh* heterozygous mutant mouse; 3, an *Mmh* homozygous mutant mouse at 9 weeks of age; and 4, an *Mmh* homozygous mutant mouse at 14 weeks of age. Arrows indicate 8-OH-dG peaks, the areas of those for homozygotes being markedly increased. (B) The measured amounts of 8-OH-G in liver tissues from *Mmh* wild-type, heterozygous mutant, and homozygous mutant mice at 9 and 14 weeks of age are summarized. Values are mean ± SD ($n = 3$). The 8-OH-G levels in *Mmh* homozygous mutant mice were significantly higher than those in *Mmh* wild-type mice at both 9 and 14 weeks of age at $P < 0.0001$ (Fisher's test). The 8-OH-G levels in 14-week-old homozygotes were significantly higher than those in 9-week-old *Mmh* homozygous mutant mice at $P < 0.0001$. (Reprinted with permission from Reference 31. Copyright 2000 National Academy of Sciences, U.S.A.)

TABLE I
Mutation Frequencies Observed in $gpt/Mmh^{+/+}$ (W) and $gpt/Mmh^{-/-}$ (O) Mouse Liver Cells

Mouse No.	Mmh	Cm^R colonies	6-TG^R and Cm^{Ra} colonies	Mutation frequency
1	W	1.3×10^6	4	3.1×10^{-6}
2	W	1.1×10^6	3	2.7×10^{-6}
3	W	1.2×10^6	5	4.2×10^{-6}
4	W	1.8×10^6	3	1.7×10^{-6}
5	W	3.9×10^5	4	10.0×10^{-6}
6	W	6.0×10^5	4	6.4×10^{-6}
Mean				$(4.7 \pm 3.1) \times 10^{-6}$
7	O	1.0×10^6	11	11.0×10^{-6}
8	O	1.7×10^6	8	4.7×10^{-6}
9	O	3.1×10^5	6	19.0×10^{-6}
10	O	8.6×10^5	7	8.1×10^{-6}
11	O	1.1×10^6	9	8.2×10^{-6}
12	O	3.9×10^5	5	13.0×10^{-6}
Mean				$(11.0 \pm 5.0) \times 10^{-6}$

[a] 6-TG^R and Cm^R stand for 6-TG-resistant and chloramphenicol-resistant, respectively. Numbers of colonies are for six independent wild-type mice 16–20 weeks old or six homozygous mice of the same age. On average, Mmh homozygotes show a 2.3-fold elevation of mutation frequency relative to the wild-type counterparts ($p < 0.03$, Fisher's test).

Reprinted with permission from Reference 31. Copyright 2000 National Academy of Sciences, U.S.A.

The mutations found were all single-base substitution. In *Mmh*-deficient mice, 26 out of a total of 46 base substitutions were GT or CA transversions; in the remainder, 15 were G to A transitions and five were T to A transversions. In the wild-type, 8 G to T or C to A transversions out of 23 single-base substitutions (35%) were found. This mutation spectrum implies that more than 70% of the increase in mutations in *Mmh* homozygous mice were accounted for by G to T or C to A transversions.

IV. Discussion and Future Direction

Two independent experiments described in the preceding sections clearly demonstrate that mammalian MMH is the major enzyme responsible for repair of 8-OH-G lesions. MMH accounts for 90–95% of the total nicking activity. *In vitro* study, by using human cell lines with antibody specific to type 1a isoform, indicated that type 1a is the most abundant form among several isoforms derived from the *Mmh* gene. This conclusion is reasonable since the amount of mitochondrial DNA is 1–2% of that of nuclear DNA, such that the amount of the isoforms localized in mitochondria is low even though they play a major role in repair of 8-OH-G in mitochondrial DNA. The residual activity, which is not accounted for by MMH, may be due to some other type of glycosylase/AP lyase. In fact, we previously isolated two different types of DNA repair enzymes for 8-OH-G in human cells, although their relative amounts are not known since they were isolated after extensive purification (33). S. Mitra and his colleagues recently reported the existence of OGG2 protein of which activity is less than 10% of that of OGG1 (34). There are several other reports showing that other enzymes have the capacity to cleave 8-OH-G containing DNA, such as N-methylpurine-DNA glycosylase (35), ribosomal protein S3 (36), nucleotide excision repair system for xeroderma pigmentosum, and mismatch repair MSH2 (37, 38). However, they are not likely to be major players for repair of 8-OH-G residue in DNA.

The increase of 8-OH-G in liver DNA of *Mmh* knockout mice is striking. The amounts of 8-OH-G increased linearly with the age of mice, reaching 7-fold that of wild-type mice at the age of 14 weeks. We are currently examining whether or not the level of 8-OH-G in DNA of knockout mice reaches a plateau after a certain age. T. Lindahl and his colleagues independently reported similar findings with their *Ogg1* knockout mice (39). The only difference between the two reports is that they (39) showed a higher amount of 8-OH-G in control wild-type mice, resulting in a similar extent of increase of 8-OH-G in the knockout mice. This discrepancy can be explained by an artificial increase of 8-OH-G during isolation of DNA. It is essential to isolate DNA in a way that minimizes artificial generation of 8-OH-G in order to get meaningful results. In general, the background level of 8-OH-G should be 1–2 per 10^6 guanine residues. Nakae *et al.* (40) first showed the usefulness of NaI extraction. Later, B. Ames and his colleagues endorsed this procedure with a detailed study (41). We currently used the NaI extraction method, as reported by H. Kasai and his colleagues (32): (1) DNA is extracted from the cell homogenate by WB kit (NaI method); (2) DNA is mixed with 10 mM sodium acetate and 1 mM EDTA; (3) nuclease P1 and acid phosphates are added, and the resultant mixture was incubated; (4) the DNAs are mixed with anion-exchange resin and filtrate to collect the supernatant; (5) the supernatant is injected onto HPLC for electrochemical analysis. It is likely that the background level of 8-OH-G in liver DNA of the knockout mice in our study (<28-OH-G per 10^6 G) is still overestimated owing to unavoidable artificial generation of 8-OH-G. Thus, it is possible that the relative increase of 8-OH-G in the knockout mice is much higher than the actual data.

It is rather unexpected to observe the moderate increase of mutation frequency in the knockout mice in spite of the large increase of 8-OH-G. Similar results have been reported by T. Lindahl and his colleagues (39). This discrepancy can be explained in several ways. It may be merely that fixation of mutation is limited because only a small fraction of adult liver cells are proliferating. It is also possible that mispairing of 8-OH-G with A during *in vivo* DNA replication is not as frequent as expected from the *in vitro* study using polymerase α (*9*). We previously reported that an 8-OH-G residue can also make a normal base pair with C. This conclusion was obtained by NMR analysis of double-stranded oligonucleotide containing 8-OH-G in one strand and cytosine at the paired position in the other strand (*42*). The other possibility is that the *MutY* product effectively removes adenine residues mispaired with 8-OH-G during DNA replication. It is also possible that the important lesion of 8-OH-G involved in induction of mutation is specifically repaired by other enzyme(s) rather than MMH. P. K. Cooper and her colleagues reported that transcription-coupled repair of 8-OH-G is carried out by XPG, TFIIH, and CSB, but not by MMH (*43*). Finally, no tumors have been found in the knockout mice thus far. These mice are kept under normal feeding conditions. In order to assess whether 8-OH-G generation is really involved in tumor formation, exposure of the mice to oxidative stress is an important experiment, which is currently underway.

Thirty percent of the increased mutations in *Mmh* homozygous, compared with wild-type mice, were not G to T or C to A transversions. In order to clarify whether other types of mutations observed in the mutant mice are due to 8-OH-G, it is necessary to perform a large number of mutational analyses, preferably with oxidative stress. However, it should be pointed out that we previously reported transfection experiments using the c-Ha-*ras* gene, which contains 8-OH-G at codons 12 or 61, and showed mutations other than G to T or C to A transversions (*44, 45*). The *ras* genes with 8-OH-G were transfected into NIH 3T3 cells and the induced mutations were analyzed by sequence determination of c-Ha-*ras* in the transformants. 8-OH-G at the first position of codons 12 and 61 induced mutation to T at the modified sites almost exclusively. On the other hand, DNA lesions at the second position of codon 12 induced a G to A transition to an appreciable extent, in addition to a G to T transversion (11 transitions versus 30 transversions). These results indicated that the type of mutation induced by 8-OH-G depends upon the site of the 8-OH-G residue in DNA and that other types of mutations are also elicited. It should be noted that Moriya also reported other types of mutation by 8-OH-G in a mammalian system (*46*). The question as to whether these observations are real in the *in vivo* situation should be answered by further studies using *Mmh* knockout mice.

References

1. M. Aida and S. Nishimura, *Mutat. Res.* **192**, 83–89 (1987).
2. H. Kasai, S. Nishimura, Y. Toriumi, A. Itai, and Y. Iitaka, *Bull. Chem. Soc. Jpn.* **60**, 3799–3800 (1987).
3. H. Kasai and S. Nishimura, *Nucleic Acids Res. Symp. Ser.* **No. 12**, s165–167 (1983).
4. H. Kasai and S. Nishimura, *Nucleic Acids Res.* **12**, 2137–2145 (1984).

5. R. A. Floyd, J. J. Watson, P. K. Wong, D. H. Altmiler, and R. C. Richard, *Free Radical Res. Commun.* **1,** 163–172 (1986).
6. H. Kasai, P. F. Crain, Y. Kuchino, S. Nishimura, A. Ootsuyama, and H. Tanooka, *Carcinogenesis* **7,** 1849–1851 (1986).
7. H. Kasai, S. Nishimura, Y. Kurokawa, and Y. Hayashi, *Carcinogenesis* **8,** 1959–1961 (1987).
8. Y. Kuchino, F. Mori, H. Kasai, H. Inoue, S. Iwai, K. Miura, E. Ohtsuka, and S. Nishimura, *Nature (London)* **327,** 77–79 (1987).
9. S. Shibutani, M. Takeshita, and A. P. Grollman, *Nature (London)* **349,** 431–434 (1991).
10. M. L. Wood, M. Dizdaroglu, E. Gajewski, and J. M. Essigmann, *Biochemistry* **29,** 7024–7032 (1990).
11. K. C. Cheng, D. S. Cahill, H. Kasai, S. Nishimura, and L. A. Loeb, *J. Biol. Chem.* **267,** 166–172 (1992).
12. M. Moriya, C. Cu, V. Bodepudi, F. Johnson, M. Takeshita, and A. P. Grollman, *Mutat. Res.* **254,** 281–288 (1991).
13. M. H. Chung, H. Kasai, D. S. Jones, H. Inoue, H. Ishikawa, E. Ohtsuka, and S. Nishimura, *Mutat. Res.* **254,** 1–12 (1991).
14. J. Tchou, H. Kasai, S. Shibutani, M.-H. Chung, J. Laval, A. P. Grollman, and S. Nishimura, *Proc. Natl. Acad. Sci. U.S.A.* **88,** 4690–4694 (1991).
15. M. L. Michaels, L. Pham, C. Cruz, and J. H. Miller, *Nucleic Acids Res.* **19,** 3629–3632 (1991).
16. M. L. Michaels and J. H. Miller, *J. Bacteriol.* **174,** 6321–6325 (1992).
17. H. Maki and M. Sekiguchi, *Nature (London)* **355,** 273–275 (1992).
18. M. M. Slupska, C. Baikalov, W. M. Luther, J.-H. Chaing, Y.-F. Wei, and J. H. Miller, *J. Bacteriol.* **178,** 3885–3892 (1996).
19. K. Sakumi, M. Furuichi, T. Tsuzuki, T. Kakuma, S. Kawabata, H. Maki, and M. Sekiguchi, *J. Biol. Chem.* **268,** 23524–23530 (1993).
20. P. A. van der Kemp, D. Thomas, R. Barbey, R. de Oliveira, and S. Boiteux, *Proc. Natl. Acad. Sci. U.S.A.* **93,** 5197–5202 (1996).
21. H. M. Nash, S. D. Bruner, O. D. Scharer, T. Kawate, T. A. Addona, E. Spooner, W. S. Lane, and G. L. Verdine, *Curr. Biol.* **6,** 968–980 (1996).
22. H. Aburatani, Y. Hippo, T. Ishida, R. Takashima, C. Matsuba, T. Kodama, T. Takao, A. Yasui, K. Yamamoto, M. Asano, K. Fukasawa, T. Yoshinari, H. Inoue, E. Ohtsuka, and S. Nishimura, *Cancer Res.* **57,** 2151–2156 (1997).
23. T. A. Rosenquist, D. I. Zharkov, and A. P. Grollman, *Proc. Natl. Acad. Sci. U.S.A.* **94,** 7429–7434 (1997).
24. T. Rol'dn-Arjona, Y.-F. Wei, K. C. Carter, A. Klungland, C. Anselmino, R.-P. Wang, M. Augustus, and T. Lindahl, *Proc. Natl. Acad. Sci. U.S.A.* **94,** 8016–8020 (1997).
25. J. P. Radicella, C. Dherin, C. Desmaze, M. S. Fox, and S. Boiteux, *Proc. Natl. Acad. Sci. U.S.A.* **94,** 8010–8015 (1997).
26. K. Arai, K. Morishita, K. Shinmura, T. Kohno, S.-R. Kim, T. Nohmi, M. Taniwaki, S. Ohwada, and J. Yokota, *Oncogene* **14,** 2857–2861 (1997).
27. R. Lu, H. M. Nash, and G. L. Verdine, *Curr. Biol.* **7,** 397–407 (1997).
28. M. Bjørås, L. Luna, B. Johnsen, E. Hoff, T. Haug, T. Rognes, and E. Seeberg, *EMBO J.* **16,** 6314–6322 (1997).
29. M. Takao, H. Aburatani, K. Kobayashi, and A. Yasui, *Nucleic Acids Res.* **26,** 2917–2922 (1998).
30. Y. Monden, T. Arai, M. Asano, E. Ohtsuka, H. Aburatani, and S. Nishimura, *Biochem. Biophys. Res. Commun.* **258,** 605–610 (2000).
31. O. Minowa, T. Arai, M. Hirano, Y. Monden, S. Nakai, M. Fukuda, M. Ito, H. Takano, Y. Hippou, H. Aburatani, K. Masumura, S. Nishimura, and T. Noda, *Proc. Natl. Acad. Sci. U.S.A.* **97,** 4156–4161 (2000).

32. S. Asami, T. Hirano, R. Yamaguchi, Y. Tomioka, H. Itoh, and H. Kasai, *Cancer Res.* **56,** 2546–2549 (1996).
33. T. Bessho, K. Tano, H. Kasai, E. Ohtsuka, and S. Nishimura, *J. Biol. Chem.* **268,** 19416–19421 (1993).
34. T. K. Hazra, T. Izumi, L. Maidt, R. A. Floyd, and S. Mitra, *Nucleic Acids Res.* **26,** 5116–5122 (1998).
35. T. Bessho, R. Roy, K. Yamamoto, H. Kasai, S. Nishimura, K. Tano, and S. Mitra, *Proc. Natl. Acad. Sci. U.S.A.* **90,** 8901–8904 (1993).
36. A. Yacoub, L. Augeri, M. R. Kelley, P. W. Doetsch, and W. A. Deutsch, *EMBO J.* **15,** 2306–2312 (1996).
37. J. T. Reardon, T. Bessho, H. C. Kung, P. H. Bolton, and A. Sancar, *Proc. Natl. Acad. Sci. U.S.A.* **94,** 9463–9468 (1997).
38. T. L. DeWeese, J. M. Shipman, N. A. Larrier, N. M. Buckley, L. R. Kidd, J. D. Groopman, R. G. Cutler, H. teRiele, and W. G. Nelson, *Proc. Natl. Acad. Sci. U.S.A.* **95,** 11915–11920 (1998).
39. A. Klungland, I. Rosewell, S. Hollenbach, E. Larsen, D. Daly, B. Epe, E. Seeberg, T. Lindahl, and D. E. Barnes, *Proc. Natl. Acad. Sci. U.S.A.* **96,** 13300–13305 (1999).
40. D. Nakae, Y. Mizumoto, E. Kobayashi, O. Noguchi, and Y. Konishi, *Cancer Lett.* **97,** 233–239 (1995).
41. H. J. Helbock, K. B. Beckman, M. K. Shigenaga, P. B. Walter, A. A. Woodall, H. C. Yeo, and B. N. Ames, *Proc. Natl. Acad. Sci. U.S.A.* **95,** 288–293 (1998).
42. Y. Oda, S. Uesugi, M. Ikehara, S. Nishimura, Y. Kawase, H. Ishikawa, H. Inoue, and E. Ohtsuka, *Nucleic Acids Res.* **19,** 1407–1412 (1991).
43. F. Le Page, E. E. Kwoh, A. Avrutskaya, A. Gentil, S. A. Leadon, A. Sarasin, and P. K. Cooper, *Cell (Cambridge, Mass.)* **101,** 159–171, 2000.
44. H. Kamiya, K. Miura, H. Ishikawa, H. Inoue, S. Nishimura, and E. Ohtsuka, *Cancer Res.* **52,** 3483–3485 (1992).
45. H. Kamiya, N. Murata-Kamiya, S. Koizume, H. Inoue, S. Nishimura, and E. Ohtsuka, *Carcinogenesis* **16,** 883–889 (1995).
46. M. Moriya, *Proc. Natl. Acad. Sci. U.S.A.* **90,** 1122–1126 (1993).

Session 3
Complexities of BER

MICHAEL WEINFELD

Cross Cancer Institute

As recently as five years ago the base excision repair (BER) pathway was depicted as a straightforward series of enzymatic steps: base removal, strand cleavage, removal of the deoxyribose phosphate, replacement of the missing base, and strand rejoining (*1*). The mammalian system seemed to closely mimic the BER system in *Escherichia coli*. The only complexity appeared to be in the number of DNA glycosylases that recognized different base lesions and the activities of abasic (AP) lyases vs. true AP endonucleases. In the intervening period, we have come to realize that the picture in mammalian cells is considerably more complicated than this. This chapter highlights several of the new developments, in particular the diversity of BER mechanisms, the regulatory control of BER, and the mutation frequency observed for the BER process for the removal of uracil.

It is now apparent that there are two distinct BER pathways operant in higher eukaryotes (*2*, *3*). Mechanistically, the two pathways diverge after the AP endonuclease incision. In one pathway, removal of deoxyribose phosphate is effected by DNA polymerase β, which subsequently replaces the missing base before strand ligation. In the other, the deoxyribose phosphate group is removed as part of a short oligonucleotide by FEN1 and the gap is filled in by DNA polymerase δ or ε. The latter pathway also requires the participation of proliferating cell nuclear antigen (PCNA) and replication factor C (RFC). Among the points that remain to be definitively answered at this stage are (1) the role of accessory proteins such as PARP, XRCC1, RFC, and PCNA; (2) the length of the gap generated by FEN1; and (3) whether each pathway is associated with particular glycosylases and DNA ligases. Yoshihiro Matsumoto has been able to reconstitute the PCNA-dependent pathway *in vitro*. He demonstrated that the repair of a plasmid substrate containing an AP site could be completed with six proteins: AP endonuclease, RFC, PCNA, FEN1, pol δ, and DNA ligase 1. Another accessory protein, replication protein A (RPA), was not required. With this system, Matsumoto has been able to establish the size of the repair patch. He has also examined potential interactions of PCNA with other BER proteins. Inspection of the binding of PCNA with p21 had revealed a 13-amino acid motif critical for the interaction. A search for identical or similar sequences in other BER proteins revealed a consensus motif present in four of the BER proteins presumed to interact with PCNA: namely, RFC, FEN1, pol δ, and DNA ligase I. Intriguingly, the consensus sequence was also found in two DNA glycosylases, UNG2 and MYH, leading to speculation that the AP sites generated by these two glycosylases are handled by the PCNA-dependent BER pathway.

Mammalian cells contain three distinct genes coding for DNA ligases: *LIG1*, *LIG3*, and *LIG4*. (DNA ligase III exists in two forms, α and β, as a result of alternative splicing at the C terminus.) Alan Tomkinson and colleagues present an overview of the role of each of these ligases in DNA repair. Cell lines deficient in either ligase I or III were found to be sensitive to alkylating agents; consequently, it was presumed that these ligases play a role in BER. Tomkinson details the genetic and biochemical evidence indicating that DNA ligase I participates in the PCNA-dependent pathway while DNA ligase IIIα is responsible for strand rejoining in the pol β pathway.

The development of a new sensitive immunoassay for thymine glycols (4) has provided the first opportunity to measure the induction and global removal of this lesion from the DNA of human cells irradiated with a clinical dose of ionizing radiation (2 Gy). The assay revealed that while removal is 80% complete within 4 hours, there appears to be a notable delay for the first 30–60 minutes. This delay disappears, however, if the cells are irradiated with 0.25 Gy 4 hours prior to the 2-Gy dose, suggesting that the BER response to radiation damage is inducible. Here, Weinfeld and colleagues discuss further data accrued with this assay that support the observations of Cooper *et al.* (5) and Klungland *et al.* (6) that XPG is required for optimal removal of thymine glycol. Preliminary data also indicate that quiescent cells and actively dividing cells have similar kinetics of removal of thymine glycol.

One of the major lesions acted upon by the BER pathway is uracil. This base is introduced in DNA by spontaneous deamination of cytosine or by incorporation of dUMP during DNA synthesis. Clearly, a uracil introduced opposite any base other than adenine would be a mutagenic event unless corrected, and accordingly, a diminished capacity for base excision repair of uracil-modified DNA leads to a mutator phenotype. One question addressed here by Dale Mosbaugh and his associates is the error frequency associated with proficient BER. Making use of an elegant *in vitro* reversion assay, they have closely examined the products of complete BER carried out by human cell-free extracts on a uracil specifically positioned in the M13mp2 plasmid.

REFERENCES

1. E. C. Friedberg, G. C. Walker, and W. Siede, "DNA Repair and Mutagenesis." ASM Press Washington, D.C., 1995.
2. Y. Matsumoto, K. Kim, and D. F. Bogenhagen, *Mol. Cell. Biol.* **14**, 6187–6197 (1994).

3. G. Frosina, P. Fortini, O. Rossi, F. Carrozzino, G. Raspaglio, L. S. Cox, D. P. Lane, A. Abbondandolo, and E. Dogliotti, *J. Biol. Chem.* **271,** 9573–9578 (1996).
4. X. C. Le, J. Z. Xing, J. Lee, S. A. Leadon, and M. Weinfeld, *Science* **280,** 1066–1069 (1998).
5. P. K. Cooper, T. Nouspikel, S. G. Clarkson, and S. A. Leadon, *Science* **275,** 990–993 (1997).
6. A. Klungland, M. Hoss, D. Gunz, A. Constantinou, S. G. Clarkson, P. W. Doetsch, P. H. Bolton, R. D. Wood, and T. Lindahl, *Mol. Cell* **3,** 33–42 (1999).

Molecular Mechanism of PCNA-Dependent Base Excision Repair

YOSHIHIRO MATSUMOTO

Department of Radiation Oncology
Fox Chase Cancer Center
Philadelphia, Pennsylvania 19111

I. Introduction.. 129
II. PCNA-Dependent AP Site Repair as an Alternative Pathway
   in Base Excision Repair.................................... 130
III. Interaction of PCNA with Various Proteins................. 132
IV. Replication-Coupled Base Excision Repair via Interaction
   between DNA-Glycosylases and PCNA.......................... 134
   References... 137

In higher eukaryotes, base excision repair can proceed by two alternative pathways: a DNA polymerase β-dependent pathway and a proliferating cell nuclear antigen (PCNA)-dependent pathway. Recently, we have reconstituted the PCNA-dependent AP site repair reaction with six purified human proteins: AP endonuclease, replication factor C (RFC), PCNA, flap endonuclease 1 (FEN1), DNA polymerase δ (pol δ), and DNA ligase I. In this reconstituted system, the number of nucleotides replaced during the repair reaction (patch size) was predominantly two nucleotides.

PCNA can directly interact with RFC, pol δ, FEN1 and DNA ligase I. These interactions are partly through a consensus motif, QXX(I/L/M)XX(F/H)(F/Y), found in each of the four proteins. PCNA functions as a molecular adaptor for recruiting these factors to the site of DNA repair.

Two DNA-N-glycosylases among those so far cloned from human, UNG2 and MYH, are found to have the same PCNA-binding motif. Major substrates of these enzymes, a uracil opposite an adenine for UNG2 and an adenine opposite an 8-oxoguanine for MYH, are formed during DNA replication. Therefore, UNG2 and MYH may serve for replication-coupled base excision repair through the direct interaction with PCNA in the replication machinery.   © 2001 Academic Press.

I. Introduction

Proliferating cell nuclear antigen (PCNA) is one of the auxiliary factors for DNA polymerases δ (pol δ) and ε (pol ε). PCNA forms a homotrimer in a torus structure that can be loaded on double-stranded DNA as a sliding clamp. It can directly bind to a number of proteins, and through these interactions PCNA participates in various DNA transactions, such as DNA replication, repair, and chromatin assembly, as well as cell cycle control.

The participation of PCNA in DNA repair was first demonstrated for an *in vitro* nucleotide excision repair reaction in which PCNA is one of the essential factors (*1*). Subsequently, a reconstituted system for AP site repair was developed with proteins from *Xenopus laevis* ovaries in which PCNA is required for one of the alternative pathways (*2*). It has also been demonstrated that anti-PCNA antibody can inhibit AP site repair in mammalian cell extracts (*3*). In addition to these *in vitro* data, there are several lines of data supporting the inference that the PCNA-dependent pathway of base excision repair is functional *in vivo*. The majority of lesions induced by ionizing radiation are modified bases, AP sites, and single-strand breaks, which are targets of base excision repair. It is observed that mammalian cells treated with ionizing radiation form tight complexes of PCNA and nuclear DNA, suggesting participation of PCNA in repair of radiation-induced damages (*4–6*). A budding yeast strain that is partially deficient in pol δ, one of the PCNA-dependent DNA polymerases, is hypersensitive to methyl-methanesulfonate (MMS) (*7*). Furthermore, some mutations in the PCNA gene make yeast cells hypersensitive to MMS without apparent perturbation of growth rate (*8*). It should be noted, however, that a pol β-like enzyme, POL4, is not required for base excision repair in budding yeast. Accordingly, the PCNA-dependent pathway may be the only mechanism of base excision repair in yeast.

Recently, reconstituted systems with purified human proteins have been established to repair AP sites *in vitro* (*9, 10*). A separate line of studies has demonstrated that a nuclear form of human uracil DNA-*N*-glycosylase (UNG2) can directly bind to PCNA and also to RPA, and that UNG2 appears to colocalize with these proteins in replication foci (*11*). In this chapter, I summarize the functions of PCNA in base excision repair as well as present a hypothetical model of a PCNA-dependent repair.

II. PCNA-Dependent AP Site Repair as an Alternative Pathway in Base Excision Repair

In vitro systems of the PCNA-dependent base excision repair have been developed using circular DNA containing a single AP site at a defined position as a substrate. Since PCNA is a sliding clamp that easily falls away from linear DNA, it is critical to use DNA substrates that do not have free ends, such as circular molecules, for analysis of the PCNA-dependent reactions *in vitro* (*12*). To identify all the essential factors for the PCNA-dependent base excision repair, we attempted to reconstitute AP site repair with purified human proteins. As a result, the repair of AP sites could be completed with six protein factors: AP endonuclease, replication factor C (RFC), PCNA, flap endonuclease 1 (FEN1), pol δ, and DNA ligase I (*10*). When any one of the six proteins was omitted, the AP site was not successfully repaired. Another group obtained similar results, which also demonstrated that pol ε was able to replace pol δ without a significant difference in repair efficiency (*9*). Replication protein A (RPA; also called HSSB) and a Ku 70/86 complex were also tested for their effects on this AP site repair. RPA is one of the essential factors for DNA replication and also for nucleotide excision repair. The Ku 70/86 complex is involved in the nonhomologous end-joining repair of double-strand breaks and also has an affinity to single-strand breaks of DNA. It turned out, however, that neither RPA nor the Ku complex enhanced AP site repair in the PCNA-dependent system (*10, 13*).

Based on these results, the mechanism of AP site repair by the PCNA-dependent pathway is presented as shown in Fig. 1. The AP site is first recognized and cleaved at its 5' side by AP endonuclease. The resultant 3' terminus

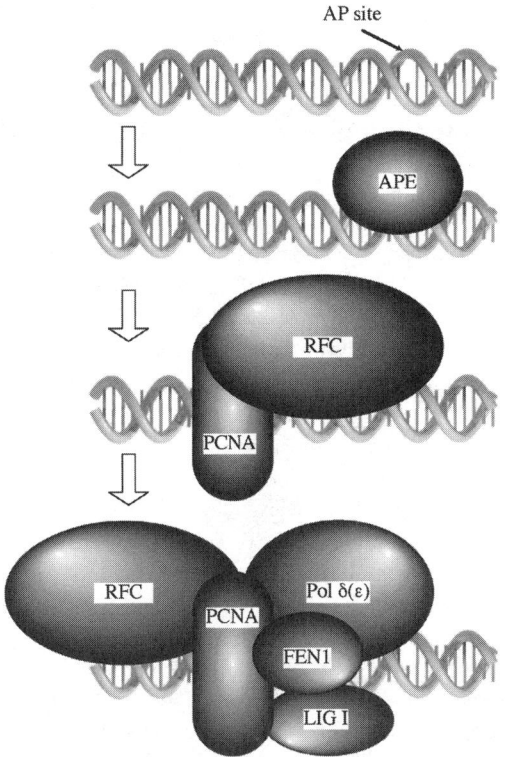

FIG. 1. A schematic model of protein assembly in PCNA-dependent AP site repair.

may serve as the target for recruiting RFC, which then assembles PCNA on DNA. Subsequently, pol δ, FEN1, and DNA ligase I are recruited to the site of repair and carry out DNA synthesis, excision of the AP site and the adjacent nucleotide(s), and ligation. As discussed in the next section, pol δ, FEN1, and DNA ligase I may form a complex with PCNA at this stage.

In base excision repair, the number of nucleotides that are replaced during the repair reaction (so-called repair patch size) is generally small. In the pol β-dependent pathway, the repair reaction predominantly completes with replacement of only one nucleotide, because pol β can remove only a dRP group to make a single-nucleotide gap. In contrast, the PCNA-dependent pathway employs FEN1 to remove a dRP group and its 3′-adjacent nucleotide(s) together. Consequently, this pathway must replace at least two nucleotides. To directly determine the major patch size of the PCNA-dependent repair, we added α-thio-dNTPs instead of regular dNTPs to the repair reaction. α-Thionucleotides incorporated into the repaired DNA were resistant to exonuclease III digestion, and provided information about the patch size. In the PCNA-dependent repair with six purified factors, a majority of repaired products had a repair patch of just two nucleotides.

III. Interaction of PCNA with Various Proteins

In recent years, a large number of proteins have been identified as physically interactive with PCNA. The list ranges from proteins directly involved in DNA transactions to those involved in cell cycle control. These proteins include p21(Cip1/Waf1) (*14*), Gadd45 (*15*), cyclin D (*16*), MSH2, MLH1 (*17*), MyD118 (*18*), DNA-(cytosine-5) methyltransferase (MCMT) (*19*), and XPG (*20*) in addition to RFC, pol δ, FEN1, and DNA ligase I as described above.

p21 is the protein most extensively characterized for its interaction with PCNA. Studies with deletion mutants and synthetic peptides derived from p21 narrowed down the domain required for interaction with PCNA to the 22 residues near the C terminus (*21–24*). Analysis of alanine-scanning mutations of p21 identified 143-RQXXMTXFYXXXR-155 (in the human protein; X, any residue) as critical amino acids for p21-PCNA interaction (*21*). Furthermore, the crystal structure of human PCNA complexed with a peptide derived from the C terminus of p21 unraveled detailed interactions between the residues of p21 and PCNA (*25*). In this structure, the p21-derived peptide interacts with a hydrophobic pocket and with the interdomain connector loop, both of which reside on the outer surface of the torus-shaped PCNA homotrimer. Accordingly, while this peptide does not directly interact either with the interfaces for homotrimer formation or with the inside hole for DNA interaction, one PCNA homotrimer molecule has three sites for binding to p21. Comparison of the

p21	141-KRRQTSMTDFYHSKRRLIFSKRKP-c
FEN1	334-GSTQGRLDDFFKVTGSLSSAKRKE-X$_{23}$-c
DNA ligase I	1-MQRSIMSFFHPKKEGKAKKPEK-X$_{897}$-c
XPG	987-QQTQLRIDSFFRLAQQEKEDAKRI-X$_{176}$-c
MCMT	40-STRQTTITSHFAKGPAKRKPQEES-X$_{1432}$-c
RFC p140	1-MDIRKFFGVIPSGKKLVSET-X$_{1128}$-c
Pol δ 3rd subunit	453-ANRQVSITGFFQRK-c
MSH6	1-MSRQSTLYSFFPKSPALSDANKAS-X$_{1336}$-c
UNG2	1-MIGQKTLYSFFSPSPARKRHAPSP-X$_{279}$-c
MYH	509-RMGQQVLDNFFRSHISTDAHSLNSAAQ-c

FIG. 2. A list of human proteins carrying the conserved PCNA-binding motif. The number attached with the left-end residue of each amino acid sequence indicates its position in the whole sequence. "c" indicates the C-terminus. The number of amino acids to the C terminus is also indicated. The most conserved four residues are marked with gray boxes.

sequence critical for p21-PCNA interaction with the amino acid sequences of several proteins that interact with PCNA uncovered a conserved motif for interaction with PCNA: QXX(I/L/M)XX(F/H)(F/Y) (26). Subsequent searches for this binding motif in the whole list of the PCNA-binding proteins revealed several proteins that may interact with PCNA in the same manner as p21 (Fig. 2). In addition, a few previously unlisted proteins were also identified as potentially interactive with PCNA. It has been demonstrated for some of these proteins that this conserved motif is essential for their binding to PCNA: FEN1 (20), XPG (20), MCMT (19), DNA ligase I (27), and UNG2 (11).

The effect of the interactions of PCNA with FEN1 and DNA ligase I on *in vitro* AP site repair has been evaluated by comparing wild-type proteins and mutant proteins in which the conserved motif was disrupted (10). For example, the F343A/F344A mutant of FEN1 could not bind to PCNA, but retained the same catalytic activity as the wild-type protein. When this mutant protein replaced the wild-type FEN1 protein in the reconstituted system, both the efficiencies of AP site excision and complete repair were significantly decreased. Likewise, the F8A/F9A mutation of DNA ligase I abolished the PCNA-binding ability, but did not affect its intrinsic catalytic activity. Furthermore, the efficiency of AP site repair in the PCNA-dependent pathway was significantly lower when the mutant DNA ligase I was used instead of the wild-type ligase in the reconstituted repair assay. However, in both cases of FEN1 and DNA ligase I, the inefficiency of the mutations could be overcome by the addition of excess amounts of those proteins. In addition, a majority of the repaired products from the reactions with either the mutant FEN1 or the mutant DNA ligase I had a patch size of two nucleotides. These results indicate that the interactions of FEN1 and DNA ligase I with PCNA are not essential for, but facilitate repair *in vitro*.

A previous study suggested that the enzymes for DNA synthesis, excision, and ligation for AP site repair may form a complex at the lesion (28). It is well established that pol δ, FEN1, and DNA ligase I can bind to PCNA. Furthermore, in addition to FEN1 and DNA ligase I, one of the subunits of pol δ has the conserved PCNA-binding motif. Therefore, it is possible that the three binding sites for this motif of the PCNA homotrimer may be occupied by pol δ, FEN1, and DNA ligase I, leading to formation of a PCNA-dependent repair complex. Such a complex formation may be critical for efficient repair by the PCNA-dependent pathway.

IV. Replication-Coupled Base Excision Repair via Interaction between DNA-Glycosylases and PCNA

In base excision repair, a wide variety of modified or mispaired bases are removed by different DNA-N-glycosylases, each of which recognizes a specific base or a specific group of bases. A large number of DNA-N-glycosylases have been cloned from a variety of organisms ranging from bacteriophages to humans. In the case of human enzymes, seven DNA-N-glycosylases have been cloned (Table I), although some are in several variant forms, resulting from different transcriptional initiation and/or alternative splicing. Interestingly, two enzymes among them are found to carry the PCNA-binding motif: UNG2 and MYH. UNG2 is a nuclear form of the major uracil DNA-N-glycosylase, and it has the PCNA-binding motif at its N-terminal region (29, 30). In contrast, the alternative form, UNG1, is a mitochondrial form, and it has a different N-terminal region which contains a mitochondrial targeting signal but not the PCNA-binding motif. MYH is a homolog of *Escherichia coli* MutY, and its human protein is found

TABLE I
HUMAN DNA-N-GLYCOSYLASES

Glycosylase	Major substrates	PCNA-binding motif
UNG2	Uracil on ssDNA and on dsDNA	N terminus
TDG	Uracil and thymine opposite guanine	—
MED1/MBD4	Thymine and uracil opposite guanine	—
MPG	N-Alkylpurine	—
NTH	Thymine glycol	—
OGG1	8-Oxoguanine opposite cytosine	—
MYH	Adenine opposite 8-oxoguanine	C terminus

in multiple forms, some of which have a mitochondrial targeting signal at their N termini and some of which lack it and seem to locate in nuclei (*31, 32*). All the forms of MYH appear to carry the PCNA-binding motif near the C terminus. The other five human DNA-*N*-glycosylases do not appear to carry the consensus motif.

Two possible mechanisms can be proposed by which the interaction of UNG2 and MYH with PCNA can facilitate specific base excision repair reactions. The first possibility is that UNG2 and MYH may preferentially recruit the PCNA-dependent pathway for the AP site repair following the removal of the impaired base. The second possibility is that UNG2 and MYH may associate with the replication machinery through binding to PCNA. A recent study of UNG2 demonstrates that it indeed binds to PCNA through the conserved motif and also to RPA, a single-strand DNA-binding protein essential for replication. UNG2 also colocalizes with RPA and PCNA in replication foci, thus supporting the second proposed mechanism (*11*), although not ruling out the first mechanism.

A uracil in DNA is a major substrate of UNG2. Such a lesion is generated by one of two distinct mechanisms: (1) incorporation of dUMP instead of dTMP during DNA replication; (2) deamination of cytosine in DNA. In the first mechanism, the misincorporated uracil forms a base pair with an adenine, and the frequency of this misincorporation seems to be related to the size of the intracellular dUTP pool (*33*). dUTP is a normal metabolite in dTTP biosynthesis *in vivo*, and its cellular level may be physiologically regulated. In the second mechanism, the uracil converted from a cytosine forms a mispair with a guanine. It is estimated that 100–500 uracil residues are generated per mammalian genome per day in this way. UNG2 can remove uracils both from the U:A pair and the U:G pair. Two other DNA glycosylases, G/T(U)-mismatch-specific thymine DNA glycosylase (TDG) and MED1 (also known as MBD4), can remove a uracil only from the U:G pair, but not from the U:A pair (*34, 35, 35a*). This suggests that the uracils incorporated during DNA replication will be processed exclusively by UNG2, while the uracils generated by the cytosine deamination irrespective of replication can be removed by TDG and MED1 as well as UNG2. Lack of the conserved PCNA-binding motif in TDG and MED1 will not matter for their function in U:G site repair. In contrast, the interaction of UNG2 with PCNA through the binding motif could serve for coupling U:A site repair with DNA replication (Fig. 3, right side).

A similar model can be proposed for the repair mechanism with MYH (Fig. 3, left side). One of the major substrates of MYH is an adenine opposite an 8-oxoguanine. This mispairing is formed by incorporation of dAMP at the opposite position of the unrepaired 8-oxoguanine during DNA replication. It has been reported that replicative DNA polymerases, pols α and δ, preferentially incorporate dAMP opposite 8-oxodG whereas pol β incorporates dCMP more frequently than dAMP (*36*). The interaction of MYH with PCNA may facilitate the repair of the mispair generated by DNA replication. Another

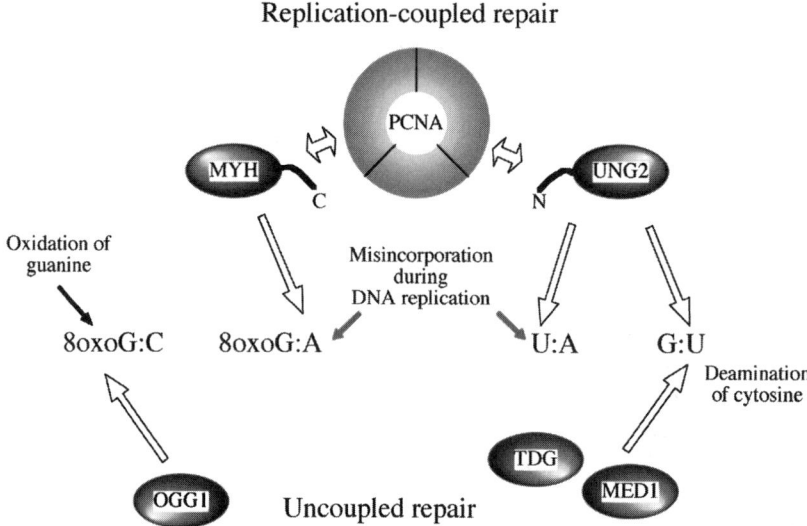

FIG. 3. A schematic model of replication-coupled base excision repair with UNG2 and MYH.

DNA-N-glycosylase serving for repair of oxidative damage is OGG1, a functional homolog of *E. coli* FPG. OGG1 removes an 8-oxoguanine opposite a cytosine, but does not remove the 8-oxoguanine base-paired with an adenine. The 8-oxoguanine opposite the cytosine results mainly from direct oxidation of the guanine in double-stranded DNA. Unlike MYH, OGG1 does not carry the PCNA-binding motif, nor does it need the interaction with PCNA for repair of 8-oxoG:C.

At this point, no experimental data are provided to indicate that MYH indeed interacts with PCNA through the conserved motif or that MYH is associated with the replication machinery. Nevertheless, the analogy of the two cases, repair of uracil DNA and repair of oxidative damage, is highly remarkable. UNG2 and MYH, which serve for repair of the replication-caused errors, carry the PCNA-binding motif, whereas TDG, MED1, and OGG1, which do not have the same binding motif, serve for removal of the damaged bases whose generation is not related to DNA replication. Thus, it is reasonable to propose that both UNG2 and MYH are involved in replication-coupled repair through their interaction with PCNA.

A few problems remain to be resolved in this model. One is how many proteins can interact with PCNA in the replication machinery. One PCNA homotrimer molecule loaded on DNA has three sites available for interaction through the conserved motif. However, there are four proteins involved in DNA replication that carry the PCNA-binding motif: RFC (the large subunit), pol δ

(the third subunit), FEN1, and DNA ligase I. Taking into account MCMT, another protein potentially associated with DNA replication, and UNG2 and MYH, a problem arises in explaining how seven proteins interact with PCNA in the replication machinery. It is possible that some proteins may bind to PCNA temporarily and may exchange their binding site with other proteins. This mechanism has yet to be elucidated.

The second problem is how the AP site is repaired following the removal of the incorrect base by UNG2 or MYH. In the UNG2-initiated repair reaction, the subsequent repair of the AP site seems straightforward. Either the PCNA-dependent pathway or the pol β-dependent pathway can complete this repair reaction by inserting dTMP. However, in the MYH-initiated repair, the subsequent AP site repair should not be catalyzed by pol δ in the PCNA-dependent pathway, since pol δ preferentially incorporates dAMP opposite 8-oxodG, leading to regeneration of the mispair. Therefore, the pol β-dependent pathway that inserts dCMP opposite the 8-oxodG should be employed in this case. It has not yet been determined what nucleotide is inserted opposite the 8-oxodG by other DNA polymerases, such as pol ε, and recently isolated translesion DNA polymerases. One of these DNA polymerases may be employed for the MYH-initiated repair, if it can incorporate dCMP.

In summary, PCNA works as a molecular adaptor in some types of base excision repair, and serves in two phases: (1) to couple the damage-scanning by UNG2 and MYH with DNA replication; (2) to recruit pol δ/ε, FEN1, and DNA ligase I to the repair site and to facilitate the coordinated reactions catalyzed by these enzymes. In both cases, the specific interactions of the repair factors with PCNA help to hold them closely at the target site for repair.

Acknowledgments

I thank A. E. McKenna for critical reading of the manuscript. This work was supported by NIH Grants CA06927, CA63154, and an appropriation from the Commonwealth of Pennsylvania.

References

1. M. K. K. Shivji, M. K. Kenny, and R. D. Wood, *Cell (Cambridge, Mass.)* **69,** 367–374 (1992).
2. Y. Matsumoto, K. Kim, and D. F. Bogenhagen, *Mol. Cell. Biol.* **14,** 6187–6197 (1994).
3. G. Frosina, P. Fortini, O. Rossi, F. Carrozzino, G. Raspaglio, L. S. Cox, D. P. Lane, A. Abbondandolo, and E. Dogliotti, *J. Biol. Chem.* **271,** 9573–9578 (1996).
4. M. Miura, M. Domon, T. Sasaki, and Y. Takasaki, *J. Cell Physiol.* **150,** 370–376 (1992).
5. L. A. Stivala, E. Prosperi, R. Rossi, and L. Bianchi, *Carcinogenesis* **14,** 2569–2573 (1993).
6. M. Miura, T. Sasaki, and Y. Takasaki, *Radiat. Res.* **145,** 75–80 (1996).
7. A. Blank, B. Kim, and L. A. Loeb, *Proc. Natl. Acad. Sci. U.S.A.* **91,** 9047–9051 (1994).

8. R. Ayyagari, K. J. Impellizzeri, B. L. Yoder, S. L. Gary, and P. M. Burgers, *Mol. Cell. Biol.* **15**, 4420–4429 (1995).
9. B. Pascucci, M. Stucki, Z. O. Jonsson, E. Dogliotti, and U. Hubscher, *J. Biol. Chem.* **274**, 33696–33702 (1999).
10. Y. Matsumoto, K. Kim, J. Hurwitz, R. Gary, D. S. Levin, A. E. Tomkinson, and M. S. Park, *J. Biol. Chem.* **274**, 33703–33708 (1999).
11. M. Otterlei, E. Warbrick, T. A. Nagelhus, T. Haug, G. Slupphaug, M. Akbari, P. A. Aas, K. Steinsbekk, O. Bakke, and H. E. Krokan, *EMBO J.* **18**, 3834–3844 (1999).
12. S. Biade, R. W. Sobol, S. H. Wilson, and Y. Matsumoto, *J. Biol. Chem.* **273**, 898–902 (1998).
13. M. Stucki, B. Pascucci, E. Parlanti, P. Fortini, S. H. Wilson, U. Hubscher, and E. Dogliotti, *Oncogene* **17**, 835–843 (1998).
14. S. Waga, G. J. Hannon, D. Beach, and B. Stillman, *Nature (London)* **369**, 574–578 (1994).
15. M. L. Smith, I. T. Chen, Q. Zhan, I. Bae, C. Y. Chen, T. M. Gilmer, M. B. Kastan, P. M. O'Connor, and A. J. Fornace Jr., *Science* **266**, 1376–1380 (1994).
16. S. Matsuoka, M. Yamaguchi, and A. Matsukage, *J. Biol. Chem.* **269**, 11030–11036 (1994).
17. A. Umar, A. B. Buermeyer, J. A. Simon, D. C. Thomas, A. B. Clark, R. M. Liskay, and T. A. Kunkel, *Cell (Cambridge, Mass.)* **87**, 65–73 (1996).
18. M. Vairapandi, A. G. Balliet, A. J. Fornace, Jr., B. Hoffman, and D. A. Liebermann, *Oncogene* **12**, 2579–2594 (1996).
19. L. S. Chuang, H. I. Ian, T. W. Koh, H. H. Ng, G. Xu, and B. F. Li, *Science* **277**, 1996–2000 (1997).
20. R. Gary, D. L. Ludwig, H. L. Cornelius, M. A. MacInnes, and M. S. Park, *J. Biol. Chem.* **272**, 24522–24529 (1997).
21. M. Nakanishi, R. S. Robetorye, O. M. Pereira-Smith, and J. R. Smith, *J. Biol. Chem.* **270**, 17060–17063 (1995).
22. Z. Q. Pan, J. T. Reardon, L. Li, H. Flores-Rozas, R. Legerski, A. Sancar, and J. Hurwitz, *J. Biol. Chem.* **270**, 22008–22016 (1995).
23. E. Warbrick, D. P. Lane, D. M. Glover, and L. S. Cox, *Curr. Biol.* **5**, 275–282 (1995).
24. J. Chen, R. Peters, P. Saha, P. Lee, A. Theodoras, M. Pagano, G. Wagner, and A. Dutta, *Nucleic Acids Res.* **24**, 1727–1733 (1996).
25. J. M. Gulbis, Z. Kelman, J. Hurwitz, M. O'Donnell, and J. Kuriyan, *Cell (Cambridge, Mass.)* **87**, 297–306 (1996).
26. E. Warbrick, *Bioessays* **20**, 195–199 (1998).
27. A. Montecucco, R. Rossi, D. S. Levin, R. Gary, M. S. Park, T. A. Motycka, G. Ciarrocchi, A. Villa, G. Biamonti, and A. E. Tomkinson, *EMBO J.* **17**, 3786–3795 (1998).
28. Y. Matsumoto and D. F. Bogenhagen, *Mol. Cell. Biol.* **11**, 4441–4447 (1991).
29. M. Otterlei, T. Haug, T. A. Nagelhus, G. Slupphaug, T. Lindmo, and H. E. Krokan, *Nucleic Acids Res.* **26**, 4611–4617 (1998).
30. E. Warbrick, W. Heatherington, D. P. Lane, and D. M. Glover, *Nucleic Acids Res.* **26**, 3925–3932 (1998).
31. M. Takao, Q. M. Zhang, S. Yonei, and A. Yasui, *Nucleic Acids Res.* **27**, 3638–3644 (1999).
32. T. Ohtsubo, K. Nishioka, Y. Imaiso, S. Iwai, H. Shimokawa, H. Oda, T. Fujiwara, and Y. Nakabeppu, *Nucleic Acids Res.* **28**, 1355–1364 (2000).
33. N. V. Tomilin and O. N. Aprelikova, *Int. Rev. Cytol.* **114**, 125–179 (1989).
34. P. Neddermann and J. Jiricny, *Proc. Natl. Acad. Sci. U.S.A.* **91**, 1642–1646 (1994).
35. B. Hendrich, U. Hardeland, H. H. Ng, J. Jiricny, and A. Bird, *Nature (London)* **401**, 301–304 (1999).
35a. F. Petronzelli, A. Riccio, G. D. Markham, S. H. Seeholzer, J. Stoerker, M. Genuardi, A. T. Yeung, Y. Matsumoto, and A. Bellacosa, *J. Biol. Chem.* **275**, 32422–32429 (2000).
36. S. Shibutani, M. Takeshita, and A. P. Grollman, *Nature (London)* **349**, 431–442 (1991).

Factors Influencing the Removal of Thymine Glycol from DNA in γ-Irradiated Human Cells

MICHAEL WEINFELD,*
JAMES Z. XING,[†] JANE LEE,*
STEVEN A. LEADON,[‡]
PRISCILLA K. COOPER,**
AND X. CHRIS LE[†]

*Experimental Oncology
Cross Cancer Institute
Edmonton, Alberta T6G 1Z2, Canada
[†]Department of Public Health Sciences
University of Alberta
Edmonton, Alberta T6G 2G3, Canada
[‡]Department of Radiation Oncology
University of North Carolina School
of Medicine
Chapel Hill, North Carolina
**Department of Radiation Biology and
DNA Repair
Lawrence Berkeley National Laboratory
Berkeley, California 94720

I. Assays for Oxidative DNA Damage 140
II. Induction of Thymine Glycol: Comparison of CE-LIF to Other Assays . . . 144
III. Cellular Removal of Thymine Glycol from DNA 145
References 148

The toxic and mutagenic effects of ionizing radiation are believed to be caused by damage to cellular DNA. We have made use of a novel immunoassay for thymine glycol to examine the removal of this lesion from the DNA of irradiated human cells. Because of the sensitivity of the assay, we have been able to keep the radiation doses at or below the standard clinical dose of 2 Gy. Our initial observations indicated that although removal of thymine glycol is >80% complete by 4 h post-irradiation with 2 Gy, there is a lag of 30–60 min before

---

Abbreviations: BrdU, bromodeoxyuridine; CE, capillary electrophoresis; CSB, Cockayne syndrome complementation group B; GC-MS, gas chromatography/mass spectrometry; LIF, laser-induced fluorescence detection; PMT, photomultiplier tube; Tg, thymine glycol; XPG, Xeroderma pigmentosum complementation group G.

repair commences. However, if cells are irradiated with 0.25 Gy 4 h prior to the 2-Gy dose, removal of the thymine glycols commences immediately after the second irradiation, suggesting that repair of thymine glycol is inducible. Our current studies are directed at two aspects of the repair process, (1) factors involved in the repair process leading up to and including glycosylase-mediated removal of thymine glycol and (2) the control of the inducible response. We have observed that mutation of the XPG gene drastically reduced the level and rate of global removal of thymine glycol (induced by 2-Gy irradiation), and there was no evidence for an inducible response. Similar results were seen with a Cockayne syndrome B (CSB) cell line. We have also examined repair in quiescent and phytohemagglutinin-stimulated human lymphocytes. Both show similar kinetics for the rate of removal of thymine glycol under induced and noninduced conditions. © 2001 Academic Press.

I. Assays for Oxidative DNA Damage

Cellular DNA is continually subject to damage. As a result, cells possess systems to repair the damage. Failure to do so is presumed to lead to mutation or cell death, depending on the nature of the DNA lesions. Sources of DNA damage include endogenous causes, such as spontaneous depurination and oxidative free radicals, and external agents, such as ionizing radiation and alkylating agents. In addition to natural exposure, these external agents are commonly used in the treatment of cancer. Ionizing radiation causes many different types of DNA lesions including single-strand breaks, double-strand breaks, DNA-protein crosslinks, and a plethora of modified base and deoxyribose residues (1), such as thymine glycol (Fig. 1) and 8-oxoguanine.

Numerous assays have been devised for the measurement of DNA lesions (see Refs. 2, 3). The most widely used for radiation-induced base damage include gas chromatography/mass spectrometry (GC-MS), HPLC-electrochemical detection, ^{32}P-postlabeling, PCR-based assays, and strand break-based assays, such as the comet assay, and immunoassays. Each assay has its advantages and disadvantages. Some, such as ^{32}P-postlabeling and GC-MS, can measure several lesions simultaneously, but the protocols are protracted, requiring either enzymatic or chemical digestion to the base or nucleotide level. This represents a particular problem when measuring oxidative DNA damage because

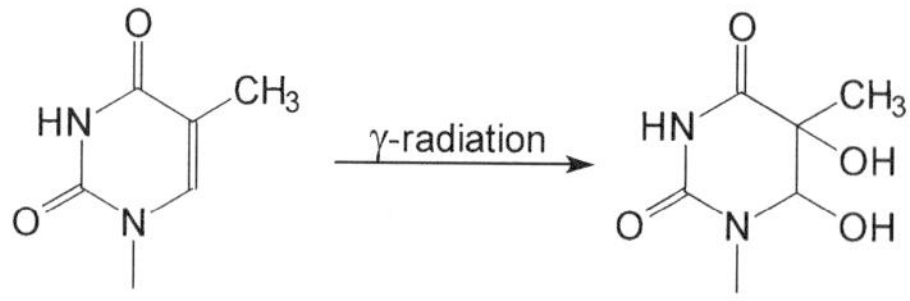

FIG. 1. Structure of thymine glycol.

of the increased possibility of introducing artifacts. A second notable problem encountered in the measurement of radiation-induced DNA damage is that of sensitivity. Until recently, it has rarely been possible to measure DNA damage induced in cells by radiation doses lower than 10 Gy, yet the standard clinical dose is 2 Gy per fraction. Since a major aim of the research in this field is to understand cellular responses to clinical treatment, it has been necessary to develop more sensitive techniques. This chapter describes a recently developed immunoassay (4) that takes advantage of the sensitivity afforded by capillary electrophoresis coupled with laser-induced fluorescence detection (CE-LIF). The assay, which has an absolute detection in the zeptomole (10^{-21} mol) range, has been used to measure the formation of thymine glycol in 2-Gy-irradiated cells and its subsequent removal.

Figure 2a shows the basic principle of the immunoassay. The lesion is recognized by a specific antibody to thymine glycols (5), which in turn is recognized by a fluorescently labeled secondary antibody. Thus, there are three fluorescent species that need to be separated and quantified: (1) the secondary antibody alone, which is used in excess; (2) the complex of the secondary antibody with unbound primary antibody, which is also in excess; and (3) the complex of the two antibodies bound to the damaged DNA. The separation is accomplished by capillary electrophoresis and the samples are detected by laser-induced fluorescence (shown schematically in Fig. 2b). A full description of the system is published elsewhere (6), so only a summary of the technique is provided here. CE separation is typically carried out in a 40-cm capillary (20 μm i.d.) at an electric field of 500 V cm^{-1}. The high-voltage power supply that drives the electrophoresis is directed by a computer that controls the separation voltage, injection voltage, and injection time. The high-voltage injection end of the capillary, together with a platinum electrode, is inserted into the sample solution (when injecting sample) or running buffer (when performing separation). The other end of the capillary is grounded and inserted into the sheath flow cuvette detection cell. For fluorescence detection, we use a 1.0-mW green helium–neon laser with a wavelength of 543.5 nm as the excitation source. The light beam is focused with a microscope objective onto a sheath flow cuvette, which acts as the fluorescence detection cell (7), and fluorescence is detected by a photomultiplier tube (PMT) after passage through a microscope objective, a bandpass filter, and a 1-mm pinhole. The output from the PMT is digitized by a data acquisition board in the computer.

To calibrate the system, we prepare a set of standard samples of plasmid DNA, containing a set number of bromodeoxyuridine (BrdU) residues. This DNA is incubated with monoclonal antibody to BrdU and tetramethylrhodamine-conjugated goat anti-mouse IgG (Calbiochem). Both antibodies are used in excess over antigen. Approximately 1 nl of this mixture is injected into the capillary. The running buffer is 1 × TBE (pH 8.3). (This change in pH from our previously published buffer, pH 10.5, has led to more consistent results.)

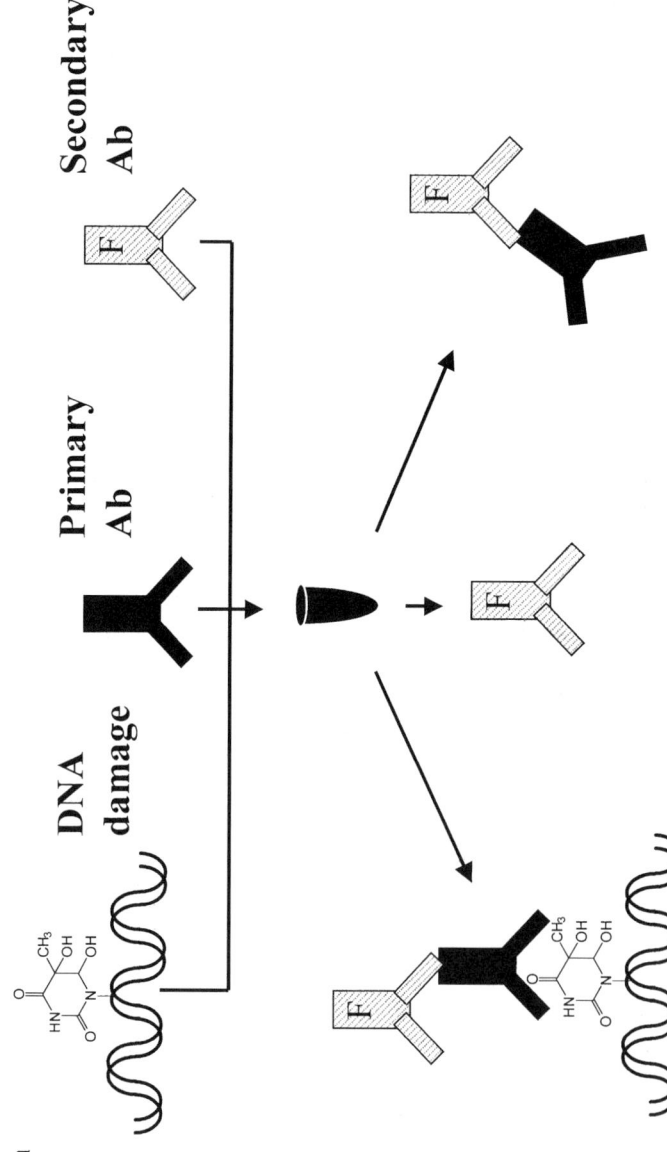

FIG. 2. Basic outline of the capillary-electrophoresis immunoassay. (a) Interaction of the primary and fluorescently tagged (F) secondary antibodies with each other and the damaged DNA. (b) The capillary electrophoresis with laser-induced fluorescence detector instrumentation.

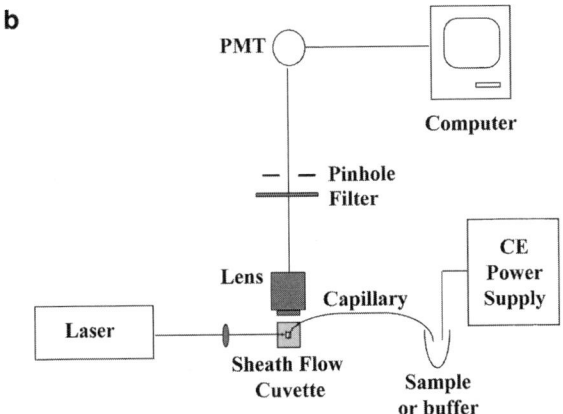

FIG. 2. (*continued*).

It is important to realize that the capillary contains only a running buffer; i.e., there is no matrix such as polyacrylamide. This allows for the easy migration of high-molecular-weight molecules such as cellular DNA.d

To measure thymine glycol, cells were irradiated and the DNA extracted, either immediately or after a period of repair, using DNAzol reagent (Gibco-BRL)

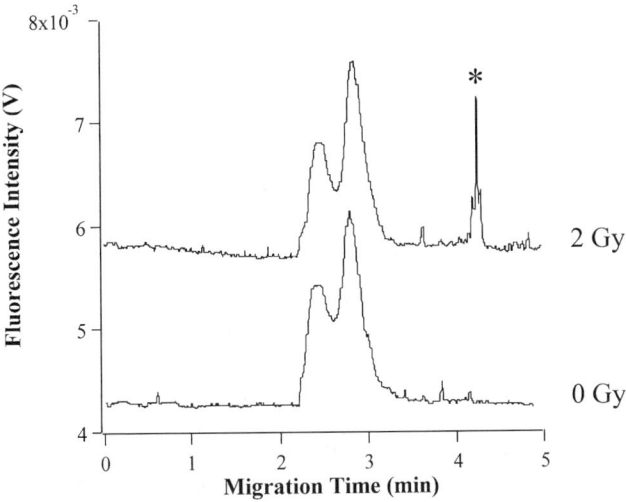

FIG. 3. Electropherograms of DNA recovered from irradiated and unirradiated A549 cells. The asterisk marks the peak of the antibody–DNA complex. The large peaks at 2.4 and 2.9 min represent the excess secondary antibody alone and secondary antibody bound to primary antibody, respectively.

as directed by the manufacturer. The DNA was then incubated with both the primary and secondary antibodies. The monoclonal antibodies to thymine glycol had been raised against an immunogen consisting of osmium tetroxide-treated poly(dT) electrostatically complexed to methylated bovine serum albumin (5). The secondary antibodies were the same as those used for the BrdU calibration. Because the primary antibody is saturated with the secondary antibody and because the fluorescence from the same secondary antibody is measured, we assume that the relative fluorescence intensity from 1 mol of thymine glycol is equal to that from 1 mol of BrdU. Figure 3 shows the electropherograms for DNA from unirradiated and 2-Gy-irradiated human A549 cells.

II. Induction of Thymine Glycol: Comparison of CE-LIF to Other Assays

We have previously shown that the CE-LIF immunoassay gives a linear dose response for thymine glycols up to and including 5 Gy, the highest dose of γ-radiation used. Table I presents a comparison of the results obtained with this assay, for both naked DNA and cellular DNA, to those of others. In the case of naked DNA, the result of the present assay, 6 thymine glycols/10^6 bases/Gy falls

TABLE I
MEASUREMENTS OF THYMINE GLYCOL (Tg)

Sample	Dose	Quantity	Assay type	Reference
Naked DNA				
Calf thymus	100 Gy	24 Tg/10^6 bases/Gy	GC/MS	26
Calf thymus	100 Gy	2.5 Tg/10^6 bases/Gy	^{32}P-Postlabeling	27
Calf thymus	50 Gy	1.5 Tg/10^6 bases/Gy	^{32}P-Postlabeling	28
ϕX174 RF	40 Gy	2.5 Tg/10^6 bases/Gy	ELISA	29
Calf thymus	1 Gy	6 Tg/10^6 bases/Gy	Immuno CE/LIF	4
Cellular DNA				
Rat liver	0 Gy	<1 Tg/10^6 bases	^{32}P-Postlabeling	27
Mouse epidermis	0 Gy	300 Tg/10^6 bases	^{3}H-Postlabeling	30
Mouse liver	0 Gy	80 Tg/10^6 bases	GC/MS	31
Human A549 cells	0 Gy	<0.005 Tg/10^6 bases	Immuno CE/LIF	4
Mouse liver	100 Gy (whole body)	80 Tg/10^6 bases	GC/MS	31
Human K562 cells	116 Gy	0.4 Tg/10^6 bases/Gy	GC/MS	32
Human fibroblasts	10 Gy	0.1 Tg/10^6 bases/Gy	ELISA	8
Human A549 cells	1 Gy	0.1 Tg/10^6 bases/Gy	Immuno CE/LIF	4

within the range of 1.5–24 thymine glycols/10^6 bases/Gy measured by others, although our result was obtained with considerably lower doses of radiation. For DNA from irradiated cells, our result of 0.1 thymine glycol/10^6 bases/Gy was very similar to that obtained by Cooper *et al.* (8), who used the same anti-thymine glycol antibody in an ELISA assay. These values are lower than those provided by GC/MS. The difference of almost two orders of magnitude for induction of thymine glycol between naked DNA and cellular DNA indicates that the cell affords considerable radioprotection to its DNA. The level of protection against thymine glycol induction is similar to that observed for strand breakage (9, 10).

Finally, it is worth noting the difference in the background levels of thymine glycol in cellular DNA as determined by several assays. Our value of <0.005 thymine glycols/10^6 bases, which is equivalent to <30 thymine glycols per cell, is significantly lower than values obtained by other methods. Although it will have to be verified by other approaches, we believe that the low background level of thymine glycol observed by CE-LIF immunoassay is probably closer to the true value for two reasons. (1) Thymine glycol formation requires the action of a highly reactive radical, the hydroxyl free radical; i.e., more common cellular oxygen radicals, such as the superoxide radical ion, are insufficiently reactive to initiate reaction with thymine (*11*). (2) While the generation of thymine glycol over the period of a day, for example, may be high, we are measuring the level at a single instant, and it is unlikely that cells with a fully functioning repair machinery would permit the buildup of high levels of this lesion.

III. Cellular Removal of Thymine Glycol from DNA

The sensitivity of the CE-LIF immunoassay allowed us to monitor the removal of thymine glycol from the DNA of A549 human lung tumor cells irradiated with 2 Gy (Fig. 4). Repair was seen to be more than 80% complete by 4 h. We have observed similar kinetics in other human cell lines including GM 38 fibroblasts. Interestingly, there appeared to be a lag of 30–60 minutes before removal started. We questioned whether this was an indication that the repair of thymine glycols may be inducible. Accordingly, we tested this idea by first irradiating the cells with a low dose (0.25 Gy) and then incubating the cells for a further 4 h before challenging the cells with 2-Gy irradiation. Under these circumstances, A549 cells displayed an immediate repair response after the challenge dose (Fig. 4). A similar phenomenon was observed with GM 38 cells. This observation provides strong support for an inducible component to the removal of thymine glycol. Adaptive responses to radiation, i.e., the ability to resist the challenge of a high dose after exposure to a low dose a few hours earlier, have been observed in a variety of mammalian cells with several endpoints, including

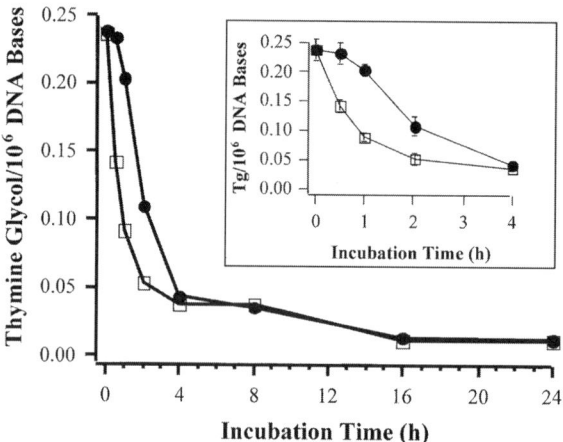

FIG. 4. Kinetics of thymine glycol (Tg) removal from human A549 cells irradiated with 2 Gy with (open squares) or without (closed circles) a priming dose of 0.25 Gy 4 h before the 2-Gy dose. (Reprinted with permission from X. C. Le, J. Z. Xing, J. Lee, S. A. Leadon, and M. Weinfeld, "Inducible Repair of Thymine Glycol Detected by an Ultrasensitive Assay for DNA Damage." *Science* 1998; 280: 1068. Copyright 1995, American Association for the Advancement of Science.)

chromosome aberrations, sister chromatid exchanges, and micronuclei (12–15). A limited number of studies have also shown an adaptive response for mutation, transformation, and cell survival *in vitro* and *in vivo* (16–18). Typically, the initial dose is $\ll 1$ Gy while the challenging dose is >1 Gy. Thus, we are in a position in followup experiments to examine aspects of the enzymology of the repair pathway responsible for the repair of thymine glycols and the inducible response in cells irradiated with a clinical dose.

The base excision repair (BER) pathway is the major mechanism used for removal of small lesions, including many of the base modifications produced by reactive oxygen species and ionizing radiation, such as thymine glycol. The process involves a series of steps: (1) The damaged base is removed by a glycosylase; (2) the sugar phosphate backbone is incised by an endonuclease or an accompanying AP lyase activity of some glycosylases; (3) the deoxyribose-phosphate group formerly attached to the damaged base is removed; (4) the resulting gap is filled in by a polymerase; (5) the strand is rejoined by a DNA ligase. It now appears that there may be two distinct pathways that act on subsets of base lesions and introduce different sized repair patches, a short-patch BER pathway involving replacement of one nucleotide and a long-patch BER pathway with gap-filling of several nucleotides (19–21). To what extent they utilize different sets of enzymes, including different polymerases and ligases, is still a matter of conjecture, but the long-patch pathway is characterized by the requirement for FEN-1 (DNase IV) to remove the residual deoxyribose-phosphate group—a task that may be fulfilled by DNA polymerase β in the short-patch pathway—and by proliferating cell nuclear antigen (PCNA) acting as an accessory protein. Other accessory proteins that are believed to play important roles in BER are poly(ADP-ribose) polymerase, XRCC1, and XPG.

Our assay detects the glycosylase step, i.e., the cleavage of the modified base from its deoxyribose. Although this is, in effect, the first step in the classical description of BER provided above, there is the possibility that in the cell, as opposed to purely *in vitro* systems with naked DNA, other proteins are required for this step to occur. These proteins could be involved upstream of the cleavage step in rendering the chromatin structure more accessible to the glycosylase or in damage recognition. Alternatively, they could be required as part of a repair complex together with the glycosylase.

The first protein we examined was XPG. This had been shown to activate removal of oxidized pyrimidines by human endonuclease III *in vitro* (22), and it was suggested that XPG serves in the assembly of a preincision complex. In addition, XPG appears to play a role in the transcription-coupled repair of oxidative damage (8). When we tested an XPG cell line for its capacity to remove thymine glycol, we observed a substantially lower level of removal 6 h after irradiation than either A549 or GM 38 cells (Table II). Furthermore, there was no sign of an inducible response following the 0.25-Gy irradiation. We also observed a fairly similar response with cells derived from a Cockayne syndrome B (CSB) patient, with only limited removal of thymine glycol after 2 Gy (with or without the prior 0.25 Gy irradiation) and a very weak inducible response (Table II). These data point to a fundamental involvement of XPG and CSB with BER following low doses of radiation. Given the links between these proteins and transcription-coupled repair (8), it is possible that the responses seen in these cell lines may be caused by impaired transcriptional bypass of thymine glycols or by failure to induce expression of repair proteins.

TABLE II
GLOBAL REMOVAL OF THYMINE GLYCOL (Tg) FROM DNA IN HUMAN CELLS

Cell line	Extent (%) Tg removal at 6 h post-irradiation	Lag time (min) prior to onset of Tg removal	Inducible response
A549	>80	30–60	+
GM 38	>80	30–60	+
GM 3021 (XPG)	~40	60	−
GM 739 (CSB)	~40	30–60	±
Resting human lymphocytes	>80	30	+
PHA-stimulated human lymphocytes	>80	30	+

We have also explored whether the proliferative status of cells might influence the rate of repair and/or the inducible response. Some studies have reported a proliferation-dependent increase in base excision repair; in particular, the expression of certain DNA repair glycosylases appear to increase with proliferation (23–25). We compared the removal of thymine glycol from the DNA of irradiated resting lymphocytes (>98% of the cells in G_0 phase) versus that in phytohemagglutinin-stimulated lymphocytes (58% of the cells in G_0/G_1 phase, 27% in S phase, and 15% in G_2 phase). Although both resting and stimulated cells efficiently removed thymine glycol and demonstrated an inducible response (Table II), the kinetics indicated faster repair in the resting lymphocytes than in the stimulated cells ($t_{1/2}$ ~60 min vs. $t_{1/2}$ ~90 min). This observation was unexpected, and further study will be required to determine if there is a marked reduction in the rate of repair during one or more of the phases of the cell cycle outside of G_0.

Acknowledgments

This work was supported by grants from the National Cancer Institute of Canada and the Natural Science and Engineering Research Council of Canada.

References

1. C. von Sonntag, "The Chemical Basis of Radiation Biology," Taylor and Francis, London, 1987.
2. J. Cadet and M. Weinfeld, *Anal. Chem.* **65**, A675–A682 (1993).
3. "Technologies for the Detection of DNA Damage and Mutation" (G. P. Pfeifer, ed.), Plenum Press, New York, 1996.
4. X. C. Le, J. Z. Xing, J. Lee, S. A. Leadon, and M. Weinfeld, *Science* **280**, 1066–1069 (1998).
5. S. A. Leadon and P. C. Hanawalt, *Mutat. Res.* **112**, 191–200 (1983).
6. J. Z. Xing, J. Lee, S. A. Leadon, M. Weinfeld, and X. C. Le, in "Methods (A Companion to Methods in Enzymology)," **22**, 157–163 (2000).
7. Y. F. Cheng and N. J. Dovichi, *Science* **242**, 562–564 (1988).
8. P. K. Cooper, T. Nouspikel, S. G. Clarkson, and S. A. Leadon, *Science* **275**, 990–993 (1997).
9. M. Ljungman, *Radiat. Res.* **126**, 58–64 (1991).
10. M. Ljungman, S. Nyberg, J. Nygren, M. Eriksson, and G. Ahnstrom, *Radiat. Res.* **127**, 171–176 (1991).
11. L. S. Myers, in "Physical Mechanisms in Radiation Biology" (R. D. Cooper and R. W. Wood, eds.), pp. 185–206. U.S. Atomic Energy Commission, 1974.
12. G. Olivieri, J. Bodycote, and S. Wolff, *Science* **223**, 594–597 (1984).
13. Z. Q. Wang, S. Saigusa, and M. S. Sasaki, *Mutat. Res.* **246**, 179–186 (1991).
14. T. Ikushima, *Mutat. Res.* **227**, 241–246 (1989).
15. I. Azzam, G. P. Raaphorst, and R. E. Mitchel, *Radiat. Res.* **138**, S28–31 (1994).
16. P. K. Zhou, X. Q. Xiang, W. Z. Sun, X. Y. Liu, Y. P. Zhang, and K. Wei, *Radiat. Environ. Biophy.* **33**, 211–217 (1994).

17. J. D. Shadley and G. Q. Dai, *Mutat. Res.* **265,** 273–281 (1992).
18. N. Yoshida, H. Imada, N. Kunugita, and T. Norimura, *J. Radiat. Res.* **34,** 269–276 (1993).
19. Y. Matsumoto, K. Kim, and D. F. Bogenhagen, *Mol. Cell Biol.* **14,** 6187–6197 (1994).
20. A. Klungland and T. Lindahl, *EMBO J.* **16,** 3341–3348 (1997).
21. S. Mitra, T. K. Hazra, R. Roy, S. Ikeda, T. Biswas, J. Lock, I. Boldogh, and T. Izumi, *Mol. Cell* **7,** 305–312 (1997).
22. A. Klungland, M. Hoss, D. Gunz, A. Constantinou, S. G. Clarkson, P. W. Doetsch, P. H. Bolton, R. D. Wood, and T. Lindahl, *Mol. Cell* **3,** 33–42 (1999).
23. L. Cool and M. A. Sirover, *Mutat. Res.* **237,** 211–220 (1990).
24. T. J. Schrader, *Mutat. Res.* **273,** 29–42 (1992).
25. G. Slupphaug, L. C. Olsen, D. Helland, R. Aasland, and H. E. Krokan, *Nucleic Acids Res.* **19,** 5131–5137 (1991).
26. A. F. Fuciarelli, B. J. Wegher, W. F. Blakely, and M. Dizdaroglu, *Int. J. Radiat. Biol.* **58,** 397–415 (1990).
27. M. E. Hegi, P. Sagelsdorff, and W. K. Lutz, *Carcinogenesis* **10,** 43–47 (1989).
28. M. Weinfeld and K.-J. Soderlind, *Biochemistry* **30,** 1091–1097 (1991).
29. K. Hubbard, H. Huang, M. F. Laspia, H. Ide, B. F. Erlanger, and S. S. Wallace, *Radiat. Res.* **118,** 257–268 (1989).
30. K. Frenkel, Z. J. Zhong, H. C. Wei, J. Karkoszka, U. Patel, K. Rashid, M. Georgescu, and J. J. Solomon, *Anal. Biochem.* **196,** 126–136 (1991).
31. T. Mori, Y. Hori, and M. Dizdaroglu, *Int. J. Radiat. Biol.* **64,** 645–650 (1993).
32. Z. Nackerdien, R. Olinski, and M. Dizdaroglu, *Free Radical Res. Commun.* **16,** 259–273 (1992).

Completion of Base Excision Repair by Mammalian DNA Ligases

ALAN E. TOMKINSON, LING CHEN, ZHIWAN DONG, JOHN B. LEPPARD, DAVID S. LEVIN, ZACHARY B. MACKEY, AND TERESA A. MOTYCKA

Department of Molecular Medicine
Institute of Biotechnology
The University of Texas Health Science
Center at San Antonio
San Antonio, Texas 78245

I. Introduction.. 152
II. Mammalian *LIG* Genes and Their Protein Products................. 153
   A. *LIG1*.. 153
   B. *LIG3*.. 154
   C. *LIG4*.. 155
III. DNA Ligase-Associated Proteins................................. 155
   A. DNA Ligase I.. 155
   B. DNA Ligase III.. 157
   C. DNA Ligase IV... 158
IV. Phenotype of DNA Ligase-Deficient Cell Lines................... 158
   A. DNA Ligase I-Deficient Cell Lines......................... 158
   B. DNA Ligase III-Deficient Cell Lines....................... 159
   C. DNA Ligase IV-Deficient Cell Lines........................ 159
V. Involvement of Mammalian DNA Ligases in BER.................... 160
VI. Concluding Remarks... 161
   References.. 162

Three mammalian genes encoding DNA ligases—*LIG1*, *LIG3*, and *LIG4*—have been identified. Genetic, biochemical, and cell biology studies indicate that the products of each of these genes play a unique role in mammalian DNA metabolism. Interestingly, cell lines deficient in either DNA ligase I (46BR.1G1) or DNA ligase III (EM9) are sensitive to simple alkylating agents. One interpretation of these observations is that DNA ligases I and III participate in functionally distinct base excision repair (BER) subpathways. In support of this idea, extracts from both DNA ligase-deficient cell lines are defective in catalyzing BER *in vitro* and both DNA ligases interact with other BER proteins. DNA ligase I interacts directly with proliferating cell nuclear antigen (PCNA) and DNA polymerase β

(Pol β), linking this enzyme with both short-patch and long-patch BER. In somatic cells, DNA ligase IIIα forms a stable complex with the DNA repair protein Xrcc1. Although Xrcc1 has no catalytic activity, it also interacts with Pol β and poly(ADP-ribose) polymerase (PARP), linking DNA ligase IIIα with BER and single-strand break repair, respectively. Biochemical studies suggest that the majority of short-patch base excision repair events are completed by the DNA ligase IIIα/Xrcc1 complex. Although there is compelling evidence for the participation of PARP in the repair of DNA single-strand breaks, the role of PARP in BER has not been established. © 2001 Academic Press.

I. Introduction

Phosphodiester bond formation is a common step in many DNA metabolic pathways, including DNA replication, DNA excision repair, and DNA strand-break repair. All the DNA ligation events in the prokaryote *Escherichia coli* are catalyzed by a single species of DNA ligase in an NAD-dependent reaction (*1*). In contrast, mammalian cells contain several distinct species of ATP-dependent DNA ligases (*2*). Except for the different nucleotide cofactor, prokaryotic and eukaryotic DNA ligases employ the same basic three-step reaction mechanism (*1, 2*). Initially, the enzyme reacts with the nucleotide cofactor to form a covalent enzyme–AMP complex, in which the AMP moiety is linked to a specific lysine residue. Next, the AMP group is transferred from the polypeptide to the 5'-phosphate terminus of a DNA nick. Finally, the nonadenylated enzyme catalyzes phosphodiester bond formation between the 3'-hydroxyl and 5'-phosphate termini of the nick, releasing AMP.

Based on their different biochemical properties, it was predicted that the mammalian DNA ligase activities were encoded by different genes (*3, 4*). This has been validated by the identification of three mammalian *LIG* genes: *LIG1*, *LIG3*, and *LIG4* (*5–7*). It seems reasonable to assume that the increased size and complexity of the eukaryotic genome provided the evolutionary driving force for multiple DNA ligases with distinct functions in cellular DNA metabolism. The presence of a conserved catalytic domain within the polypeptides encoded by the mammalian *LIG* genes supports the notion that these genes were generated by duplication events. Insights into the cellular roles of the mammalian *LIG* gene products have been gleaned both from the phenotype of DNA ligase-deficient cell lines and from the identification of proteins that specifically interact with the unique regions that flank the conserved catalytic domain.

In the yeast *Saccharomyces cerevisiae*, the *CDC9* and *DNL4* genes are homologous with the mammalian *LIG1* and *LIG4* genes, respectively (*5, 8–10*). Elegant genetic and biochemical studies have shown that, in addition to its essential role in DNA replication, Cdc9 DNA ligase completes nucleotide and base excision repair pathways (*11, 12*). The assignment of DNA ligases to excision repair pathways in mammalian cells is complicated by the presence of the *LIG3* gene. Here we will focus on the products of the mammalian *LIG* genes with particular emphasis on the contribution of these enzymes to base excision repair (BER).

II. Mammalian *LIG* Genes and Their Protein Products

The relationship between the mammalian *LIG* genes and their products is shown in Table I and described below.

A. *LIG1*

A full-length cDNA encoding DNA ligase I was identified by screening for human cDNAs that complemented the conditional lethal phenotype of a S. cerevisiae cdc9 DNA ligase mutant (5). The relationship between the 919-amino acid polypeptide encoded by the cDNA and DNA ligase I was confirmed by the alignment of peptide sequences from purified bovine DNA ligase I with sequences within the open reading frame encoded by the cDNA (5). A notable feature of the DNA ligase I polypeptide is its high proline content, which causes aberrant mobility during SDS-PAGE. Thus, DNA ligase I has a molecular mass of 125 kDa when measured by SDS-PAGE compared with a calculated molecular weight of 102,000 (5, 13).

When compared with other eukaryotic DNA ligases, mammalian DNA ligase I is most closely related to the replicative DNA ligases of S. cerevisiae and *Schizosaccharomyces pombe* encoded by the *CDC9* and *CDC17* genes, respectively (5). However, the amino acid homology is restricted to the C-terminal catalytic domains of these enzymes. The noncatalytic N-terminal domain of DNA ligase I contains sequences that target this polypeptide to the nucleus and to specific subnuclear locations (14, 15). In addition, this domain is phosphorylated *in vivo* and mediates specific protein–protein interactions (16–18).

The association of DNA ligase I with DNA replication is further supported by several other observations: (1) The level of DNA ligase I correlates with cell proliferation (19, 20); (2) DNA ligase I colocalizes with other replication

TABLE I
MAMMALIAN *LIG* GENES AND THEIR PRODUCTS

Gene	Enzyme
LIG1	DNA ligase I
LIG3	DNA ligase IIIα-nucl
	DNA ligase IIIα-mito
	DNA ligase IIIβ
LIG4	DNA ligase IV

proteins at the sites of DNA synthesis, termed replication foci, in replicating cells (21); (3) DNA ligase I copurifies with a high-molecular-weight replication complex (22).

B. LIG3

This gene was cloned at about the same time by two different approaches—searching an EST database with a motif conserved among eukaryotic DNA ligases, and a conventional cDNA cloning strategy based on partial amino acid sequences from purified bovine DNA ligase III (6, 7). The absence of *LIG3* homologs in the genomes of *S. cerevisiae, Drosophila melanogaster,* and *Caenorhabditis elegans* suggests that this gene was a relatively recent addition to the genome of mammals during evolution. In contrast to the *LIG1* and *LIG4* genes, the *LIG3* gene encodes several polypeptides that appear to have distinct cellular functions. Two forms of DNA ligase III, α and β, are produced by an alternative splicing event that results in proteins with different C termini (23). The C-terminal 77 amino acids of human DNA ligase IIIα are replaced by a unique 17-amino acid sequence in human DNA ligase IIIβ. The sequence of the C-terminal 100 amino acids of DNA ligase IIIα exhibits homology with the BRCT motif, a putative protein–protein interaction domain that was first identified in the product of the breast cancer susceptibility gene, *BRCA1* (24, 25). DNA ligase IIIα mRNA is expressed in all tissues and cells, whereas DNA ligase IIIβ mRNA expression is restricted to male germ cells, specifically pachytene spermatocytes and round spermatids (23). The differential expression profiles suggest that DNA ligase IIIα is a housekeeping enzyme, whereas DNA ligase IIIβ participates either in meiotic recombination or postmeiotic DNA repair.

Although the human DNA ligase IIIα open reading frame encodes a 949-amino acid polypeptide, the second in-frame ATG was chosen as the most likely *in vivo* translation initiation site based on homology with the Kozak consensus sequence (6, 7). More recently, it was noticed that the preceding 87-amino acid sequence contained a putative mitochondrial targeting signal (26). It was assumed that DNA ligase IIIα was a nuclear protein, as this enzyme was purified from nuclear extracts and immunocytochemistry experiments indicated that the majority of DNA ligase IIIα protein is located within the nucleus (23, 27). However, fusion of the putative targeting sequence to a heterologous protein resulted in mitochondrial localization of the fusion protein (26). Therefore, it appears that mitochondrial and nuclear forms of DNA ligase IIIα are produced in somatic cells by alternative translation initiation. Presumably, mitochondrial DNA ligase IIIα participates in the replication and repair of the mitochondrial genome. At present the mechanism underlying the nuclear localization of DNA ligase IIIα is unknown.

A zinc finger motif at the N terminus of DNA ligase III distinguishes the products of the *LIG3* gene from other eukaryotic DNA ligases. Intriguingly, the DNA ligase III zinc finger is most closely related to the two zinc fingers located at the N terminus of poly(ADP-ribose) polymerase (PARP) (7). PARP is a relatively abundant nuclear protein whose avid binding to both single- and double-strand breaks is mediated by its zinc fingers. DNA binding activates the polymerase activity of PARP, resulting in the ADP-ribosylation of itself and other nuclear proteins (28). The single zinc finger of DNA ligase III has similar DNA binding properties in that it mediates the binding of this enzyme to DNA strand breaks with a preference for single-strand breaks (29). Although the zinc finger is not required for catalytic activity *in vitro* or for functional complementation of a temperature-sensitive *E. coli lig* mutant, it has a significant effect on the DNA-binding and DNA-joining activities of DNA ligase III at physiological salt concentrations (29).

C. LIG4

This gene, which was cloned by searching an EST database with a motif conserved among eukaryotic DNA ligases, was identified prior to the detection of the enzyme activity in mammalian cell extracts (7, 30). There are several reasons why DNA ligase IV activity escaped detection in the earlier enzyme purification studies: (1) In extracts from proliferating cells, DNA ligase IV is a minor contributor to cellular DNA ligase activity (30); (2) both the *LIG3* and *LIG4* genes encode polypeptides with similar electrophoretic mobility (7, 30); (3) unlike DNA ligases I and III, a significant fraction of DNA ligase IV remains adenylated during fractionation, making it difficult to label with $[\alpha^{32}P]ATP$ in the adenylation assay (30). A unique feature of the *LIG4* gene product compared with the *LIG1* and *LIG3* gene products is its long C-terminal extension beyond the conserved catalytic domain. Furthermore, this C-terminal region contains two BRCT motifs (7).

III. DNA Ligase-Associated Proteins

The identification of specific protein partners for each of the DNA ligases (Table II) has provided insights into both the molecular mechanisms by which these enzymes recognize interruptions in the phosphodiester backbone and their cellular functions.

A. DNA Ligase I

Using affinity chromatography with either DNA Pol β antibody or Pol β as the ligand, a multiprotein complex that catalyzed the repair of a uracil-containing DNA substrate was partially purified from a bovine testis nuclear extract (31). Subsequent studies revealed that DNA ligase I was a component of this BER complex and that DNA ligase I interacts directly with Pol β (18, 31). This interaction occurs between the noncatalytic N-terminal domain of DNA ligase I and

TABLE II
DNA LIGASE INTERACTING PROTEINS [a]

Enzyme	Interacting protein
DNA ligase I	PCNA
	DNA Pol β
DNA ligase IIIα	XRCC1
DNA ligase IIIβ	?
DNA ligase IV	XRCC4

[a]Biochemical and genetic studies indicate that the majority of short-patch BER events are completed by the DNA ligase IIIα/Xrcc1 complex. The contribution of DNA ligase I to short-patch BER via its interaction with Pol β is not known. The interaction of DNA ligase I with PCNA suggests that this enzyme completes the PCNA-dependent long-patch BER subpathway.

the 8-kDa N-terminal domain of Pol β that has DNA binding and 5′-deoxyribose phosphate diesterase activities (32, 33).

Intuitively, an association between two enzymes that catalyze consecutive steps in a reaction pathway suggests that the interaction should enhance the overall rate of the reaction. In Figure 1, we show that gap-filling synthesis and ligation of a DNA substrate with a single nucleotide gap by the concerted action of Pol β and DNA ligase I occurs more rapidly than ligation of a DNA substrate with a single ligatable nick. It should be noted that this experiment does not demonstrate that the increased ligation rate is a direct consequence of the binding of DNA ligase I to Pol β. Nonetheless, it is intriguing that DNA ligase I, which transfers the AMP group to the 5′-phosphate terminus at a nick, interacts with the domain of Pol β that binds to and can modify the 5′ terminus at a nick or short gap in duplex DNA (32, 33).

An association between DNA ligase I and proliferating cell nuclear antigen (PCNA) was also detected by affinity chromatography, in this case using DNA ligase I as the ligand (17). The direct interaction of these proteins is mediated by the noncatalytic N-terminal domain of DNA ligase I. Cell biology studies had demonstrated that this region of the protein was required both for nuclear localization and for subnuclear targeting to replication foci (14). Subsequent studies mapped these targeting activities to two distinct regions within the N-terminal region of DNA ligase I (15). Interestingly, the sequence required for targeting to replication foci was coincident with that required for PCNA binding (15). The replication foci targeting/PCNA binding sequence of DNA ligase I exhibits homology with a PCNA binding motif initially identified in the cell division kinase inhibitor p21 and subsequently found in several PCNA binding proteins (34). Although these observations indicate that the stable association of

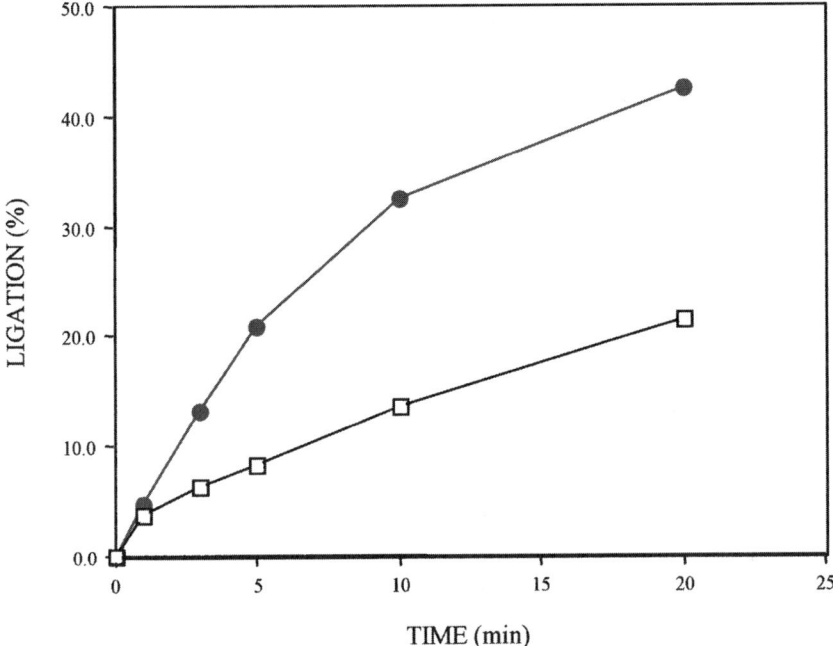

FIG. 1. Repair of a one-nucleotide gap by Pol β and DNA ligase I. DNA ligase I (0.06 pmol) and Pol β (0.12 pmol) were incubated with a labeled linear duplex DNA substrate (4 pmol) containing a one-nucleotide gap in the presence of the 4 dNTPs (filled circles). DNA ligase I (0.06 pmol) was incubated with a labeled linear duplex DNA substrate (4 pmol) containing a single ligatable nick in the presence of the 4 dNTPs (open squares). Aliquots were removed at the times indicated. After separation by denaturing gel electrophoresis, ligated product was detected and quantitated by PhosphorImager analysis. The inclusion of Pol β in assays with the nicked DNA substrate had no effect on the rate of ligation.

DNA ligase I with replication foci is mediated by PCNA binding, they do not address the direct role of PCNA binding in cellular ligation events catalyzed by DNA ligase I. In this regard, the inclusion of PCNA in DNA ligase assays does not stimulate DNA joining by DNA ligase I. However, DNA ligase I will bind to a PCNA molecule that is topologically linked to a circular DNA duplex so PCNA binding could serve to tether DNA ligase I to DNA substrates (17).

B. DNA Ligase III

DNA ligase III was identified as a protein that bound specifically to a tagged version of the DNA repair protein Xrcc1 (35). The human XRCC1 gene was identified by its ability to complement the DNA damage hypersensitivity of a mutant Chinese hamster cell line, EM9 (36), which is described in more detail below. Following the discovery of the different forms of DNA ligase III generated by alternative splicing, it was demonstrated that DNA ligase IIIα but not DNA ligase IIIβ formed a complex with Xrcc1 (23, 37). Interestingly, complex formation is mediated by an interaction between BRCT motifs present at the C termini of both Xrcc1 and DNA ligase IIIα (37). Although Xrcc1 has no catalytic activity, the reduced level of DNA ligase IIIα in the *xrcc1* mutant cell line indicates that Xrcc1 is important for the stability of DNA ligase IIIα (38). Moreover, protein–protein interactions between Xrcc1 and both Pol β and PARP have been characterized (39, 40). Each of these binding events involves different regions of Xrcc1 that are also distinct from the DNA ligase IIIα binding site, suggesting that Xrcc1 may act as a scaffold for the assembly of a multiprotein complex (39, 40).

C. DNA Ligase IV

DNA ligase IV was identified as a protein that specifically coimmunoprecipitated with a tagged version of the DNA repair protein Xrcc4 (41). The human *XRCC4* gene was identified by its ability to complement the DNA damage hypersensitivity of a mutant Chinese hamster cell line XR1 (42). In contrast to the interaction between DNA ligase IIIα and Xrcc1, the interaction between DNA ligase IV and Xrcc4 is not mediated by BRCT motifs, but instead Xrcc4 binds to the region between the two C-terminal BRCT motifs of DNA ligase IV (43). The reduced level of DNA ligase IV protein in *xrcc4* mutant cells suggests that the interaction with Xrcc4 stabilizes DNA ligase IV (44). Since Xrcc4 binds to DNA, it is possible that this protein mediates the binding of DNA ligase IV to its DNA substrate (45).

IV. Phenotype of DNA Ligase-Deficient Cell Lines

Our understanding of the cellular functions of the mammalian DNA ligases has been greatly enhanced by the isolation of mammalian cell lines deficient in each of the DNA ligases. The phenotypes of these cell lines are summarized below.

A. DNA Ligase I-Deficient Cell Lines

The fibroblast cell line 46BR and an SV40-immortalized derivative 46BR.1G1 were established from an immunodeficient individual who inherited different point mutations in each *LIG1* allele (46, 47). One of the mutations abolishes DNA ligase activity, whereas the other mutant allele encodes a polypeptide that is 10- to 20-fold less active than the wild-type enzyme. As expected, the DNA ligase I deficiency results in abnormal joining of Okazaki fragments (48, 49). The cell lines are also hypersensitive to the cytotoxic effects of monofunctional

DNA alkylating agents and are moderately sensitive to ultraviolet and ionizing radiation (50). These observations are consistent with DNA ligase I participating in strand-break and excision-repair pathways. In support of this idea, 46BR.1G1 extracts are defective in the repair of circular duplex DNA containing a single uracil residue (48). Specifically, the DNA ligase I deficiency resulteds in reduced levels of repaired circular product and increased levels of DNA repair synthesis.

B. DNA Ligase III-Deficient Cell Lines

The mutant Chinese hamster ovary cell line EM9 was isolated based on its hypersensitivity to DNA alkylating agents (51). Further analysis revealed that this cell line was also moderately sensitive to ionizing radiation and exhibited elevated levels of sister chromatid exchange when grown in the presence of bromodeoxyuridine (52). As mentioned previously, expression of the human *XRCC1* gene complemented the DNA damage-sensitive phenotype of the mutant CHO cell line (36). Analysis of DNA repair in EM9 cells revealed a defect in events after DNA damage-incision. The interactions of Xrcc1 with both Pol β and DNA ligase IIIα provide a molecular explanation for the participation of Xrcc1 in the DNA synthesis and ligation reactions that complete base excision repair (35, 40). In further support of the idea that Xrcc1 plays an important role in BER, extracts from EM9 cells are defective in this type of repair (53).

It should be noted that Xrcc1 may function independently of DNA ligase IIIα. As mentioned previously, Xrcc1 interacts with PARP, an enzyme that is not required for the reconstitution of BER *in vitro* (39). In addition, a recent study with *xrcc1* mutant cells reported evidence that the DNA ligase IIIα/Xrcc1 complex plays an important role in DNA repair within the G_1 phase of the cell cycle whereas Xrcc1 alone mediates DNA repair events within the DNA synthesis phase of the cell cycle (54). The construction and characterization of *lig3* mutant cell lines will facilitate further analysis of the cellular functions of DNA ligase III and Xrcc1.

C. DNA Ligase IV-Deficient Cell Lines

The mutant Chinese hamster cell line XR1 is hypersensitive to ionizing radiation and defective in V(D)J recombination (42). Interestingly, the DNA damage hypersensitivity is most pronounced in the G_1 phase of the cell cycle. As noted above, both the repair and V(D)J recombination defects in XR1 cells are complemented by the human *XRCC4* gene whose product forms a stable complex with the *LIG4* gene product (41, 55). More recently, the *XRCC4* gene has been deleted in mouse embryonic cells by gene targeting and embryonic fibroblast cell lines established from the resultant *xrcc4*$^{-/-}$ embryos. These cell lines reiterate the phenotype of the XR1 cells (56).

Cell lines with mutations in the *LIG4* gene are also available. A radiosensitive human cell line 180BR was established from a leukemia patient who had severe reactions to both radiotherapy and chemotherapy (57). Subsequently, a point mutation was identified within the *LIG4* gene of the 180BR cells (58). Interestingly, the leukemia patient did not appear to be immunodeficient, and the 180BR cells, when activated for V(D)J recombination, had no obvious defect in this type of site-specific recombination (58). It is possible that the mutant allele encodes a product that retains sufficient activity to complete V(D)J recombination, but not for the repair of DNA damage-induced double-strand breaks. The *LIG4* gene has been deleted in mouse embryonic cells by gene targeting, and embryonic fibroblast cell lines have been established from the resultant $lig4^{-/-}$ embryos. These cell lines exhibit the same phenotype as the *xrcc4* mutant cell line in that they are hypersensitive to ionizing radiation and defective in V(D)J recombination (59). Based on the results described above and comparable studies in the yeast *S. cerevisiae* (8–10), it appears that DNA ligase IV plays a major role in the repair of DNA double-strand breaks by nonhomologous end joining but does not contribute significantly to BER and other excision repair pathways.

V. Involvement of Mammalian DNA Ligases in BER

Biochemical studies using cell-free extracts and purified proteins have identified two pathways of BER that can be distinguished both by the extent of DNA repair synthesis and the requirement for PCNA (40, 60–63). At the present time, the spectrum of DNA base lesions repaired *in vivo* by the short-patch (single-nucleotide repair patch, PCNA-independent) and long-patch (repair patches of 2–11 nucleotides, PCNA-dependent) BER subpathways is not known. Thus it is possible that certain base lesions can be repaired by only one of the BER subpathways, whereas other lesions can be effectively repaired by either subpathway.

As mentioned above, cell lines defective in either DNA ligase I or DNA ligase III are hypersensitive to killing by monofunctional alkylating agents such as methyl methanesulfonate (MMS) (50, 52). Since the base lesions caused by MMS are thought to be repaired by BER, one explanation of these observations is that there are two functionally distinct BER subpathways, one involving DNA ligase I and the other involving DNA ligase III, that repair different cytotoxic lesions induced by MMS exposure (Fig. 2). In support of this model, extracts from cell lines defective in either DNA ligase I or DNA ligase III exhibit abnormalities in BER assays (48, 53). The extracts with reduced DNA ligase III activity are defective in short-patch BER but proficient in long-patch BER (53). This suggests that the interaction between Pol β and the Xrcc1 subunit of the Xrcc1/DNA ligase IIIα complex is critical for the completion of short-patch BER (Fig. 2).

Although the interactions of DNA ligase I with both Pol β and PCNA suggest that this enzyme could function in both short- and long-patch BER (17, 18), the

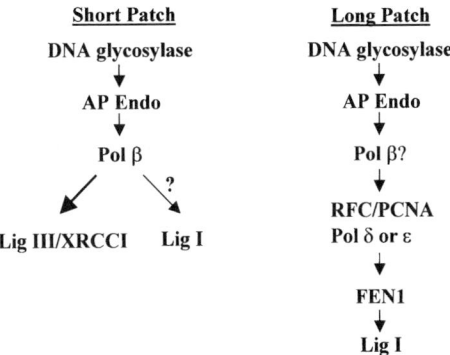

FIG. 2. A model illustrating the participation of DNA ligases I and III in BER.

effect of DNA ligase I deficiency on these subpathways has not been determined. Since DNA ligase III deficiency effects short- but not long-patch BER, it is conceivable that DNA ligase I deficiency will have the opposite effect, causing a defect in long-patch, but not short-patch BER (Fig. 2). If this is the case, the DNA ligase III- and DNA ligase I-deficient cell lines should be useful reagents to determine the spectrum of base lesions repaired *in vivo* by short- and long-patch BER.

VI. Concluding Remarks

Genetic and biochemical studies suggest that both the DNA ligase IIIα/Xrcc1 complex and DNA ligase I play biologically significant roles in BER. Based on currently available evidence, it appears that the DNA ligase IIIα/Xrcc1 complex completes the majority of short-patch BER events (*40, 53*). The interaction of DNA ligase I with PCNA makes this species of DNA ligase the most likely candidate for the long-patch BER subpathway (*17, 60, 63*). Further studies are required to determine the *in vivo* roles of these BER subpathways. At the present time, the biological significance of the interaction between DNA ligase I and Pol β is unclear. Since the 180-kDa BER complex containing these proteins was isolated from extracts of bovine testes (*31*), it is possible that the involvement of DNA ligase I and DNA ligase IIIα/Xrcc1 in short-patch BER is cell type-specific. Further studies are required to elucidate the functional consequence of the interactions between the DNA ligases and Pol β and/or PCNA on the coordination of the reactions that complete BER.

The role of the nuclear protein PARP in BER is controversial. Although both the short- and long-patch BER pathways can be reconstituted *in vitro* without PARP (*40, 60, 61*), the results of BER assays with extracts from parp$^{-/-}$

cells and the interaction between PARP and Xrcc1 suggest that PARP may contribute to the completion of BER (39, 64). The involvement of PARP in BER does not appear to be compatible with the current model of BER, in which pairwise interactions between BER proteins mediate the handover of repair intermediates. In fact, this model predicts that PARP would be actively excluded from DNA undergoing BER by the coordinate action of the BER proteins. It is possible that the DNA ligase IIIα/Xrcc1 complex participates in both PARP-dependent repair of DNA single-strand breaks and PARP-independent short-patch BER. Further biochemical and molecular genetic studies are required to clarify the complex relationships between these DNA repair proteins.

Acknowledgments

Studies in A.E.T.'s laboratory were supported by grants from the Department of Health and Human Services (GM47251 and GM57479), the Nathan Shock Aging Center, and the San Antonio Cancer Institute. D.S.L. and Z.B.M. were supported by the Training Program in the Molecular Basis of Breast Cancer. D.S.L. is supported by the Training Program in the Molecular Basis of Aging. Z.B.M is a UNCF-MERCK fellow.

References

1. I. R. Lehman, *Science* **186**, 790–797 (1974).
2. A. E. Tomkinson and Z. B. Mackey, *Mutat. Res.* **407**, 1–9 (1998).
3. A. E. Tomkinson, E. Roberts, G. Daly, N. F. Totty, and T. Lindahl, *J. Biol. Chem.* **286**, 21728–21735 (1991).
4. J. E. Arrand, A. E. Willis, I. Goldsmith, and T. Lindahl, *J. Biol. Chem.* **261**, 9079–9082 (1986).
5. D. E. Barnes, L. H. Johnston, K. Kodama, A. E. Tomkinson, D. D. Lasko, and T. Lindahl, *Proc. Natl. Acad. Sci. U.S.A.* **87**, 6679–6683 (1990).
6. J. Chen, A. E. Tomkinson, W. Ramos, Z. B. Mackey, S. Danehower, C. A. Walter, R. A. Schultz, J. M. Besterman, and I. Husain, *Mol. Cell. Biol.* **15**, 5412–5422 (1995).
7. Y.-F. Wei, P. Robins, K. Carter, K. Caldecott, D. J. C. Pappin, G.-L. Yu, R.-P. Wang, B. K. Shell, R. A. Nash, P. Schär, D. E. Barnes, W. A. Haseltine, and T. Lindahl, *Mol. Cell. Biol.* **15**, 3206–3216 (1995).
8. P. Schär, G. Herrman, G. Daly, and T. Lindahl, *Genes Dev.* **11**, 1912–1924 (1997).
9. S. H. Teo and S. P. Jackson, *EMBO J.* **16**, 4788–4795 (1997).
10. T. E. Wilson, U. Grawunder, and M. R. Lieber, *Nature (London)* **388**, 495–498 (1997).
11. X. Wu, E. Braithwaite, and Z. Wang, *Biochemistry* **38**, 2628–2635 (1999).
12. D. R. Wilcox and L. Prakash, *J. Bacteriol.* **148**, 618–623 (1981).
13. A. E. Tomkinson, D. D. Lasko, G. Daly, and T. Lindahl, *J. Biol. Chem.* **265**, 12611–12617 (1990).
14. A. Montecucco, E. Savini, F. Weighardt, R. Rossi, G. Ciarrocchi, A. Villa, and G. Biamonti, *EMBO J.* **14**, 5379–5386 (1995).
15. A. Montecucco, R. Rossi, D. S. Levin, R. Gary, M. S. Park, T. A. Motycka, G. Ciarrocchi, A. Villa, G. Biamonti, and A. E. Tomkinson, *EMBO J.* **17**, 3786–3795 (1998).

16. C. Prigent, D. D. Lasko, K. Kodama, J. R. Woodgett, and T. Lindahl, *EMBO J.* **11**, 2925–2933 (1994).
17. D. S. Levin, W. Bai, N. Yao, M. O'Donnell, and A. E. Tomkinson, *Proc. Natl. Acad. Sci. U.S.A.* **94**, 12863–12868 (1997).
18. E. K. Dimitriadis, R. Prasad, M. K. Vaske, L. Chen, A. E. Tomkinson, M. S. Lewis, and S. H. Wilson, *J. Biol.Chem.* **273**, 20540–20550 (1998).
19. J. H. J. Petrini, K. G. Huwiler, and D. T. Weaver, *Proc. Natl. Acad. Sci. U.S.A.* **88**, 7615–7619 (1991).
20. S. Soderhall, *Nature (London)* **260**, 640–642 (1976).
21. D. Wilcock and D. P. Lane, *Nature (London)* **349**, 429–431 (1991).
22. C. Li, J. Goodchild, and E. F. Baril, *Nucleic Acids Res.* **22**, 632–638 (1994).
23. Z. B. Mackey, W. Ramos, D. S. Levin, C. A. Walter, J. R. McCarrey, and A. E. Tomkinson, *Mol. Cell. Biol.* **17**, 989–998 (1996).
24. I. Callebaut and J. P. Mornon, *FEBS Lett.* **400**, 25–30 (1997).
25. E. V. Koonin, S. F. Alschul, and P. Bork, *Nature Genet.* **13**, 266–267 (1996).
26. U. Lakshmipathy and C. Campbell, *Mol. Cell. Biol.* **19**, 3869–3876 (1999).
27. I. Husain, A. E. Tomkinson, W. A. Burkhart, M. B. Moyer, W. Ramos, Z. B. Mackey, J. M. Besterman, and J. Chen, *J. Biol. Chem.* **270**, 9683–9690 (1995).
28. G. de Murcia and J. M. de Murcia, *Trends Biochem. Sci.* **19**, 172–176 (1994).
29. Z. B. Mackey, C. Niedergang, J. M. Murcia, J. Leppard, K. Au, J. Chen, G. de Murcia, and A. E. Tomkinson, *J. Biol. Chem.* **274**, 21679–21687 (1999).
30. P. Robins and T. Lindahl, *J. Biol. Chem.* **271**, 24257–24261 (1996).
31. R. Prasad, R. K. Singhal, D. K. Srivastava, J. T. Molina, A. E. Tomkinson, and S. H. Wilson, *J. Biol. Chem.* **271**, 16000–16007 (1996).
32. Y. Matsumoto and K. Kim, *Science* **269**, 699–702 (1995).
33. R. Prasad, W. A. Beard, and S. H. Wilson, *J. Biol. Chem.* **269**, 18096–18101 (1994).
34. E. Warbrick, *Bioessays* **20**, 195–199 (1998).
35. K. W. Caldecott, C. K. McKeown, J. D. Tucker, S. Ljunquist, and L. H. Thompson, *Mol. Cell. Biol.* **14**, 68–76 (1994).
36. L. H. Thompson, K. W. Brookman, N. J. Jones, S. A. Allen, and A. V. Carrano, *Mol. Cell. Biol.* **10**, 6160–6171 (1990).
37. R. A. Nash, K. Caldecott, D. E. Barnes, and T. Lindahl, *Biochemistry* **36**, 5207–5211 (1997).
38. S. Ljungquist, K. Kenne, L. Olsson, and M. Sandstrom, *Mutat. Res.* **314**, 177–186 (1994).
39. M. Masson, C. Niedergang, V. Schreiber, S. Muller, J. M. de Murcia, and G. de Murcia, *Mol. Cell. Biol.* **18**, 3563–3571 (1998).
40. Y. Kubota, R. A. Nash, A. Klungland, P. Schär, D. E. Barnes, and T. Lindahl, *EMBO J.* **15**, 6662–6670 (1996).
41. U. Grawunder, M. Wilm, X. Wu, P. Kulesza, T. E. Wilson, M. Mann, and M. R. Lieber, *Nature (London)* **388**, 492–495 (1997).
42. Z. Li, T. Otevrel, Y. Gao, H.-L. Cheng, B. Seed, T. D. Stamato, G. E. Taccioli, and F. W. Alt, *Cell (Cambridge, Mass.)* **83**, 1079–1089 (1995).
43. U. Grawunder, D. Zimmer, and M. R. Leiber, *Curr. Biol.* **8**, 873–876 (1998).
44. M. Bryans, M. C. Valenzano, and T. Stamato, *Mutat. Res.* **433**, 53–58 (1999).
45. M. Modesti, J. E. Hesse, and M. Gellert, *EMBO J* **18**, 2008–2017 (1999).
46. D. E. Barnes, A. E. Tomkinson, A. R. Lehmann, A. D. B. Webster, and T. Lindahl, *Cell (Cambridge, Mass.)* **69**, 495–503 (1992).
47. A. D. B. Webster, D. E. Barnes, C. F. Arlett, A. R. Lehmann, and T. Lindahl, *Lancet* **339**, 1508–1509 (1992).
48. C. Prigent, M. S. Satoh, G. Daly, D. E. Barnes, and T. Lindahl, *Mol. Cell. Biol.* **14**, 310–317 (1994).

49. V. J. Mackenney, D. E. Barnes, and T. Lindahl, *J. Biol. Chem.* **272**, 11550–11556 (1997).
50. I. A. Teo, C. F. Arlett, S. A. Harcourt, A. Priestly, and B. C. Broughton, *Mutat. Res.* **107**, 371–386 (1983).
51. L. H. Thompson, T. Shiomi, E. P. Salazar, and S. A. Stewart, *Somat. Cell Mol. Genet.* **14**, 605–612 (1988).
52. L. H. Thompson, K. W. Brookman, L. E. Dillehay, A. V. Carrano, J. A. Mazrimas, C. L. Mooney, and J. L. Minkler, *Mutat. Res.* **95**, 247–254 (1982).
53. E. Cappelli, R. Taylor, M. Cevasco, A. Abbondandolo, K. Caldecott, and G. Frosina, *J. Biol. Chem.* **272**, 23970–23975 (1997).
54. R. M. Taylor, D. J. Moore, J. Whitehouse, P. Johnson, and K. W. Caldecott, *Mol. Cell. Biol.* **20**, 735–740 (2000).
55. S. E. Critchlow, R. P. Bowater, and S. P. Jackson, *Curr. Biol.* **7**, 588–598 (1997).
56. Y. Gao, Y. Sun, K. M. Frank, P. Dikkes, Y. Fujiwara, K. J. Seidl, J. M. Sekiguchi, G. A. Rathbun, W. Swat, J. Wang, R. T. Bronson, B. A. Malynn, M. Bryans, C. Zhu, J. Chaudhuri, L, Davidson, R. Ferrini, T. Stamato, S. H. Orkin, M. E. Greenberg, and F. W. Alt, *Cell (Cambridge, Mass.)* **95**, 891–902 (1998).
57. P. N. Plowman, B. A. Bridges, C. F. Arlett, A. Hinney, and J. E. Kingston, *Br. J. Radiol.* **63**, 624–628 (1990).
58. E. Riballo, S. E. Critchlow, S.-H. Teo, A. J. Doherty, A. Priestly, B. Broughton, B. Kysela, H. Beamish, N. Plowman, C. F. Arlett, A. R. Lehmann, S. P. Jackson, and P. A. Jeggo, *Curr. Biol.* **9**, 699–702 (1999).
59. K. M. Frank, J. M. Sekiguchi, K. J. Seidl, W. Swat, G. A. Rathbun, H. L. Cheng, L. Davidson, L. Kangaloo, and F. W. Alt, *Nature (London)* **396**, 173–176 (1998).
60. A. Klungland and T. Lindahl, *EMBO J.* **16**, 3341–3348 (1997).
61. Y. Matsumoto, K. Kim, J. Hurwitz, R. Gary, D. S. Levin, A. E. Tomkinson, and M. Park, *J. Biol. Chem.* **274**, 33703–33708 (1999).
62. D. K. Srivastava, B. J. Vande Berg, R. Prasad, J. T. Molina, W. A. Beard, A. E. Tomkinson, and S. H. Wilson, *J. Biol. Chem.* **273**, 21203–21209 (1998).
63. G. Frosina, P. Fortini, O. Rossi, F. Carrozzino, G. Raspaglio, L. S. Cox, D. P. Dane, A. Abbondandolo, and E. Dogliotti, *J. Biol. Chem.* **271**, 9573–9578 (1996).
64. C. Trucco, F. J. Oliver, G. de Murcia, and J. Menissier-de Murcia, *Nucleic Acids Res.* **26**, 2644–2649 (1998).

Uracil-Initiated Base Excision DNA Repair Synthesis Fidelity in Human Colon Adenocarcinoma LoVo and *Escherichia coli* Cell Extracts

Russell J. Sanderson,[†]
Samuel E. Bennett,[*]
Jung-Suk Sung,[*] and
Dale W. Mosbaugh[*,†,‡]

*Departments of Environmental and
Molecular Toxicology and
[†]Biochemistry and Biophysics
[‡]Environmental Health Science Center
Oregon State University
Corvallis, Oregon 97331

I. Introduction.	166
II. Uracil-Initiated Base Excision DNA Repair Assay.	167
A. Molecular Logic of the Uracil-Initiated *lacZα* DNA Reversion Assay.	167
B. Blueprint for Measuring the Fidelity of Nucleotide Incorporation at the Uracil Target.	169
III. Detection of Uracil-Initiated DNA Repair in Human Colon Adenocarcinoma LoVo Cell Extracts.	171
A. Uracil-DNA Glycosylase-Initiated DNA Repair.	171
B. Ugi-Resistant Uracil-DNA Glycosylase-Initiated DNA Repair.	172
IV. Uracil-DNA Repair Patch Size Produced by BER in LoVo Cell Extracts.	175
A. Effect of Aphidicolin on Uracil-DNA Repair Synthesis.	177
B. Effect of Ugi on Uracil-DNA Repair Synthesis.	180
V. Mutations Produced by Uracil-Initiated Base Excision DNA Repair Synthesis.	181
A. Mutation Frequency.	181
B. Mutational Spectrum.	184
VI. Concluding Remarks.	186
References.	187

Abbreviations: BER, base excision repair; AP, apurinic/apyrimidinic; dRP, 2′-deoxyribose 5′-phosphate; dRPase, 2′-deoxyribose 5′-phosphatase; PCNA, proliferating cell nuclear antigen; FEN1, flap endonuclease 1; IPTG, isopropyl β-D-thiogalactopyranoside; X-gal, bromo-4-chloro-3-indolyl β-D-galactopyranoside; Ung, *E. coli* uracil-DNA glycosylase; UDGΔ84, human uracil-DNA glycosylase (UNG) in which the N-terminal 84 amino acids encoded by the human *UNG* gene were replaced by three amino acids from the pTRC expression vector; Ugi, uracil-DNA glycosylase inhibitor protein; Exo III, *E. coli* exonuclease III; Endo IV, *E. coli* endonuclease IV; Pol β, DNA polymerase beta; Pol δ, DNA polymerase delta; Pol ε, DNA polymerase epsilon; Pol η, DNA polymerase eta; Pol ζ, DNA polymerase zeta; and dNMP[α-S]s, 2′-deoxynucleoside α-thiolmonophosphates.

The error frequency of uracil-initiated base excision repair (BER) DNA synthesis in human and *Escherichia coli* cell-free extracts was determined by an M13mp2 *lacZα* DNA-based reversion assay. Heteroduplex M13mp2 DNA was constructed that contained a site-specific uracil target located opposite the first nucleotide position of opal codon 14 in the *lacZα* gene. Human glioblastoma U251 and colon adenocarcinoma LoVo whole-cell extracts repaired the uracil residue to produce form I DNA that was resistant to subsequent *in vitro* cleavage by *E. coli* uracil-DNA glycosylase (Ung) and endonuclease IV, indicating that complete uracil-initiated BER repair had occurred. Characterization of the BER reactions revealed that (1) the majority of uracil-DNA repair was initiated by a uracil-DNA glycosylase-sensitive to Ugi (uracil-DNA glycosylase inhibitor protein), (2) the addition of aphidicolin did not significantly inhibit BER DNA synthesis, and (3) the BER patch size ranged from 1 to 8 nucleotides. The misincorporation frequency of BER DNA synthesis at the target site was 5.2×10^{-4} in U251 extracts and 5.4×10^{-4} in LoVo extracts. The most frequent base substitution errors in the U251 and LoVo mutational spectrum were T to G > T to A ≫ T to C. Uracil-initiated BER DNA synthesis in extracts of *E. coli* BH156 (*ung*) BH157 (*dug*), and BH158 (*ung, dug*) was also examined. Efficient BER occurred in extracts of the BH157 strain with a misincorporation frequency of 5.6×10^{-4}. A reduced, but detectable level of BER was observed in extracts of *E. coli* BH156 cells; however, the mutation frequency of BER DNA synthesis was elevated 6.4-fold.   © 2001 Academic Press.

I. Introduction

In both *Escherichia coli* and human cells, the presence of persistent uracil residues in DNA have been shown to mediate cytotoxic, mutagenic, and lethal consequences (1–4). Uracil-initiated DNA base excision repair (BER) is a cellular defense mechanism that acts to maintain genetic integrity by removing uracil residues that accumulate in DNA following deamination of cytosine residues or incorporation of dUMP during DNA synthesis (1, 3, 5, 6). *E. coli* (2, 7), yeast (8, 9), and human cells that are defective in uracil-DNA BER exhibit a mutator phenotype characterized by a predisposition toward C to T base substitutions (Radany and Mosbaugh, unpublished observations).

The fundamental series of chemomechanical steps that comprise uracil-initiated BER appear to be evolutionarily conserved from *E. coli* to humans. BER is initiated when uracil-DNA glycosylase recognizes a uracil residue in DNA and catalyzes the hydrolytic cleavage of the *N*-glycosylic bond that links the uracil base to the deoxyribose phosphate DNA backbone (10). This action releases free uracil and creates an apyrimidinic site in the DNA (10). Incision by a class II AP endonuclease cleaves the phosphodiester bond 5' to the AP site to generate a terminal 3'-hydroxyl-containing nucleotide and a deoxyribose 5'-phosphate (dRP) residue (11, 12). At this point, BER may proceed by one of two pathways that may be distinguished by the extent of DNA synthesis associated with the repair patch size, at least in mammalian cells (13–16).

In the short-patch BER pathway, DNA synthesis at the repair site is confined to one nucleotide. During short-patch repair, DNA polymerase β incorporates a single deoxyribonucleoside monophosphate to fill the one-nucleotide gap left by AP endonuclease, and by means of an associated deoxyribose 5′-phosphatase (dRPase) activity, removes the dRP moiety (15, 17, 18). Removal of dRP appears to constitute the rate-limiting step in short-patch BER and may play an important role in dictating the BER pathway taken (19, 20). Following DNA repair synthesis, DNA ligase activity completes the BER process by restoring the covalent phosphodiester bond (21, 22).

In the other BER pathway, referred to as long-patch BER, DNA repair synthesis may involve incorporation of 2–8 nucleotides per repair event (23, 24). During long-patch repair, after uracil removal and incision by AP endonuclease, the dRP residue is removed along with several downstream nucleotides by DNase IV (also referred to as FEN1) (25, 26). In concert with strand displacement DNA synthesis by DNA polymerase δ, the action of DNase IV is reported to produce predominantly a 5′-dRP-trinucleotide (20, 27). Long-patch BER was initially reported to require proliferating cell nuclear antigen (PCNA), circular double-stranded DNA, and either DNA polymerase δ or ε (24). More recently, PCNA has been reported to promote the DNA polymerase β-dependent long-patch pathway on a double-stranded oligodeoxynucleotide 60-mer through stimulation of DNase IV (25). Several investigations have shown that a circular DNA BER substrate appears to greatly facilitate long-patch BER, whereas duplex oligodeoxynucleotide substrates favor short-patch BER (14, 28). In addition, it has been reported that Pol δ, as well Pol ε, can substitute for Pol β in short-patch repair (29). However, it is thought by many that Pol β is responsible for the large majority of short-patch BER DNA repair synthesis (14, 25, 30). The relative contributions of DNA polymerases β; δ, and ε to uracil-initiated long-patch BER *in vivo* remain to be determined.

While our knowledge of the basic mechanisms of short- and long-patch BER is substantial, we still know very little about the fidelity and error specificity of uracil-initiated BER. To address these issues, we have developed an M13mp2 *lacZα* DNA-based reversion assay for detecting base substitution errors produced during BER in human and *E. coli* cell extracts. In this chapter, we present measurements of the fidelity of uracil-initiated BER and mutational spectra of the misincorporation events.

II. Uracil-Initiated Base Excision DNA Repair Assay

A. Molecular Logic of the Uracil-Initiated BER *lacZα* DNA Reversion Assay

In order to detect DNA base excision repair initiated by uracil excision and to determine the base substitution error frequency associated with completed

BER, we have developed a M13mp2 *lacZα* DNA-based reversion assay (Fig. 1). Site-directed mutagenesis was used to construct a DNA substrate (M13mp2op14) in a two-step modification of M13mp2 DNA. First, the arginine codon (CGT) at *lacZα* nucleotide positions 78–80, which codes amino acid 14 of the *lacZα* gene product, was replaced with an opal codon (TGA); this substitution precluded α-complementation when M13mp2op14 DNA was introduced into the *E. coli* indicator strain, CSH50 (Δ(*lacZ*)M15). Second, a silent mutation in the proline 20 codon at nucleotide position 98 (CC<u>T</u> → CC<u>G</u>) was introduced to create a novel *Sma*I restriction endonuclease recognition site 3′ to opal codon 14, whereas a preexisting *Eco*RI recognition site was located 5′ to the stop codon (Fig. 1A).

To construct the substrate for the BER reaction, single-stranded M13mp2op14 DNA (+ strand) was purified and an oligodeoxynucleotide (5′-CCC AGT CAC GTC UTT GTA AAA CG-3′) was annealed that contained a

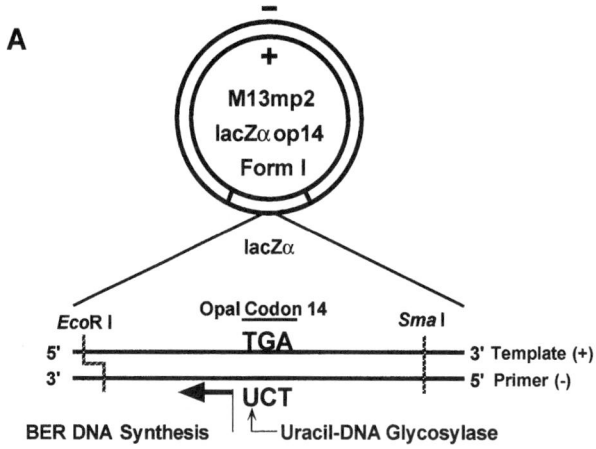

FIG. 1. Scheme for detecting faithful and unfaithful nucleotide incorporation at the uracil target site during base excision DNA repair synthesis.

site-specific uracil residue. Following a primer extension and ligation reaction, the 23-mer oligodeoxynucleotide was incorporated into the (−) strand DNA, and form I heteroduplex M13mp2op14 DNA was isolated as previously described (31). This DNA substrate contained a U·T base mispair at the first position of the opal codon and served as the uracil target for repair. Excision of the uracil residue by uracil-DNA glycosylase initiates the BER pathway, which ultimately concludes with ligation of the newly synthesized (−) strand DNA to form the repair patch (Fig. 1A). The uracil residue was strategically located so that faithful and unfaithful uracil-initiated DNA repair synthesis could be distinguished by the *lacZα* complementation phenotype of the M13mp2 phage genome.

If faithful DNA synthesis occurs during BER opposite the template thymine residue at position 78, a dAMP nucleotide will be incorporated into the (−) strand, and the opal codon will be reestablished in both DNA strands. As the (−) strand DNA serves as the template for production of single-stranded M13 DNA in *E. coli*, the resulting phage are defective in α-complementation, and the plaques they produce are colorless when grown on medium containing isopropyl β-D-thiogalactopyranoside (IPTG) and 5-bromo-4-chloro-3-indolyl β-D-galactopyranoside (X-gal) (Fig. 1B). However, if BER DNA synthesis is unfaithful, dCMP, dGMP, or dTMP may be misincorporated. Each of these base substitutions restores (reverts to wild-type) the α-complementation phenotype; therefore, these mutant phages will produce blue-colored plaques. Incorporation of either dGMP or dTMP restores the wild-type arginine codon (dark blue plaque phenotype), whereas dCMP incorporation results in glycine codon, shown to generate a light blue plaque phenotype (31). Only reversion at the opal codon results in a blue plaque phenotype; misincorporation at other positions in the *lacZα* gene is not detected.

B. Blueprint for Measuring the Fidelity of Nucleotide Incorporation at the Uracil Target

Detection of uracil-initiated base excision DNA repair begins with incubation of form I (U·T) heteroduplex M13mp2op14 DNA (Fig. 2A) with cell extracts that have been supplemented with dNTPs and the components of an ATP regenerating system (31). Various other additions may be made to the cell extract, such as ^{32}P-labeled or 2′-deoxyribonucleoside α-thioltriphosphate DNA precursors, and chemical or protein inhibitors of any of the several steps of BER (31). Initially, a time course of incubation is carried out in order to optimize BER reaction conditions and determine the extent of repair. Aliquots of the reaction products are then analyzed by agarose gel electrophoresis and visualized by ethidium bromide staining in order to distinguish form I (repaired) from form II (unrepaired) DNA (Fig. 2B). The M13mp2op14 DNA substrate is rapidly converted from form I to form II DNA following uracil excision and

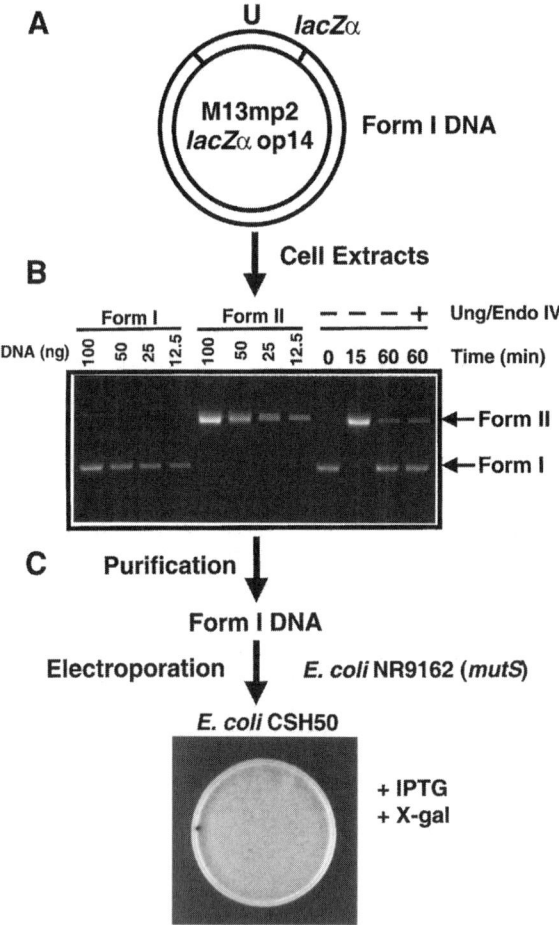

FIG. 2. Procedure for measuring uracil-initiated base excision DNA repair synthesis fidelity in cell extracts.

incision of the resulting apyrimidinic site. As the BER reaction progresses, the form II DNA will be converted back to form I DNA upon conclusion of DNA repair synthesis and ligation of the final intermediate.

After the M13mp2op14 DNA is recovered from the cell extract, it is treated *in vitro* with excess *E. coli* uracil-DNA glycosylase (Ung) and endonuclease IV (Endo IV), as indicated in Fig. 2B. The rationale for this treatment is as follows. The extent of the BER reaction is contingent on the degree of uracil excision. After incubation of the substrate DNA with the cell extract, it is possible that some M13mp2op14 DNA molecules may still contain the target uracil (unreacted substrate). Both unreacted substrate DNA and fully repaired M13mp2op14

DNA migrate as form I molecules during agarose gel electrophoresis, whereas M13mp2op14 DNA that is nicked and/or missing one or more nucleotides (incomplete or aberrant repair) migrates as form II DNA. Fully repaired M13mp2op14 DNA no longer contains dUMP or a nick, as the uracil residue has been excised, the resultant AP site incised, the deoxyribose phosphate moiety removed, and the gap subjected to DNA repair synthesis and ligation. Thus, treatment of uracil-containing M13mp2op14 DNA with excess Ung and Endo IV removes unreacted substrate from the pool of repaired molecules by creating a nicked AP-site-containing form II DNA molecule. In order to quantify the amount of form I and form II product DNA in the agarose gel, form I and form II DNA standards are separated by electrophoresis, and the relative fluorescence of each species determined. This step is necessary, since form I DNA binds significantly less ethidium bromide than form II DNA (*31*).

In order to distinguish faithful from unfaithful DNA repair synthesis events, the Ung/Endo IV-resistant form I DNA is extracted from the agarose gel by electroelution (*31*) and used to electroporate competent *E. coli* NR9162 (*mutS*) cells (Fig. 2C). This strain, which is defective for methyl-directed DNA mismatch repair (*32*, *33*), is used in order to avoid *E. coli*-mediated correction of base–base mispairs that may arise during the uracil-initiated base excision DNA repair reaction (*31*). Transfected NR9162 cells are then combined with the indicator *E. coli* strain CSH50 and plated with top agar containing IPTG and X-gal (*31*). Following overnight incubation at 37°C, the M13mp2op14 plaques may be scored for colorless or blue phenotype. As discussed in Section II,A, faithfully repaired M13mp2op14 DNA produces a clear (colorless) plaque phenotype owing to maintenance of the opal codon 14 and the failure of the truncated *lacZα* gene product to perform α-complementation. In contrast, the unfaithful incorporation of a noncomplementary nucleotide in place of the uracil residue restores the reading frame of the *lacZα* gene and through α-complementation results in a blue plaque phenotype. The frequency of unfaithful DNA synthesis may be calculated by dividing the number of blue plaques produced by the sum of clear and blue plaques. Omission of the Ung/Endo IV treatment and/or failure to separate form II from form I DNA negatively impacts any measurement of the fidelity of uracil-initiated base excision repair DNA synthesis.

III. Detection of Uracil-Initiated DNA Repair in Human Colon Adenocarcinoma LoVo Cell Extracts

A. Uracil-DNA Glycosylase-Initiated DNA Repair

Initial experiments conducted to detect uracil-initiated DNA repair were carried out in extracts of human colon adenocarcinoma LoVo cells using the assay outlined in Fig. 2. LoVo cells were selected because they contain a defective hMSH2 gene and are deficient in DNA mismatch repair (32, 33). Thus, the base substitution errors generated by unfaithful M13mp2op14 DNA repair synthesis in LoVo cell extracts should persist owing to the absence of DNA mismatch repair. The M13mp2op14 DNA substrate was incubated for various times (0–90 min) in LoVo whole-cell extracts prepared as described by Wood et al. (34) and supplemented with [α-^{32}P]dATP. Following product recovery, the M13mp2op14 DNA was treated with excess Ung/Endo IV and the reaction products were resolved by 0.8% agarose gel electrophoresis (Fig. 3). A control experiment was also conducted in which M13mp2op14 DNA was mock-incubated in the absence of LoVo cell extract and subjected (or not) to Ung/Endo IV treatment. Prior to Ung/Endo IV treatment, >95% of the M13mp2op14 DNA migrated as form I molecules, whereas after Ung/Endo IV treatment, no detectable form I DNA was observed (Fig. 3, lanes 2 and 3). These results established that the Ung/Endo IV treatment was sufficient to convert the form I M13mp2op14 uracil-DNA substrate to form II molecules. Examination of the DNA reaction products following incubation in the LoVo cell extract revealed a time-dependent increase in Ung/Endo IV-resistant form I DNA (Fig. 3, lanes 4–9). Maximal repair was observed between 60 and 90 min when ~45% of the recovered M13mp2op14 DNA became resistant to Ung/Endo IV treatment.

B. Ugi-Resistant Uracil-DNA Glycosylase-Initiated DNA Repair

To determine whether uracil-initiated base excision repair was responsible for the production of Ung/Endo IV-resistant form I M13mp2op14 DNA, the Ugi protein was added to the LoVo cell extract. Ugi is an irreversible inhibitor of uracil-DNA glycosylase from many biological sources, including human (35–37), and would therefore be expected to block the BER reaction. As anticipated, in the reaction conducted in the presence of Ugi, a significantly reduced level of Ung/Endo IV-resistant form I DNA was observed (Fig. 3, lanes 10–15). This finding established that the majority of repair events were indeed initiated by a Ugi-sensitive uracil-DNA glycosylase and implied the involvement of the BER pathway. However, several lines of evidence suggested that a secondary uracil-DNA repair system was operant in human cell extracts.

As observed in Fig. 3, and as previously reported by Sanderson and Mosbaugh (31) for human glioblastoma U251 cell extracts, inclusion of excess Ugi in the M13mp2op14 DNA-based BER assay did not completely eliminate the formation of repaired form I DNA. Interestingly, in a standard *in vitro* uracil-DNA glycosylase assay conducted with [^{3}H-uracil] DNA containing U·A target sites, the addition of Ugi reduced the level of endogenous uracil-DNA glycosylase activity below the limit of detection (31). In contrast, similar Ugi treatment of

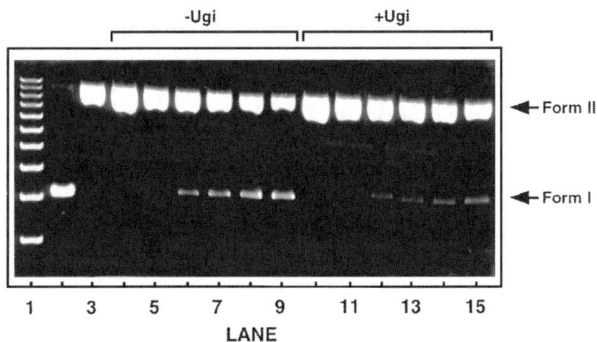

FIG. 3. Agarose gel electrophoresis of M13mp2op14 DNA following uracil-initiated base excision repair in human LoVo cell extracts. Two standard BER reaction mixtures (600 μl) containing 6 μg of M13mp2op14 (U·T) DNA, 1.2 mg of LoVo cell extract protein, and 120 μCi [^{32}P]dATP were prepared as previously described (31). One reaction mixture was supplemented with 1000 units of Ugi (+) and the other contained no Ugi (−) prior to the addition of substrate as indicated. Samples (100 μl) were removed after 0, 15, 30, 45, 60, and 90 min at 30°C from BER reaction mixtures incubated in the absence of Ugi (lanes 4–9, respectively) and the presence of Ugi (lanes 10–15, respectively). Reactions were terminated, and the DNA was isolated, treated with *E. coli* Ung and Endo IV, and prepared for 0.8% agarose gel electrophoresis. As a control, M13mp2op14 (U·T) DNA (1 μg) was mock-treated without cell extract and incubated with Ung and Endo IV (lane 3). Untreated M13mp2op14 (U·T) DNA (100 ng) and a sample containing 2.5 μg of a 1-kb DNA ladder (Gibco BRL) were analyzed as reference standards (lanes 2 and 1, respectively). The location of ethidium bromide-stained form I and II DNA bands are indicated by arrows.

either LoVo or U251 cell extracts significantly diminished, but did not eliminate uracil-DNA repair as detected by the M13mp2op14 DNA-based reversion assay. Thus, in both cell extracts Ung/Endo IV-resistant form I DNA was produced, albeit at a much reduced level (Fig. 3) (31).

The addition of exogenous human uracil-DNA glycosylase [UDGΔ84 (38)] to the Ugi-containing reaction in subsaturating amounts did not stimulate production of repaired form I M13mp2op14 DNA (data not shown). Therefore, it seemed unlikely that the uracil-DNA repair detected in the presence of Ugi resulted from uracil-DNA glycosylase that had escaped inhibition by Ugi. These results suggested that an alternative enzyme may recognize the uracil target residue and initiate repair, an enzyme that was resistant to inactivation by Ugi. In support of this proposition, a novel Ugi-resistant uracil-DNA glycosylase activity has been detected in human glioblastoma U251 mugi-17 (Radany and Mosbaugh, unpublished observations) and HeLa cells (39). Taken together, these observations imply that some repair of M13mp2op14 DNA occurred without the involvement of a Ugi-sensitive uracil-DNA glycosylase.

To ascertain the location of repair synthesis on the M13mp2op14 DNA substrate, [α-^{32}P]dATP was included in the standard BER reaction described above.

Using extracts from both LoVo and U251 cells, reactions were conducted with M13mp2op14 DNA that contained either an A·T base pair or a U·T mispair at position 78 of the *lacZα* gene. The two DNA substrates were used to distinguish uracil target-specific from nonspecific DNA synthesis. Following the BER reactions, the [^{32}P]DNA was isolated and digested with the restriction endonuclease *Hin*fI. M13mp2op14 DNA contains 26 *Hin*fI recognition sites distributed about the circular molecule; however, only the unique, 529-bp fragment encompasses the uracil target site. The DNA fragments produced by *Hin*fI digestion were resolved by nondenaturing polyacrylamide gel electrophoresis, and those fragments into which [^{32}P]dAMP was incorporated were subsequently detected by autoradiography (Fig. 4).

Analysis of the autoradiogram showed that in both U251 and LoVo cell extracts, preferential incorporation of [^{32}P]dAMP occurred specifically in the 529-bp DNA fragment of M13mp2op14 (U·T) DNA (Fig. 4, lanes 2 and 4).

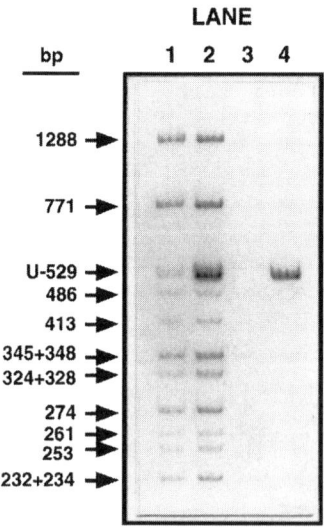

FIG. 4. Analysis of uracil-initiated base excision repair-specific DNA synthesis in human U251 and LoVo cell extracts. BER reaction mixtures (200 µl) containing 0.4 mg of human cell extract protein (U251, lanes 1 and 2; LoVo, lanes 3 and 4), 2 µg of M13mp2op14 (A·T) DNA (lanes 1 and 3) or M13mp2op14 (U·T) DNA (lane 2 and 4), and 40 µCi of [^{32}P]dATP were incubated at 30°C for 60 min (U251) or 45 min (LoVo). Each reaction was terminated with 2000 units of Ugi protein and adjusted to 20 m*M* EDTA. DNA was isolated, and samples (2.5 µl, ~50 ng) were removed for digestion with 5 units of *Hin*fI for 1 h at 37°C. The resulting DNA restriction fragments were resolved by 5% nondenaturing polyacrylamide gel electrophoresis and the ^{32}P-labeled DNA was detected by autoradiography. The location of the uracil-containing fragment (U-529 bp) and other *Hin*fI DNA fragments are indicated by arrows.

In contrast, HinfI fragment-specific [^{32}P]dAMP incorporation was not seen in M13mp2op14 (A·T) DNA (Fig. 4, lanes 1 and 3), and the overall level of [^{32}P]dAMP incorporation into this substrate was significantly reduced relative to M13mp2op14 (U·T) DNA. Uracil-dependent, target fragment-specific [^{32}P]dAMP incorporation was observed in both LoVo and U251 cell extracts, and the extent of [^{32}P]dAMP incorporation into other HinfI DNA fragments was essentially constant for both DNA substrates. These results were consistent with a uracil-initiated BER mechanism.

Additional observations further supported the interpretation that the [^{32}P]dAMP incorporation detected in the 529-bp DNA fragment was attributable to BER DNA synthesis. Sanderson and Mosbaugh (*31*) demonstrated that in reactions containing U251 cell extracts (1) 5.2- to 7.6-fold more [^{32}P]dAMP incorporation occurred in the 529-bp uracil target fragment relative to the similar-sized 486-bp fragment located on the opposite side of the M13mp2op14 DNA molecule; (2) the two DNA fragments (253-bp and 261-bp) flanking the uracil target fragment showed low levels of [^{32}P]dAMP incorporation; (3) DNA synthesis in the 529-bp fragment was confined specifically to the (−) strand DNA, where the uracil target was originally located; and (4) the addition of Ugi to U251 cell extracts significantly inhibited DNA repair synthesis. Similar observations were made for BER reactions conducted with LoVo cell extracts (data not shown). While the amount of [^{32}P]dAMP incorporation in the 529-bp fragment containing the uracil target was similar for reactions conducted with U251 or LoVo cell extracts, the extent of nonspecific DNA synthesis was obviously different (Fig. 4). HinfI fragments that did not contain uracil exhibited a 4.8- to 6.6-fold reduced level of [^{32}P]dAMP incorporation in reactions with U251 cell extract, whereas a 16.2- to 42.6-fold reduction in nonspecific [^{32}P]dAMP incorporation was observed for BER reactions conducted with LoVo cell extracts. Whether this relative reduction of nonspecific DNA synthesis resulted as a consequence of the hMSH2 defect in LoVo cells remains to be determined.

IV. Uracil-DNA Repair Patch Size Produced by BER in LoVo Cell Extracts

We have developed an assay to determine the patch size of DNA repair synthesis associated with uracil-initiated BER that is based on the resistance of DNA containing 2′-deoxyribonucleoside α-thiolmonophosphates to degradation by *E. coli* exonuclease III (Exo III). For this purpose, the M13mp2op14 (U·T) DNA substrate was modified to include a site-specific [^{32}P]dCMP residue located at nucleotide position 90 on the (−) strand situated between the uracil target and the SmaI restriction endonuclease recognition site (Fig. 5A). Using

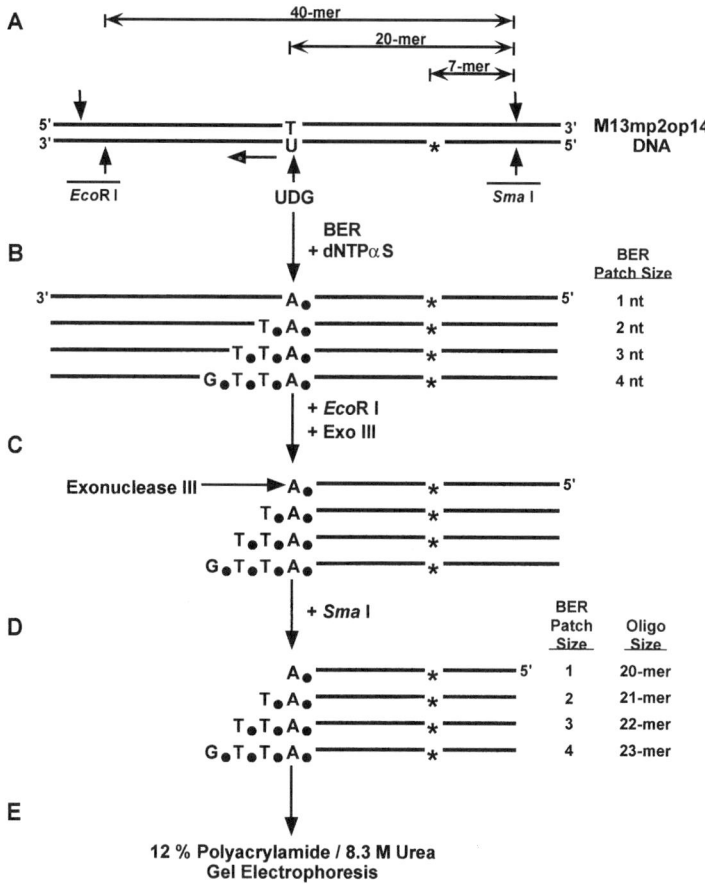

FIG. 5. Scheme for determining uracil-initiated base excision DNA repair synthesis patch size.

this M13mp2op14 [^{32}P]DNA substrate, standard BER reactions were conducted with LoVo cell extracts, except that the four complementary 2′-deoxyribonucleoside triphosphates were replaced with the corresponding 2′-deoxyribonucleoside α-thioltriphosphates. As a result, various amounts of dNMP[α-S] were incorporated into the [^{32}P]DNA during DNA repair synthesis, and the extent of dNMP[α-S] incorporation corresponded to the repair patch size (Fig. 5B).

In order to locate the 3′ boundary of the repair patch produced during the uracil-initiated BER DNA synthesis, the recovered M13mp2op14 [^{32}P]DNA was cleaved with *Eco*RI to generate linear DNA with a recessed 3' terminus located on the (−) strand 20 nucleotides downstream from the uracil target residue. The (−) strand of the linearized [^{32}P]DNA was then digested in the 3′ to 5′ direction by Exo III (Fig. 5C). We anticipated that the 3′ to 5′ degradation of the (−) strand would terminate at the 3′ border of the repair patch because it had been reported that Exo III does not hydrolyze phosphorothioate DNA linkages (*40*).

Subsequent to Exo III treatment, the [^{32}P]DNA was restricted with SmaI to generate a DNA fragment containing a ^{32}P-labeled oligodeoxynucleotide whose length was diagnostic of the BER patch size (Fig. 5D). For example, a repair patch size of one nucleotide would be expected to produce a [^{32}P]oligodeoxynucleotide 20-mer, since the incorporation of a single dAMP[α-S] at the uracil target site would block digestion by Exo III at a location 20 nucleotides from the 5′ end created by SmaI restriction. Similarly, repair patches consisting of two or more dNMP[α-S] incorporations were expected to produce ^{32}P-labeled oligodeoxynucleotides of increasing size (i.e., 21-, 22-, 23-mer). The various ^{32}P-labeled DNA fragments produced in the patch size assay were resolved by denaturing 12% polyacrylamide gel electrophoresis and detected by autoradiography (Fig. 5E). This assay was used to (1) establish the nature of the BER patch size created by uracil-initiated DNA repair synthesis, (2) investigate the effect of aphidicolin on uracil-DNA repair synthesis, (3) study the influence of Ugi-mediated inhibition of uracil-DNA repair on patch size, and (4) elucidate the effect of the combined treatment of aphidicolin and Ugi on patch size.

A. Effect of Aphidicolin on Uracil-DNA Repair Synthesis

Prior to examining the effect of aphidicolin on repair patch size, the general influences of aphidicolin on DNA polymerase activity and BER efficiency in human cell extracts were investigated. *In vitro* DNA synthesis reactions utilizing calf thymus DNA and LoVo cell extract were conducted to assess the influence of the fungal antibiotic on the class B (α-like) family of DNA polymerases (*41*). As anticipated, a significant fraction of the DNA polymerase activity in the LoVo cell extract was sensitive to inhibition by aphidicolin (Fig. 6A). Maximum inhibition, approximately 80% reduction of DNA polymerase activity, was achieved at ~25 μM aphidicolin, and the residual activity was attributed to aphidicolin-resistant DNA polymerases, such as DNA polymerase β, ζ, or η (*41–43*).

The influence of aphidicolin on the extent of uracil-initiated BER was examined in LoVo cell extracts using the M13mp2op14 DNA substrate described in Fig. 3. Quantitative analysis of the form I and II DNA reaction products, resolved by agarose gel electrophoresis and stained with ethidium bromide, revealed that the percentage of repaired form I DNA did not vary significantly when LoVo cell extracts were supplemented with aphidicolin from 1 to 50 μM (Fig. 6B). For each aphidicolin concentration tested, 48–55% of the M13mp2op14 DNA substrate was recovered as Ung/Endo IV-resistant form I molecules. Moreover, the extent of uracil-initiated BER in the presence or absence of aphidicolin was very similar. Thus, the efficiency of uracil-initiated BER was not particularly influenced by aphidicolin-induced inhibition of α-like DNA polymerases.

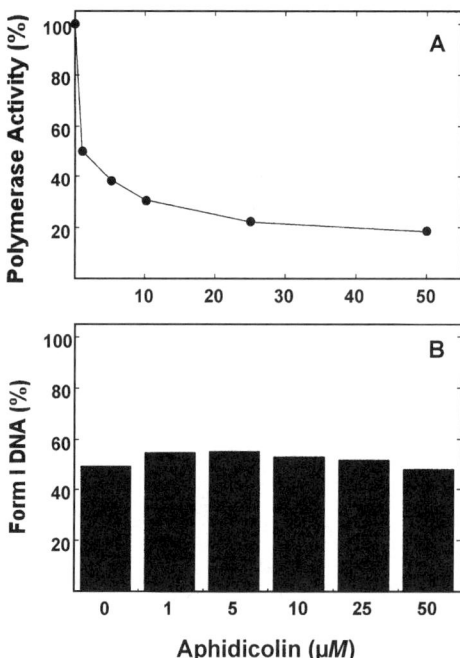

FIG. 6. Effect of aphidicolin on DNA polymerase activity and uracil-DNA base excision repair in human LoVo cell extracts. (A) DNA polymerase activity was assayed under standard BER reaction conditions in reaction mixtures (100 μl) containing 20 μg of LoVo cell extract protein, 10 μg of activated calf thymus DNA, and [^{3}H]dTTP (1700 cpm/pmol). Reaction mixtures were supplemented with 0, 1, 5, 10, 25, and 50 μM aphidicolin in DMSO prior to the addition of substrate and incubated for 45 min at 30°C. Reactions were terminated on ice, DNA precipitated, and acid-insoluble DNA was analyzed for [^{3}H]dTMP incorporation. Incorporation of 9.8 pmol of [^{3}H]dTMP corresponded to 100% DNA polymerase activity. (B) Standard BER reaction mixtures (100 μl) containing 1 μg of M13mp2op14 (U·T) DNA, 200 μg of LoVo cell extract protein, and 20 μCi of [^{32}P]dATP were supplemented with aphidicolin as indicated above. Samples were incubated for 45 min at 30°C and terminated by addition of 2000 units of Ugi and 20 mM EDTA. DNA was isolated, treated with *E. coli* Ung and Endo IV, and analyzed by 0.8% agarose gel electrophoresis as described in Fig. 3. Form I and form II DNA detected by ethidium bromide staining were quantified with a Gel Documentation System, and the percentage of form I DNA was determined. The amount of form I and II DNA was measured relative to standards (6.3–100 ng) analyzed on the same gel.

Experiments were then conducted to establish the size and distribution of the DNA repair synthesis patch produced during uracil-initiated BER. The patch size substrate, M13mp2op14 [^{32}P]DNA, was characterized in two control reactions. Unreacted M13mp2op14 [^{32}P]DNA was either (1) restricted with *Eco*RI and *Sma*I, or (2) treated with Exo III after *Eco*RI digestion but prior to *Sma*I restriction. Inspection of the autoradiogram shows that restriction of

M13mp2op14 [^{32}P]DNA with EcoRI and SmaI produces the expected ^{32}P-labeled oligodeoxynucleotide 40-mer product (Fig. 7, lane 1). Furthermore, in the absence of dNMP[α-S] incorporation during BER, the M13mp2op14 [^{32}P]DNA substrate was completely susceptible to degradation by Exo III (Fig. 7, lane 2).

When all four 2′-deoxyribonucleoside α-thioltriphosphate precursors were included in the BER reaction with LoVo cell extract, restriction of the M13mp2op14 [^{32}P]DNA with EcoRI and SmaI again produced the anticipated 40-mer [^{32}P]DNA fragment (Fig. 7, lane 3). However, when linearized by EcoRI cleavage and subjected to degradation by Exo III, the M13mp2op14 [^{32}P]DNA proved refractory to complete digestion (Fig. 7, lanes 4–6). As a consequence of dNMP[α-S] incorporation, a series of [^{32}P]DNA fragments 20–27 nucleotides

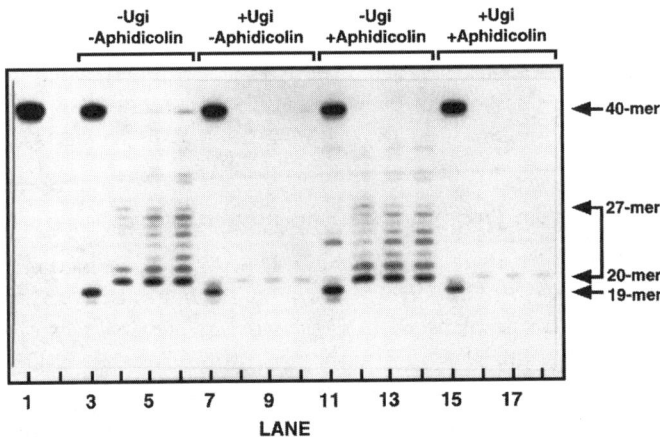

FIG. 7. Effect of Ugi and aphidicolin on the 3′ boundary of DNA synthesis introduced by uracil-initiated base excision repair in human LoVo cell extracts. Standard BER reaction mixtures (500 μl) containing 5 μg of M13mp2op14 (U·T) [^{32}P]DNA, 1 mg of human LoVo cell extract protein, and 20 μM each of dATP[α-S], dTTP[α-S], dCTP[α-S], and dGTP[α-S] were supplemented with no addition (lanes 3–6), 2000 units of Ugi (lanes 7–10), 50 μM aphidicolin (lanes 11–14), or 2000 units of Ugi and 50 μM aphidicolin (lanes 15–18). Reactions were incubated at 30°C for 45 min and DNA reaction products were isolated. Samples (8 μl, ~500 ng) were removed for digestion with 25 units of EcoRI for 1 h at 25°C. The linearized DNA was then supplemented with no addition (lanes 3, 7, 11, 15) or the addition of 20 units (lanes 4, 8, 12, 16), 2 units (lanes 5, 9, 13, 17), or 0.2 units (lanes 6, 10, 14, 18) of E. coli Exo III for 30 min at 37°C. Following Exo III digestion, reaction products were restricted with 25 units of SmaI for 1 h at 25°C. As controls, unreacted M13mp2op14 (U·T) [^{32}P]DNA (~500 ng) was incubated with EcoRI and SmaI (lane 1) or EcoRI, Exo III (0.2 units), and SmaI (lane 2). Following these reactions, DNA reaction products were resolved by 12% polyacrylamide/8.3 M urea gel electrophoresis. The location of the 40-mer corresponding to the (−) strand EcoRI/SmaI restriction fragment and the 19-mer through 27-mer BER reaction products are located by arrows on the autoradiogram.

in length were detected; these fragments correspond to DNA repair patches of 1–8 nucleotides, respectively. Under the reaction conditions used, ~32% of repair synthesis events resulted in a DNA repair patch of one nucleotide. The 19-mer [^{32}P]DNA fragment observed in Fig. 7 (lane 3) most likely corresponds to the BER intermediate following uracil excision and AP-site incision, as this fragment was sensitive to Exo III digestion.

To assess the influence of aphidicolin on the repair patch size, we included aphidicolin in addition to the dNTP[α-S] precursors in the BER reaction with LoVo cell extract, and repeated the experiment described above. Restriction of the [^{32}P]DNA reaction product yielded predominantly a [^{32}P]oligodeoxynucleotide 40-mer (Fig. 7, lane 11), and the amount of ^{32}P-labeled 40-mer detected was approximately equal to that detected in the reaction lacking aphidicolin (Fig. 7, lane 3). These results were consistent with our previous observation that aphidicolin did not appreciably inhibit the BER reaction (Fig. 6B). Significantly, the addition of aphidicolin did not appear to alter the size or distribution of the DNA repair patches produced by BER DNA synthesis (Fig. 7, lanes 12–14). In the presence of aphidicolin, ~30% of the BER was attributed to one-nucleotide DNA repair synthesis. We infer from this result that α-like DNA polymerases are not specifically required to carry out long-patch BER synthesis of 2–8 nucleotides.

B. Effect of Ugi on Uracil-DNA Repair Synthesis

Although the majority of uracil-initiated DNA repair observed in LoVo cell extracts was sensitive to inhibition by Ugi, a minor fraction of BER was apparently initiated by a Ugi-resistant uracil-DNA glycosylase activity. This observation prompted us to examine the BER patch size attributed to uracil-DNA repair conducted in the presence of Ugi. Following the addition of excess Ugi to the LoVo cell extract, uracil-DNA repair patch size assays were conducted using the M13mp2op14 [^{32}P]DNA substrate as described earlier (Figs. 5 and 7). Treatment of the BER reaction product with EcoRI and SmaI successfully released the 40-mer [^{32}P]DNA fragment (Fig. 7, lane 7). However, the vast majority of this DNA had not undergone BER since it was in large part sensitive to complete digestion by Exo III (Fig. 7, lanes 8–10). These results are consistent with those obtained in Fig. 3.

Upon detailed examination of the BER patch size distribution using various autoradiographic exposures of the polyacrylamide gel, we determined that ~46% of the Ugi-resistant BER events generated a ^{32}P-labeled 20-mer product that resulted from a one-nucleotide repair patch (Fig. 7, lanes 8–10). In addition to this short-patch BER product, ~54% of the repair events involved long-patch BER and consisted of repair patches 2–8 nucleotides long. Thus, the distribution of short- and long-patch BER products was quite similar in the Ugi-sensitive and Ugi-resistant BER pathways.

When BER reactions were conducted in the presence of both Ugi and aphidicolin, the results were similar to those obtained with Ugi alone (Fig. 7, lanes 15–18). However, the overall amount of uracil-DNA repair was slightly reduced as was the fraction of short-patch (one-nucleotide) BER, as short- and long-patch repair constituted ~27% and ~73% of the total BER events detected. Taken together, these results imply that an aphidicolin-resistant DNA polymerase acts in both short- and long-patch DNA repair synthesis, and that both a Ugi-sensitive and Ugi-resistant uracil-DNA glycosylase activity can initiate a uracil-DNA BER pathway.

V. Mutations Produced by Uracil-Initiated Base Excision DNA Repair Synthesis

The fidelity of uracil-initiated base excision DNA repair was examined in human U251 and LoVo cell extracts, as well as in extracts of *E. coli* strains BH156, BH157, and BH158, using the M13mp2op14 *lacZα* DNA-based reversion assay, as discussed above. The mutation frequency at opal codon 14 was determined by analysis of the M13 phage plaque phenotypes that resulted from proficient or deficient α-complementation on host indicator plates (Fig. 2). The presence of a light or dark blue plaque phenotype indicated that an error had occurred during base excision DNA repair. Thus, the ratio of blue plaques to total plaques produced corresponded to the error frequency of BER at the uracil target. Hence, the reversion frequency of *lacZα*-complementation directly reflected the error frequency of complete BER.

A. Mutation Frequency

We first examined the background reversion frequency associated with the M13mp2op14 *lacZα* DNA-based reversion assay caused by nonspecific DNA synthesis and degradation of the DNA substrate in the cell extracts. To this end, we incubated the M13mp2op14 (A·T) DNA substrate with the U251 and LoVo cell extracts and subsequently determined the reversion frequency. The M13mp2op14 (A·T) DNA substrate contained a dAMP residue opposite the first nucleotide of the opal codon (nucleotide position 78), and therefore did not present an opportunity for uracil-DNA glycosylase-mediated initiation of BER. The reversion frequency of M13mp2op14 (A·T) DNA in U251 and LoVo cell extracts was determined to be 0.24×10^{-4} and 0.31×10^{-4}, respectively (Table I). These values compared favorably with the reversion frequency obtained (0.079×10^{-4}) when reaction buffer was substituted for cell extract during incubation (*31*), indicating that a minimal amount of nonspecific processing occurred in the absence of the uracil target.

TABLE I
FREQUENCY OF MUTATIONS PRODUCED BY URACIL-INITIATED BER IN HUMAN U251 AND LoVo CELL EXTRACTS[a]

Human CE	DNA substrate $(-/+)$[c]	Protein addition	Plaques scored		Reversion frequency[b] $(\times 10^{-4})$
			Total	Blue	
U251	A·T	—	532,480	13	0.24
	U·T	—	150,705	78	5.2
	U·T	Ugi	350,428	993	28.3
LoVo	A·T	—	222,820	7	0.31
	U·T	—	121,290	66	5.4
	U·T	Ugi	163,670	191	11.7

[a]Standard base excision DNA repair reaction mixtures (500 μl) were prepared containing 1 mg of human U251 or LoVo cell extract (CE) protein and either 5 μg of M13mp2op14 (U·T) or (A·T) DNA. After incubation at 30°C for 60 min (U251) or 45 min (LoVo), reactions were terminated, DNA products recovered, and form I DNA that was resistant to *E. coli* Ung/Endo IV treatment was isolated by 0.8% agarose gel electrophoresis as described in Fig. 3. Form I DNA was then transfected into *E. coli* NR9162 cells and the M13mp2 *lacZα* DNA-based reversion assay was performed as described by Sanderson and Mosbaugh (*31*).

[b]Reversion frequencies were calculated by dividing the number of blue plaques scored by the total number of blue plus colorless plaques. Revertants included dark blue and light blue phenotypes.

[c]The (−) and (+) strand nucleotide at the target site.

When M13mp2op14 (U·T) DNA was incubated with U251 cell extracts and processed, a reversion frequency of 5.2×10^{-4} was observed (Table I). This value is considerably greater (~22-fold) than that obtained for M13mp2op14 (A·T) DNA, and is indicative of the level of misincorporation events associated with uracil-initiated base excision repair. When M13mp2op14 (U·T) DNA was incubated with U251 cell extract in the presence of Ugi, the reversion frequency determined was 28.3×10^{-4} (Table I). Therefore, Ugi addition elevates the reversion frequency associated with U251 cells by more than 5-fold, demonstrating that the Ugi-resistant repair pathway is more error-prone than the uracil-DNA glycosylase-initiated repair pathway in this cell line.

To address the question whether DNA mismatch repair capacity might influence the fidelity of uracil-initiated base excision DNA repair, we utilized the human colon adenocarcinoma LoVo cell line. LoVo cells are homozygous for a deletion in the hMSH2 gene and are defective in base–base DNA mismatch repair (*32, 33*). The reversion frequency of M13mp2op14 (U·T) DNA repaired in LoVo cell extracts was determined to be 5.4×10^{-4} (Table I). This value was essentially indistinguishable from that obtained for U251 cells (5.2×10^{-4}). Therefore, it would appear that any influence DNA mismatch repair may have on the frequency of misincorporation events associated with uracil-initiated base excision DNA repair in this system is below the limit of detection of the assay.

We also determined the fidelity of Ugi-resistant uracil-DNA repair in LoVo cell extracts by supplementing the BER repair reaction with Ugi as described above for U251 extracts: the reversion frequency observed was 11.7×10^{-4} (Table I). This value is ~2-fold greater than that obtained in the absence of Ugi for LoVo cells. In both cases, the presence of Ugi provoked a more error-prone uracil-initiated BER process. The reason why Ugi appears to have less of an impact on the fidelity of BER in LoVo cell extracts is unclear, and may or may not be related to this cell line's defect in DNA mismatch repair.

We also examined the fidelity of uracil-initiated base excision DNA repair in the *E. coli* strains BH156, BH157, and BH158, using M13mp2op14 (U·T) DNA (Table II). These *E. coli* strains are isogenic except for inactivating mutations in either or both of the *ung* and *dug* genes (44). The *ung* gene codes for uracil-DNA glycosylase, which this laboratory (37, 45–47) and others (10, 48, 49) have extensively characterized. We suggest that the gene encoding double-strand uracil-DNA glycosylase be given the acronym *dug*, in accordance with accepted *E. coli* nomenclature, rather than the designation *mug* (for mismatch-specific uracil-DNA glycosylase), as the latter term does not accurately describe the enzymatic properties of the gene product (J.-S. Sung and D. W. Mosbaugh, unpublished observations).

The reversion frequency obtained for *E. coli* BH157, which is wild-type for the *ung* gene, was 5.6×10^{-4} (Table II). This value is very similar to the mutation

TABLE II
FREQUENCY OF MUTATIONS PRODUCED BY URACIL-INITIATED BER IN VARIOUS *E. coli* CELL-FREE EXTRACTS[a]

E. coli extract	Allele	DNA substrate $(-/+)$[c]	Plaques scored		Reversion frequency[b] $(\times 10^{-4})$
			Total	Blue	
BH156	$ung^- dug^+$	U·T	239,154	873	36
BH157	$ung^+ dug^-$	U·T	1,229,600	690	5.6
BH158	$ung^- dug^-$	U·T	ND[d]	ND	ND

[a] Base excision DNA repair reaction mixtures (100 μl) were prepared containing 0.1 mg of *E. coli* BH156, BH157, or BH158 cell-free extract and 1 μg of M13mp2op14 (U·T) DNA. After incubation at 30°C for 60 min, reactions were terminated. DNA products were recovered and form I DNA that was resistant to *E. coli* Ung/Endo IV treatment was isolated by 0.8% agarose gel electrophoresis. Form I DNA was then transfected into *E. coli* NR 9162 cells and the M13mp2op14 *lacZα* DNA-based reversion assay was performed essentially as described in Table I.

[b] Reversion frequencies were calculated by dividing the number of blue plaques scored by the total number of blue plus colorless plaques. Revertants included dark blue and light blue phenotypes.

[c] The $(-)$ and $(+)$ strand nucleotide at the target site.

[d] ND indicates not determined since form I DNA was not detected following the base excision DNA repair reaction containing the *E. coli* BH158 cell extract.

frequencies obtained for the human U251 (5.2×10^{-4}) and LoVo (5.4×10^{-4}) cells, which also contain a Ugi-sensitive uracil-DNA glycosylase activity. In contrast, the reversion frequency observed in extracts of BH156 was 36×10^{-4} (Table II). The genotype of BH156, as it concerns uracil excision, is $ung^- \, dug^+$. Since dug encodes a Ugi-resistant uracil-DNA glycosylase activity, the uracil-excision activity profile of BH156 extracts is similar to that of the human cell extracts supplemented with Ugi (J.-S. Sung and D. W. Mosbaugh, unpublished observation). This comparison is supported by the observation that the mutation frequency in BH156, like those observed in human cell extracts containing Ugi, is substantially greater (6.4-fold) than in the uracil-DNA glycosylase-proficient strain, BH157 (Table II). Interestingly, we were unable to detect uracil-DNA repair in BH158, in which both ung and dug are defective. This observation suggests that Ung and Dug may be the only activities in E. coli capable of initiating uracil-excision DNA repair in this system.

B. Mutational Spectrum

To define the type of misincorporation event that produced the individual revertant M13 phage, 78 blue plaques containing M13mp2op14 (U·T) DNA repaired in U251 cell extracts were subjected to DNA sequence analysis of the reversion target, opal codon 14. The distribution of single-base substitutions located within the target sequence is shown in Fig. 8A. Seventy-four (95%) of the mutations occurred at the first nucleotide position of the opal codon, one mutation (1%) was detected at the second nucleotide position, and three (4%) were observed at the third nucleotide position. Clearly, almost all base substitutions occurred at the location of the uracil target. Transversion mutations represented 94% of the errors at this site, and were divided nearly equally between T to G and T to A changes in the (+) strand template; four sequences scored as T to C transition mutations.

In an analogous fashion, DNA sequence information was acquired from 59 of the 66 blue plaques produced in BER reactions containing LoVo cell extract. The distribution and specificity of these base substitution mutations are shown in Fig. 8B. Fifty-four (90%) of the mutations occurred at the first nucleotide position of the opal codon, and six (10%) were detected at the third nucleotide position; one double mutant with base substitutions at the first and third nucleotide positions was detected. As was the case with U251 cell extracts, almost all of the errors observed in M13mp2op14 (U·T) DNA repaired in LoVo cell extracts occurred at the uracil target site. Of the base substitutions detected at this position, 91% promoted transversion mutations of which 61% were T to G, and 39% T to A, changes in the (+) strand template; five mutations (9%) scored as T to C transitions.

URACIL-INITIATED BER SYNTHESIS FIDELITY 185

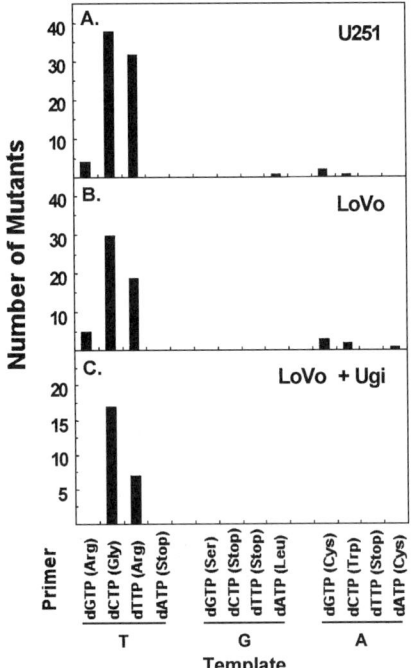

FIG. 8. Mutation spectrum of uracil-initiated base excision DNA repair synthesis. Standard BER reactions were carried out using M13mp2op14 (U·T) DNA and human LoVo cell extract as described in Table I. Following transfection of Ung/Endo IV resistant form I DNA into *E. coli* NR9162 cells, blue plaques were isolated, single-stranded DNA purified, and DNA sequenced over the *lacZα* gene. The nucleotide sequence (TGA) for the opal codon opposite the transcribed (−) strand serves as the template strand for uracil-initiated base excision DNA repair synthesis and is indicated as "Template." Above each template nucleotide, the four possible incoming deoxyribonucleoside triphosphate precursors are indicated together with the coded amino acid in parenthesis. The numbers of individual base-substitution mutations detected by DNA sequence analysis are plotted for reaction mixtures containing either U251 (A) or LoVo (B) cell extract, or LoVo cell extract incubated in the presence of Ugi (C), as indicated.

A limited investigation of mutations introduced during the Ugi-resistant uracil-DNA repair pathway in LoVo cell extracts was also conducted. DNA sequence information was obtained from 24 of the 191 blue plaques detected, and the distribution and specificity of mutations is shown in Fig. 8C. All mutations occurred at the first nucleotide position and were scored as transversion mutations, of which 71% were T to G changes, and 29% T to A changes in the (+) strand template. Thus, similar mutational spectra were observed for completed uracil-initiated BER conducted in both U251 and LoVo cell extracts, as well as for Ugi-sensitive and Ugi-resistant uracil-DNA repair in LoVo cell extracts. One possible interpretation of these results would suggest that the same BER activities are utilized in each system for mutation fixation. This interpretation does not necessarily imply that only one pathway is operational, but rather that the contribution(s) from one or more BER pathways and/or DNA polymerases are represented approximately equally in each system.

VI. Concluding Remarks

It has been known for many years that a deficiency in cellular uracil-initiated BER confers a mutator phenotype resulting in an elevated spontaneous mutation rate. However, an understanding of the error frequency associated with proficient uracil-initiated BER has only recently been considered. We have determined the fidelity of completed uracil-initiated base excision DNA repair in extracts of human and bacterial cells by using a covalently closed circular duplex DNA substrate containing a defined uracil target. We found that human cells carried out uracil-initiated BER along both the short- and long-patch repair pathways. However, the fidelity of short-patch versus long-patch BER is an issue that remains to be addressed. The fidelity assay developed herein detects only misincorporations located at the uracil-substituted *lacZα* reversion target, as errors that occur downstream of the uracil excision site on the (−) strand are not observed. As designed, the BER fidelity assay measured the occurrence of errors that were the end product of the DNA synthesis and ligation steps. What remains to be elucidated is the relative contribution, *in vivo*, of specific DNA polymerases and DNA ligases to mutation fixation in both the short- and long-patch BER pathways.

We also observed that the initiation of uracil-mediated BER may proceed via a Ugi-sensitive or a Ugi-resistant avenue in both human and bacterial cells, and that the fidelity associated with the Ugi-resistant mode is considerably lower than that observed for Ugi-sensitive BER. Whether the Ugi-resistant avenue of uracil-initiated DNA repair represents a novel BER pathway is a question that awaits further investigation, as does determination of the underlying causes of the increased error frequency associated with this form of BER. In addition to the natural accumulation of uracil-DNA, the cell must also deal with a wide variety of spontaneous and environmentally induced DNA damage. Is the fidelity of non-uracil-initiated BER comparable to that of uracil-initiated BER? If not, what is the impact on genomic stability? Finally, the effect of sequence context in determining the type and frequency of mutation made during BER awaits examination. These and many more stimulating questions await biochemists in the twenty-first century.

Acknowledgments

We thank Christine Mosbaugh for technical assistance and Nanci Adair for conducting the DNA sequence analysis through the Center for Gene Research and Biotechnology at Oregon State University. We also thank Dr. A. S. Bhagwat at Wayne State University for his generous gift of the *E. coli* strains BH156, BH157, and BH158. This work was supported by the National Institutes of Health Grants GM32823 and ES00210.

References

1. T. Lindahl, *Prog. Nucleic Acid Res. Mol. Biol.* **22**, 135–192 (1979).
2. B. K. Duncan and B. Weiss, *J. Bacteriol.* **151**, 750–755 (1982).
3. D. W. Mosbaugh and S. E. Bennett, *Prog. Nucleic Acid Res. Mol. Biol.* **48**, 315–370 (1994).
4. H. A. Ingraham, L. Dickey, and M. Goulian, *Biochemistry* **25**, 3223–3230 (1986).
5. H. Hyatsu, *Prog. Nucleic Acid Res. Mol. Biol.* **16**, 75–124 (1976).
6. L. A. Frederico, T. A. Kunkel, and B. R. Shaw, *Biochemistry* **29**, 2532–2537 (1990).
7. B. K. Duncan and J. H. Miller, *Nature (London)* **287**, 560–561 (1980).
8. Z. Wang, X. Wu, and E. C. Friedberg, *Mol. Cell. Biol.* **13**, 1051–1058 (1993).
9. P. M. Burgers and M. B. Klein, *J. Bacteriol.* **166**, 905–913 (1986).
10. T. Lindahl, S. Ljungquist, W. Siegert, B. Nyberg, and B. Sperens, *J. Biol. Chem.* **252**, 3286–3294 (1977).
11. D. W. Mosbaugh and S. Linn, *J. Biol. Chem.* **255**, 11743–11752 (1980).
12. J. D. Levin and B. Demple, *Nucleic Acids Res.* **18**, 5069–5075 (1990).
13. G. Frosina, P. Fortini, O. Rossi, F. Carrozzino, G. Raspaglio, L. S. Cox, D. P. Lane, A. Abbondandolo, and E. Dogliotti, *J. Biol. Chem.* **271**, 9573–9578 (1996).
14. P. Fortini, B. Pascucci, E. Parlanti, R. W. Sobol, S. H. Wilson, and E. Dogliotti, *Biochemistry* **37**, 3575–3580 (1998).
15. D. W. Mosbaugh and S. Linn, *J. Biol. Chem.* **258**, 108–118 (1983).
16. D. W. Mosbaugh and S. Linn, *J. Biol. Chem.* **259**, 10247–10251 (1984).
17. Y. Matsumoto and K. Kim, *Science* **269**, 699–702 (1995).
18. H. Pelletier, M. R. Sawaya, W. Wolfle, S. H. Wilson, and J. Kraut, *Biochemistry* **35**, 12742–12761 (1996).
19. D. K. Srivastava, B. J. V. Berg, R. Prasad, J. T. Molina, W. A. Beard, A. E. Tomkinson, and S. H. Wilson, *J. Biol. Chem.* **273**, 21203–21209 (1998).
20. R. Prasad, G. L. Dianov, V. A. Bohr, and S. H. Wilson, *J. Biol. Chem.* **275**, 4460–4466 (2000).
21. I. Husain, A. E. Tomkinson, W. A. Burkhart, M. B. Moyer, W. Ramos, Z. B. Mackey, J. M. Besterman, and J. Chen, *J. Biol. Chem.* **270**, 9683–9690 (1995).
22. R. Prasad, R. K. Singhal, D. K. Srivastava, J. T. Molina, A. E. Tomkinson, and S. H. Wilson, *J. Biol. Chem.* **271**, 16000–16007 (1996).
23. Z. Wang, X. Wu, and E. C. Friedberg, *J. Biol. Chem.* **272**, 24064–24071 (1997).
24. G. Frosina, P. Fortini, O. Rossi, F. Carrozzino, A. Abbondandolo, and E. Dogliotti, *Biochem. J.* **304**, 699–705 (1994).

25. A. Klungland and T. Lindahl, *EMBO J.* **16,** 3341–3348 (1997).
26. M. S. DeMott, S. Zigman, and R. A. Bambara, *J. Biol. Chem.* **273,** 27492–27498 (1998).
27. G. L. Dianov, R. Prasad, S. H. Wilson, and V. A. Bohr, *J. Biol. Chem.* **274,** 13741–13743 (1999).
28. S. Biade, R. W. Sobol, S. H. Wilson, and Y. Matsumoto, *J. Biol. Chem.* **273,** 898–902 (1998).
29. M. Stucki, B. Pascucci, E. Parlanti, P. Fortini, S. H. Wilson, U. Hubscher, and E. Dogliotti, *Oncogene* **17,** 835–843 (1998).
30. K. Nealon, I. D. Nicholl, and M. K. Kenny, *Nucleic Acids Res.* **24,** 3763–3770 (1996).
31. R. J. Sanderson and D. W. Mosbaugh, *J. Biol. Chem.* **273,** 24822–24831 (1998).
32. P. Branch, R. Hampson, and P. Karran, *Cancer Res.* **55,** 2304–2309 (1995).
33. A. Umar, J. C. Boyer, D. C. Thomas, D. C. Nguyen, J. I. Risinger, J. Boyd, Y. Ionov, M. Perucho, and T. A. Kunkel, *J. Biol. Chem.* **269,** 14367–14370 (1994).
34. R. D. Wood, M. Biggerstaff, and M. K. K. Shivji, *in* "Methods: A Companion to Methods in Enzymology," Vol. 7, pp. 163–175. Academic Press, San Diego, 1995.
35. C. D. Mol, A. S. Arvai, R. J. Sanderson, G. Slupphaug, B. Kavli, H. E. Krokan, D. W. Mosbaugh, and J. A. Tainer, *Cell (Cambridge, Mass.)* **82,** 701–708 (1995).
36. Z. Wang and D. W. Mosbaugh, *J. Biol. Chem.* **264,** 1163–1171 (1989).
37. S. E. Bennett, M. I. Schimerlik, and D. W. Mosbaugh, *J. Biol. Chem.* **268,** 26879–26885 (1993).
38. G. Slupphaug, I. Eftedal, B. Kavli, S. Bharati, N. M. Helle, T. Haug, D. W. Levine, and H. E. Krokan, *Biochemistry* **34,** 128–138 (1995).
39. P. Neddermann and J. Jiricny, *Proc. Natl. Acad. Sci. U.S.A.* **91,** 1642–1646 (1994).
40. S. D. Putney, S. J. Benkovic, and P. R. Schimmel, *Proc. Natl. Acad. Sci. U.S.A.* **78,** 7350–7354 (1981).
41. T. S.-F. Wang, *Annu. Rev. Biochem.* **60,** 513–552 (1991).
42. J. R. Nelson, C. W. Lawrence, and D. C. Hinkle, *Science* **272,** 1646–1649 (1996).
43. C. Masutani, M. Araki, A. Yamada, R. Kusumoto, T. Nogimori, T. Maekawa, S. Iwai, and F. Hanaoka, *EMBO J.* **18,** 3491–3501 (1999).
44. E. Lutsenko and A. S. Bhagwat, *J. Biol. Chem.* **274,** 31034 (1999).
45. S. E. Bennett and D. W. Mosbaugh, *J. Biol. Chem.* **267,** 22512–22521 (1992).
46. S. E. Bennett, O. N. Jensen, D. F. Barofsky, and D. W. Mosbaugh, *J. Biol. Chem.* **269,** 21870–21879 (1994).
47. S. E. Bennett, R. J. Sanderson, and D. W. Mosbaugh, *Biochemistry* **34,** 6109–6119 (1995).
48. U. Varshney and J. H. van de Sande, *Biochemistry* **30,** 4055–4061 (1991).
49. J. T. Stivers, K. W. Pankeiwicz, and K. A. Watanabe, *Biochemistry* **38,** 952–963 (1999).

Session 4
DNA Glycosylases: Specificity and Mechanisms

| SANKAR MITRA
| *University of Texas Medical Branch*

Since the discovery of uracil-DNA glycosylase (UDG) and the elucidation of its mechanism more than two decades ago (*1*), many DNA glycosylases have been identified in bacteria, yeast, and mammalian cells. Early studies involved purification of endogenous enzymes for their biochemical characterization. Subsequently, rapid development of recombinant DNA methodologies allowed cloning of the coding sequences of glycosylases and construction of expression vectors. As a result, active recombinant enzymes could now be produced in significant quantities in most cases, and these have been utilized for biochemical and structural investigation.

It is remarkable that the structure, sequence, and activity of DNA glycosylases are conserved to a significant extent in all organisms, ranging from prokaryotes to complex eukaryotes. This may, however, be expected because of the simple reaction of N-glycosylic bond cleavage that is carried out by these enzymes. Several general properties of these enzymes as a group are revealed from a large number of studies. The session on specificity and mechanisms of DNA glycosylases was limited to DNA glycosylases responsible for repair of 8-oxoguanine and thymine (U) in G·T (U) pairs. The mechanism and structural basis of enzymatic activity of several other DNA glycosylases have been extensively reviewed elsewhere in this volume.

Table I lists the classes of major DNA glycosylases characterized so far in both eukaryotes and prokaryotes. Although some of their general properties have been discussed in recent reviews (*2*) as well as in some of the articles in this volume, it may be worthwhile to recapitulate some salient features.

1. Most DNA glycosylases, with the notable exception of UDG, have strong preference for duplex DNA even though their initial substrate recognition may require subtle structural perturbation or base unpairing at the site of lesion. As discussed extensively in Session 6 (this volume), base or nucleotide flipping appears to be the common mechanism of substrate recognition and catalysis by all DNA glycosylases.
2. DNA glycosylases require no cofactor including exogenous divalent metal ions.

TABLE I
DNA GLYCOSYLASE: SUBSTRATE BASES

Class of enzyme	AP lyase activity	Substrate base
UDG (UNG)	−	Uracil
MPG (mammals)/MAG (yeast) AlkA (*E. coli*)	−	Alkylated bases
OGGs (eukaryotes)/MutM (*E. coli*)	+	8-Oxoguanine, Fapy
NTH (*E. coli* and mammals)/Nei (*E. coli*)/Ntgs (yeast)	+	Oxidized pyrimidines, DHU
MutY (*E. coli*)/MYH (mammals)	−	A in 8-oxoG·A pair
G·T(U)-DNA glycosylase (*E. coli*, eukaryotes)	−	T(U) in G·T (U) pair
Pyrimidine dimer-DNA glycosylase	+	*cis–syn* Cyclobutane pyrimidine dimers

3. While DNA glycosylases have molecular masses in the range of 25–40 kDa, the eukaryotic enzymes tend to be larger than their bacterial paralogs owing to the presence of additional polypeptide domains at one or both termini. These nonconserved regions may be involved in interaction with other proteins or in organelle targeting.

4. Three oxidized base lesion-specific DNA glycosylases in *Escherichia coli*, namely, endonuclease III (Nth), endonuclease VIII (Nei), and MutM/Fapy-DNA glycosylase (Fpg), and their overlapping substrate range are reviewed by Ide and colleagues in this session. Although early studies indicated that Nth and Nei are specific for oxidized pyrimidine derivatives, including thymine glycol and 5-hydroxycytosine, MutM was shown to be specific for 8-oxoguanine and the ring-opened formamidopyrimidine (Fapy) (3). Thus, the presence of a purine ring is not a determinant for MutM, although Nei and Nth appeared to prefer pyrimidine derivatives (4). However, more recent studies show that Nei actively excises 8-oxoG from DNA (5), and that MutM is capable of excising pyrimidine base lesions (6). Thus the mechanistic basis for substrate recognition is a major challenge for future study.

To reconcile these observations, we propose that the active site pockets of DNA glycosylases are somewhat plastic and may assume multiple conformations to accommodate substrate base lesions of different structures. One notable exception to this is UDG, which may have a rather rigid active site pocket and recognizes only U and a few other uracil derivatives (7).

5. As discussed elsewhere in this volume, the simple DNA glycosylase and the DNA glycosylases with AP lyase activity have distinct reaction mechanisms. For most simple DNA glycosylases, a water molecule, activated by an active site acidic residue, acts as the nucleophile in an S_N2

mechanism for glycosyl bond breakage without formation of a covalent enzyme substrate intermediate. For a DNA glycosylase/AP lyase, an amino group (of Lys) or imino group (of Pro) is the nucleophile that forms a Schiff base with the free deoxyribose after removal of the damaged base (8, 9). In the subsequent lyase reaction, the DNA strand is cleaved due to β (β/δ) elimination concomitant with release of the enzyme. The subtle mechanism of the AP lyase reaction of MutM is discussed in Kow and colleague's article in this session.

It is interesting to note that all oxidized damaged bases are repaired by DNA glycosylase/AP lyases. Nevertheless, the teleological basis for the lyase activity is not clear because the AP site generated after excision of the substrate base could be repaired quite efficiently by AP endonucleases, and no significant advantage is evident for repair of the AP lyase product over that of the AP site itself. In fact, APE is required for repair in both cases, with its dual function as an endonuclease or as a 3'-phosphodiesterase. Interestingly, the major human APE (hAPE1) is much more active in repair of AP sites than of the 3'-$\alpha\beta$-unsaturated aldehyde generated by AP lyases.

This phosphodiesterase activity of APE appears to be the rate-limiting step in repair of AP lyase products (10). It is noteworthy that 5'-phosphodeoxyribose (dRP) is generated from an AP site by the 5'-endonuclease activity of APE. Subsequent cleavage of this dRP group by lyase activity of DNA pol-β is the rate-limiting step (10). We should caution, however, that the above studies have been carried out in *in vitro* systems. Understanding the potential benefit of AP lyase activity in *in vivo* BER poses a future challenge. It may be pointed out in this context, as reviewed by Hazra and colleagues, that human OGG1 turns over extremely poorly, and the AP site product remains bound to this enzyme with weak AP lyase activity.

6. One of the unresolved issues regarding DNA glycosylases is the presence of multiple enzymes in the same organism with similar substrate preference and their physiological significance. Do these provide simple back-up systems or have distinct *in vivo* functions? The first example of this was observed in *E. coli* in which Tag, a DNA glycosylase with strong specificity for 3-methyladenine, is constitutively expressed while AlkA, which acts on 3-methyladenine and many other alkyl base adducts, is expressed as a component of the *ada* regulon (11). More recently, Nei and Nth in *E. coli* and Ntg1 and Ntg2 in *Saccharomyces cerevisiae* were shown to have a strong overlap in their substrate preference (12, 13). Similarly, G·T specific thymine-DNA glycosylase (TDG) could also function as a UDG (14). The *in vivo* function of TDG and its regulation, including potential involvement in bridging DNA repair and other cellular functions, is reviewed in the article by Hardeland and colleagues.

We and others have proposed a rationale for the presence of these enzymes with similar substrate specificity in that these act as distinct DNA strand-specific (and possibly cell cycle-specific) enzymes. The possibility of distinct *in vivo* functions of the two OGGs is discussed by Hazra *et al.* and Krokan *et al.* in this volume. Thus, UDG (UNG2) or Nei (Ntg1) may prefer nascent DNA strand while TDG or Nth may be active in repair of the parental DNA strand. In other words, TDG or Nth will remove DNA damage produced *in situ* while UDG or Nei acts on the nascent strand to prevent incorporation of modified or abnormal bases. Recently Nilsen *et al.* have shown that UNG protein's main role is to counteract U·A base pairs formed by dUMP incorporation during DNA synthesis (*15*).

Finally, one general feature of DNA glycosylases is their intracellular distribution in the cytoplasm, nucleus and mitochondria. Whether and how these enzymes are targeted to their sites of action in the nucleus or mitochondria in response to genotoxic stress should be a future topic of study.

REFERENCES

1. T. Lindahl, *Prog. Nucleic Acid Res. Mol. Biol.* **22**, 135–192 (1979).
2. H. E. Krokan, R. Standal, and G. Slupphaug, *Biochem. J.* **325**, 1–16 (1997).
3. S. Boiteux, T. R. O'Connor, F. Lederer, A. Gouyette, and J. Laval, *J. Biol. Chem.* **265**, 3916–3922 (1990).
4. Y. Saito, F. Uraki, S. Nakajima, A. Asaeda, K. Ono, K. Kubo, and K. Yamamoto, *J. Bacteriol.* **179**, 3783–3785 (1997).
5. T. K. Hazra, T. Izumi, R. Venkataraman, Y. W. Kow, M. Dizdaroglu, and S. Mitra, *J. Biol. Chem.* **275**, 27762–27767 (2000).
6. Z. Hatahet, Y. W. Kow, A. A. Purmal, R. P. Cunningham, and S. S. Wallace, *J. Biol. Chem.* **269**, 18814–18820 (1994).
7. C. D. Mol, A. S. Arvai, G. Slupphaug, B. Kavli, I. Alseth, H. E. Krokan, and J. A. Tainer, *Cell (Cambridge, Mass.)* **80**, 869–878 (1995).
8. M. L. Dodson, M. L. Michaels, and R. S. Lloyd, *J. Biol. Chem.* **269**, 32709–32712 (1994).
9. H. M. Nash, R. Lu, W. S. Lane, and G. L. Verdine, *Chem. Biol.* **4**, 693–702 (1997).
10. T. Izumi, T. K. Hazra, I. Boldogh, A. E. Tomkinson, M. S. Park, S. Ikeda, and S. Mitra, *Carcinogenesis* **21**, 1329–1334 (2000).
11. T. Lindahl, B. Sedgwick, M. Sekiguchi, and Y. Nakabeppu, *Annu. Rev. Biochem.* **57**, 133–157 (1988).
12. D. Jiang, Z. Hatahet, J. O. Blaisdell, R. J. Melamede, and S. S. Wallace, *J. Bacteriol.* **179**, 3773–3782 (1997).
13. S. Senturker, P. Auffret van der Kemp, H. J. You, P. W. Doetsch, M. Dizdaroglu, and S. Boiteux, *Nucleic Acids Res.* **26**, 5270–5276 (1998).
14. P. Gallinari and J. Jiricny, *Nature (London)* **383**, 735–738 (1996).
15. H. Nilsen, I. Rosewell, P. Robins, C. F. Skjelbred, S. Andersen, G. Slupphaug, G. Daly, H. E. Krokan, T. Lindahl, and D. E. Barnes, *Mol. Cell* **5**, 1059–1065 (2000).

Multiple DNA Glycosylases for Repair of 8-Oxoguanine and Their Potential in Vivo Functions

TAPAS K. HAZRA, JEFF W. HILL,
TADAHIDE IZUMI, AND
SANKAR MITRA

*Sealy Center for Molecular Science and
Department of Human Biological
Chemistry and Genetics
The University of Texas Medical Branch
Galveston, Texas 77555*

I. Repair of 8-Oxoguanine via the BER Pathway	194
II. Antimutagenic Processing of 8-OxoG: GO Model	194
III. Presence of Multiple OGGs in Bacteria and Eukaryotes	195
IV. Bipartite Antimutagenic Processing of 8-OxoG	197
V. Nei Is the OGG2 Paralog in *E. coli*	198
VI. Preferential Repair of 8-OxoG from the 8-OxoG·G Pair by OGG2 and Its Potential *in Vivo* Implication in Replication-Coupled Repair	199
VII. Mutual Interference of MutY and OGGs	200
VIII. Potential Role of OGG2 in Transcription-Coupled Repair	202
IX. Enzymatic Reaction of OGG1: The Handoff Process in Repair of 8-OxoG	203
References	204

8-Oxoguanine (8-oxoG) is a critical mutagenic lesion because of its propensity to mispair with A during DNA replication. All organisms, from bacteria to mammals, express at least two types of 8-oxoguanine-DNA glycosylase (OGG) for repair of 8-oxoG. The major enzyme class (OGG1), first identified in *Escherichia coli* as MutM (Fpg), and later in yeast and humans, excises 8-oxoG when paired with C, T, and G but rarely with A. In contrast, a distinct and less abundant OGG, OGG2, prefers 8-oxoG when paired with G and A as a substrate, and has been characterized in yeast and human cells. Recently, OGG2 activity was detected in *E. coli* which was subsequently identified to be Nei (Endo VIII). In view of the ubiquity of OGG2, we have proposed a model named "bipartite antimutagenic processing of 8-oxoguanine" and is an extension of the original "GO model." The GO model explains the presence of OGG1 (MutM) that excises 8-oxoG from nonreplicated DNA. If 8-oxoG mispairs with A during replication, MutY excises A and provides an opportunity for insertion of C opposite 8-oxoG during subsequent repair replication. Our model postulates that whereas OGG1 (MutM) is responsible for global repair of 8-oxoG in the nonreplicating genome, OGG2 (Nei) repairs

8-oxoG in nascent or transcriptionally active DNA. Interestingly, we observed that MutY and MutM reciprocally inhibited each other's catalytic activity but observed no mutual interference between Nei and MutY. This suggests that the recognition sites on the same substrate for Nei and MutY are nonoverlapping.

Human OGG1 is distinct from other oxidized base-specific DNA glycosylases because of its extremely low turnover, weak AP lyase activity, and nonproductive affinity for the abasic (AP) site, its first reaction product. OGG1 is activated nearly 5-fold in the presence of AP-endonuclease (APE) as a result of its displacement by the latter. These results support the "handoff" mechanism of BER in which the enzymatic steps are coordinated as a result of displacement of the DNA glycosylase by APE, the next enzyme in the pathway. The physiological significance of multiple OGGs and their *in vivo* reaction mechanisms remain to be elucidated by further studies.   © 2001 Academic Press.

I. Repair of 8-Oxoguanine via the BER Pathway

8-Oxoguanine (8-oxoG or G*), which is arguably the most important mutagenic and procarcinogenic lesion induced in cellular genomes by most types of reactive oxygen species (ROS), is often used as a marker of cellular oxidative stress (*1*). As first observed in *Escherichia coli,* 8-oxoG, like other oxidatively generated DNA lesions and small DNA base adducts, is repaired via the base excision repair (BER) pathway, in which the base lesion is excised from DNA by a specific DNA glycosylase, namely, 8-oxoguanine-DNA glycosylase (OGG). OGG1, the major OGG, was discovered first in *E. coli* and named MutM (Fpg). Enzymes with similar activities have been described subsequently in many organisms. Recent *in vivo* studies suggest that, at least in yeast, there is an overlap among various excision processes for repair of oxidized bases and other lesions in DNA (*2*). However, studies with bacterial and mammalian cell mutants lacking BER enzymes show that the removal of 8-oxoG by OGG1 in the BER process is the predominant mechanism for repair of this oxidized lesion (*3*). Commensurate with the critical need for repair of this mutagenic lesion, the contribution of other repair processes in the absence of OGG is perhaps expected.

II. Antimutagenic Processing of 8-OxoG: GO Model

In view of the importance of 8-oxoG as a mutagenic lesion, a mutation avoidance strategy has evolved in all organisms ranging from bacteria to mammals. This involves enzymes that are functionally, and to some extent structurally, conserved in all organisms. The mutation avoidance model, first proposed in *E. coli* and named the GO system (*4*), involves three enzymes. MutM (OGG1) is primarily responsible for repairing 8-oxoG once formed in DNA. However, if

8-oxoG remains unrepaired and then mispairs with A during subsequent DNA replication, excision of 8-oxoG would itself be mutagenic. In that case, a second enzyme, MutY, excises A and thus provides a second chance for incorporation of a nonmutagenic C opposite 8-oxoG by a DNA polymerase. MutM and the major eukaryotic OGGs (OGG1) do not excise 8-oxoG to a significant extent from an 8-oxoG·A pair, and their preferred substrate is the 8-oxoG·C pair. Thus, the two enzymes, MutM (OGG1) and MutY (MYH in eukaryotes), acting in concert, could prevent 8-oxoG-induced mutation before and after replication.

However, 8-oxoG is induced not only in DNA, but also in the deoxynucleoside triphosphate pool, from which it could be incorporated into DNA. Thus, the third enzyme in the GO system, Mut T (MTH in eukaryotes), an 8-oxodGTPase, hydrolyzes 8-oxodGTP and prevents its incorporation into DNA (5).

Although *in vitro* studies showed that 8-oxodGTP is not a good substrate for DNA polymerases (6), the fact that the spontaneous mutation frequency of *mutT* $^-$ *E. coli* is much higher than that of *mutM* and *mutY* mutants strongly suggests that 8-oxodGMP can be incorporated in DNA to a significant extent *in vivo*. It may be pointed out in this context that *mutM, mutY,* and *mutT* genes were identified in *E. coli* on the basis of enhanced spontaneous mutagenesis in the absence of their gene products. These observations underscore the role of 8-oxoG as a major, endogenous mutagenic lesion.

III. Presence of Multiple OGGs in Bacteria and Eukaryotes

After the discovery of MutM (OGG1) in *E. coli,* its homologs were subsequently identified and characterized in other organisms. The first eukaryotic OGG1 gene was cloned from yeast on the basis of phenotypic suppression of a GC→TA transversion mutation (7). The mammalian OGG1 orthologs were subsequently cloned on the basis of sequence homology with yeast OGG1 (8–10). It is interesting to note that while bacterial MutM is nearly identical to eukaryotic OGG1 functionally, it is structurally distinct, specifically in regard to its active site residues, as discussed in detail in other articles in this volume. In fact, MutM of *E. coli* and other bacteria, stands out as a unique class of DNA glycosylase because of the N-terminal Pro serving as the active site nucleophile (11). The eukaryotic OGGs share common structural motifs of the prototype endonuclease III (Nth) of *E. coli,* a DNA glycosylase/AP lyase that prefers oxidized pyrimidines as substrates. It is also important to point out that endonuclease VIII (Nei) is another DNA glycosylase that is predicted to be structurally similar to MutM.

Although the yeast and mammalian OGG1 genes and cDNAs were cloned and the recombinant enzymes were expressed and biochemically characterized,

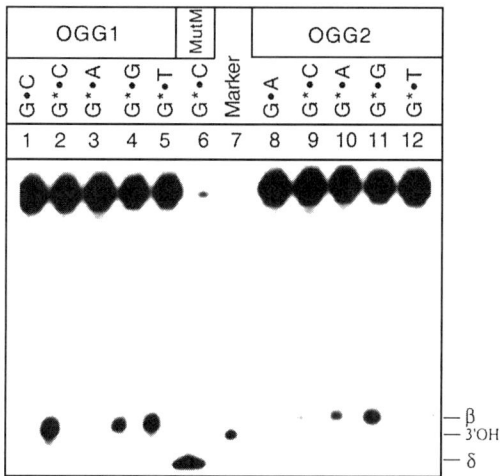

FIG. 1. Substrate specificity and AP lyase activity of OGG1 and OGG2 partially purified from HeLa cells. Lanes 1–5: Incubation of OGG1 (50 ng) with different substrates; lane 6, control G*·C oligo with MutM; lane7, 15-mer marker; lanes 8–12, reaction of OGG2 (150 ng) with different substrates. Reaction buffer (50 µl) contained 25 mM HEPES, pH 7.9, 50 mM KCl, 2.5 mM EDTA, 2 mM DTT, and 2.5% glycerol. After termination of the reaction with phenol/chloroform and ethanol precipitation, the oligonucleotides were separated by denaturing polyacrylamide (15%) gel electrophoresis.

no studies on purification of endogenous OGG1 from a eukaryotic source had been reported. Thus, we carried out systematic purification of OGG1 from HeLa cells and purified the enzyme several thousandfold (13). Multiple mRNA species of human OGG1 (hOGG1) have been recently identified, which suggests the presence of multiple isoforms of OGG1 in human cells (12). Our results showed that the major OGG1 species in HeLa cells with a molecular mass of 38 kDa corresponds to OGG1-1a, the predominant gene product identified by others (12, 13). We observed a minor 42-kDa species, which could correspond to a mitochondrial-specific isoform of OGG1 (12, 13).

During the purification process, we discovered the presence of a second OGG, which we named OGG2. OGG2 is immunologically distinct from OGG1, and these enzymes could be separated by FPLC. The first evidence for the presence of OGG2 was obtained from yeast (14). Human OGG2 and OGG1 were found to have overlapping but distinct substrate preferences in regard to 8-oxoG-containing base pairs. Specifically, OGG2, unlike OGG1s of bacteria and eukaryotes, was found to excise 8-oxoG from an 8-oxoG·G pair more efficiently than from an 8-oxoG·C pair (13). Subsequent studies showed that yeast OGG2 was identical to Ntg1, the yeast paralog of Nei of *E. coli*, with a similar preference for oxidized pyrimidines as substrates (14). The total OGG2 activity was about 10% of the OGG1 activity in HeLa extract with an 8-oxoG·G oligonucleotide substrate. However, the unique feature of human OGG2 is its preference for 8-oxoG·G and 8-oxoG·A base pairs as substrates. In contrast, we showed, in confirmation of earlier results, that OGG1 did not significantly excise 8-oxoG from an 8-oxoG·A pair, and was only weakly active in excising the oxidized lesion from the 8-oxoG·G pair (Fig. 1).

IV. Bipartite Antimutagenic Processing of 8-OxoG

Although a comparative substrate preference study has not been reported for yeast OGG1 and OGG2 as for the human OGGs, we predict that the yeast enzymes would show a pattern of substrate specificity similar to that of their human counterparts. The distinct substrate specificities of human OGGs led us to propose a hypothesis by extending the original GO model and postulating distinct *in vivo* functions of the two OGGs. We named our model the "bipartite antimutagenic processing of 8-oxoguanine" (Fig. 2). The key feature of this model is that OGG2 functions primarily in excising 8-oxoG when it is incorporated in the nascent DNA strand opposite A in the parental strand. As discussed earlier, such incorporation should be largely prevented by MutT (MTH). However,

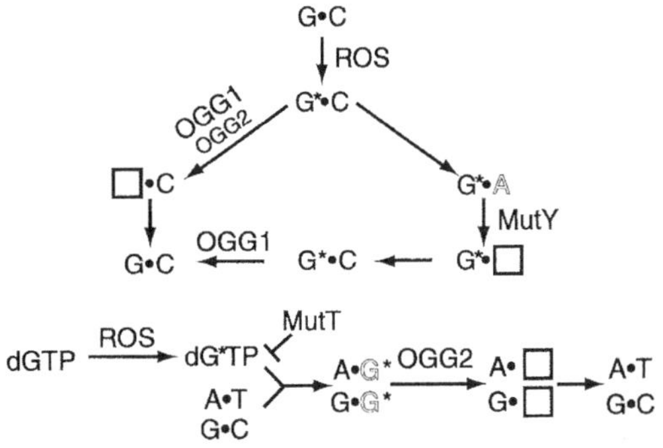

FIG. 2. Bipartite antimutagenic processing of 8-oxoguanine (G*) model showing differential functions of OGG1 and OGG2 in the repair of 8-oxoguanine. Upper panel: G* is paired with C when induced in DNA by ROS. Both OGG1 and OGG2 excise G* leaving an AP site (~). Unrepaired G* may pair with A during replication. MutY excises nascent strand-specific A (lightly tinted) and provides a second chance for repairing G*, as postulated in the GO model. Lower panel: G* could be incorporated from the dG*TP pool (if not degraded by MutT) in the nascent strand (lightly tinted) opposite A or G. OGG2, but not OGG1 removes nascent strand-specific G*. This model predicts that nascent strand specificity of MutY and OGG2 prevents G*-induced mutation.

hydrolysis of 8-oxodGTP by MutT may not always be complete. MutY should excise A from the nascent strand only when it is incorporated opposite 8-oxoG (or G) in the template strand. Thus, both MutY (and its mammalian ortholog MYH) and OGG2 should be specific for their respective substrate in the nascent DNA strand. While the nascent strand specificity could be conferred on these enzymes in a variety of ways, we have proposed that these DNA glycosylases utilize the nascent strand recognition mechanism of the DNA mismatch repair pathway (13).

V. Nei Is the OGG2 Paralog in *E. coli*

A critical requirement for validating our hypothesis is to show that an 8-oxoG·A-specific, OGG2-type activity is ubiquitous, and it should thus be possible to identify and characterize such an OGG2 in *E. coli*. As indicated earlier, there were no published reports about the presence of a second OGG in *E. coli*, particularly one that preferentially excises 8-oxoG from an 8-oxoG·A pair. We looked for OGG activity in cell-free extract of a *mutM* deletion strain of *E. coli*, and observed a surprisingly robust OGG activity. More importantly, the crude extract showed higher 8-oxoG-excision activity from an 8-oxoG·A pair than from an 8-oxoG·C pair (15). This reinforced our conclusion that this unidentified enzyme is distinct from MutM, which was, of course, absent in the mutant *E. coli* used for making the extract.

We named the enzyme OGG2 and purified the enzyme-substrate trapped complex to a point where we could identify the protein on the basis of peptide sequences. The N terminus of the protein was blocked. However, on the basis of an internal peptide sequence, we identified it to be Nei (15). As mentioned before, Nei was identified as the second oxidized pyrimidine-specific DNA glycosylase in *E. coli*, after Nth (16). However, its primary sequence and sequence motif is much closer to that of MutM. In particular, the N-terminal region, including the N-terminal Pro, the active site nucleophile of MutM, is highly conserved between the two enzymes. It is therefore surprising that there was no report describing 8-oxoG excision activity of Nei until recently (15, 17).

We purified to homogeneity recombinant wild-type Nei expressed in a $mutM^-$ nei^- *E. coli* strain and examined its substrate range (15). It was evident that dihydrouracil (DHU), a known substrate for Nei and Nth, is a better substrate for Nei than 8-oxoG. However, the rather small difference in the K_m and k_{cat} of Nei for the two substrates suggests that 8-oxoG functions as a substrate of Nei *in vivo*, particularly considering that the endogenous concentration of 8-oxoG may be significantly higher than that of DHU and most other ROS-induced lesions. More importantly, there is a remarkable difference between MutM and Nei with respect to excision of 8-oxoG when paired with

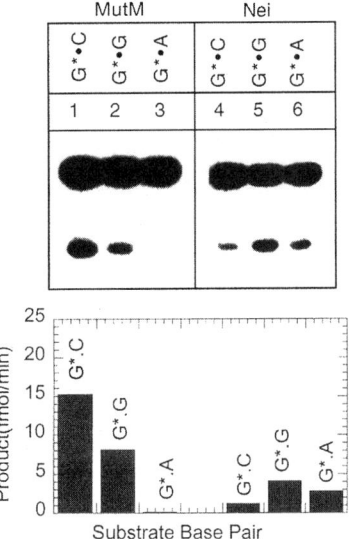

FIG. 3. Substrate specificity of MutM and Nei with 8-oxoG containing duplex oligos. 5 nM MutM (lanes 1–3) or Nei (lanes 4–6) was incubated at 37°C for 20 min with 200 nM duplex oligos. Reaction buffer (20 μl) contained 20 mM Tris–HCl, pH 7.5, 50 mM NaCl, and 1 mM EDTA. Lower panel, quantitative comparison of activity.

various bases in duplex oligonucleotide substrates (Fig. 3). As expected, Mut M did not excise 8-oxoG when paired with A, and did so poorly when paired with G. Its favored substrate was 8-oxoG·C. In contrast, Nei cleaved 8-oxoG from an 8-oxoG·G pair most efficiently, followed by the 8-oxoG·A pair. The 8-oxoG·C pair was a poor substrate. It is extremely interesting that the relative 8-oxoG base pair preferences of MutM and Nei closely resemble that of human OGG1 and OGG2 (13). These results strongly support our model of nascent strand-specific repair of 8-oxoG.

VI. Preferential Repair of 8-OxoG from the 8-OxoG·G Pair by OGG2 and Its Potential *in Vivo* Implication in Replication-Coupled Repair

We were surprised by the finding that the best substrate among 8-oxoG base pairs in DNA for both human OGG2 and its *E. coli* counterpart, Nei, is the 8-oxoG·G pair (15). It should be noted that in the 8-oxoG·A pair, 8-oxoG is present as a syn enantiomer, while in both 8-oxoG·C and 8-oxoG·G pairs, the base lesion is present in the *anti* conformation, the characteristic of all normal base pairs in B-DNA (*18*). The fact that the 8-oxoG·G pair is a preferred substrate for OGG2 suggests that such base pairing can occur *in vivo*, when 8-oxoG is incorporated in the nascent strand opposite G. Recent results showing that MutY can excise G from an 8-oxoG·G pair, although with a lower efficiency than A from an 8-oxoG·A pair, further support the existence of the 8-oxoG·G pair in DNA *in vivo* (*19*). Additionally, GC→CG transversion mutations, often observed after ROS treatment, are most likely to result from mispairing of G with a G oxidation product (rather than C pairing with an oxidation product of C). Although some early *in vitro* studies indicated that 8-oxoG·G pairing could occur during *in vitro* DNA synthesis (*20*), more recent studies, which are generally accepted now, showed that 8-oxoG preferentially paired with C or A and not G during *in vitro* DNA synthesis with DNA polymerases (*21*). In spite of these later studies, our results warrant a reexamination of 8-oxoG pairing during its incorporation in *in vitro* studies, preferably using a complete DNA replication system rather than purified DNA polymerases.

VII. Mutual Interference of MutY and OGGs

Even before the discovery of OGG2, the nascent and template strand specificity of MutY and MutM toward 8-oxoG in DNA was apparent. Recently, Li *et al.* showed that MutY, which binds 8-oxoG·C, 8-oxoG·G, and 8-oxoG·A pairs but utilizes only the latter two as substrates, inhibits MutM from excising 8-oxoG from an 8-oxoG·C pair (*22*). This is somewhat unexpected, because 8-oxoG·C is not a substrate for MutY. More importantly, however, we investigated the possibility of mutual interference between MutY, MutM, and Nei. Figure 4 shows that MutY inhibited MutM in its excision of 8-oxoG from both 8-oxoG·C and 8-oxoG·G pairs. In contrast, Nei activity was not affected by MutY with 8-oxoG·A, 8-oxoG·C, or 8-oxoG·G pair-containing DNA substrates.

These results could be reconciled with a scenario, consistent with our modified GO model, in which both MutY and Nei recognize the same 8-oxoG·A pair as a substrate, but have largely nonoverlapping interactions with the lesion pair itself. Whether the enzymes can simultaneously interact with the substrate remains to be determined. In contrast, because MutM acts on 8-oxoG generated *in situ* in DNA, while MutY is specific for the nascent strand, the latter has the priority in protecting DNA from the mutagenic process of excision of 8-oxoG from the progeny strand by inhibiting MutM.

These preliminary results do not shed light on how MutY or Nei recognizes the nascent strand. Once such a mechanism is elucidated, interference among these enzymes needs to be examined using a more complete, reconstituted assay system.

DNA GLYCOSYLASES FOR REPAIR OF 8-OXOGUANINE   201

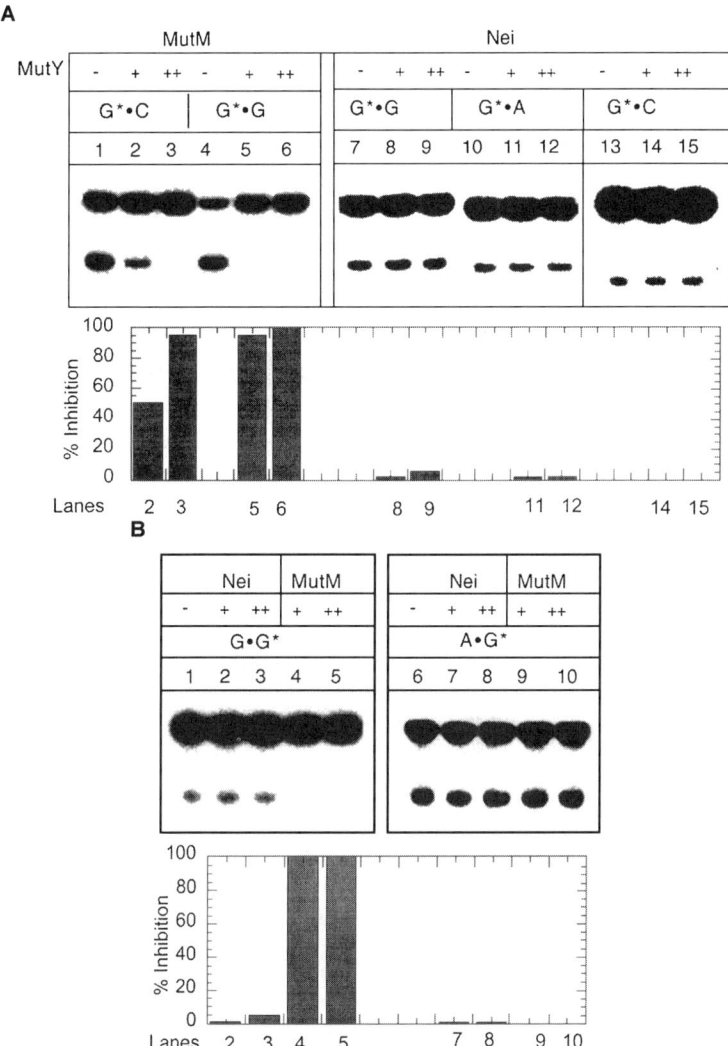

FIG. 4. Mutual interference of MutY and MutM but not Nei. (A) Effect of MutY on the activity of MutM and Nei. Upper panel: Analysis of cleaved products of G*·C and G*·G pair-containing oligo with 5 nM MutM (lanes 1–6) or G*·C, G*·G and G*·A with 5 nM Nei (lanes 7–15) in the presence of 10 (+) or 100 (++) nM of MutY. Substrate oligos (5 nM) were preincubated with MutY at room temperature for 5 min followed by incubation with MutM or Nei for an additional 15 min. Lower panel: Quantitative comparison of inhibition. (B) Effect of MutM and Nei on the activity of MutY. 5 (+) or 25 (++) nM of MutM or Nei was preincubated with 5 nM of G*·A or G*·G duplex oligo for 5 min at room temperature followed by further incubation with MutY (25 nM) for another 15 min. Lower panel: Quantitative comparison of inhibition.

VIII. Potential Role of OGG2 in Transcription-Coupled Repair

Direct evidence for the existence and function of a second OGG in mammals was provided from studies with OGG1 null mouse cells. Targeted mutagenesis of OGG1 was performed in mouse embryonic stem cells, leading to generation of OGG1 knockout mutant mice, in such a way that all variants of OGG1 were deleted in the animals. As expected, the level of endogenous 8-oxoG in the cellular genome was significantly elevated in homozygous mutant cells compared to that in wild-type or OGG1 heterozygous mutants (3). However, the mutant animals, at least at birth or after several months of growth, have no discernible phenotype. This observation was quite unexpected, because it raises the possibility that 8-oxoG has no significant etiologic role in pathophysiology, or at least that an increase in the level of genomic 8-oxoG may not be immediately harmful to the organism. Because of the difficulties in accurate quantitation of 8-oxoG in DNA [when 8-oxoG can be generated artifactually during the DNA preparation procedure (23)], it was not clear whether 8-oxoG is indeed repaired in OGG1 knockout mutant cells. At the same time, it was clear that OGG1 is responsible for bulk repair of 8-oxoG.

A more refined assay for repair of 8-oxoG, placed at a defined site in a plasmid, which was introduced into wild-type and OGG1-negative mouse cells, allowed for quantitative determination of 8-oxoG repair in the absence of OGG1. It is evident that 8-oxoG was indeed repaired to a small extent in OGG1-deficient cells (3). Although the possibility has not been directly tested, we suggest that OGG2 was responsible for this repair. More importantly, this repair occurred in the transcribed strand but not in the nontranscribed strand (24, 25). Transcription-coupled repair (TCR) was first elucidated during investigation of repair of bulky base adducts via the nucleotide excision repair pathway (26). However, unlike these adducts which block transcription, the oxidized base lesions in general allow transcription to proceed, although sometimes with a pause. Thus, the observation of repair of oxidized lesions, e.g., thymine glycol, via the TCR pathway was unexpected, and led to identification of a more complex function of XPG, a site-specific 3'-endonuclease in NER (27). XPG appears to function in recruitment of DNA glycosylases, such as OGG2, and thus activates repair of lesions in the transcribed strand (24). Whether XPG may also help recruit OGG1 under certain circumstances is not known. In any event, such interactions need to be experimentally demonstrated.

Taken together, the above results suggest that 8-oxoG repair *in vivo* and its regulation are quite complex, and that OGG2 interacts with other proteins *in vivo* carrying out both replication- and transcription-coupled repair.

IX. Enzymatic Reaction of OGG1: The Handoff Process in Repair of 8-OxoG

The distinct preference of OGG1 and OGG2 for 8-oxoG when paired with different bases must reflect their distinct substrate recognition mechanisms, specifically their interaction with the complementary base. Human OGG2 has not yet been available in recombinant, homogeneous form for enzymatic and structural studies. Recently, a detailed study on the enzymatic mechanism of mouse OGG1 has been published (28). We have carried out similar studies on the optimum reaction conditions for human OGG1 and obtained very similar results (28a). It is generally believed that OGG1 acts like type I restriction enzymes, and remains bound to its product. Our results confirmed this conclusion. We observed a very low rate of turnover with the 38-kDa OGG1 (OGG1-1α), the major isoform of the enzyme. However, Michaelis–Menten kinetics could not be observed for the enzyme because the reaction rate did not approach a plateau at high substrate concentrations.

We then observed that the major human AP-endonuclease, APE1, stimulated the 8-oxoG excision activity of OGG1 about 5-fold. More importantly, the reaction showed typical Michaelis–Menten kinetics [28a]. Our observations are consistent with the "handoff model" of BER, at least in regard to the early enzymes in the pathway, namely DNA glycosylases and AP-endonuclease (29, 30). Thus, OGG1 (and other DNA glycosylases) remain bound to their reaction product, namely AP sites, until they are displaced by AP-endonucleases. Figure 5 provides direct evidence for such displacement by using an electrophoretic mobility shift assay. In this experiment, we used an AP site-containing oligonucleotide and an inactive OGG1 mutant (K249Q) in which the active site nucleophile, Lys, is substituted with Gln. The mutant and wild-type OGG1 have similar substrate affinities. The enzyme bound to the AP site was displaced by

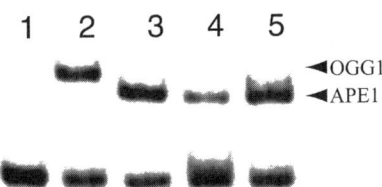

FIG. 5. Displacement of OGG1 from AP·C by APE1. A ^{32}P-labeled, AP site-containing oligo was incubated for 5 min with no protein (lane 1) or 5 nM of K249Q OGG1 (lanes 2, 4, and 5), then APE1 was added in lanes 3 (5 nM), 4 (2.5 nM), and 5 (10 nM) for an additional 5 min. The complexes were then analyzed by EMSA. Lanes 1, 2, and 4: With 1 mM Mg^{2+}; lanes 3 and 5: 1 mM EDTA. Arrows: Positions of OGG1-DNA and APE1-DNA complex.

wild-type APE1 in the absence of its activity in the presence of EDTA. However, active APE1 in the presence of Mg^{2+} was more effective in displacing OGG1 from its AP site product.

These results underscore the potential problem of studying the enzymatic characteristics of DNA glycosylases like OGG1 in isolation, because it would be difficult to extrapolate these results to the *in vivo* situation. Furthermore, these results raise questions regarding the physiological relevance of the intrinsic AP lyase activity in DNA glycosylases, in particular those of OGG1 and NTH1, which are involved in repair of oxidized base damage. It appears that the AP lyase activity of OGG1 is very weak so that the enzyme remains unproductively bound to its first product, i.e., an AP site, without carrying out the subsequent β-elimination reaction. A possible physiological significance of OGG1 affinity for the AP site is the need for coordination in the BER process. This may be particularly important for OGG1 because 8-oxoG is generated by all types of ROS, and may be more abundant than other base lesions *in vivo*. Cleavage of AP sites by the AP lyase activity of OGG1 would cause DNA strand breaks, which in turn may induce many signaling responses, including cell cycle arrest and apoptosis, in the absence of APE1 and other BER enzymes. The tight binding of OGG1 to AP sites may help prevent such strand breaks. Once APE1 and other components of the repair replication complex are recruited at the AP site, OGG1 is displaced for completion of repair. We have recently observed physical interaction among APE1, FEN1, and PCNA using purified proteins and HeLa cell-free extracts [Roy *et al.*, submitted]. These results suggest that repair of 8-oxoG *in vivo* could involve a complex of APE1 with FEN1, PCNA, and the replicative DNA polymerase, pol δ (or ε).

In conclusion, studies of repair of 8-oxoG have now entered an exciting phase in which it will soon be possible to dissect the complex machinery involved in the high-precision repair process of 8-oxoG and other oxidative base damage.

Acknowledgments

The studies described in this article were funded by the U.S. Public Health Service Grants CA81063, CA53791, and AG10514. The authors thank Drs. Y. W. Kow and M. Dizdaroglu for their suggestions and collaborative efforts, and Dr. David Konkel for critically reading the manuscript. We gratefully acknowledge Dr. R. Stephen Lloyd and Mr. Raymond Manuel for a gift of purified MutY and Dr. Cynthia J. Burrows for providing the 8-oxoguanine-containing oligonucleotide used in this study.

References

1. H. J. Helbock, K. B. Beckman, and B. N. Ames, *Methods Enzymol.* **300**, 156–166 (1999).
2. R. L. Swanson, N. J. Morey, P. W. Doetsch, and S. Jinks-Robertson, *Mol. Cell. Biol.* **19**, 2929–2935 (1999).

3. A. Klungland, I. Roswell, S. Hollenbach, E. Larsen, G. Daly, B. Epe, E. Seeberg, T. Lindahl, and D. E. Barnes, *Proc. Natl. Acad. Sci. U.S.A.* **96,** 13300–13305 (1999).
4. M. L. Michaels and J. H. Miller, *J. Bacteriol.* **174,** 6321–6325 (1993).
5. H. Maki and M. Sekiguchi, *Nature (London)* **355,** 273–275 (1992).
6. H. J. Einolf, N. Schnetz-Boutaud, and F. P. Guengerich, *Biochemistry* **37,** 13300–13312 (1998).
7. P. A. van der Kemp, D. Thomas, R. Barbey, R. de Oliveira, and S. Boiteux, *Proc. Natl. Acad. Sci. U.S.A.* **93,** 5197–5202 (1996).
8. H. Aburatani, Y. Hippo, T. Ishida, R. Takashima, C. Matsuba, T. Kodama, M. Takao, A. Yasui, K. Yamamoto, M. Asano, K. Fukasawa, T. Yoshinari, H. Inoue, E. Ohtsuka, and S. Nishimura, *Cancer Res.* **57,** 2151–2156 (1997).
9. R. Lu, H. M. Nash, and G. L. Verdine, *Curr. Biol.* **7,** 397–407 (1997).
10. J. P. Radicella, C. Dherin, C. Desmaze, M. S. Fox, and S. Boiteux, *Proc. Natl. Acad. Sci. U.S.A.* **94,** 8010–8015 (1997).
11. D. O. Zharkov, R. A. Rieger, C. R. Iden, and A. P. Grollman, *J. Biol. Chem.* **272,** 5335–5341 (1997).
12. K. Nishioka, T. Ohtsubo, H. Oda, T. Fujiwara, D. Kang, K. Sugimachi, and Y. Nakabeppu, *Mol. Biol. Cell* **10,** 1637–1652 (1999).
13. T. K. Hazra, T. Izumi, L. Maidt, R. A. Floyd, and S. Mitra, *Nucleic Acid Res.* **26,** 5116–5122 (1998).
14. S. D. Bruner, H. M. Nash, W. S. Lane, and G. L. Verdine, *Curr. Biol.* **8,** 393–403 (1998).
15. T. K. Hazra, T. Izumi, R. Venkataraman, Y. W. Kow, M. Dizdaroglu, and S. Mitra, *J. Biol. Chem.* **275,** 27762–27767 (2000).
16. R. J. Melamede, Z. Hatahet, Y. W. Kow, H. Ide, and S. S. Wallace, *Biochemistry* **33,** 1255–1264 (1994).
17. J. O. Blaisdell, Z. Hatahet, and S. S. Wallace, *J. Bacteriol.* **181,** 6396–6402 (1999).
18. S. J. Culp, B. P. Cho, F. F. Kadlubar, and F. E. Evans, *Chem. Res. Toxicol.* **2,** 416–422 (1989).
19. Q. M. Zhang, N. Ishikawa, T. Nakahara, and S. Yonei, *Nucleic Acids Res.* **26,** 4669–4675 (1998).
20. Y. Kuchino, F. Mori, H. Kasai, H. Inoue, S. Iwai, K. Miura, E. Ohtsuka, and S. Nishimura, *Nature (London)* **327,** 77–79 (1987).
21. S. Shibutani, M. Takeshita, and A. P. Grollman, *Nature (London)* **349,** 431–434 (1991).
22. X. Li, P. M. Wright, and A-L. Lu, *J. Biol. Chem.* **275,** 8448–8455 (2000).
23. H. J. Helbock, K. B. Beckman, M. K. Shigenaga, P. B. Walter, A. A. Woodall, H. C. Yeo, and B. N. Ames, *Proc. Natl. Acad. Sci. U.S.A.* **95,** 288–293 (1998).
24. F. Le Page, E. E. Kwoh, A. Avrutskaya, A. Gentil, S. A. Leadon, A. Sarasin, and P. I. Cooper, *Cell (Cambridge, Mass.)* **101,** 159–171 (2000).
25. F. Le Page, A. Klungland, D. E. Barnes, A. Sarasin, and S. Boiteux, *Proc. Natl. Acad. Sci. U.S.A.* **97,** 8397–8402 (2000).
26. I. Mellon, G. Spivak, and P. C. Hanawalt, *Cell (Cambridge, Mass.)* **51,** 241–249 (1987).
27. A. Sancar, *Annu. Rev. Biochem.* **65,** 43–81 (1996).
28. D. O. Zharkov, T. A. Rosenquist, S. E. Gerchman, and A. P. Grollman, *J. Biol. Chem.* **275,** 28607–28617 (2000).
28a. J. W. Hill, T. K. Hazra, T. Izumi, and S. Mitra, *Nucleic Acids Res.* **29,** 430–438 (2001).
29. C. D. Mol, D. J. Hosfield, and J. A. Tainer, *Mutat. Res.* **460,** 211–229 (2000).
30. S. H. Wilson and T. A. Kunkel, *Nat. Struct. Biol.* **7,** 176–178 (2000).

DNA Substrates Containing Defined Oxidative Base Lesions and Their Application to Study Substrate Specificities of Base Excision Repair Enzymes

Hiroshi Ide

*Department of Mathematical
and Life Sciences
Graduate School of Science
Hiroshima University
Higashi-Hiroshima 739-8526, Japan*

I. DNA Containing Defined Oxidative Base Lesions	208
II. Substrate Specificity of Endo III and Fpg Homologs	210
A. Endo III Homologs	210
B. Fpg Homologs	215
III. Sequence Context Effects on Damage Recognition	216
References	219

Reactive oxygen species generate structurally diverse base lesions in DNA. These lesions are primarily removed by base excision repair (BER) enzymes in prokaryotic and eukaryotic cells. Biochemical properties of BER enzymes such as substrate specificity, enzymatic parameters, and action mechanisms can be best studied by employing defined oligonucleotide and DNA substrates. Currently available methods are listed to prepare defined DNA substrates containing oxidative base damage and analogs.

BER enzymes for oxidative base damage are classified into two subgroups that recognize pyrimidine lesions (Endo III homologs) and purine lesions (Fpg homologs), though *E. coli* Fpg exhibits weak repair activity for certain pyrimidine damage. Recently, several interesting findings have been reported in relation to the substrate specificity of BER enzymes. *Saccharomyces cerevisiae* Endo III homologs (NTG1 and NTG2) have been shown to recognize formamidopyrimidine (Fapy) lesions that are derived from purine. Endo III and Endo VIII have a very weak activity to dihydrothymine in comparison with thymine glycol. Excision of 7,8-dihydro-8-oxoguanine by Fpg and human OGG1 is paired-base-dependent, whereas that of Fapy is essentially paired-base-independent. The repair efficiency of BER enzymes is affected by surrounding sequence contexts. In general, the sequence context effect appears to be more pronounced for Fpg homologs than Endo III homologs. © 2001 Academic Press.

I. DNA Containing Defined Oxidative Base Lesions

Reactive oxygen species generated by ionizing radiation or normal aerobic metabolism induce oxidative base lesions in DNA. If unrepaired, these lesions inhibit DNA replication or induce replication errors, thereby exerting cytotoxic and/or genotoxic effects (1, 2). Unlike DNA damages formed by ultraviolet light or alkylating agents, reactive oxygen species generate structurally diverse DNA lesions (3, 4). Nearly a hundred base and sugar damages have been identified to date. The oxidative base damages induce relatively localized structural perturbations and are primarily restored by the base excision repair (BER) pathway initiated by DNA glycosylases (1, 5, 6). Although DNA substrates damaged by ionizing irradiation or Fenton-type reactions would be ultimate substrates to study the biochemistry of BER enzymes, they contain too many types of base lesions, thereby not allowing detailed and reliable biochemical analyses of the property of BER enzymes. The presence of diverse lesions in DNA particularly interferes with the analyses of enzymatic parameters and damage recognition/catalytic mechanisms of BER enzymes. The most suitable substrates for these types of study are oligonucleotides containing a unique damage at a defined site.

Table I summarizes oxidative base damages that have been specifically incorporated into DNA (mostly in oligonucleotides) (7–50). Many of these substrates have been used to characterize the substrate specificity and action mechanisms of BER enzymes. The most powerful method to prepare the defined substrate is chemical synthesis using the phosphoramidite method. A limitation of this method comes from the availability of the phosphoramidite monomer containing damage. The commercially available phosphoramidite units are listed without references in Table I. Precautions are also necessary in deprotection of synthesized oligonucleotides since many of the listed base damages are alkali-labile. If deprotection is performed under the standard conditions (conc. NH_4OH, $60°C$ for 6 h), the introduced damage will be completely destroyed. Therefore, very mild conditions are used for deprotection and special amino protecting groups are employed for normal A, G, C phosphoramidite units (T bears no protecting groups). This information is usually provided by the suppliers of phosphoramidite units.

Although it is impossible to go through every detail of the data in Table I, some of the more recently reported are noteworthy. For fairly alkali-labile 5-formyluracil, two approaches have been used to introduce this lesion. In the first method, deoxyribonucleoside 5'-triphosphate of 5-formyluracil was synthesized (26) and incorporated into oligonucleotides by DNA polymerase to study its repair and base pairing properties (27, 28). In the second method, a phosphoramidite unit containing an alkali-stable precursor of 5-formyluracil was used to prepare oligonucleotides (24). After standard deprotection, the precursor

TABLE I
OXIDATIVE BASE LESIONS AND ANALOGS SPECIFICALLY INCORPORATED INTO DNA SUBSTRATES

Base	Damage	Method of introduction (references)
Thymine	Thymine glycol	Treatment with OsO_4 (7, 8) or $KMnO_4$ (pH 8) (9–11)
	5-Hydroxy-5,6-dihydrothymine	Phosphoramidite (12, 13)
	5,6-Dihydrothymine	Phosphoramidite[a]; enzymatic (14–16) and in vivo (17) incorporation
	5-Hydroxy-5-methylhydantoin	Phosphoramidite (18)
	Urea	Phosphoramidite (19, 20); alkali treatment of thymine glycol (8, 11, 21)
	Formamide	Phosphoramidite (19)
	β-Ureidoisobutyric acid	Alkali treatment of 5,6-dihydrothymine (17)
	Hydroxy-β-ureidoisobutyric acid	Alkali treatment of 5-hydroxy-5,6-dihydrothymine (22, 23)
	5-Formyluracil	Phosphoramidite (24, 25); enzymatic incorporation (26–28)
	5-Hydroxymethyluracil	Phosphoramidite (29); enzymatic incorporation (30), SPO1 phage (31)
Cytosine	Uracil glycol	Enzymatic incorporation (32, 33)
	5-Hydroxycytosine	Phosphoramidite[a]; enzymatic incorporation (34, 35)
	5-Hydroxyuracil	Phosphoramidite[a]; enzymatic incorporation (34, 35)
	Uracil	Phosphoramidite[a], PBS phage (36)
	5,6-Dihydrouracil	Phosphoramidite[a]
Guanine	7,8-Dihydro-8-oxoguanine	Phosphoramidite[a]
	me-Fapy[b]	Enzymatic incorporation (37)
	Xanthine	Phosphoramidite (38–41)
	Oxanine	Treatment with HNO_2 or NO (42); enzymatic incorporation (43)
	2-Aminoimidazolone	Photooxidation of 8-methoxyguanine (44)
Adenine	7,8-Dihydro-8-oxoadenine	Phosphoramidite[a]
	2-Hydroxyadenine (iso-guanine)	Phosphoramidite[a]; enzymatic incorporation (45, 46)
	Hypoxanthine	Phosphoramidite[a]
	α-Deoxyadenosine	Phosphoramidite (47–50)
Abasic site	Tetrahydrofuran	Phosphoramidite[a]

[a] Phosphoramidite units are commercially available.
[b] 2,6-Diamino-4-hydroxy-5-N-methylformamidopyrimidine.

was converted to 5-formyluracil by a chemical treatment. These oligonucleotides were further used to analyze the stability of DNA duplexes (25). 5-Hydroxycytosine as well as 5-hydroxyuracil and uracil glycol (both are deaminated cytosine lesions) were introduced into oligonucleotides by the phosphoramidite method (51) or enzymatic incorporation (32–35), and their base pairing properties and repair have been studied (32–35, 52–54). The phosphoramidite monomers of 5-hydroxycytosine and 5-hydroxyuracil are now commercially available. The advantage of the enzymatic incorporation over the phosphoramidite method is that only a deoxyribonucleoside 5'-triphosphate of a damaged base is necessary. Thus, the chemistry involved is relatively simple and the synthesis of the triphosphate is less laborious, although it must serve as a substrate for DNA polymerase or terminal deoxynucleotidyl transferase (TdT). 7,8-Dihydro-8-oxopurines (both A and G) and formamidopyrimidine (Fapy) are major oxidative purine lesions. Phosphoramidite monomers of 7,8-dihydro-8-oxopurines (both A and G) are commercially available, but Fapy lesions are rather difficult to be specifically introduced. Therefore, N-methylated analogs of G derived-Fapy (2,6-diamino-4-hydroxy-5-N-methylformamidopyrimidine, me-Fapy) has been mostly used in enzymatic studies. In the conventional method, me-Fapy was introduced into DNA by methylation followed by alkali treatment (55). Methylation of DNA results in 7-methylguanine as a major product, but inevitably induces other undefined lesions at the same time. This problem was overcome by introducing 7-methylguanine by the DNA polymerase reaction, and the oligonucleotide substrates containing me-Fapy were used to characterize the biochemical properties of Fpg and human OGG1 (37). 2-Aminoimidazolone (56, 57) and oxanine (58) have been identified as major products of G in photooxidation and treatments with nitrogen oxides (NO, NO_2^-), respectively. The stability of duplexes containing oxanine and related properties has been investigated (42, 43, 59). In addition, it has been shown that 2-aminoimidazolone can be selectively introduced by photooxidation of 8-methoxyguanine in oligonucleotides (44). The repair enzymes for these two lesions remain to be identified.

II. Substrate Specificity of Endo III and Fpg Homologs

A. Endo III Homologs

The oxidative base damages are mostly repaired by bifunctional DNA N-glycosylases. Repair activities of these enzymes for oxidative damages of thymine, cytosine, and purines are summarized in Tables II–IV, respectively (23, 33, 53–55, 60–87). In E. coli, these enzymes are classified into two subgroups. One subgroup (Endo III and Endo VIII) recognizes oxidized pyrimidines and analogs (Tables II and III), and the other subclass (Fpg) recognizes oxidized purine lesions (Table IV).

TABLE II
REPAIR ACTIVITY OF BER ENZYMES FOR OXIDATIVE THYMINE LESIONS AND ANALOGS

		E. coli			Yeast			Mammal		E. coli	Yeast	Mammal
Damage	Substrate	Endo III	Endo VIII	Nth-Spo	NTG1	NTG2	mNTH1	hNTH1	Fpg	yOGG1	h(m)OGG1
Thymine glycol	Oligo	33, 65, 73, 79	33, 73, 79		79, 80	79, 80	11		33		
	OsO$_4$–DNA	62	72	67							
	KMnO$_4$–DNA	64					11	71, 81, 82			
	γ-DNA	60, 63		66	68	68		70			
	γ-DNA	60, 63			68	68		70			
5-Hydroxy-5,6-dihydrothymine	Oligo	65, 79	79		79	79			65		
	γ-DNA	60									
	PM2–DNA		72								
5,6-Dihydrothymine											
5-Hydroxy-5-methylhydantoin	KMnO$_4$–DNA	64									
	γ-DNA	60, 63			68	68					
	Oligo	79	79		79	79					
Urea	OsO$_4$/base-DNA	62	72				11				
	KMnO$_4$–DNA	64		67				71			
Methyltartonylurea	γ-DNA	63									
β-Ureidoisobutyric acid	KMnO$_4$–DNA	64	72								
Hydroxy-β-ureidoisobutyric acid	Oligo	23							23		

TABLE III
REPAIR ACTIVITY OF BER ENZYMES FOR OXIDATIVE CYTOSINE LESIONS AND ANALOGS

		E. coli			Yeast		Mammal		E. coli	Yeast	Mammal
Damage	Substrate	Endo III	Endo VIII	Nth-Spo	NTG1	NTG2	mNTH1	hNTH1	Fpg	yOGG1	h(m)OGG1
Uracil glycol	Oligo	33, 54, 79	33, 79						33		
	γ-DNA	60, 61									
5-Hydroxycytosine	Oligo	53, 54, 65, 73, 79	79		79	79			53, 65		
5-Hydroxyuracil	γ-DNA	61		66	68	68			53		
	Oligo	53, 54, 73, 79	79		79, 80	79		70			
5,6-Dihydrouracil	γ-DNA	61		66	68	68					
	Oligo				69, 79	69, 79	70	82			
5-Hydroxyhydantoin	γ-DNA	60									
	γ-DNA	61									
5,6-Dihydroxycytosine	γ-DNA			66							
5-Hydroxy-5,6-dihydrocytosine	γ-DNA							70			
5-Hydroxy-5,6-dihydrouracil	γ-DNA	60			68	68					
Alloxan	γ-DNA	60									
Dihydroxyimidazolidine	γ-DNA	61									

TABLE IV
REPAIR ACTIVITY OF BER ENZYMES FOR OXIDATIVE PURINE LESIONS AND ANALOGS

		E. coli			Yeast		Mammal		E. coli	Yeast	Mammal
Damage	Substrate	Endo III	Endo VIII	Nth-Spo	NTG1	NTG2	mNTH1	hNTH1	Fpg	yOGG1	h(m)OGG1
7,8-Dihydro-8-oxoguanine:C	Oligo								75	77, 87	37, 78, 83–86
	γ-DNA								74	76	
7,8-Dihydro-8-oxoguanine:G	Oligo				68						
7,8-Dihydro-8-oxoadenine	Oligo								75		
Formamido-pyrimidine-A	γ-DNA				68	68			74		
Formamido-pyrimidine-G	γ-DNA				68	68			74	76	
Methylformamido-pyrimidine-G	Oligo Poly(dG/dC)				68, 69, 80	68, 69, 80			55	77	37

Endo III and Endo VIII exhibit little amino acid sequence homology (88, 89), but their substrate specificities are considerably overlapping as shown in Tables II and III. The *E. coli* mutant deficient in Endo III (*nth*) or Endo VIII (*nei*) exhibit neither apparent mutator phenotype nor an increased sensitivity to X-rays and hydrogen peroxide. However, the *nth nei* double mutant shows mutator phenotype and an increased sensitivity to X-rays and hydrogen peroxide (88, 89), suggesting redundant repair pathways for oxidized pyrimidine lesions. Like Endo III, eukaryotic Endo III homologs from *S. pombe* (Nth-Spo), *S. cerevisiae* (NTG1 and NTG2), and mouse and human NTH1 have the DNA recognition helix–hairpin–helix motif, two catalytically essential amino acids (corresponding to Lys-120 and Asp-138 in Endo III), and the 4Fe-4S cluster (except NTG1) (90). They recognize a variety of pyrimidine lesions that have been already identified as substrates of Endo III. Interestingly, however, *S. cerevisiae* NTG1 and NTG2 also excise formamidopyrimidines (Fapy) derived from purines (68, 69, 80). In addition, NTG1 but not NTG2 recognizes 8-oxoG paired with G, albeit weakly (68). Previously, the activities of Endo III (60) and human NTH1 (70) have been examined using γ-irradiated DNA that contained Fapy-A and Fapy-G lesions. However, release of free Fapy-A and Fapy-G was not detected. Therefore, the activity for Fapy might be peculiar to the *S. cerevisiae* Endo III homologs. Alternatively, the activity has been somehow overlooked in the previous studies. Since defined oligonucleotides containing me-Fapy are now available (37), this question can be asked by employing the substrates.

Another interesting finding on the substrate specificity of Endo III homologs is a large activity difference for thymine glycol (Tg) and dihydrothymine (DHT) (65). Based on the V_{max}/K_m values of Endo III, DHT is an approximately 10-fold poorer substrate than Tg embedded in the same site of an oligonucleotide substrate. This appears to also be the case for NTG1 and NTG2 (79). The activities of Endo VIII for DHT and Tg were also compared (Ide *et al.* unpublished data). Duplex oligonucleotide substrates (19-mer) containing DHT and Tg (both paired with A) at the same site were treated with Endo VIII and nicked products were analyzed by gel electrophoresis. DHT was a very poor substrate for Endo VIII and the activity for DHT was 22-fold lower than that for Tg (Fig. 1). When DHT was placed opposite G, C, T to form noncanonical base pairs, the repair activity was markedly enhanced. The enhancement ratio of the activity relative to a DHT:A pair was 39 (DHT:G), 10 (DHT:C), and 13 (DHT:T), thus particularly high for a DHT:G pair. The repair activity for Tg was also enhanced moderately for a Tg:G pair (3.6-fold) but not for Tg:C and Tg:T pairs (Fig. 1). When paired with A, DHT will stack into the helix fairly well, but Tg will not because it has polar and larger hydroxyl groups. Unlike DHT, which induces a minimum perturbation, Tg induces localized but significant distortion in the helix and can adopt both intrahelical and extrahelical positions

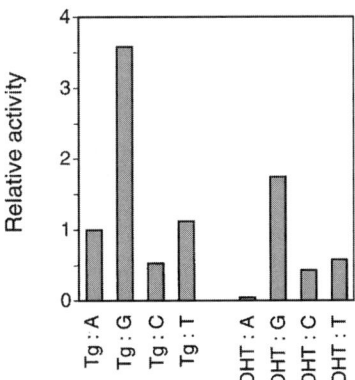

FIG. 1. Repair activity of Endo VIII for Tg and DHT. Duplex oligonucleotide substrates (19-mer) containing DHT and Tg (both paired with A, G, C, T) at the same site were treated with Endo VIII and nicked products were quantified by polyacrylamide gel electrophoresis (PAGE). The activity was standardized relative to that for the Tg:A pair. (Ide et al. unpublished data).

(9, 10). Concerning the paired effects, two factors are likely to be involved in the enhancement of the repair efficiency, i.e., the loss of hydrogen bonding and steric crash in the helix core. Destabilization of DHT by the loss of proper hydrogen bonding in the mispairs (G, C, T) facilitates the recognition by Endo VIII. The steric crash of DHT with G but not pyrimidine bases (C and T) in the helix core can further promote this process by altering the disposition of the base pair. Since Tg exerts a destabilizing effect by itself, the steric crash occurring with G becomes a dominating factor in the repair.

B. Fpg Homologs

Fpg recognizes mostly purine lesions. Although it has been reported that Fpg recognizes certain pyrimidine lesions such as Tg, DHT, uracil glycol, 5-hydroxycytosine, and 5-hydroxyuracil (Tables II and III), the activity for these substrates is notably lower than that of its intrinsic substrate 8-oxoG or that of Endo III (33, 53, 65). Therefore, Fpg is indeed characterized as a repair enzyme for purine lesions. Fpg and its eukaryotic functional homologs (yOGG1, mOGG1, hOGG1) recognize 8-oxoG, 7,8-dihydro-8-oxoadenine (8-oxoA) and Fapy analogs, but the activity for 8-oxoA seems to be weaker than other two substrates (75, 76). Fpg and eukaryotic OGG1 are functionally similar, but their amino acid sequences show no homology (78, 83–87). OGG1 has the helix–hairpin–helix motif and two key amino acids that are conserved in Endo III homologs, thus belonging to Endo III superfamily in this sense. A unique feature associated with the repair activity of Fpg and OGG1 is paired base effects. Fpg and OGG1 efficiently acts on 8-oxoG:C pairs but not 8-oxoG:A pairs to avoid mutagenic repair of the latter pairs (91, 92). The stringent discrimination of paired base also occurs when OGG1 acts on abasic sites (85, 87). yOGG1 and hOGG1 preferentially incise abasic sites paired with C. Conversely, Fpg show no preference of the base opposite an abasic site (87). We have found that hOGG1 also acts on a urea residue in a paired base-dependent manner (Ide *et al.*, unpublished data). A urea residue was an excellent substrate when paired with C, but other pairs were very poor substrates. It has recently been reported that the stringent discrimination of paired base by Fpg and hOGG1 is almost abolished when acting on me-Fapy (37). The discrimination of the paired base may not be essential for me-Fapy because me-Fapy (originally derived from G) does not form any mispairs during DNA replication and me-Fapy paired with C is only subjected to repair by the enzymes in cells.

III. Sequence Context Effects on Damage Recognition

Heterogeneity in DNA repair has been reported by a number of studies. One class of the heterogeneous repair is related to chromatin structure and transcriptional activity, and hence is domain-dependent (see Ref. 93 and references cited therein). The other class is observed as a variation of the repair rate at the nucleotide level, and is therefore sequence context-dependent (see Ref. 94 and references cited therein). Both domain-dependent and sequence context-dependent repair have been observed for nucleotide excision repair, whereas base excision repair is mostly related to sequence context-dependent repair. Variation of the *in vitro* repair rate from position to position or sequence to sequence was up to 17-fold for uracil glycosylase (95), 6.6-fold (me-Fapy) and 33-fold (8-oxoG) for Fpg (37, 96), 17.2-fold (me-Fapy) for hOGG1 (37), 5-fold ($1,N^6$-ethenoadenine), and 185-fold (m^7G) for human methylpurine glycosylase (MPG/APNG) (93, 94). Efficiently and inefficiently repaired consensus sequences for Fpg and hOGG1 have been also reported (37, 96), but the two enzymes exhibit quite different sequence specificities, probably reflecting the difference in the architecture of the active site and and its proximal regions.

We compared sequence context effects on the repair of Tg and urea residues by Endo III and mouse NTH1 (mNTH1) (Ide *et al.*, unpublished data). To prepare substrates, single-stranded M13 DNA was primed by a ^{32}P-labeled primer and replicated by *E. coli* DNA polymerase I (Klenow fragment). The purified DNA was treated with 0.1% OsO_4 (70°C for 30 min) to introduce Tg (8). DNA containing urea residues was prepared by the alkali treatment of Tg-DNA at pH 11.4 (8). M13 DNA containing Tg or urea residues was treated with Endo III or mNTH1, and the bands arising from excision of damage were analyzed by PAGE (Fig. 2). The piperidine treatment of DNA containing Tg revealed that introduction of the damage was relatively uniform and the variation of the

SUBSTRATE SPECIFICITIES OF BER ENZYMES

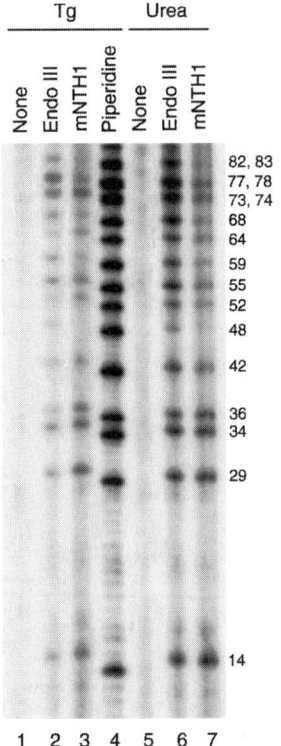

FIG. 2. PAGE analysis of the sequence context effects on the repair of Tg and urea residues by Endo III and its mouse homolog (mNHT1). Duplex M13 DNA containing randomly distributed Tg or urea residues was treated by Endo III or mNTH1. Products were analyzed by 8% PAGE. The substrates and enzymes used are indicated on the top. For the sequence and position (numbers on the right), see Fig. 3. (Ide *et al.* unpublished data).

band intensity was less than 2-fold (lane 4). For quantitative analysis, individual band intensities were quantified. The intensity of the individual band formed by Endo III (lanes 2, 6) or mNTH1 (lanes 3, 7) was divided by that of the corresponding piperidine-generated band (lane 4) to correct the site-dependent variation of the lesion frequency in the substrate DNA. To compare the repair efficiency of the individual site by Endo III and mNTH1, a normalized repair efficiency (nicked % per sum of nicked % for positions 14–83) was calculated and plotted against the sequence (Fig. 3). The normalized repair efficiency represents the distribution of a repair event when enzymes remove the same amount of lesions from the region of interest. The repair efficiency of Endo III and mNTH1 for the individual Tg site varied 3.5- and 2.7-fold, respectively (Fig. 3A), and that for urea residues was 3.1-fold (Endo III) and 3.9-fold (mNTH1)

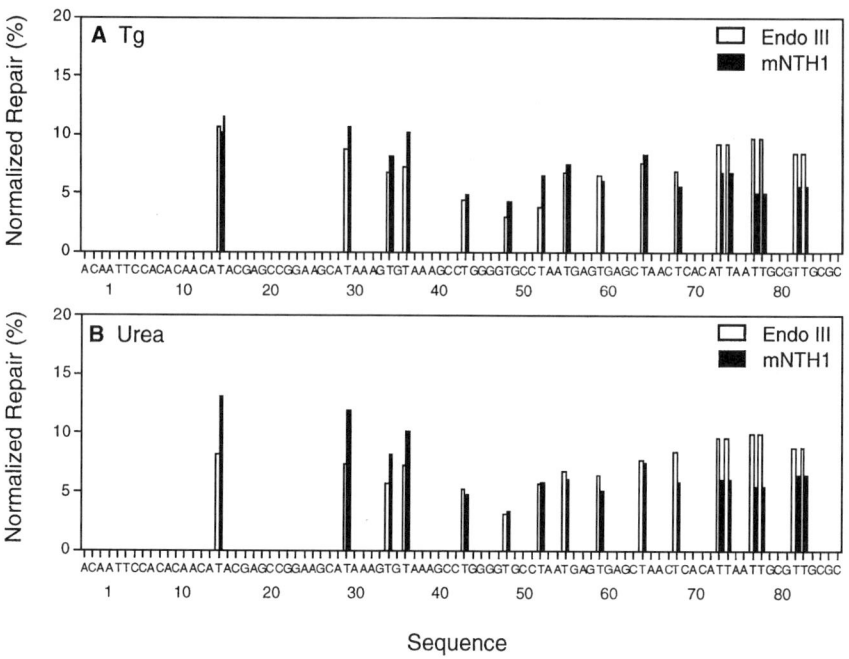

FIG. 3. Normalized repair efficiencies of individual Tg and urea sites by Endo III and its mouse homolog (mNTH1). The intensity of each band formed by Endo III or mNTH1 was divided by that of the corresponding piperidine-generated band to determine the repair efficiency at individual sites. The normalized repair efficiency (nicked % per sum of nicked % for positions 14–83) was calculated and plotted against the sequence for (A) Tg and (B) urea residues. The orientation of the sequence shown below was 5' (left) to 3' (right). The bands for the T doublets at positions 2 and 3 were obscured by the intense primer band so that these data was not included in the analysis. (Ide *et al.* unpublished data).

(Fig. 3B). Therefore, the sequence-dependent variations of the repair efficiency of Endo III and mNTH1 were smaller than those for Fpg, hOGG1 and other enzymes mentioned above. Furthermore, Endo III and mNTH1 showed similar repair patterns with respect to the distribution of preferred and unpreferred sites, suggesting similar context effects on the repair.

ACKNOWLEDGMENTS

The author thanks Kenjiro Asagoshi, Takao Yamada, and Hiroaki Odawara in his laboratory for providing their data for this review. Special thanks are extended to Yoshihiko Ohyama and Hiroaki Terato for helpful discussions and comments on this manuscript.

References

1. E. C. Friedberg, G. C. Walker, and W. Siede, "DNA Repair and Mutagenesis" ASM Press, Washington, D.C., 1995.
2. Z. Hatahet and S. S. Wallace, in "DNA Damage and Repair" (J. A. Nickoloff and M. F. Hoekstra, eds.), Vol. 1, pp. 229–262. Humana Press, Totowa, NJ, 1998.
3. C. von Sonntag, "The Chemical Basis of Radiation Biology." Taylor & Francis, New York, 1987.
4. A. P. Breen and J. A. Murphy, *Free Radic. Biol. Med.* **18,** 1033–1077 (1995).
5. I. D. Hickson, "Base Excision Repair of DNA Damage" Landes Bioscience, Austin, TX, 1997.
6. S. S. Wallace, in "Oxidative Stress and the Molecular Biology of Antioxidant Defenses" (J. Scandalios, ed.), pp. 49–90. Cold Spring Harbor Laboratory, Cold Spring Harbor, NY, 1997.
7. J. M. Clark and G. P. Beardsley, *Biochemistry* **26,** 5398–5403 (1987).
8. H. Ide, Y. W. Kow, and S. S. Wallace, *Nucleic Acids Res.* **13,** 8035–8052 (1985).
9. J. Y. Kao, I. Goljer, T. A. Phan, and P. H. Bolton, *J. Biol. Chem.* **268,** 17787–17793 (1993).
10. H. C. Kung and P. H. Bolton, *J. Biol. Chem.* **272,** 9227–9236 (1997).
11. A. H. Sarker, S. Ikeda, H. Nakano, H. Terato, H. Ide, K. Imai, K. Akiyama, K. Tsutsui, Z. Bo, K. Kubo, K. Yamamoto, A. Yasui, M. C. Yoshida, and S. Seki, *J. Mol. Biol.* **282,** 761–774 (1998).
12. T. J. Matray and M. M. Greenberg, *J. Amer. Chem. Soc.* **116,** 6931–6932 (1994).
13. M. M. Greenberg and T. J. Matray, *Biochemistry* **18,** 14071–14079 (1997).
14. H. Ide, R. J. Melamede, and S. S. Wallace, *Biochemistry* **26,** 964–969 (1987).
15. H. Ide and S. S. Wallace, *Nucleic Acids Res.* **16,** 11339–11354 (1988).
16. M. A. Chaudhry and M. Weinfeld, *J. Mol. Biol.* **249,** 914–922 (1995).
17. H. Ide, L. A. Petrullo, Z. Hatahet, and S. S. Wallace, *J. Biol. Chem.* **266,** 1467–1477 (1991).
18. A. Guy, J. Dubet, and R. Teoule, *Tetrahedron Lett.* **34,** 8101–8102 (1993).
19. A. Guy, H. Bazin, and R. Teoule, *Nucleosides Nucleotides* **10,** 565–566 (1991).
20. V. Gervais, J. A. H. Cognet, A. Guy, J. Cadet, R. Teoule, and G. V. Fazakerley, *Biochemistry* **37,** 1083–1093 (1998).
21. Y. W. Kow and S. S. Wallace, *Proc. Natl. Acad. Sci. U.S.A.* **82,** 8354–8358 (1985).
22. T. J. Matray, K. J. Haxton, and M. M. Greenberg, *Nucleic Acids Res.* **23,** 4642–4648 (1995).
23. J. Jurado, M. Saparbaev, T. J. Matray, M. M. Greenberg, and J. Laval, *Biochemistry* **37,** 7757–7763 (1998).
24. H. Sugiyama, S. Matsuda, K. Kino, Q. M. Zhang, S. Yonei, and I. Saito, *Tetrahedron Lett.* **37,** 9067–9070 (1996).
25. Q. M. Zhang, H. Sugiyama, I. Miyabe, S. Matsuda, K. Kino, I. Saito, and S. Yonei, *Int. J. Radiat. Biol.* **75,** 59–65 (1999).
26. M. Yoshida, K. Makino, H. Morita, H. Terato, Y. Ohyama, and H. Ide, *Nucleic Acids Res.* **25,** 1570–1577 (1997).
27. A. Masaoka, H. Terato, M. Kobayashi, A. Honsho, Y. Ohyama, and H. Ide, *J. Biol. Chem.* **274,** 25136–25143 (1999).
28. H. Terato, A. Masaoka, M. Kobayashi, S. Fukushima, Y. Ohyama, M. Yoshida, and H. Ide, *J. Biol. Chem.* **274,** 25144–25150 (1999).
29. L. C. Sowers and G. P. Beardsley, *J. Org. Chem.* **58,** 1664–1665 (1993).
30. D. D. Levy and G. W. Teebor, *Nucleic Acids Res.* **19,** 3337–3343 (1991).
31. R. G. Kallen, M. Simon, and J. Marmur, *J. Mol. Biol.* **5,** 248–250 (1962).
32. A. A. Purmal, J. P. Bond, B. A. Lyons, Y. W. Kow, and S. S. Wallace, *Biochemistry* **37,** 330–338 (1998).
33. A. A. Purmal, G. W. Lampman, J. P. Bond, Z. Hatahet, and S. S. Wallace, *J. Biol. Chem.* **273,** 10026–10035 (1998).
34. Z. Hatahet, A. A. Purmal, and S. S. Wallace, *Nucleic Acids Res.* **21,** 1563–1568 (1993).
35. A. A. Purmal, Y. W. Kow, and S. S. Wallace, *Nucleic Acids Res.* **22,** 72–78 (1994).

36. I. Takahashi and J. Marmur, *Nature (London)* **197**, 794–795 (1963).
37. K. Asagoshi, T. Yamada, H. Terato, Y. Ohyama, Y. Monden, T. Arai, S. Nishimura, H. Aburatani, T. Lindahl, and H. Ide, *J. Biol. Chem.* **275**, 4956–4964 (2000).
38. A. Van Aerschot, M. Mag, P. Herdewijn, and H. Vanderhaeghe, *Nucleosides Nucleotides* **8**, 159–178 (1989).
39. R. Eritja, D. M. Horowitz, P. A. Walker, J. P. Ziehler-Martin, M. S. Boosalis, M. F. Goodman, K. Itakura, and B. Kaplan, *Nucleic Acids Res.* **14**, 8135–8153 (1986).
40. M. J. Lutz, H. A. Held, M. Hottiger, U. Hubscher, and S. Benner, *Nucleic Acids Res.* **24**, 1308–1313 (1996).
41. H. Kamiya, M. Shimizu, M. Suzuki, H. Inoue, and E. Ohtsuka, *Nucleosides Nucleotides* **11**, 247–260 (1992).
42. T. Suzuki, Y. Matsumura, H. Ide, K. Kanaori, K. Tajima, and K. Makino, *Biochemistry* **36**, 8013–8019 (1997).
43. T. Suzuki, M. Yoshida, M. Yamada, H. Ide, M. Kobayashi, K. Kanaori, K. Tajima, and K. Makino, *Biochemistry* **37**, 11592–11598 (1998).
44. H. Ikeda and I. Saito, *J. Amer. Chem. Soc.* **121**, 10836–10837 (1999).
45. C. Y. Switzer, S. E. Moroney, and S. A. Bonner, *Biochemistry* **32**, 10489–10496 (1993).
46. H. Kamiya and H. Kasai, *J. Biol. Chem.* **270**, 19446–19450 (1995).
47. H. Ide, M. Okagami, H. Murayama, Y. Kimura, and K. Makino, *Biochem. Mol. Biol. Int.* **31**, 485–491 (1993).
48. H. Ide, K. Tedzuka, H. Shimzu, Y. Kimura, A. A. Purmal, S. S. Wallace, and Y. W. Kow, *Biochemistry* **33**, 7842–7847 (1994).
49. H. Ide, H. Shimizu, Y. Kimura, S. Sakamoto, K. Makino, M. Glackin, S. S. Wallace, H. Nakamuta, M. Sasaki, and N. Sugimoto, *Biochemistry* **34**, 6947–6955 (1995).
50. H. Shimizu, R. Yagi, Y. Kimura, K. Makino, H. Terato, Y. Ohyama, and H. Ide, *Nucleic Acids Res.* **25**, 597–603 (1997).
51. M. L. Morningstar, D. A. Kreutzer, and J. M. Essigmann, *Chem. Res. Toxicol.* **10**, 1345–1350 (1997).
52. D. A. Kreutzer and J. M. Essigmann, *Proc. Natl. Acad. Sci. U.S.A.* **95**, 3578–3582 (1998).
53. Z. Hatahet, Y. W. Kow, A. A. Purmal, R. P. Cunningham, and S. S. Wallace, *J. Biol. Chem.* **269**, 18814–18820 (1994).
54. D. Wang and J. M. Essigmann, *Biochemistry* **36**, 8628–8633 (1997).
55. C. J. Chetsanga and T. Lindahl, *Nucleic Acids Res.* **5**, 3673–3684 (1979).
56. J. Cadet, M. Berger, G. W. Buchko, P. C. Joshi, S. Raoul, and J.-L. Ravanat, *J. Amer. Chem. Soc.* **116**, 7403–7404 (1994).
57. K. Kino, I. Saito, and H. Sugiyama, *J. Amer. Chem. Soc.* **120**, 7373–7374 (1998).
58. T. Suzuki, R. Yamaoka, M. Nishi, H. Ide, and K. Makino, *J. Amer. Chem. Soc.* **118**, 2515–2516 (1996).
59. T. Suzuki, H. Ide, M. Yamada, N. Endo, K. Kanaori, K. Tajima, T. Morii, and K. Makino, *Nucleic Acids Res.* **28**, 544–551 (2000).
60. M. Dizdaroglu, J. Laval, and S. Boiteux, *Biochemistry* **32**, 12105–12111 (1993).
61. J. R. Wagner, B. Blount, and C. M. Weinfeld, *Anal. Biochem.* **233**, 76–86 (1996).
62. H. L. Katcher and S. S. Wallace, *Biochemistry.* **22**, 4071–4081 (1983).
63. L. H. Breimer and T. Lindahl, *Biochemistry* **24**, 4018–4022 (1985).
64. L.H. Breimer and T. Lindahl, *J. Biol. Chem.* **259**, 5543–5548 (1984).
65. C. D'Ham, A. Romieu, M. Jaquinod, D. Gasparutto, and J. Cadet, *Biochemistry* **38**, 3335–3344 (1999).
66. B. Karahalil, T. Roldan-Arjona, and M. Dizdaroglu, *Biochemistry* **37**, 590–595 (1998).
67. T. Roldan-Arjona, C. Anselmino, and T. Lindahl, *Nucleic Acids Res.* **24**, 3307–3312 (1996).
68. S. Senturker, P. Auffret-van-der-Kemp, H. J. You, P. W. Doetsch, M. Dizdaroglu, and S. Boiteux, *Nucleic Acids Res.* **26**, 5270–5276 (1998).

69. H. J. You, R. L. Swanson, and P. W. Doetsch, *Biochemistry* **37**, 6033–6040 (1998).
70. M. Dizdaroglu, B. Karahalil, S. Senturker, T. J. Buckley, and T. Roldan-Arjona, *Biochemistry* **38**, 243–246 (1999).
71. R. Aspinwall, D. G. Rothwell, T. Roldan-Arjona, C. Anselmino, C. J. Ward, J. P. Cheadle, J. R. Sampson, T. Lindahl, P. C. Harris, and I. D. Hickson, *Proc. Natl. Acad. Sci. U.S.A.* **94**, 109–114 (1997).
72. R. J. Melamede, Z. Hatahet, Y. W. Kow, H. Ide, and S. S. Wallace, *Biochemistry* **33**, 1255–1264 (1994).
73. D. Jiang, Z. Hatahet, R. J. Melamede, Y. W. Kow, and S. S. Wallace, *J. Biol. Chem.* **272**, 32230–32239 (1997).
74. A. Karakaya, P. Jaruga, V. A. Bohr, A. P. Grollman, and M. Dizdaroglu, *Nucleic Acids Res.* **25**, 474–479 (1997).
75. J. Tchou, V. Bodepudi, S. Shibutani, I. Antoshechkin, J. Miller, A. P. Grollman, and F. Johnson, *J. Biol. Chem.* **269**, 15318–15324 (1994).
76. B. Karahalil, P. M. Girard, S. Boiteux, and M. Dizdaroglu, *Nucleic Acids Res.* **26**, 1228–1233 (1998).
77. P. Auffret-van-der-Kemp, D. Thomas, R. Barbey, R. de Oliveira, and S. Boiteux, *Proc. Natl. Acad. Sci. U.S.A.* **93**, 5197–5202 (1996).
78. H. Aburatani, Y. Hippo, T. Ishida, R. Takashima, C. Matsuba, C. Kodama, M. Takao, A. Yasui, K. Yamamoto, M. Asano, K. Fukasawa, T. Yoshinari, H. Inoue, E. Ohtsuka, and S. Nishimura, *Cancer Res.* **57**, 2151–2156 (1997).
79. H. J. You, R. L. Swanson, C. Harrington, A. H. Corbett, S. Jinks-Robertson, S. Senturker, S. S. Wallace, S. Boiteux, M. Dizdaroglu, and P. W. Doetsch, *Biochemistry* **38**, 11298–11306 (1999).
80. I. Alseth, L. Eide, M. Pirovano, T. Rognes, E. Seeberg, and M. Bjoras, *Mol. Cell. Biol.* **19**, 3779–3787 (1999).
81. T. P. Hilbert, W. Chaung, R. J. Boorstein, R. P. Cunningham, and G. W. Teebor, *J. Biol. Chem.* **272**, 6733–6740 (1997).
82. S. Ikeda, T. Biswas, R. Roy, T. Izumi, I. Boldogh, A. Kurosky, A. H. Sarker, S. Seki, and S. Mitra, *J. Biol. Chem.* **273**, 21585–21593 (1998).
83. J. P. Radicella, C. Dherin, C. Desmaze, M. S. Fox, and S. Boiteux, *Proc. Natl. Acad. Sci. U.S.A.* **94**, 8010–8015 (1997).
84. T. Roldan-Arjona, Y. F. Wei, K. C. Carter, A. Klungland, C. Anselmino, R. P. Wang, M. Augustus, and T. Lindah, *Proc. Natl. Acad. Sci. U.S.A.* **94**, 8016–8020 (1997).
85. M. Bjoras, L. Luna, B. Johnsen, E. Hoff, T. Haug, T. Rognes, and E. Seeberg, *EMBO J.* **16**, 6314–6322 (1997).
86. T. A. Rosenquist, D. O. Zharkov, and A. P. Grollman, *Proc. Natl. Acad. Sci. U.S.A.* **94**, 7429–7434 (1997).
87. P. M. Girard, N. Guibourt, and S. Boiteux, *Nucleic Acids Res.* **25**, 3204–3211 (1997).
88. D. Jiang, Z. Hatahet, J. O. Blaisdell, R. J. Melamede, and S. S. Wallace, *J. Bacteriol.* **179**, 3773–3782 (1997).
89. Y. Sato, F. Uraki, S. Nakajima, A. Asaeda, K. Ono, K. Kubo, and K. Yamamoto, *J. Bacteriol.* **179**, 3783–3785 (1997).
90. S. S. David and S. D. Williams, *Chem. Rev.* **98**, 1221–1261 (1998).
91. M. L. Michaels and J. H. Miller, *J. Bacteriol.* **174**, 6321–6325 (1992).
92. R. P. Cunningham, *Curr. Biol.* **6**, 1230–1233 (1996).
93. N. Ye, G. P. Holmquist, and T. R. O'Connor, *J. Mol. Biol.* **284**, 269–285 (1998).
94. B. Hang, J. Sagi, and B. Singer, *J. Biol. Chem.* **273**, 33406–33413 (1998).
95. H. Nilsen, S. P. Yazdankhah, J. Eftedal, and H. E. Krokan, *FEBS Lett.* **362**, 205–209 (1995).
96. Z. Hatahet, M. Zhou, L. J. Reha-Krantz, S. W. Morrical, and S. S. Wallace, *Proc. Natl. Acad. Sci. U.S.A.* **95**, 8556–8561 (1998).

Mechanism of Action of *Escherichia coli* Formamidopyrimidine *N*-Glycosylase: Role of K155 in Substrate Binding and Product Release

Lois Rabow, Radhika Venkataraman, and Yoke W. Kow

Division of Cancer Biology
Department of Radiation Oncology
Emory University School of Medicine
Atlanta, Georgia 30335

I. Introduction...	224
II. Materials and Methods..	226
A. Oligonucleotides and DNA Substrates	226
B. Enzymes..	228
C. Analysis of fpg-Induced Cleavage Activity	228
III. Results..	228
A. Effect of DNA Structures on the Yield of β,δ-Elimination Products ..	228
B. Effect of Sequence Context on the Yield of β,δ-Elimination Products...	230
C. Pre–Steady-State Burst Kinetics	231
IV. Conclusions...	232
References...	233

Escherichia coli formamidopyrimidine *N*-glycosylase (fpg) is a DNA glycosylase with an associated β,δ-lyase activity. We have recently shown that the highly conserved lysine residue K155 is important for base recognition. Incubation of a double-stranded DNA containing an abasic site with the wild-type fpg protein generated only β,δ-product. However, incubation of a double-stranded DNA containing an abasic site opposite a small gap with fpg protein generated predominantly β-product. These data suggested that the induction of a double-strand break by fpg led to the destabilization of the protein–DNA covalent intermediate, causing the fpg protein to prematurely dissociate from the DNA substrate. Furthermore, when a double-stranded DNA containing an abasic site opposite an A was used as a substrate, K155A mutant fpg protein yielded a mixture of β- and

---

Abbreviations: 8-oxoG, 7-hydro-8-oxoguanine; AP, apyrimidinic/apurinic site; fpg, formamidopyrimidine *N*-glycosylase.

β,δ-products. These data suggested that K155 is essential for maintaining the stability of the intermediary protein–DNA covalent complex. Pre–steady-state burst kinetics showed that mutation in K155 led to the apparent disappearance of the initial burst, suggesting that the rate of product release from K155A is much greater than the rate of chemical reaction catalyzed by the mutant enzyme. This is consistent with the idea that K155A dissociates prematurely from the covalent complex, leading to a higher turnover number observed for K155A for DNA substrate containing an AP site. © 2001 Academic Press.

I. Introduction

Formamidopyrimidine N-glycosylase (fpg) from *Escherichia coli* recognizes diverse purine products including 7-hydro-8-oxoguanine (8-oxoG), formamidopyrimidines (FapyG and FapyA, imidazole ring–opened products of guanine and adenine, respectively), and N^7-methylformamidopyrimidines (1–5). Interestingly, the enzyme is unable to recognize 7-hydro-8-oxoadenine, the hydroxylation product of deoxyadenosine (4). Recently, it was shown that in addition to the oxidative purine products, fpg also recognizes some oxidative pyrimidine products including urea residue, 5-hydroxycytosine, 5-hydroxyuracil, uracil glycol, and thymine glycol (6, 7) as well as the lesion homologs generated by the action of alkoxyamine on abasic (AP) sites (8). In addition to the N-glycosylase activity, fpg has an associated β,δ-AP lyase activity, catalyzing successive β- and δ-elimination reactions (3, 8–12) (the δ-elimination reaction is actually a second β-elimination reaction, Fig. 1). The combined action of the N-glycosylase and AP lyase activity of fpg leads to DNA strand scission 3′ and 5′ to the damaged base, resulting in the release of the base lesion, a monomeric five-carbon fragment derived from deoxyribose (4-oxo-2-pentenal) and a one base-gap DNA terminated by 3′ and 5′ phosphates (9).

Like fpg protein, endonuclease III from *E. coli* is an N-glycosylase with an associated β-AP lyase activity (10, 13–16). In contrast to the β,δ-AP lyase activity of fpg, the β-AP lyase activity of endonuclease III catalyzes only a single β-elimination reaction. The lytic reaction carried out by both AP lyases requires the formation of a covalent protein–DNA complex (β-complex, Fig. 1), an imine intermediate formed between the active site amino acid and the sugar moiety (14–16). Since the initial course of the reaction carried out by both AP lyases is the same, it is interesting to find out what factors are important in allowing fpg to carry out a second β-elimination reaction, leading to the observed β,δ-elimination reaction. The proposed reaction scheme presented in Fig. 1 suggests that in order for the β,δ-elimination to occur, the β-complex has to persist long enough for the second elimination reaction to occur. This is corroborated by the finding that rebinding of fpg to the β-elimination product is rather poor (17). Therefore, it is likely that the differences between reactions catalyzed

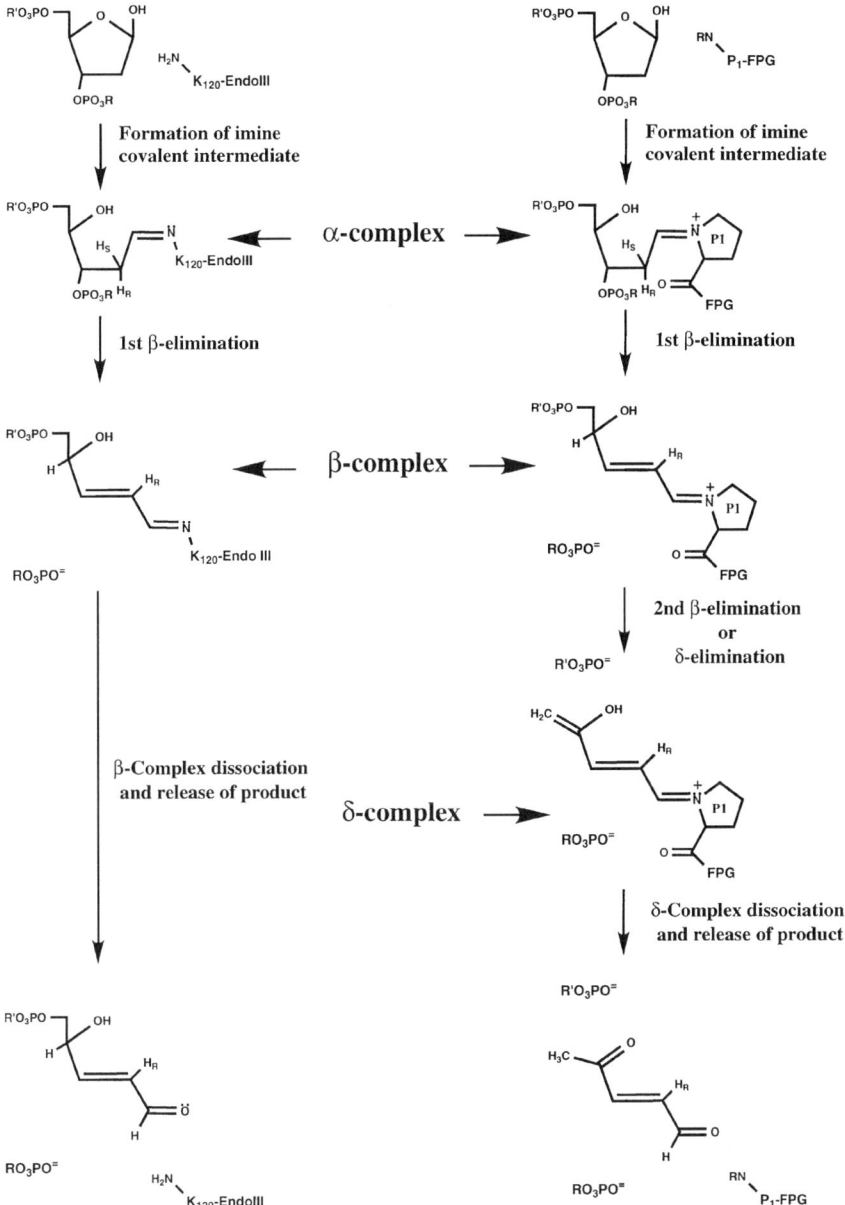

FIG. 1. Postulated intermediates in AP lyase reactions.

by these two AP lyases might reflect the relative stability of the intermediary β-complex (Fig. 1). Incidentally, it is interesting that fpg has a much higher affinity for its DNA substrates as compared to endonuclease III, suggesting that the β-complex derived from fpg is much more stable than the β-complex derived from endonuclease III (4, 18, 19). In order to examine the possibility that the course of the AP lyase reaction can be modified by affecting the stability of the β-complex, several approaches were taken. These include varying the sequence context surrounding the AP sites and the use of mutant fpg protein.

Lysine 57 and 155 are two highly conserved lysine residues in fpg protein. Earlier studies from this laboratory showed that K155A mutant fpg protein exhibited a large decrease in activity for 8-oxoG substrate but a significant increase in the AP lyase activity (20). The reduction in the activity against 8-oxoG led us to suggest that lysine 155 interacts directly with the 8-oxo group of the 8-oxoG lesion. However, the mechanism that led to the increase in AP lyase activity in the K155A mutant fpg is not clear. In this chapter, we show that in addition to interacting directly with 8-oxoG, K155 plays an important role in maintaining the stability of the protein–DNA complex and that the stability of the covalent intermediate is essential for the β,δ-elimination reaction.

II. Materials and Methods

A. Oligonucleotides and DNA Substrates

The sequences of oligonucleotides used in this study were as follows: U-60mer, 5'-AATTCAGCGTACTGATCTTAGACATGCC**U**GGTCAGACTCAGC GAACTGCACTGTCAGCTA; UComp-60mer (complementary strand for U-60mer), 5'-AGCTTAGCTTGACAGTGCAGTTCGCTGAGTCTGACCAGGC ATGTCTAAGATCAGTACGCTG; GAP1-I, 5'-ACAGTGCAGTTCGCTGAGT CTGACC; GAP1-II, 5'-pGGCATGTCTAAGATCAG–TACGCTG; GAP3-I, 5'-ACAGTGCAGTTCGCTGAGTCTGAC; GAP3-II, 5'-GCATGTCTAAGATCA GTACGCTG; GAP17-I, 5'-ACAGTGCAGTTCGCTGA; GAP17-II, 5'-TAAGA TCAGTACGCTG; U-24mer, 5'-GAACTAGTG**U**ATCCCC–CGGGCTGC; CComp-24mer (complementary strand for U-24mer with C opposite U when hybridized with U-24mer), 5'-GCAGCCCGGGGGAT**C**CACTAGTTC; AComp-24mer (complementary strand for U-24mer with A opposite U when hybridized with U-24mer), 5'-GCAGCCCGGGGGAT**A**CACTAGTTC; U-27mer, 5'-GGTC GACT**U**AG–GAGGATCCCCGGGTAC; CComp-27mer (complementary strand for U-27mer with C opposite U when hybridized with U-27mer), 5'-CCGGGGATCCTCCT**C**AGTCGACC; AComp-27mer (complementary strand for U-27mer with A opposite U when hybridized with U-27mer), 5′-CCGGGGA TCCTCCTAAGTCGACC. U-24mer containing a unique uracil residue was 5′-end-labeled with [γ-^{32}P]ATP (DuPont NEN) using T4 polynucleotide kinase (US Biochemical Corp.). Excess [γ-^{32}P]ATP was removed by gel filtration through Sephadex G50 (Pharmacia). The labeled oligonucleotide was annealed to the complementary strand (CComp-24mer) in a 1 : 1.5 ratio in a buffer containing 10 mM Tris–HCl, pH 7.5, and 0.1 M NaCl by heating to 90°C and cooling down gradually to room temperature, giving rise to duplex oligonucleotide containing a unique U/C pair. Single-stranded AP-24mer or duplex AP-24mer containing an AP/C pair was then prepared by incubating single-stranded U-24mer or duplex U-24mer with a U/C pair with excess amount of uracil DNA N-glycosylase (1 unit of enzyme per pmol of substrate) for 30 min at 37°C. The resulting DNA containing the AP site was used as a substrate for fpg without removal of uracil DNA N-glycosylase. Other duplex substrates were prepared in a similar manner. Gapped DNA substrates were prepared by hybridizing the respective three strands of oligonucleotides in a 1 : 1.5 : 1.5 ratio. For example, one-nucleotide gap DNA substrate was prepared by hybridizing 5′-labeled U-60mer with GAP-I and GAP-II in a 1 : 1.5 : 1.5 ratio following hybridizing conditions as described earlier. The resulting one-nucleotide gap uracil containing duplex was then treated with excess uracil DNA N-glycosylase to generate one-nucleotide gap duplex containing an AP site opposite the one-nucleotide gap. Other gapped DNA substrates were prepared in a similar manner. Table I lists the structures of the DNA substrates used in this study.

TABLE I
DNA SUBSTRATES

AP-24mer/C[a]	5′-GAACTAGTGSATCCCCCGGGCTGC
	CTTGATCACCTAGGGGGCCCGACG-5′
AP-24mer/A	5′-GAACTAGTGSATCCCCCGGGCTGC
	CTTGATCACATAGGGGGCCCGACG-5′
AP-27mer/C	5′-GGTCGACTSAGGAGGATCCCCGGGTAC
	CCAGCTGACTCCTCCTAGGGGCC-5′
AP-27mer/A	5′-GGTCGACTSAGGAGGATCCCCGGGTAC
	CCAGCTGAATCCTCCTAGGGGCC-5′
AP-60mer	5′-AATTCAGCGTACTGATCTTAGACATGCCSGGTCAGACTCAGCGAACTGCACTGTCAGCTA
	GTCGCATGACTAGAATCTGTACGGACCAGTCTGAGTCGCTTGACGTGACAGTTCGATTCGA-5′
GAP1	5′-AATTCAGCGTACTGATCTTAGACATGCCSGGTCAGACTCAGCGAACTGCACTGTCAGCTA
	GTCGCATGACTAGAATCTGTACGG CCAGTCTGAGTCGCTTGACGTGACA-5′
GAP3	5′-AATTCAGCGTACTGATCTTAGACATGCCSGGTCAGACTCAGCGAACTGCACTGTCAGCTA
	GTCGCATGACTAGAATCTGTACG CAGTCTGAGTCGCTTGACGTGACA-5′
GAP17	5′-AATTCAGCGTACTGATCTTAGACATGCCSGGTCAGACTCAGCGAACTGCACTGTCAGCTA
	GTCGCATGACTAGAAT AGTCGCTTGACGTGACAC-5′

[a]Substituted deoxynucleotides: S, abasic site.

B. Enzymes

Wild-type fpg protein was prepared according to procedures published earlier using an *E. coli* strain harboring an overproducing plasmid, pLRfpg16, that codes for an fpg protein modified to contain six C-terminal histidines (20). Similarly, mutant fpg protein (K155A) was also prepared as described earlier (20). Mutant protein K57A was also prepared using oligonucleotide-directed mutagenesis as described earlier (primer: 5′-p AGGCAGCTCGAGCAGCAGCA-GATATGCAGCCCGCCG) (20).

C. Analysis of fpg-Induced Cleavage Activity

The cleavage activity of fpg enzymes was assayed in a standard reaction buffer (10 μl) containing 20 fmol of DNA substrates, 10 mM Tris–HCl, pH 7.5, 50 mM NaCl, and an appropriate amount of enzyme at 37°C for 10–15 min. The reaction was stopped by the addition of an equal volume of formamide loading buffer (0.05% bromphenol blue, 0.05% xylene cyanol, and 10 mM EDTA in 98% formamide). The reaction substrates and products were separated on a 12.5% denaturing polyacrylamide gel. The polyacrylamide gel was then dried under vacuum, analyzed by autoradiography, and quantified with a Fuji BAS1000 phosphorImager system.

III. Results

A. Effect of DNA Structures on the Yield of β,δ-Elimination Products

As we and others have shown, under most reaction conditions, wild-type fpg protein yields only β,δ-elimination product (DNA terminated with 3′ phosphoryl ends) from duplex DNA containing either an AP or 8-oxoG site (3, 8–12). It is therefore interesting to find out whether the course of the reaction will be affected if the lesion (AP site) is situated directly opposite a nick or a gap in DNA. The rationale is that after the first β-elimination reaction catalyzed by fpg protein, the product generated will contain a double-strand break. The formation of a double-strand break should substantially increase the flexibility at the break and thus weaken the stability of the β-complex and lead to its dissociation. To examine this possibility, DNA substrates containing an AP site located directly opposite a nick or a small gap were incubated with fpg protein. Figure 2 shows the result of such an experiment. The cleavage products of endonuclease III and fpg on double-stranded DNA containing an AP site were shown to be predominantly β-elimination and β,δ-elimination products, respectively. The migration of the β-elimination product on a 12.5% SDS denaturing polyacrylamide gel is approximately one and half nucleotide distance slower than the

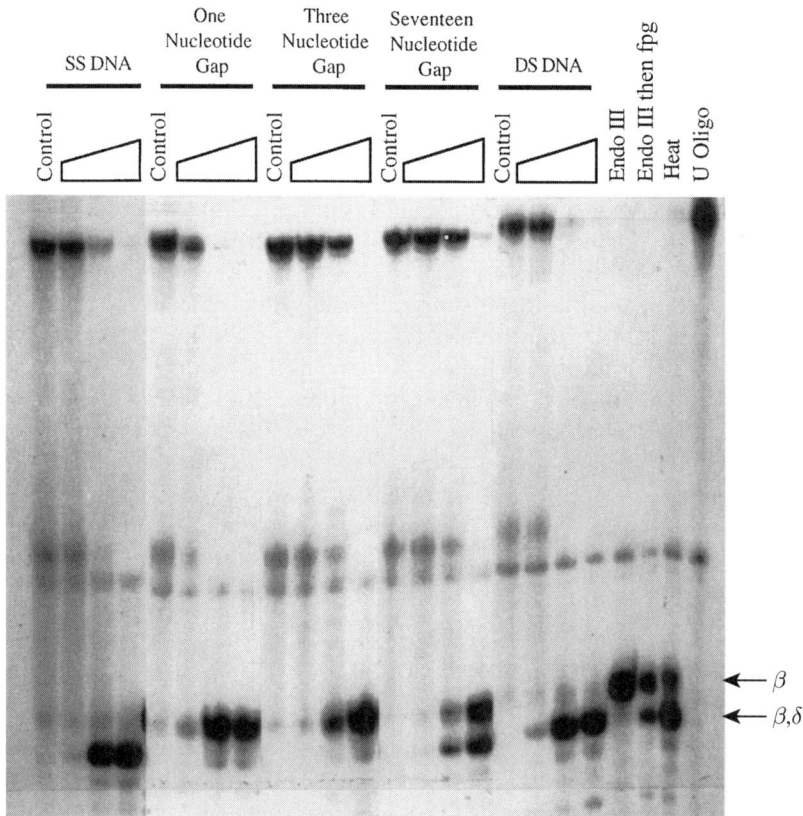

FIG. 2. Small gap in DNA promotes formation of β-elimination products by fpg. [^{32}P]5′-End-labeled AP-60mer (20 fmol, 10 μl reaction volume) containing an AP site was treated with various amounts of fpg in a reaction mixture containing 10 mM Tris–HCl, pH 7.5, 50 mM NaCl at 37°C for 10 min. DNA was treated with no fpg (lanes 1, 5, 9, 13, 17); 0.4 ng fpg (lanes 2, 6, 10, 14, 18); 4 ng fpg (lanes 3, 7, 11, 15, 19); and 40 ng fpg (lanes 4, 8, 12, 16, 20). Lane 21: Double-stranded AP-60-mer containing an AP site was treated with excess endonuclease III (200 ng) for 30 min. Lane 22: Double-stranded AP-60mer containing an AP site was first treated with endonuclease III (30 min) and then with fpg (15 min), showing the conversion of β-elimination product to β,δ-elimination product by fpg protein. Lane 23: Double-stranded AP-60-mer containing an AP site was heated at 90°C for 1 h to fully cleave the DNA at the AP site, generating both β- and β,δ-elimination products (*arrows*). Lane 24: Double-stranded U-24mer containing a uracil. β, position of β-elimination product generated by endonuclease III reaction; β,δ, position of β,δ-elimination product generated by fpg.

β,δ-elimination product generated by fpg protein (see Refs. 8 and 13; Fig. 2, lanes 20 and 21). Furthermore, prolonged heating of an AP site is known to generate a mixture of β- and β,δ-elimination products (Fig. 2, lane 23). Based on the rationale discussed above, it is believed that the slower migrating products generated by fpg on gapped DNA substrate were β-elimination products (Fig. 2, lanes 7–8, 11–12, and 17–18). As expected, the major product generated by fpg with double-stranded DNA containing a unique AP site was predominantly β,δ-elimination product (Fig. 2, lanes 19–20). Interestingly, when a one-nucleotide or three-nucleotide gapped DNA substrate was treated with fpg, the enzyme generated predominantly β,δ-elimination product (Fig. 2, lanes 7–8 and lanes 11–12). However, as the gap size of the substrate DNA was increased to 17, a mixture of β- and β,δ-elimination products was observed (Fig. 2, lanes 15–16). Since β-elimination product can be formed only when fpg dissociates from the β-complex, the data therefore suggest that fpg dissociates from the β-complex before the δ-elimination reaction can occur. When single-stranded DNA containing an AP site was used as a substrate, fpg yielded predominantly a major cleavage product that migrated to the same distance as the β,δ-elimination product and a minor amount of slower migrating species (Fig. 2, lanes 3–4).

B. Effect of Sequence Context on the Yield of β,δ-Elimination Products

It is known that the enzymatic reaction constant (K_m) as well as the apparent binding constant (K_d) of fpg toward its substrate is affected by the base opposite the lesion. The affinity of fpg for DNA substrate containing either an oxoG/C or AP/C pair is 10-fold higher than the DNA substrate containing an oxoG/A or AP/A pair (4, 21). A weaker K_d or K_m might reflect a weaker β-complex formed between fpg with DNA containing an AP/A pair as compared to DNA containing an AP/C pair. It is thus expected that an AP/A pair promotes premature dissociation of the β-complex. To examine the role of sequence context, DNA containing either an AP/C or AP/A pair was incubated with either wild-type or K155A mutant fpg protein. Mutant K155A fpg protein was shown to have a substantial decrease in the N-glycosylase activity, but an increase in the AP lyase activity (20). When the DNA substrate contained an AP/C pair, the reaction product generated by both the wild-type and K155A fpg consisted mainly of β,δ-elimination products (Table II). However, when the DNA substrate contained an AP/A pair, K155A mutant fpg protein yielded a substantial amount of β-elimination product (Table II), indicative of a premature dissociation of the β-complex. Little or no β-elimination product was observed with the wild-type protein. With all four DNA substrates containing an AP site, either opposite to an A or C, little or no β-product was observed with wild-type fpg protein. For

TABLE II
ACTIVITY OF WILD-TYPE, K57A, AND K155A MUTANT fpg WITH VARIOUS
AP-CONTAINING OLIGONUCLEOTIDES

	Specific activity (nmol/min/mg protein) (% of total product)								
	Wild-type			K155A			K57A		
Substrates	β	δ	Total	β	δ	Total	β	δ	Total
AP-24mer/C	0.21 (0.6)	34.7 (99.4)	34.9	2.83 (9.3)	27.7 (90.7)	30.5	0.47 (5)	8.62 (95)	9.09
AP-24mer/A	0.19 (4.4)	4.09 (95.6)	4.28	1.4 (26.7)	3.85 (73.3)	5.25	0.25 (15)	1.42 (85)	1.67
AP-27mer/C	0 (0)	32.5 (100)	32.5	0.41 (0.65)	63.5 (99.4)	63.8	0 (0)	14 (100)	14
AP-27mer/A	0 (0)	6.02 (100)	6.02	2.96 (51)	2.8 (49)	5.76	0.21 (11)	1.75 (89)	1.96

K155A mutant fpg protein, as much as 50% of the product was identified as β-product. It is also interesting that for K57A mutant fpg protein, about 10–15% of the reaction products were found to be β-product for DNA substrates containing an AP/A pair, while little or no β-product was observed for DNA substrates containing an AP/C pair. Changes in the structural conformation of the K57A protein have been observed, and the mutant protein exhibits only 15% of the wild-type activity in both the N-glycosylase and the AP lyase activity. However, recent data from the Laval laboratory showed that K57 might have a catalytic role in the fpg-catalyzed reaction.

C. Pre–Steady-State Burst Kinetics

In order to further understand the role of lysine 155, a transient kinetics assay was performed for both the wild-type and K155A mutant fpg proteins. Pre–steady-state burst kinetics were performed with 10 nM of either the wild-type or K155A mutant fpg proteins and incubated with 100 nM of 5'-end-labeled 24-mer containing an AP site. The points were taken and the reaction was stopped by the addition of gel loading buffer and heated to 50°C for 5 min before loading onto a 15% gel. It is interesting to observe that for the wild-type fpg protein, the progress curve showed a distinct burst followed by a steady-state kinetics (data not shown). Interestingly, mutation in K155 led to an apparent disappearance of the initial burst, suggesting that the rate of product released by the K155A mutant fpg is much greater than the rate of chemical reaction catalyzed by the mutant protein.

IV. Conclusions

Fpg is an N-glycosylase with an associated β,δ-AP lyase. It has been shown by many investigators that fpg forms a covalent imine intermediate with the DNA substrate, and the intermediary complex can be stabilized by reduction with sodium borohydride (11, 12, 20, 22, 23). Based on the reaction mechanism depicted in Fig. 1, three possible sequential imine protein–DNA complexes are involved before the enzyme dissociates from the final reaction product. The α-complex is the initial covalent intermediate that is formed between fpg and the C1-sugar moiety. For endonuclease III, the α-complex is generated by the reaction of lysine-120 (K120) of the active site with the C1-aldehyde of the ring-opened deoxyribose moiety (24). For fpg, the α-complex is formed by the nucleophilic reaction of the N-terminal proline (P1) residue (11, 12, 22). The β-complex is the covalent protein–DNA complex produced after the β-elimination reaction catalyzed by the α-complex (the abstraction of the C2-proS proton, H_S), followed by the cleavage of the 3′-phosphodiester bond by a β-elimination reaction (15). The δ-complex is the covalent complex of 4-oxo-2-pentenal with fpg protein (the abstraction of the C4 proton followed by the cleavage of the 5′-phosphodiester bond by a second β-elimination reaction) (9, 20).

Both α- and β-complexes have been demonstrated in our earlier experiments using borohydride reduction (20). Using a 5′-end-labeled double-stranded DNA substrate containing an AP site, the α-complex was identified as a protein–DNA crosslink with the initial DNA substrate (the DNA substrate has not been nicked yet) and β-complex was identified as fpg protein covalently linked to the nicked substrate (20). The presence of a δ-complex has yet to be demonstrated. It is therefore, conceivable that in order to proceed from the β-complex to the δ-complex to yield the β,δ-elimination product, the β-complex must be stable long enough for the second β-elimination reaction to occur. It is important to point out that β-elimination product is normally not observed for fpg-catalyzed reactions. Factors that affect the stability of β-complex might lead to premature dissociation of fpg from the β-complex, leading to the generation of β-elimination product (an abortive β,δ-elimination reaction). In this chapter, we show that the sequence context surrounding the lesion (AP site), the DNA structure in the vicinity of the lesion (i.e., the size of the gap), as well as the active site environment of the enzyme (effect of K155A mutation) are factors that can affect the outcome of the fpg-catalyzed reactions. It is probably safe to assume that if the lesion is opposite to a small gap, the formation of β-complex will lead to a double-strand break. The generation of a double-strand break might lead to a large decrease in the affinity (or an increase in K_d) of the β-complex, probably due to changes in the DNA conformation at the substrate-binding site. Alternatively, the generation of a double-strand break might lead to a large increase in the flexibility at the break junction, thus causing the enzyme to dissociate prematurely from the DNA–protein complex. The dissociation of the protein–DNA complex should involve hydrolysis of the imine intermediate, followed by a physical dissociation of fpg from the DNA, thus resulting in the generation of the observed β-product. It is interesting to note that if the gap size became larger (the region where the lesion is located is becoming more single-stranded–like) or the substrate is a single-stranded DNA, the product generated by fpg is mostly β,δ-product. However, unlike the situation with double-stranded DNA, the single-stranded and 17-nucleotide gapped double-stranded DNA have low but significant levels of β-products. These data suggest that the binding of fpg to double-stranded DNA is different from single-stranded DNA or single-stranded region. This is conceivable because single-stranded DNA adopts a random coil structure while double-stranded DNA adopts a double helical structure. Alternatively, a different mechanism might be involved in the generation of β,δ-product.

The generation of β-elimination product by fpg is thus highly indicative of an abortive β,δ-lyase reaction, and that the fpg protein dissociates prematurely from the protein–DNA complex. Mutation in K155 was observed to lead to an increased production of β-elimination product. Thus, the data suggest that in addition to playing a direct role in lesion recognition, K155 is also involved in maintaining the stability of the β-complex, possibly by interacting with the phosphodiester backbone after the β-elimination reaction. The additional role of K155 was corroborated by the preliminary pre–steady-state kinetic study. The pre–steady-state kinetic progress curve for K155 protein did not exhibit any observable burst kinetics. The lack of a burst in the pre–steady-state progress curve is thus highly suggestive that K155A mutant protein dissociates rapidly from the cleavage product at a rate much faster than the cleavage reaction. This observation is consistent with the idea that lysine 155 helps the wild-type protein to maintain a highly stable DNA–protein complex, a condition that is essential for the subsequent δ-elimination reaction to occur.

ACKNOWLEDGMENT

This work was supported by the National Institutes of Health Grant GM 37216.

REFERENCES

1. L. H. Breimer, *Nucleic Acids Res.* **12,** 6359–6367 (1984).
2. C. J. Chetsanga, M. Lozon, C. Makaroff, and L. Savage, *Biochemistry* **20,** 5201–5207 (1981).

3. J. Tchou, H. Kasai, S. Shibutani, M. H. Chung, J. Laval, A. P. Grollman, and S. Nishimura, *Proc. Natl. Acad. Sci. U.S.A.* **88,** 4690–5694 (1991).
4. J. Tchou, V. Bodepudi, S. Shibutani, I. Antoshechkin, J. Miller, A. P. Grollman, and F. Johnson, *J. Biol. Chem.* **269,** 15318–15324 (1991).
5. B. Tudek, S. Boiteux, and J. Laval, *Nucleic Acids Res.* **20,** 3079–3084 (1992).
6. Z. Hatahet, Y. W. Kow, A. A. Purmal, R. P. Cunningham, and S. S. Wallace, *J. Biol. Chem.* **269,** 18814–18820 (1994).
7. A. A. Purmal, G. W. Lampman, J. P. Bond, Z. Hatahet, and S. S. Wallace, *J. Biol. Chem.* **273,** 10026–10035 (1998).
8. V. Bailly, M. Derydt, and W. G. Verly, *Biochem. J.* **261,** 707–713 (1989).
9. M. Bhagwat and J. A. Gerlt, *Biochemistry* **35,** 659–665 (1996).
10. A. A. Purmal, L. E. Rabow, G. W. Lampman, R. P. Cunningham, and Y. W. Kow, *Mutat. Res.* **364,** 193–207 (1996).
11. J. Tchou and A. P. Grollman, *J. Biol. Chem.* **270,** 11671–11677 (1995).
12. D. O. Zharkov, R. A. Rieger, C. R. Iden, and A. P. Grollman, *J. Biol. Chem.* **272,** 5335–5341 (1997).
13. V. Bailly and W. G. Verly, *Biochem. J.* **242,** 565–572 (1987).
14. Y. W. Kow and S. S. Wallace, *Biochemistry* **26,** 8200–8206 (1987).
15. A. Mazumder, J. A. Gerlt, M. Absalon, J. Stubbe, R. P. Cunningham, J. Withka, and P. H. Bolton, *Biochemistry* **30,** 1119–1126 (1991).
16. J. Kim and S. Linn, *Nucleic Acids Res.* **16,** 1135–1141 (1988).
17. V. Bailly, W. G. Verly, T. O'Connor, and J. Laval, *Biochem. J.* **262,** 581–589 (1989).
18. S. O'Handley, C. P. Scholes, and R. P. Cunningham, *Biochemistry* **34,** 2528–2536 (1995).
19. D. Xing, R. Dorr, R. P. Cunningham, and C. P. Scholes, *Biochemistry* **34,** 2537–2544 (1995).
20. L. E. Rabow and Y. W. Kow, *Biochemistry* **36,** 5084–5096 (1997).
21. B. Castaing, A. Geiger, H. Seliger, P. Nehls, J. Laval, C. Zelwer, and S. Boiteux, *Nucleic Acids Res.* **21,** 2899–2905 (1993).
22. S. V. Kuznetsov, O. M. Sidorkina, J. Jurado, M. Bazin, P. Tauc, J. C. Brochon, J. Laval, and R. Santus, *Eur. J. Biochem.* **253,** 413–420 (1998).
23. B. Tudek, A. A. Van Zeeland, J. T. Kusmierek, and J. Laval, *Mutat. Res.* **407,** 169–176 (1998).
24. C. F. Kuo, D. E. McRee, C. L. Fisher, F. S. O'Handley, R. P. Cunningham, and J. A. Tainer, *Science* **258,** 434–440 (1992).

Thymine DNA Glycosylase

ULRIKE HARDELAND,
MARC BENTELE, TERESA
LETTIERI, ROLAND
STEINACHER, JOSEF JIRICNY, AND
PRIMO SCHÄR

Institute of Medical Radiobiology
University of Zürich and the Paul Scherrer
Institute
August-Forel Strasse 7, CH-8008 Zürich,
Switzerland

I. Evidence for Short-Patch Mismatch Correction and Discovery of
   Thymine DNA Glycosylase   236
II. The TDG Protein: Its Structure and Molecular Functions   237
   A. Primary Structure, Sequence Conservation, and Interacting Proteins. . .   237
   B. Three-Dimensional Structure and Mechanistic Implications   239
III. The Biochemistry of Thymine DNA Glycosylase.   241
   A. Enzymatic Activities of TDG.   241
   B. Substrate Specificity of Thymine DNA Glycosylase   243
IV. The Biology of Thymine DNA Glycosylase   246
   A. G·T Mismatch Correction   247
   B. Correction of Mismatched Uracil in DNA.   248
   C. Correction of Damaged Pyrimidine Bases   249
   D. Transcription-Associated DNA Repair   250
   E. TDG and Human Cancer   250
V. Conclusions   251
   References.   252

More than 50% of colon cancer-associated mutations in the p53 tumor suppressor gene are C → T transitions. The majority of them locate in CpG dinucleotides and are thought to have arisen through spontaneous hydrolytic deamination of 5-methylcytosine. This deamination process gives rise to G·T mispairs that need to be repaired to G·C in order to avoid C → T mutation. Similarly, deamination of cytosine generates G·U mispairs that also produce C → T transitions if not repaired. Restoration of both G·T and G·U mismatches was shown to be mediated by a short-patch excision repair pathway, and one principal player implicated in this process may be thymine DNA glycosylase (TDG). Human TDG was discovered as an enzyme that has the potential to specifically remove thymine and uracil bases mispaired with guanine through hydrolysis of their N-glycosidic bond, thereby generating abasic sites in DNA and initiating a base excision repair reaction. The same protein was later found to interact physically and functionally with the retinoid receptors RAR and RXR, and this implicated an unexpected function of TDG in nuclear receptor-mediated transcriptional activation of gene expression. The objective of this chapter is to put together the results of different lines of experimentation that have explored the thymine DNA glycosylase since its discovery and to critically evaluate their implications for possible physiological roles of this enzyme.   © 2001 Academic Press.

I. Evidence for Short-Patch Mismatch Correction and Discovery of Thymine DNA Glycosylase

The first evidence for efficient and specific G·T mismatch repair in mammalian cells emerged from heteroduplex transfection experiments designed by Brown and Jiricny (1, 2). They found that, upon transfection of SV40 DNA containing a single, strategically positioned mismatch, African green monkey kidney cells corrected G·T mispairs with very high efficiency (96%), predominantly by replacing the thymine with a cytosine (92%). By contrast, the same cells corrected A·C substrates with lower efficiency and without an apparent strand or base preference. Later, these observations were corroborated in human cells (3), and G·T heteroduplex correction was found to be equally efficient in normal fibroblasts and in several mutant cell lines originating from patients with DNA repair deficiencies, namely xeroderma pigmentosum, ataxia telangiectasia, and Bloom's syndrome. These data were interpreted as evidence for the existence of a mismatch repair activity in mammalian cells that has evolved specifically to counteract the mutagenic potential of hydrolytic deamination of 5-methylcytosine.

In an attempt to identify enzymes involved in this homoduplex restoration process, Wiebauer and Jiricny (4) developed an *in vitro* assay that allowed the correction of a synthetic G·T-mismatch containing substrate to be monitored after incubation with nuclear extracts of HeLa cells. This biochemical approach led to the discovery of a cellular activity that specifically incised the T-containing strand at the site of the mispaired thymine. Further examination of this reaction revealed that it could be accounted for by the concerted action of at least two enzymes—a mismatch-specific thymine DNA glycosylase, generating an apyrimidinic (AP) site opposite the guanine, and an AP-endonuclease, incising the DNA strand 5′ to the AP-site. Later, the same authors produced evidence for the fact that DNA polymerase β may be in charge of filling in the single-nucleotide gaps arising by processing of the AP sites (5).

Finally, a 55-kDa protein with thymine DNA glycosylase activity was purified from HeLa cells by affinity chromatography using a G·U derivatized matrix (6), and the enzyme was named after its biochemical activity—mismatch-dependent thymine DNA glycosylase (TDG). Microsequencing of proteolytic peptides then provided the structural information for isolation of the TDG encoding cDNA (7). Two species of cDNAs were identified, differing in the lengths of the 3′ untranslated region but not in their coding sequences. The full-length cDNA encodes a protein of 410 amino acids with a calculated molecular mass of 46 kDa and, upon heterologous expression in bacteria, yielded a recombinant enzyme with biochemical properties indistinguishable from those of the native TDG.

II. The TDG Protein: Its Structure and Molecular Functions

A. Primary Structure, Sequence Conservation, and Interacting Proteins

The revelation of the primary structure for the human TDG protein enabled the search for relatives in DNA and protein databases. Initially, two putative bacterial homologs were identified. The proteins from *Escherichia coli* and *Serratia marcescens* share about 37% amino acid sequence identity with the human enzyme and align with its central part (Fig. 1). The *E. coli* enzyme was expressed *in vitro* and shown to be active as a G·U but not as a G·T mismatch-specific uracil glycosylase. This protein was therefore named Mug for mismatch-specific uracil DNA glycosylase (8). Deletion analysis of the human enzyme confirmed that its conserved central domain (residues 112–360) is indeed sufficient for processing of G·U substrate, while hydrolysis of thymine from G·T substrate requires additional N-terminal sequence located between amino acid residues 56–112 (Fig. 1). Recently, we identified and cloned TDG homologs of *Drosophila melanogaster* and *Schizosaccharomyces pombe* (Marc Bentele and Primo Schär, unpublished results). A comparison of the primary structures of all TDG proteins suggests a common domain organization for the eukaryotic homologs. They are all composed of a conserved core domain and nonconserved, species-specific N-terminal and C-terminal extensions of variable length. Within the core domains, the insect and the fission yeast homologs share around 52% and 36% amino acid sequence identity with the human protein, respectively (Figs. 1 and 2). Strikingly, the *Drosophila* TDG has uniquely long N- and C-terminal sequences that account for its much higher molecular mass (>191 kDa) compared to that predicted for the other eukaryotic counterparts (37–46 kDa) (Fig. 1). The function of these N- and C-terminal domains is largely unknown, and future investigation will have to address this question. It seems likely, however, that they fulfill a role in establishing species- or pathway-specific protein–protein interactions that may be required to modulate its activity and/or to target the glycosylase activity to sites where its action is needed.

Thus far, two interacting proteins have been reported for TDG (Fig. 1B). The laboratory of Pierre Chambon published an observation that, at first encounter, seemed totally unrelated to any of the physiological roles that had been envisaged for TDG (9). In a yeast two-hybrid screen, they identified the murine TDG as an interacting partner of the retinoid acid receptor and the

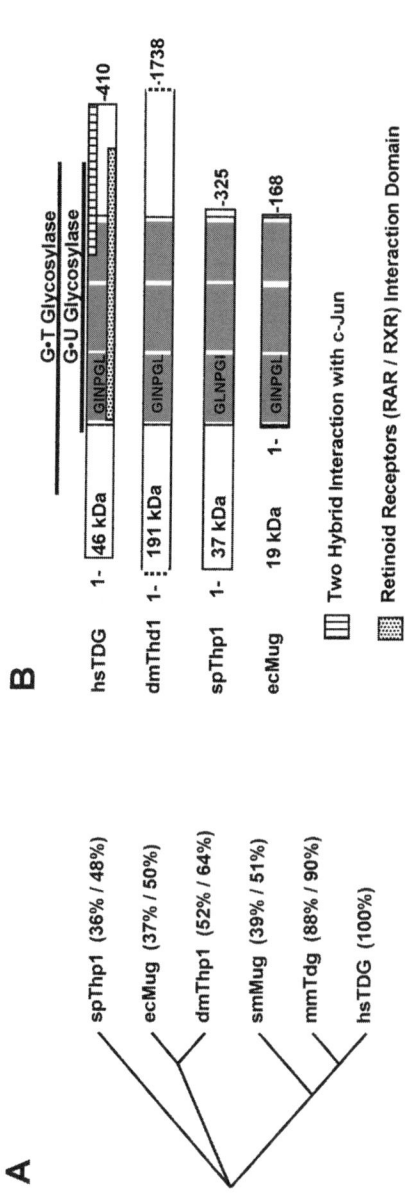

FIG. 1. Evolutionary conservation of thymine DNA glycosylase. (A) Dendrogram representing the clustering relationships between TDG homologs of different species as calculated with the PILEUP program of the Genetics Computer Group, Inc., software package (Version 10). The TDG homologs included are from *Escherichia coli* (ecMug; Swissprot No. P43342), *Serratia marcescens* (smMug; Swissprot No. P43343), *Schizosaccharomyces pombe* (spThp1; Swissprot No. P56581), *Drosophila melanogaster* (dmThd1, EMBL No. AJ277789), *Mus musculus* (mmTDG; Swissprot No. AJ277958), and *Homo sapiens* (hsTDG, Accession No. U51166). Percentages in brackets represent overall identities/similarities of a respective homolog to the human protein. (B) Schematic alignment of TDG homologs from representative bacterial, yeast, insect, and mammalian species. Shaded are three distinct blocks that are characterized by highest sequence conservation within the catalytic core domains of the proteins. The conserved position of the active site motif G(I/L)NPG(L/I) in the N-terminal part of the catalytic domain is indicated. The nonconserved N- and C-terminal extensions are shown by open boxes and the predicted molecular mass of each homolog as well as the length of the amino acid sequences are also indicated. Solid bars on top of the human protein illustrate the minimal structural requirements for catalytic activity on G·T and G·U substrates as determined by deletion analysis.

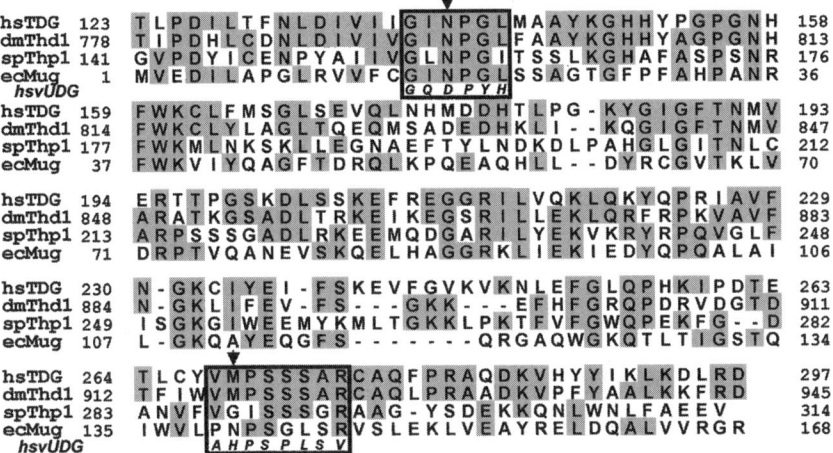

FIG. 2. Conservation of the catalytic domain in TDG homologs. A partial amino acid sequence alignment spanning the core domains of the TDG homologs of human (hsTDG), *Drosophila melanogaster* (dmThd1), *Schizosaccharomyces pombe* (spThp1), and *Escherichia coli* (ecMug). Identical residues are shaded and the amino acid motifs forming the essential parts of the proposed active site pocket are framed. Residues in italics indicate the structural-equivalent motifs present in the *Herpes simplex* virus uracil DNA glycosylase (hsvUDG; Swissprot No. P10186). Arrows indicate the positions of the active site residue N140 and the M269 that are essential for stable substrate interaction, as suggested by the crystal structure of substrate-bound Mug and confirmed by mutational analysis of the human enzyme.

retinoid X receptor (RAR and RXR). RAR and RXR are transcription factors of the family of ligand-activated nuclear receptors. This unexpected interaction was corroborated by functional evidence in the form that TDG stimulated binding of RAR/RXR homo- and heterodimers to their cognate responsive elements in *in vitro* studies and enhanced RAR/RXR-dependent transactivation of a reporter gene in transient transfection experiments. The RAR/RXR-interacting domain of TDG was mapped to its central region which also harbors the active site. These findings remind us of an earlier reported, but not further examined, "two-hybrid interaction" between the C terminus of TDG and the leucine zipper of another transcription factor, c-Jun (*10*).

B. Three-Dimensional Structure and Mechanistic Implications

The resolution of the crystal structures of both, free and substrate-bound *E. coli* Mug protein by the laboratory of Laurence Pearl brought about a major advance toward understanding the structure–function correlations for this family of DNA glycosylases (11, 12). This effort revealed striking structural similarities between Mug and the functionally related uracil DNA glycosylases (UDG) despite very limited conservation at the amino acid sequence level (<10% identity). Significant functional features of UDGs, such as the combined intercalation nucleotide-flipping mechanism, are also present in Mug, although the catalytic residues are only poorly conserved (Fig. 2). Like UDGs, Mug forms an active site pocket, and the postulated catalytic mechanism suggests that the mispaired uracil is flipped out of the DNA double helix and accommodated within this pocket in a manner that allows the N-glycosidic bond to be hydrolytically attacked by a strategically positioned, activated water molecule. Positioning of the water molecule is coordinated by the Asn40 of the highly conserved active site motif GINPGL. Base flipping by Mug is accompanied by intercalation of a three-amino acid wedge into the double helix. The less conserved residues around Gly143 occupy the abandoned space opposite the guanine and thereby maintain the base stacking interactions with the bases flanking the flipped-out uracil. A notable difference from the mechanism employed by uracil DNA glycosylase (13) is that the Mug residues involved in helix intercalation make specific contacts to the Watson–Crick face of the widowed guanine, providing specificity for a base pair rather than a single base. Ser23 of Mug is part of a small, flexible helix contributing to the fold of the active site and appears to interfere with the accommodation of the hydrophobic thymine within the active site pocket. The equivalent residue in the human enzyme is the smaller and less polar Ala145, a substitution that has been proposed to account for the human enzyme's ability to process thymine-containing substrate (12).

To test the functional predictions from the Mug crystal structure on the human TDG, we performed site-directed mutagenesis at critical residues and examined the biochemical activities of the mutant TDG variants (12a). Mutation of the postulated catalytic site asparagine to alanine (N140A) resulted in an enzyme that binds G·T and G·U substrates but is unable to catalyze base removal. Mutation of M259 in a motif with an inferred role in helix intercalation selectively inactivated stable binding of the enzyme to the mismatched substrates but not its glycosylase activity. These findings suggest that the structure–function model postulated for the E. coli enzyme is largely applicable to human TDG and its eukaryotic counterparts. It is noteworthy, however, that mutation of Ala145 to Ser did not alter the human enzyme's ability to process the G·T mismatched substrate. Thus, contrary to predictions based on the Mug crystal structure, evolutionary conversion of the equivalent of Ser23 to Ala in the human enzyme cannot adequately account for its acquired ability to process G·T mismatches.

Obviously, the bacterial model is inadequate when it comes to explaining properties specific for the larger eukaryotic enzymes, e.g., the wider substrate spectrum of the human TDG and the activity modulating role of its amino- and carboxy-terminal domains. Therefore, the ultimate understanding of the molecular mechanisms involved in substrate recognition, binding, and hydrolysis by eukaryotic TDG homologs will have to await structural analysis of the enzymes in interaction with their substrates and possibly other proteins.

III. The Biochemistry of Thymine DNA Glycosylase

A. Enzymatic Activities of TDG

If TDG evolved to counteract the mutagenic threat of cytosine and 5-methylcytosine deamination, it must be able to specifically recognize and remove from DNA uracil and thymine bases that are mispaired with guanine. While the specificity of detection and removal of uracil by a DNA glycosylase is structurally facilitated by its being a foreign base in DNA, recognition and repair of G·T mismatches requires an enzyme that is able to discriminate between a mutagenic thymine in a CpG·T sequence context and a normal thymine base-paired with adenine. Thus, G·T mismatch recognition can occur only in the context of double-stranded DNA.

Making use of oligonucleotide synthesis and labeling techniques, Josef Jiricny and colleagues developed versatile *in vitro* assays that enabled the monitoring of both mismatch binding and the generation of alkali-sensitive abasic sites (AP sites) by electrophoretic separation of enzymatic reaction products in nondenaturing and denaturing gels, respectively (Fig. 3). These assays facilitated the biochemical investigation into the catalytic properties of human TDG and revealed that, indeed, this enzyme is able to hydrolyze uracil and thymine from the biological relevant G·U and G·T substrates in a double-strand specific manner. The initial step in this reaction is mismatch binding followed by base flipping and rapid hydrolysis of the N-glycosidic bond, the last and rate-limiting step is dissociation of the enzyme from the AP-site. With the exception of the base flipping, which is inferred from the crystal structure of substrate-bound Mug, all other reaction intermediates have been demonstrated in different biochemical studies (*11*, *12*, *14*). One interesting mechanistic aspect that emerged for TDG is its property to bind to the product of its reaction, an AP site opposite guanine, with an affinity higher than that for the G·T and G·U substrates, and thus its failure to turn over under standard reactions condition (Fig. 3B) (*12a*, *14*). This product inhibition through specific AP-site binding finds a consistent explanation in the structural model for the substrate interaction of the *E. coli* Mug protein, where the helix intercalating residues of the enzyme establish specific and rigid contacts to the widowed guanine in the complementary strand (*11*). Therefore, the question arises whether TDG needs a releasing factor to stimulate its dissociation from the AP site and thus to promote enzymatic turnover. One obvious candidate for this function would be the enzyme catalyzing

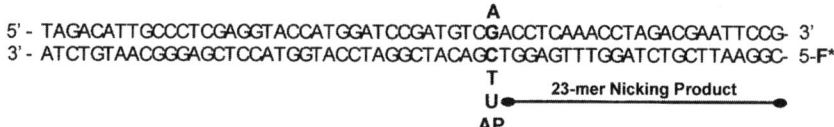
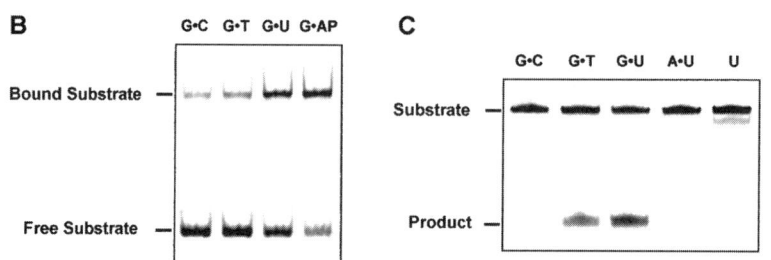

FIG. 3. Measurable enzymatic activities of thymine DNA glycosylase. The repertoire of biochemical techniques that have been used in studies with TDG includes electrophoretic mobility shift assays (EMSA, panel B) for the tracing of substrate binding and "nicking assays" (panel C) for the monitoring of abasic site (AP site) formation. (A) Shown is a 60-base-pair synthetic oligonucleotide substrate as used for the assays presented here. Different homo- and heteroduplex substrates can be obtained by annealing of appropriately designed complementary oligonucleotides (see boldface letters) and fluorescent (F*) or radioactive labeling enables detection of the reaction products after gel electrophoresis. (B) Substrate binding preferences of the human TDG. Labeled substrate DNA as indicated in panel A (1 pmol) was incubated with TDG enzyme (4 pmol) in the presence of a 10-fold excess of unlabeled competitor DNA. In the example shown here, a nonhydrolytic TDG variant (N140A, see text) was used in order to reveal the genuine mismatch binding preference of the enzyme. After incubation, enzyme-bound substrate was separated from free substrate in a native polyacrylamide gel and monitored by fluorescent scanning. The result illustrates that the human TDG binds the substrates shown with the following order of decreasing affinity: G·AP > G·U > G·T > G·C. (C) Mismatch nicking by the human TDG. Labeled substrate as indicated (1 pmol) was incubated with wild-type TDG protein (1 pmol). The ability of the enzyme to generate alkali-sensitive abasic (AP) sites at the position of the base to be removed was monitored after hot alkaline treatment and denaturing polyacrylamide gel electrophoresis. The presence of nicked product bands in the lanes with G·T and G·U substrate indicates that these are processed by TDG. No processing is evident in the lanes with the G·C, the A·U, and the single-stranded U substrates.

the subsequent step in the inferred base excision repair process, the AP-endonuclease. Peter Swann and his colleagues addressed this very problem in a series of biochemical assays with purified proteins. They could indeed show that human AP-endonuclease 1, HAP1, stimulates some turnover of the glycosylase acting on a G·T substrate, but it does so only when present in a molar excess over TDG in the reaction (15). Since the stimulatory effect of HAP1 on TDG turnover was not stoichiometric and clearly concentration-dependent and since both enzymes bind to double-stranded AP-sites with high affinities, one is tempted to argue that the HAP1 effect is accounted for by competition with TDG at the AP site substrate level rather than by an active displacement of TDG from the AP site by HAP1. In a reaction with TDG only, the enzyme would process the mismatch and remain bound to the AP site. It may well dissociate with a low rate, but then immediately bind again, and so on. In the presence of an excess of HAP1, however, spontaneous dissociation of TDG from the AP site would allow for HAP1 to come into competition, bind, and attack the AP site. If so, TDG would be forced to find a new substrate that can be either an unprocessed mismatch or an unoccupied AP site. This scenario is perfectly compatible with a concentration-dependent effect of HAP1 on TDG turnover and with the fact that different laboratories were not able to produce evidence for a physical interaction between HAP1 and TDG. Certainly, the final resolution of this issue will provide a lead to understanding the physiological role of TDG. The search for and the study of releasing factors for TDG must go on, following the example set by the group of Peter Swann.

B. Substrate Specificity of Thymine DNA Glycosylase

Studies by a number of groups using a series of different oligonucleotide substrates in assays with purified TDG protein yielded insight into its substrate preferences. The human enzyme was shown to process G·U, C·U, and T·U containing heteroduplex oligonucleotides better than G·T mismatched substrates, with the apparent k_{cat} for G·U processing being about 10-fold higher than that for G·T processing. TDG was also found to remove thymine from C·T and T·T mispairs, although with a rate that appears too low to suggest biological significance (*14, 16*). After we learned that the human enzyme readily processes the mismatched substrates under electrophoretic mobility shift assay conditions (*14*; our unpublished results), we used a nonhydrolytic mutant of TDG (N140A) to investigate its genuine substrate preferences at the recognition and binding step. The result was the following order of decreasing binding affinities: G·AP > G·U > G·T > G·C (Fig. 2A). This is consistent with the repeated observation that the wild-type enzyme processes G·T mismatches with a lower rate than G·U mismatches (*12a*).

Regarding neighboring sequence-context effects, Griffin and Karran (*17*) reported that the G·T mispair in a methylated or unmethylated CpG context was the preferred substrate for a processing activity present in HeLa cell extracts, and the same preference was observed with the purified native enzyme as well as recombinant TDG (*7, 14, 18*). These studies led to the currently established order of preferred nucleotides 3' to the mismatched thymine which decreases along the series CpG·TpG > TpG·TpA > GpG·TpC > ApG·TpT (*14, 16, 19*).

In addition to these standard base to base mismatches, both human TDG and the bacterial Mug enzyme were reported to efficiently process the mutagenic cytosine adduct 3, N^4-ethenocytosine (εC) (20). Ethenocytosine arises in DNA by the action of metabolites of environmental pollutants such as vinyl chloride and by endogenous products of lipid peroxidation, and may therefore be a biologically relevant mutagen (21). Recently, the group of Jean-Pierre Jost reported processing of 5-methylcytosine by a chicken homolog of human TDG. The enzyme copurified with a 5-methylcytosine DNA glycosylase activity from extracts of chicken embryos and also showed weak 5-methylcytosine processing in a substrate with an asymmetrically methylated mCpG/CpG sequence when purified as a recombinant protein from overexpressing bacteria (22). Using an unrelated oligonucleotide substrate with hemimethylated and fully methylated CpG sites, we were not able to detect significant 5-methylcytosine glycosylase activity with the recombinant human enzyme or in extracts of human cell lines under conditions where G·T and G·U substrates were efficiently processed (Ulrike Hardeland and Primo Schär, unpublished results). Given the importance of the issue, the resolution of this apparent discrepancy warrants careful investigation.

The groups of Karran and Day examined the enzyme's ability to remove thymine from mispairs with modified purines and purine analogs. They could show that an activity in extracts of HeLa or A1235 human malignant glioma lines was able to process duplexes containing mispairs of thymine with 6-O-methylguanine, 6-thioguanine, 2,6-diaminopurine, and 6-aminomethyl-2-aminopurine (18, 23, 24). However, in contrast to the cell extracts, purified recombinant TDG failed to act on 2,6-diaminopurine·T and 2-aminopurine·T heteroduplex substrates (19). Regarding the biologically relevant 6-O-methylguanine·T and 6-thioguanine·T mispairs, it will be important to examine whether TDG-dependent base excision repair contributes to their processing *in vivo*. If it does so with an appreciable efficiency, it might contribute to the cytotoxicity of 6-O-methylguanine- or 6-thioguanine-inducing drugs which are being used in cancer therapy.

Gregory Verdine and colleagues investigated the substrate interaction of TDG using synthetic substrate analogs that are recognized and bound but not cleaved by the enzyme. They demonstrated that replacement of protons at the 2′ position of a mispaired deoxyuridine with fluorine atoms strongly inhibited the cleaving but not the binding step of the TDG reaction. This opened the possibility of forming reasonably stable complexes between G·2′-fluorouracil mispaired substrate and TDG, and thus enabled the analysis of the protein DNA interaction by DNase I footprinting and methylation interference studies. Bound TDG was found to protect a 14-nucleotide region on the U-containing strand and a 19-nucleotide stretch on the G strand. The same study also revealed that TDG does not make specific contact with the N7 position of the mispaired guanine; rather, a specific interaction could be detected with the N7 position of the guanine immediately flanking the mispaired uracil on its 3' side (25). This is consistent with the finding that G·U or G·T mismatches are most efficiently processed by TDG if they are located within CpG sequences. It should be remembered at this point that specific contacts to the guanine in the complementary strand were evident in the crystal structure of substrate-bound Mug (11); however, these contacts did not involve the N7 position of the guanine. The strong binding of the human TDG to double-stranded substrate with an AP site opposite guanine suggests formation of specific complementary base contacts also for this enzyme.

We found that recombinant human TDG was able to process 5-fluorouracil (5-FU) with a higher efficiency than any other substrate that had been tested. This attracted our attention because 5-FU was a good substrate not only when mismatched with G but also when it was opposite A or even present in a single-strand oligonucleotide. From comparative kinetic analyses of G·T, G·U, and G·5-FU processing by the fully active TDG and by the nonhydrolytic mutant variant (see Section II,B), we learned that destabilizing the *N*-glycosidic bond through substitution of the 5-carbon hydrogen of uracil with an electron-withdrawing fluorine can enhance the activity of TDG and obviate its need for stable interaction with the substrate. This implies for the opposite case of an energetically less favorable substrate such as a thymine opposite guanine, that stable complementary strand interactions by TDG are essential to compensate for its comparably poor hydrolytic potential (12a). Given these observations, one might speculate that evolution has limited the catalytic power of TDG in order to ensure the formation of stable enzyme–substrate interactions before base hydrolysis can take place. This seems a reasonable strategy for an enzyme that has the ability to hydrolyze normal bases from DNA. In the case of thymine release by human TDG, a sufficiently stable DNA interaction is established only if the opposite base is a guanine, thus providing the substrate discrimination needed to avoid inappropriate and nonspecific base hydrolysis. Opposite-base specificity may be less important for DNA glycosylases that release damaged bases from DNA, because the specificity in these cases is achieved through physical recognition of structural base irregularities.

In another line of investigation Gallinari and Jiricny (8) employed a deletion mutagenesis strategy to identify the minimal catalytic domain of TDG. They deleted up to 264 amino acids from the N terminus and up to 201 amino acids from the C terminus and tested the mutant enzymes for G·T and G·U processing activity. Deletion of 56 N-terminal amino acids had no measurable effect on enzymatic activity *in vitro*, but deletion of a further 56 amino acids produced a mutant enzyme that was active as an uracil glycosylase on the G·U substrate but no longer able to process the G·T mispair. At least three hypotheses can be put forward to explain this remarkable change in substrate specificity. (1) TDG may contain two separate active sites, one near the N terminus responsible for G·T processing and one toward the C terminus for G·U processing. The idea is contradicted and thus falsified by the fact that a single amino acid substitution (N140A) at the active site within the conserved catalytic domain abolishes both the G·T and G·U processing activities of the enzyme (*12a*). (2) Removal of the N terminus could change the conformation of the active site pocket of the enzyme. Structural studies of the human uracil DNA glycosylase revealed that its specificity for uracil is achieved by folding of a very tight active site pocket in which a tyrosine residue is juxtaposed close to the 5-carbon of the base, blocking entrance of thymine or other uracil analogs with bulky substituents at this position (*26, 27*). No such function is apparent in the crystal structure of Mug, but molecular crowding at the 5-carbon position of uracil was proposed to explain the thymine discrimination by the enzyme (*12*). In analogy, the human TDG might have a similarly tight active site pocket, but there the addition of N-terminal amino acid residues might alter the overall structure of the pocket in a way that allows accommodation of thymine. (3) A third possibility is that the N terminus contributes to stabilizing the enzyme–substrate interaction in a way that facilitates hydrolysis of the energetically less favorable G·T mispair. This finds some support in the nonconserved primary structure of the N-terminal domain of TDG. This domain is composed of clusters of basic amino acid residues and has weak similarity to histone H1 and HMG1 proteins that have characteristic DNA-binding properties. It is therefore conceivable that the N terminus of TDG has a DNA-binding capacity that could stabilize substrate interactions *in vitro*. Clearly, this substrate-specificity modulating role of the N terminus needs further investigation. It is hoped that the resolution of this phenomenon will lead to the identification of the physiological substrate of TDG, and thus to more precise predictions as to the function of the enzyme in cells.

IV. The Biology of Thymine DNA Glycosylase

The existence of TDG homologs in organisms ranging from bacteria to man testifies to the fact that the enzyme is of ancient origin and has a conserved structure and function (Fig. 1A). Considering the ability of the human TDG to recognize and remove uracil and thymine from G·U and G·T mispairs, one might predict a function for this enzyme in the avoidance of mutation through spontaneous hydrolytic deamination of cytosine and 5-methylcytosine. However, the inability of the bacterial Mug and the conserved catalytic domain of the human enzyme to process G·T mispairs would suggest that the family of TDG enzymes has evolved as uracil DNA glycosylases specific for G·U mispairs. Then, the fact that Mug and human TDG process εC·G adducts with appreciable efficiency might indicate a role for these enzymes in repair of a wider spectrum of damaged pyrimidine bases, and, finally, the interaction of the human TDG with retinoid receptors would support a function of at least this homolog in regulation of transcriptional processes. Let us consider the evidence for each of these cases individually.

A. G·T Mismatch Correction

The extra capacity of mammalian TDG to process G·T mismatches may be a specialized function of the enzyme required only in organisms with appreciable degrees of DNA methylation at CpG sites. If we thus consider a specific role for the mammalian TDG in repair of G·T mispairs generated by deamination of methylated cytosine residues, we have to acknowledge a recent discovery that MBD4 has enzymatic activities and properties very similar to those of TDG. This mammalian protein is composed of a methylated DNA-binding domain and a DNA glycosylase domain that is related to those of the bacterial Endo III and MutY proteins but not to the core domain of TDG. *In vitro*, MBD4 processes nearly the same range of mismatched substrates as TDG and they are therefore expected to be functionally redundant, in particular for the correction of G·T mispairs in CpG sequences (28). However, as for TDG, the role of MBD4 in the repair of G·T mismatches, and thus in mutation avoidance, is hypothetical. If the biochemical properties of both enzymes do reflect their physiological functions, we have to find an explanation for the apparently conflicting observation that, despite the presence of specialized repair systems, methylated CpG sites are mutational hotspots in the mammalian genome. One possible explanation is that the actions of TDG and MBD4 might be confined to certain areas in the genome or to specific DNA metabolic processes such as transcription from selected promoters. If this were true, only some genomic regions would benefit from protection by TDG or MBD4 whereas others would be more prone to mutagenesis by 5-methylcytosine deamination. However, at present there is no evidence in support of this hypothesis. Similarly, and not mutually exclusive, the relative inefficiency of both TDG and MBD4 in G·T processing could indicate that these restoration pathways are not sufficiently efficient to completely avoid interference with the postreplicative mismatch repair system controlled by the conserved MutS and MutL proteins (29). Postreplicative mismatch repair, in the absence of a strand-discriminating signal, would then correct G·T mismatches without appreciable preference to either G·C or A·T base pairs, and thus assimilate C to T transitions in approximately 50% of the events. Indeed, the SV40 transfection experiments mentioned earlier in this review revealed 4% correction of G·T to A·T (1). Assuming random strand discrimination, this would imply a contribution to G·T correction by the postreplicative mismatch repair system in 8% of the events, which would be sufficient to account for the observation that sites of cytosine methylation are mutagenic hotspots. These are interesting observations, correlations, and hypotheses that make clear that an ultimate resolution of the issue of G·T repair will have to await the availability of appropriate TDG knockout models that will allow for testing the specific predictions associated with this function.

B. Correction of Mismatched Uracil in DNA

The problem of functional redundancy also arises when we consider a role for TDG homologs in the processing of uracil in mutagenic G·U mispairs. Why should a comparably inefficient mismatch-dependent uracil DNA glycosylase evolve side-by-side with the very potent uracil DNA glycosylase (UDG) and, in the case of the mammalian situation, with at least two other enzymes that process the same substrate with comparable efficiency. These are the recently discovered SMUG1 protein (30) and, again, the MBD4 protein. For UDG it could be envisaged that, despite being a potent repair enzyme, it may not be able to remove all uracil from DNA with equal efficiency. Its primary function may be the replication-associated processing of uracil misincorporated opposite adenine during DNA synthesis and not the processing of G·U mispairs arising in nonreplicating DNA after cytosine deamination. Supporting evidence for such a scenario was reported by Hans Krokan and his group, who established a physical and functional association of the nuclear uracil DNA glycosylase (UNG2) with DNA replication in mammalian cells (31). Further consistent with this idea are the phenotypes of a recently published mouse $ung^{-/-}$ knockout strain (32) and an udg^- mutant of Schizosaccharomyces pombe (Marc Bentele and Primo Schär, in preparation). In both organisms, the inactivation of the major uracil DNA glycosylase activity leads to an increased steady-state level of uracil in DNA but does not significantly increase the spontaneous mutation rate. Thus, the function of G·U repair in these organisms appears to be redundant, with at least four enzymes—UNG2, TDG, MBD4, and SMUG1—having the potential to compensate for each other in mammalian cells and UDG and TDG or more in S. pombe. However, a strict limitation to the correction of misincorporated uracil opposite adenine would not explain the mutator phenotype displayed by uracil DNA glycosylase mutants of two other organisms, E. coli (33) and S. cerevisiae (34, 35). In budding yeast, the mutator phenotype of ung1 mutant strains could be accounted for by the apparent lack of TDG, MBD4, and SMUG1 homologs in the genome and thus, by the absence of additional uracil glycosylase activities. However, regarding the bacterial situation where no redundant MBD4- or SMUG1-like activities are discernible, the laboratory of Ashok Bhagwat showed that inactivation of the mug gene in both wild-type and uracil DNA glycosylase deficient (ung^-) strain backgrounds had no effect on the rates of C→T or 5-methylC→T transition mutations (33). This would argue against a redundant role of a TDG homolog in mutation avoidance, at least in E. coli.

Interesting in this respect is Drosophila melanogaster. Its genome is devoid of a uracil DNA glycosylase gene but encodes a TDG homolog and a SMUG1 homolog (Marc Bentele and Primo Schär, unpublished data). A uracil-processing activity was discovered in *Drosophila*, but its expression was restricted to late larval stages of development (*36, 37*). Gallinary and Jiricny (*8*) analyzed extracts of different cell lines from pupating insects, among them S2 from *D. melanogaster*. All extracts possessed G·U mismatch-specific uracil glycosylase activity but neither G·T nor single-stranded uracil processing activity was observed. Since a UDG-like activity did indeed appear to be absent from these extracts, the existence of an enzyme capable of processing G·U mispairs supports the concept that organisms need a specific defense mechanism against cytosine deamination. As uracil in DNA is mutagenic only when present in the mismatch G·U after a cytosine deamination event, organisms could conceivably dispense with UDG as long as a G·U-specific uracil glycosylase is present. However, it is unclear at this point whether the TDG or the SMUG1 homolog of *Drosophila* or yet another protein is responsible for the uracil glycosylase activities observed. The fact that no G·T processing activity was found in any of the insect cell extracts is consistent with the absence in these organisms of DNA methylation and, thus, of the premutagenic G·T mismatches occurring by deamination of 5-methylcytosine.

One particularly interesting question arises from these reflections. Why have different organisms evolved with different sets of enzymes that can deal with uracil bases in DNA? The fact that none of the four proteins with uracil glycosylase activity—UDG, TDG, MBD4, and SMUG1—is present in the four organisms considered herein suggests that none of them is absolutely essential for life. On the other hand, the fact that no organism has been reported to be devoid of any uracil glycosylase activity suggests that uracil processing in some form is important for most organisms. Future studies addressing the role of uracil processing in organisms with different sets of uracil glycosylases may provide useful insight into the specific physiological functions of the individual enzymes.

C. Correction of Damaged Pyrimidine Bases

Both the human TDG and *E. coli* Mug protein are able to process ethenocytosine, and it appears that, at least in *E. coli*, Mug is the only enzyme to remove this lesion from DNA (*33*). Ethenocytosine adducts are present in mammalian DNA (1–3 per 10^7 bases in DNA of human liver) and they are mutagenic and genotoxic. The mutations most frequently caused by the adduct are C→A transversions and C→T transitions (*38, 39*). The observation that TDG processes substrates other than the classical G·U and G·T mismatches suggested a more general function of the enzyme in the repair of DNA base damage. We therefore tested a number of oxidized derivatives of cytosine and 5-methylcytosine and found that TDG was able to efficiently hydrolyze 5-hydroxyuracil and 5-hydroxymethyluracil mispaired with guanine. In agreement with these biochemical results, we observed that HeLa cells overexpressing TDG protein became hypersensitive to treatment with oxidizing agents (Teresa Lettieri, Josef Jiricny, and Primo Schär, unpublished results). Thus, in view of these findings, we would predict that TDG acts in cells to repair damaged cytosine or 5-methylcytosine residues in DNA. The range of substrates could include the mutagenic and cytotoxic products of cytosine deamination, alkylation, and oxidation and possibly also oxidized thymine residues. If so, TDG might share common substrates with yet another family of pyrimidine DNA glycosylases, the homologs of the *E. coli* nth protein.

D. Transcription-Associated DNA Repair

A separate line of investigation has implied a role for TDG in transcription-related processes, mainly by the evidence of a physical and functional interaction of the human and mouse TDG proteins with the retinoid receptors. This appears to contrast with the observations suggesting a DNA glycosylase function for TDG and therefore raises the question of whether the two diverging sets of experimental evidence reflect unrelated actions of TDG in DNA repair and transcription or whether they can be reconciled in a unifying functional model. There is little evidence upon which to draw any reasonably safe conclusions. However, studying active site mutations in human TDG, we found that amino acid substitutions affecting DNA glycosylase function were at the same time reducing the potential of the protein to stimulate RAR/RXR-dependent transcription in co-transfection experiments (Arndt Bennecke, Ulrike Hardeland, Pierre Chambon, Josef Jiricny, and Primo Schär, in preparation). We could therefore argue that the transcription and DNA repair-associated functions of TDG are related as they both share the same active site within the enzyme. Given this, we can put forward two hypotheses for possible transcription-related actions of TDG: one that proposes an involvement of TDG in transcription-coupled repair of pyrimidine base damage, and another that takes into account a possible 5-methylcytosine DNA glycosylase activity of TDG and postulates a role for the enzyme in conjunction with transcription factors in regulating the methylation status at critical CpG sequences in activated promoters. At this point, this is purely speculative, but certainly these hypotheses will require careful examination.

E. TDG and Human Cancer

The mutational spectrum found in DNA from tumor cells can be diagnostic for the underlying mutagenic events. A compilation of reported colon cancer-associated mutations in the p53 tumor suppressor gene revealed that nearly 50% of tumors display C to T mutations in CpG dinucleotides (40). The same type of mutation is also predominating in tumors of other tissues, and the fact that the CpG sites frequently mutated in the colon tumors were found to be methylated in normal tissue (41) suggests deamination of 5-methylcytosine in combination with malfunction of G·T repair to be the principal underlying mutagenic events. As mentioned before, the lack of TDG activity or even minor inefficiency of the enzyme causing disregulation of mismatch repair, e.g., channeling the G·T mismatches toward postreplicative mismatch repair, would be sufficient to account for the CpG mutations found in colon tumors. Interestingly, the colon is also the most frequent locus of hereditary nonpolyposis colon cancer (HNPCC) which is linked to defects in postreplicative mismatch repair, indicating that this epithelium may be particularly prone to disregulation of mismatch repair. Similarly, the substrate specificity of TDG suggests that the antimutagenic effect of the enzyme is not limited to methylated CpG sites and that mutations arising as a consequence of cytosine deamination, alkylation, or oxidation would also be more frequent in cells without functional TDG. These are very suggestive reflections combining biochemical and biological observations and culminating in an exciting anticipation as to a possible physiological role of TDG in mutation avoidance. Yet, these hypotheses require careful testing as, at present, there is no direct evidence either for TDG being associated with a G·T, G·U, or G·εC specific repair process *in vivo* or for malfunction of TDG-dependent base excision repair being linked to increased mutagenesis in general and, thus, to human malignancy in particular.

V. Conclusions

At the beginning there was a hypothesis saying that organisms that methylate cytosine in DNA need a specific repair enzyme that can recognize and process the products of spontaneous hydrolytic deamination of 5-methylcytosine in order to prevent excessive mutagenesis. Then the human TDG was discovered as an enzymatic activity that could do just that. The biochemical investigation of this protein over the past decade has provided a great deal of information about its enzymatic properties and identified a number of highly interesting DNA substrates that would indeed suggest an important role of TDG in mutation avoidance. Structural studies of the homologous Mug protein of *E. coli* have significantly advanced our understanding of the molecular details of the enzyme's interaction with the DNA substrate and the mechanism of base hydrolysis. On the other side, this appreciable biochemical and structural insight is contrasted by a total lack of physiological evidence substantiating any of the functions postulated for the enzyme. Therefore, it will be the most important challenge in future studies to explore the physiological role of TDG—in studies that will be greatly facilitated by the availability and the use of suitable yeast, insect, and mouse models. Only after these efforts have established solid connections between biochemical observations and functional speculations can we return to postulating roles for thymine DNA glycosylase in human disease with reasonable accuracy.

Acknowledgments

We are grateful to Giancarlo Marra and Orlando Schärer for helpful discussions and critical reading of the manuscript. Our own work mentioned herein was supported by research grants from the Swiss National Science Foundation and the Schweizerische Krebsliga.

References

1. T. C. Brown and J. Jiricny, *Cell (Cambridge, Mass.)* **50**, 945–950 (1987).
2. T. C. Brown and J. Jiricny, *Cell (Cambridge, Mass.)* **54**, 705–711 (1988).
3. T. C. Brown, I. Zbinden, P. A. Cerutti, and J. Jiricny, *Mutat. Res.* **220**, 115–123 (1989).
4. K Wiebauer and J. Jiricny, *Nature (London)* **339**, 234–236 (1989).
5. K. Wiebauer and J. Jiricny, *Proc. Natl. Acad. Sci. U.S.A.* **87**, 5842–5845 (1990).
6. P. Neddermann and J. Jiricny, *J. Biol. Chem.* **268**, 21218–21224 (1993).
7. P. Neddermann, P. Gallinary, T. Lettieri, D. Schmid, O. Truong, J. Justin Hsuan, K. Wiebauer, and Josef Jiricny, *J. Biol. Chem.* **271**, 12767–12774 (1996).
8. P. Gallinari and J. Jiricny, *Nature (London)* **383**, 735–738 (1996).
9. S. Um, M. Harbers, A. Benecke, B. Pierrat, R. Losson, *et al., J. Biol. Chem.* **273**, 20728–20736 (1998).
10. P. M. Chevray and D. Nathans, *Proc. Natl. Acad. Sci. U.S.A.* **89**, 5789–5793 (1992).
11. T. B. Barrett, R. Savva, G. Panayotou, T. Barlow, T. Brown, J. Jiricny, and L. H. Pearl, *Cell (Cambridge, Mass.)* **92**, 117–129 (1998).
12. T. E. Barrett, O. D. Schärer, R. Savva, T. Brown, J. Jiricny, G. L. Verdine, and L. H. Pearl, *EMBO J.* **18**, 6599–6609 (1999).
12a. U. Hardeland, M. Bentele, J. Jiricny, and P. Schär, *J. Biol. Chem.* **275**, 33449–33456.
13. G. Slupphaug, C. D. Mol, B. Kavli, A. S. Arvai, H. E. Krokan, and J. A. Tainer, *Nature (London)* **384**, 87–92 (1996).
14. T. R. Waters and P. F. Swann, *J. Biol. Chem.* **273**, 20007–20014 (1998).
15. T. R. Waters, P. Gallinari, J. Jiricny, and P. F. Swann, *J. Biol. Chem.* **274**, 67–74 (1999).
16. P. Neddermann and J. Jiricny, *Proc. Natl. Acad. Sci. U.S.A.* **91**, 1642–1646 (1994).
17. S. Griffin and P. Karran, *Biochemistry* **32**, 13032–13039 (1993).
18. U. Sibghat and R. S. Day, *Biochemistry* **34**, 6869–6875 (1995).
19. U. Sibghat, P. Gallinari, Y. Z. Xu, M. F. Goodman, L. B. Bloom, J. Jiricny, and R. S. Day, *Biochemistry* **35**, 12926–12932 (1996).
20. M. Saparbaev and J. Laval, *Proc. Natl. Acad. Sci. U.S.A.* **95**, 8508–8513 (1998).
21. F. L. Chung, H. J. Chen, and R. G. Nath, *Carcinogenesis* **17**, 2105–2111 (1996).
22. B. Zhu, Y. Zheng, D. Hess, H. Angliker, S. Schwarz, M. Siegmann, S. Thiry, and J. P. Jost, *Proc. Natl. Acad. Sci. U.S.A.* **97**, 5135–5139 (2000).
23. S. Griffin, P. Branch, Y. Z. Xu, and P. Karran, *Biochemistry* **33**, 4787–4793 (1994).
24. U. Sibghat, Y. Z. Xu, and R. S. Day, *Biochemistry* **34**, 7438–7442 (1995).
25. O. D. Schärer, T. Kawate, P. Gallinari, J. Jiricny, and G. L. Verdine, *Proc. Natl. Acad. Sci. U.S.A.* **94**, 4878–4883 (1997).
26. C. D. Mol, A. S. Arvai, R. J. Sanderson, G. Slupphaug, B. Kavli, H. E. Krokan, D. W. Mosbaugh, and J. A. Tainer, *Cell (Cambridge, Mass.)* **82**, 701–708 (1995).
27. R. Savva, H. K. McAuley, T. Brown, and L. Pearl, *Nature (London)* **373**, 487–493 (1995).
28. B. Hendrich, U. Hardeland, H. H. Ng, J. Jiricny, and A. Bird, [published erratum appears in *Nature* **404**(6777), 525, (2000)]. *Nature (London)*, **401**, 301–304 (1999).
29. P. Schär and J. Jiricny, *in* "Nucleic Acids and Molecular Biology" (F. Eckstein and D. M. J. Lilley, eds.), Vol. 12, pp. 199–247. Springer-Verlag, Berlin, Heidelberg, 1998.

30. K. A. Haushalter, M. W. Todd Stukenberg, M. W. Kirschner, and G. L. Verdine, *Curr. Biol.* **9,** 174–185 (1999).
31. M. Otterlei, E. Warbrick, T. A. Nagelhus, T. Haug, G. Slupphaug, M. Akbari, P. A. Aas , K. Steinsbekk, O. Bakke, and H. E. Krokan, *EMBO J.* **18,** 3834–3844 (1999).
32. H. Nilsen, I. Rosewell, P. Robins, C. Skjelbred, S. Andersen, G. Slupphaug, G. Daly, H. E. Krokan, T. Lindahl, and D. E. Barnes, *Mol. Cell* **5,** 1059–1065 (2000).
33. E. Lutsenko and A. S. Bhagwat, *J. Biol. Chem.* **274,** 31034–31038 (1999).
34. K. J. Impellizzeri, B. Anderson, and P. M. Burgers, *J. Bacteriol.* **173,** 6807–6810 (1991).
35. N. J. Morey, C. N. Greene, and S. Jinks-Robertson, *Genetics* **154,** 109–120 (2000).
36. A. R. Morgan and J. Chlebek, *J. Biol. Chem.* **264,** 9911–9914 (1989).
37. W. A. Deutsch, *Insect Mol. Biol.* **4,** 1–5 (1995).
38. A. K. Basu, M. L. Wood, L. J. Niedernhofer, L. A. Ramos, and J. M. Essigmann, *Biochemistry* **32,** 12793–12801 (1993).
39. M. Moriya, W. Zhang, F. Johnson, and A. P. Grollman, *Proc. Natl. Acad. Sci. U.S.A.* **91,** 11899–11903 (1994).
40. M. S. Greenblatt, W. P. Bennett, M. Hollstein, and C. C. Harris, *Cancer Res.* **54,** 4855–4878 (1994).
41. S. Tornaletti and G. P. Pfeifer, *Oncogene* **10,** 1493–1499 (1995).

Session 5
Mitochondrial DNA Repair

VILHELM A. BOHR

Laboratory of Molecular Genetics
National Institutes on Aging, NIH

Mitochondria are the energy stations for cells. Within these organelles, oxidative phosphorylation occurs, generating ATP. In this process, reactive oxygen species are formed at high frequencies, and the mitochondrial DNA (mtDNA) is directly exposed. The mitochondrial DNA does not have a recognized chromatin structure and is thus particularly exposed to oxidative DNA base lesion formation.

There are about 1000 mitochondria per mammalian cell, and each mitochondrion has 4–5 DNA plasmids, each about 16 kb long. This means that about 2% of total cellular DNA is in the mitochondria. All mtDNA is transcribed, whereas only about 1% of the nuclear DNA is transcribed; thus, the mtDNA makes up more than half of the total transcribed DNA in a mammalian cell.

Mitochondrial DNA does not code for any DNA damage processing enzymes, but only for the oxidative phosphorylation proteins and some tRNAs. Thus, all repair enzymes functioning in the mitochondria need to be transported in. It was shown more than 20 years ago that mitochondria do not repair ultraviolet (UV)-induced pyrimidine dimers, and this observation provided the basis for the assumption that there is no DNA repair capacity for mtDNA. Since then, there have been many observations of base excision repair (BER) in mitochondria. BER enzymes have been identified and characterized, and studies have shown that a number of simple monofunctional alkylating agents and oxidative DNA base lesions are efficiently removed from mtDNA. Other repair processes have also been detected, and this is reviewed extensively in Ref. (1). An important question remaining is whether mitochondria possess any capabilities to repair bulky lesions via the nucleotide excision repair pathway.

A limitation in the study of mitochondrial DNA repair has been the lack of techniques. Whereas *in vitro* repair studies have been performed with great success in nuclear or whole-cell extracts from cells, this kind of biochemical approach has only very recently become available for mtDNA. Recent advances suggest that mtDNA repair can now be studied using more sophisticated biochemical analyses as seen in this session, and this should provide great advances in the near future.

In this session, Bogenhagen described the identification of lyase/dRpase activities in mitochondrial polymerase γ and ligase; LeDoux and Wilson examined the substrates for mtDNA repair, as well as the cell specific differences and biological consequences of mtDNA repair; and Bohr examined oxidative DNA damage repair in mtDNA and the correlation of increased repair capacity with age.

REFERENCE

1. D. L. Croteau, R. H. Stierum, and V. A. Bohr, *Mutat. Res.* **434**, 137–148 (1999).

Enzymology of Mitochondrial Base Excision Repair

DANIEL F. BOGENHAGEN
KEVIN G. PINZ, AND
ROMINA M. PEREZ-JANNOTTI

Department of Pharmacological Sciences
SUNY at Stony Brook
Stony Brook, New York 11794

I. mtDNA Damage... 258
II. Base Excision Repair of mtDNA....................................... 259
III. What Nuclear Genes Encode Mitochondrial DNA Repair Enzymes?.... 261
   A. Initiation of BER by DNA Glycosylases............................ 261
   B. AP Endonuclease 263
   C. DNA Polymerase... 265
   D. mtDNA Ligase 267
   E. Is That All There Is?.. 269
   References... 270

A number of laboratories have shown that those types of DNA damage that are generally reparable by base excision repair are efficiently repaired in mtDNA. In contrast, most types of damage that require other sorts of repair machinery are not effectively repaired in mtDNA. We have shown that a set of highly purified mitochondrial proteins, including AP endonuclease (APE), DNA polymerase γ, and mtDNA ligase, is capable of efficiently repairing abasic (AP) sites in mtDNA. These three enzymes appear to conduct all four steps in a conventional BER mechanism: incision, removal of the 5′-deoxyribosephosphate by dRP lyase, polymerization, and ligation. Both DNA polymerase γ and mtDNA ligase possess some dRP lyase activity. DNA polymerase γ is a member of the family A of DNA polymerases, with clear homology to DNA pol I of *E. coli*, while mtDNA ligase is an alternatively expressed form of DNA ligase III. The dRP lyase activities discovered in these mitochondrial enzymes are not unique, but are found in all representatives tested of the family-A DNA polymerases and of the ATP-dependent DNA ligases. These dRP lyase activities have low turnover rates that may have important implications for the overall process of BER. All proteins involved in maintenance of mtDNA are encoded in the nuclear genome and must be directed to mitochondria in order to act on mtDNA. Thus, it is evident that the scope of DNA repair activities undertaken within mitochondria is determined by the set of nucleus-encoded DNA repair enzymes that are capable

Abbreviations: AP, apurinic–apyrimidinic; APE, AP endonuclease; BER, base excision repair; dRP, deoxyribose phosphate; MLS, mitochondrial localization signal; UDG, uracil DNA glycosylase.

of being imported into the organelle. A review of DNA repair proteins that may be imported into mitochondria in various organisms will be presented.

© 2001 Academic Press.

I. mtDNA Damage

The mitochondrial genome, like all DNA, is subject to continuous damage from endogenous and exogenous sources. Understanding the consequences of this damage requires an appreciation of a few key aspects of mitochondrial molecular biology. A typical cell in a higher eukaryote contains a few thousand 16–17-kb circular mtDNA molecules organized into a somewhat smaller number of heritable physical units referred to as nucleoids. MtDNA replicates throughout the cell cycle and in resting cells, as required to maintain a constant complement of mtDNA in the face of a continuous turnover of whole organelles. The mean lifetime of a mtDNA molecule has been estimated as 2–4 weeks in rat liver and brain cells, respectively (1). If these numbers can be extrapolated to humans with a life expectancy of over 70 years, mtDNA molecules in tissues such as brain, muscle, or primary oocytes may experience approximately 1000 more cycles of replication than the nuclear genome. Thus, unrepaired or poorly repaired damage to mtDNA may help drive the accumulation of mtDNA point mutations or deletions in the female germline or in somatic cells that can give rise to a variety of mitochondrially inherited diseases (2). Point mutations in any one of the 13 structural genes encoded in the mtDNA affect only a single gene product. In contrast, deletions or point mutations that inactivate tRNA genes may have pleiotropic effects on mitochondrial protein synthesis since, in general, mtDNA encodes only a single tRNA for each of the 20 amino acids. A single nascent mtDNA mutation initially has little deleterious effect since the high copy number of mtDNA provides gene products in *trans*. Mutations in mtDNA are phenotypically silent until their frequency reaches approximately 85% of the cell's mtDNA.

The incidence of damage to mtDNA can often exceed that inflicted on the nuclear genome. A large number of bulky carcinogens have been shown to cause adducts to mtDNA at higher frequency than nuclear DNA. This may reflect the fact that many chemical carcinogens have a lipophilic hydrocarbon structure or require activation by mixed-function oxidases in mitochondria or other cytoplasmic microsomal compartments. The activated carcinogens may have easy access to mtDNA owing to its proximity to mitochondrial membranes.

MtDNA is subjected to a higher rate of oxidative damage than nuclear DNA owing to its exposure to the enzymatic apparatus for oxidative phosphorylation. Mitochondria are clearly the main consumers of oxygen and the principal generators of reactive oxygen species (ROS) in cells (3). Some studies have revealed increased rates of oxidative damage to mtDNA (4), while others have not (5). Although additional work is needed to resolve this controversy, it may be concluded that oxidative damage is at least as extensive for mtDNA as for nuclear DNA. Mitochondrial ROS promote formation of a complex array of different types of chemical damage to mtDNA, as has been observed in other settings (5).

II. Base Excision Repair of mtDNA

The mtDNA repair field has been reviewed extensively in the past year (6–9). These reviews have summarized a general consensus that many types of bulky damage to mtDNA that would be processed by nucleotide excision repair in nuclear DNA are not efficiently repaired when they occur in mtDNA. In contrast, many types of damage amenable to base excision repair are processed efficiently by mitochondrial enzymes. The experiments documenting BER of mtDNA have been of two general types. First, as summarized in reviews listed above, exposure of cells to DNA oxidative stress or to simple alkylating agents has been shown to induce rapid mtDNA damage as documented by Southern blotting or PCR analysis. This damage is removed over time. Second, *in vitro* repair studies using site-specific DNA damage have shown that mitochondria contain a full set of enzymes sufficient for complete BER (10) that are capable of completing the short-patch repair documented in crude mitochondrial extracts (11).

The overall pathway for mitochondrial BER (10) is strikingly similar to that observed for the short-patch pathway of BER in the eukaryotic nucleus. This common pathway is depicted in Fig. 1 to show the enzymes identified in short-patch repair in both cellular compartments. The AP site is first attacked by a class II AP endonuclease to create 3'-OH and 5'-deoxyribosephosphate (dRP) groups flanking the incision. The 5'-dRP group can be removed by deoxyribophosphodiesterase or by dRP lyase. For the case of nuclear BER conducted by DNA pol β, the action of a separate enzyme to remove the 5'-dRP group is not necessary since DNA pol β contains an intrinsic dRP lyase (12). In this case, a single repair enzyme conducts two successive reactions in the BER scheme, elimination of the 5'-dRP blocking group and insertion of a correct single-base repair patch. Repair is completed as DNA ligase reseals the nicked DNA. Both DNA ligase I and DNA ligase III have been implicated in BER in the nucleus (13, 14).

How extensive are the similarities between mitochondrial and nuclear BER? While the overall process appears to be highly conserved, the characterization of most of the enzymes involved in mitochondrial BER is incomplete. A major source of uncertainty is the lack of complete understanding of the nature of DNA repair proteins that may be imported into mitochondria. Investigation

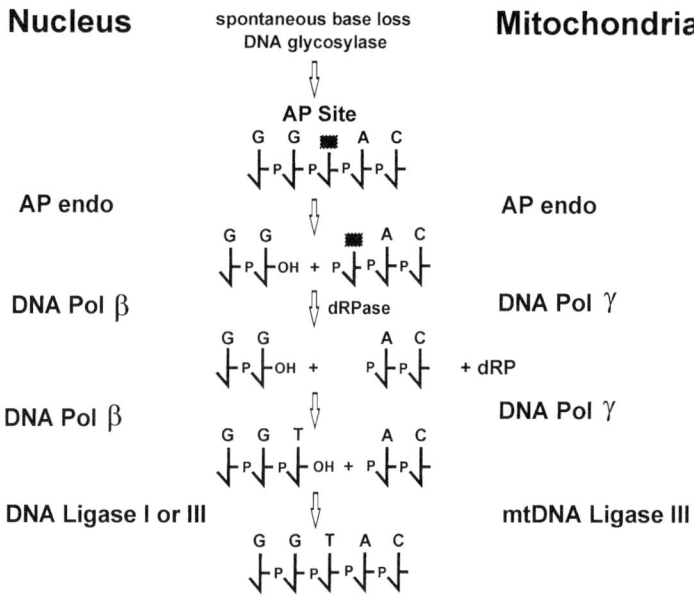

FIG. 1. Common pathway for mitochondrial and nuclear short-patch BER.

of enzymes involved in mitochondrial nucleic acid metabolism always bears the burden of proof that the enzymes in question are authentic mitochondrial components, not simply nuclear or cytoplasmic proteins that may adventitiously copurify with the mitochondrial fraction. Indeed, it is difficult to draw any firm conclusions regarding the nature of mtDNA repair proteins using biochemical approaches alone without substantiating genetic analysis. Regrettably, mtDNA repair has received little attention in genetically tractable organisms such as yeast. However, the recent expansion in genomic and proteomic databases provides a new approach to the identification of nuclear gene products that may be imported into mitochondria to function in BER. This chapter reviews the current status of our understanding of proteins that may be imported into mitochondria to function in BER. This collection depends on both biochemical characterization of repair proteins and on analysis of potential mitochondrial proteins in sequence databases. Most (but not all) proteins imported into mitochondria have N-terminal mitochondrial localization signals (MLS) that are removed upon import into mitochondria. We have used the MITOPROT algorithm (15) to assess the likelihood that a candidate protein may contain a functional MLS. It is important to recognize that a high probability score is not a guarantee that a protein is capable of being imported into mitochondria. A functional test of the activity of an apparent MLS sequence can be performed by using recombinant DNA methods to fuse the N-terminal sequence to green fluorescent protein (GFP) to test its ability to direct a fusion protein to mitochondria (16). Even this functional test is not entirely reliable, since false negative results can be obtained.

III. What Nuclear Genes Encode Mitochondrial DNA Repair Enzymes?

A. Initiation of BER by DNA Glycosylases

Both Pinz and Bogenhagen (10) and Stierum et al. (11) employed uracil-containing DNA substrates that were treated with UDG, a known mitochondrial glycosylase (17, 18), to produce AP sites. Repair of other sorts of base damage would require the activity of other glycosylases. It is not known whether BER in mtDNA initiated by the action of other glycosylases would necessarily follow the same pathway. Recent work has suggested that the nature of the initiating glycosylase may influence the subsequent pathway of BER in nuclear DNA (14).

UDG is an example of a simple glycosylase with no other associated repair activities. Some other glycosylases, including 8-oxoguanine glycosylase, also contain an AP lyase activity that acts through a β-elimination mechanism to produce a $3'$-dRP end group and a $5'$-phosphoryl moiety flanking the lesion. The distinction between these two types of activities is illustrated in Fig. 2. The action of a simple glycosylase leaves an intact phosphodiester backbone that requires action of AP endonuclease to generate a $5'$-dRP group. This $5'$-dRP can be removed by a dRP lyase that promotes β-elimination, as shown in Fig. 2. The term dRP lyase describes an AP lyase that strongly prefers to act on an exposed dRP group following incision by AP endonuclease. The action of a glycosylase with associated AP lyase activity results in a $3'$-dRP that is a potential block to the action of DNA polymerase. AP endonuclease is capable of hydrolytic excision of the $3'$-dRP moiety. Thus, following removal of an aberrant base by a simple DNA glycosylase, the activities of AP endonuclease and either AP lyase or dRP lyase are required to create a one-basepair gap in the DNA strand. In some cases, as in long-patch BER in the nucleus, a $5'$ exonuclease, like Fen1, can act to remove a $5'$-dRP group as part of an oligonucleotide (19).

Mitochondria are capable of repairing several types of mtDNA base damage, including oxidative and alkylation damage. The spectrum of damage that is repaired must correspond with the collection of DNA glycosylases available in mitochondria, as reviewed recently. These reviews have cited evidence that

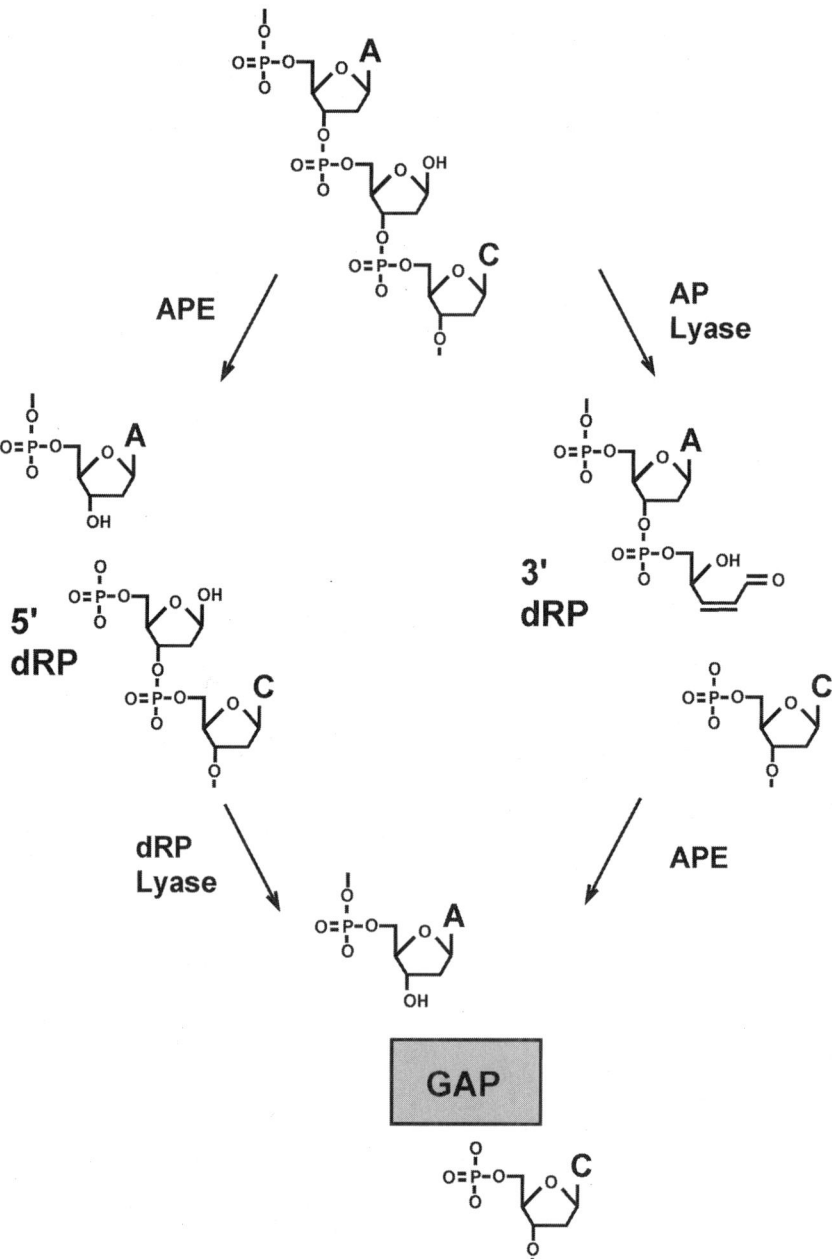

FIG. 2. AP endonuclease combines with AP lyase or dRP lyase to create a one-basepair gap. Two pathways for processing of an AP site to a one-base gap are diagrammed as described in the text.

isoforms of 8-oxoguanine glycosylase (OGG1), endonuclease III homologs (e.g., hNTH), and AG glycosylase (MutY homologs) may be imported into mitochondria to initiate repair of 8-oxoguanine, thymine glycol, and A·G or A·8-oxoG, respectively. In all of these examples, it is likely to be the case that the mitochondrial enzyme is produced by alternative splicing or alternative translation initiation from the same gene that provides enzyme for nuclear DNA repair. Indeed, there is no clear example to date of a nuclear gene encoding a glycosylase that is exclusively targeted to mitochondria. This situation may simply reflect the fact that characterization of mitochondrial repair enzymes has only just begun.

B. AP Endonuclease

AP endonucleases are ubiquitous enzymes required for BER in all organisms. *Escherichia coli* contains two major enzymes with AP endonuclease activity—exonuclease III and endonuclease IV. Although these enzymes are both class II AP endonucleases, they share only 11.6% sequence identity. Eukaryotic AP endonucleases are characterized by their relationship to one or another of these *E. coli* enzymes. Two of the first eukaryotic enzymes to be studied, human APE1 and yeast APN1 (scAPN1), are most closely related to exonuclease III and endonuclease IV, respectively.

The gene or genes encoding mitochondrial AP endonuclease have not been characterized in any organism, and this enzyme has been studied in only a few reports in the scientific literature. Pinz and Bogenhagen (10) purified a mitochondrial class II AP endonuclease from *Xenopus* oocytes with properties similar to those of the major cellular AP endonuclease in *Xenopus*, xlAPE1, first characterized by Matsumoto *et al.* (20). XlAPE1 has been cloned (Pinz and Bogenhagen, unpublished data) and is 64% identical to human APE1. The *Xenopus* mitochondrial AP endonuclease and xlAPE1 are similar in size and share immunological cross-reactivity to human APE1. The possibility that the *Xenopus* mitochondrial APE is an alternative product of the xlAPE1 gene is under active study in our lab. It is interesting to note that the *Xenopus* mitochondrial enzyme appears to be smaller than the mouse mitochondrial AP endonuclease studied by Tomkinson *et al.* (21). These workers reported partial purification of a class II AP endonuclease activity from mouse mitochondria that appeared to copurify with a 65-kDa protein immunologically related to human nuclear APE1. The mitochondrial AP endonuclease in mammals has not been further characterized since the initial report (21). Thus, its relationship to other known AP endonuclease genes has remained obscure.

In order to search for genes that may be candidates for mitochondrial AP endonucleases, we have collected a number of protein sequences and surveyed them for the possible presence of an N-terminal mitochondrial localization signal using the MITOPROT algorithm (15). The apparent evolutionary relationships

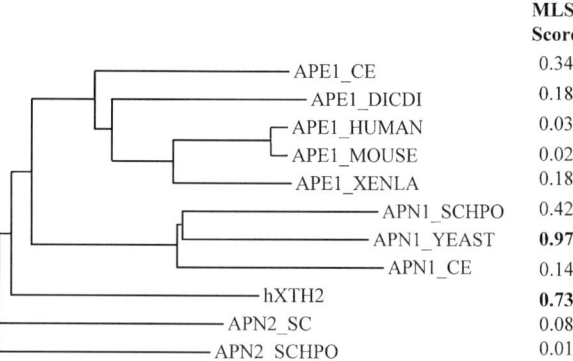

FIG. 3. Sequence relationships among AP endonucleases. Sequences for AP endonucleases from *C. elegans* (APE1, AF034258.1; APN1, T05H10), *D. discoideum* (P51173), human (SW:APE1_human; hXTH2, AJ011311.1), mouse (SW:APE1_mouse), *X. laevis* (Pinz and Bogenhagen, in preparation), *S. pombe* (SW:APN1_SCHPO; APN2, Z95620.1), and *S. cerevisiae* (SW:APN1_yeast; APN2, YBL0443) were aligned using Vector NTI to produce the dendrogram shown. The N-terminal protein sequences were evaluated for the likelihood that they might encode a mitochondrial localization signal (MLS) using the MITOPROT program cited in the text.

shared by these sequences are shown in Fig. 3. These proteins fall into three general categories. First, mouse, human and *Xenopus* APE are examples of a set of proteins 300–320 amino acids long with a high degree of homology to *E. coli* exonuclease III. A second set of AP endonucleases includes the recently described scAPN2 and may be characterized as a relative of exonuclease III with a long C-terminal extension. Very recently, the sequence of a novel human AP endonuclease gene in this category, hXTH2, has been submitted to the NCBI database (accession number AJ011311). hXth2 appears to be a good candidate for the mitochondrial enzyme characterized by Tomkinson *et al.* (21), since it is a larger relative of APE1 and has an N-terminal sequence with features characteristic of a mitochondrial localization signal (MLS). The third class of AP endonuclease proteins is related to yeast APN1 and endonuclease IV. This class includes enzymes in the fission yeast, *S. pombe,* and in the worm, *C. elegans,* but has not yet been reported in higher eukaryotes. Unfortunately, no reports of a mitochondrial AP endonuclease have appeared in any of these organisms. As shown in Fig. 3, some proteins in this group have N-terminal sequences with a significant potential ability to serve as a mitochondrial localization signal. We have performed experiments to fuse the potential mitochondrial localization signals of the three endoIV-like proteins to green fluorescent protein (GFP) in order to determine whether these candidate MLS sequences are functional.

Preliminary experiments suggest that the *S. pombe* signal sequence is capable of directing GFP to mitochondria (Connelly, Perez-Jannotti, and Bogenhagen, unpublished results).

C. DNA Polymerase

As shown in Fig. 1, after AP endonuclease incision at an AP site, repair requires both dRP lyase and DNA pol activities to prepare the DNA for ligation. The need for a DNA pol activity is unquestioned. There has been some uncertainty in the field regarding the requirement for a dRP lyase since the β-elimination reaction occurs "spontaneously" due to catalysis by solvent or small molecules. However, the slow rate of its spontaneous release is not sufficient to support the rapid overall rate of BER in cells, arguing that enzymatic catalysis is required.

In short-patch BER in the nucleus, pol β is capable of acting as a dRP lyase as well as a DNA polymerase. The 39-kDa enzyme contains two distinct domains, a 31-kDa domain containing the DNA polymerase activity and an 8-kDa domain, known as the finger domain, containing the dRP lyase activity. The dRP lyase mechanism proceeds with attack by Lys72 of pol β (22) on the C1' residue of the dRP site to form a transient Schiff base intermediate. This intermediate can be documented by treating the Schiff base intermediate with sodium borohydride, effectively crosslinking the enzyme to the DNA substrate. It appears that mitochondrial pol γ may similarly act as both a dRP lyase and a DNA polymerase in short-patch BER in mitochondria, as discussed below.

DNA pol γ is the only well-characterized DNA polymerase in mitochondria. This aphidicolin-resistant, NEM-sensitive enzyme serves as the replicative DNA polymerase in mitochondria. The enzyme has been purified to near homogeneity from several sources. In higher eukaryotes, the enzyme contains a 140-kDa catalytic subunit related to the family-A DNA polymerases, such as *E. coli* DNA pol I, and a small subunit of approximately 50 kDa that serves as a processivity factor (23). Both the large (A) and small (B) subunits have been cloned in several organisms. Interestingly, a small subunit has not been characterized in yeast DNA pol γ. DNA pol γ purified from *Xenopus* oocyte mitochondria and the recombinant human pol γA are competent to function in short-patch BER (10, 24). The pol γB subunit is dispensable for short-patch repair *in vitro*, but it is not known whether the pol γA subunit is free to engage in repair on its own *in vivo*, i.e., without the small subunit. It should be noted that two reports have appeared recently of lower-molecular-weight pol β-type polymerases in mitochondria from the protist, *Crithidia fasciculata* (25), and in cows (26). While it is tempting to think that these low-molecular-weight polymerases may act in BER in mitochondria, this remains to be tested.

The possibility that pol γ may act as a dRP lyase as well as a DNA polymerase was suggested by the fact that it was active in a borohydride trapping reaction on a substrate with an incised AP site (10). This preliminary observation has been confirmed by more detailed studies with *Xenopus* pol γ (27) and with the recombinant human pol γ (24). The finding of a dRP lyase activity in pol γ was a considerable surprise, since this activity had not previously been found in other members of the family-A DNA polymerases. We have recently shown that several family-A DNA polymerases, including the Klenow fragment of *E. coli* DNA pol I, contain dRP lyase activity using both borohydride trapping assays and dRP product release assays (27). The family-A DNA polymerases, Klenow pol and pol γ, bind avidly to a nicked AP site, bringing basic lysine residues into close contact with the 5'-dRP group. The attack of this lysine on the C1' residue of deoxyribose creates the Schiff base intermediate that is stabilized in a borohydride trapping reaction. To this point, the reaction catalyzed by family-A DNA polymerases resembles the action of pol β. However, for the case of pol β, the reaction proceeds quickly with elimination of the DNA and release of the dRP group. The family-A DNA polymerases tested to date are very slow to complete this reaction, so that intermediates with either the DNA or dRP group bound to enzyme persist for as long as one hour. The slow rate of resolution of the enzyme–dRP intermediate helps to explain why the dRP lyase activity intrinsic to family-A DNA polymerases is much less active than that of pol β. Many of the details of this reaction remain to be explored. The active site lysine residues have not been mapped in any family-A polymerases and the kinetic constants for each step in the reaction have not been determined. It will be of interest to determine whether the long-lived enzyme–DNA or enzyme–dRP complex may contribute to the toxicity of abasic sites.

These *in vitro* experiments by Pinz and Bogenhagen (27) show that family-A polymerases are capable of acting as dRP lyases in BER, but do not prove that these are the enzymes that perform this reaction *in vivo*. This would require a genetic analysis that is virtually impossible at the present time for mitochondria. However, a genetic analysis can be informative for *E. coli*. Dianov *et al.* (28) have reported that the FPG and RecJ proteins are major sources of dRP lyase in *E. coli,* but that a *fpg recj* double mutant retains BER activity. It may be that DNA pol I or other enzymes with less active sources of dRP lyase activity provide a backup for this activity in bacteria. Since family-A DNA polymerases have a high affinity for binding to incised AP sites, it seems likely that they may sometimes be the first enzyme to bind *in vivo* and may initiate the dRP lyase reaction. Due to the slow rate of resolution of the enzyme-dRP intermediate, we would not expect the dRP lyase activity of family-A DNA polymerases to be capable of dealing with high rates of base damage.

D. mtDNA Ligase

1. Identification of mtDNA Ligase

DNA ligases are responsible for the final step in both BER and DNA replication processes. All DNA ligases studied so far share the same catalytic mechanism. These enzymes react with a high-energy nucleotide substrate, either ATP or, for many bacterial enzymes, NAD, to form an adenylated enzyme. The enzyme transfers the AMP moiety to the 5′-phosphate of nicked DNA, so that subsequent nucleophilic attack by the 3′-hydroxyl can reform the phosphate backbone of a double-stranded DNA molecule.

The existence of a mtDNA ligase was predicted by the finding that mtDNA is a covalently closed circular genome and confirmed by an early report by Levin and Zimmerman (29). This study did not provide a detailed characterization of the mtDNA ligase, but did indicate that it was an ATP-dependent enzyme, more similar to eukaryotic nuclear DNA ligase and bacteriophage DNA ligases than to NAD-dependent bacterial DNA ligases.

Five ATP-dependent DNA ligase activities have been studied in eukaryotes, although at least one of them, DNA ligase II, is thought to be a proteolytic fragment of another, DNA ligase III. cDNAs corresponding to three mammalian enzymes—DNA ligase I, III, and IV—have been obtained from several vertebrate organisms including human, mouse, and *Xenopus* (30–33; Perez-Jannotti, Klein, and Bogenhagen, unpublished observations). Altogether, these genes show somewhat limited primary sequence homology, sharing several conserved motifs including an AMP active site, a DNA ligase signature, and a conserved peptide. Budding and fission yeasts appear to possess only DNA ligase I and IV homologs. The sequence relationships between eukaryotic DNA ligases are shown in Fig. 4, along with an analysis of the predicted capacity of the N-terminal sequences to serve as mitochondrial localization signals.

The first clue as to the identity of mtDNA ligase was the extensive purification of a 100–110-kDa DNA ligase from *Xenopus* mitochondria (10). The activity purified in our laboratory closely resembles mammalian DNA ligase III and IV in terms of its size, ATP dependence, and ability to ligate the synthetic homopolymer pairs oligo(dT)·poly(dA) and oligo(dT)·poly(rA). The immunological relationship of *Xenopus* mtDNA ligase to DNA ligase III suggests that it is related to this nuclear DNA ligase (10). Data confirming the mitochondrial role of mammalian DNA ligase III were also recently obtained by green fluorescent protein and antisense strategies (34).

Several eukaryotic DNA ligase genes appear to be capable of generating two polypeptides by a common mechanism of alternative translation initiation. DNA ligase I from *S. cerevisiae, S. pombe,* and *C. elegans,* and DNA ligase III and IV from *Xenopus,* mouse, and human contain multiple in-frame translation initiation

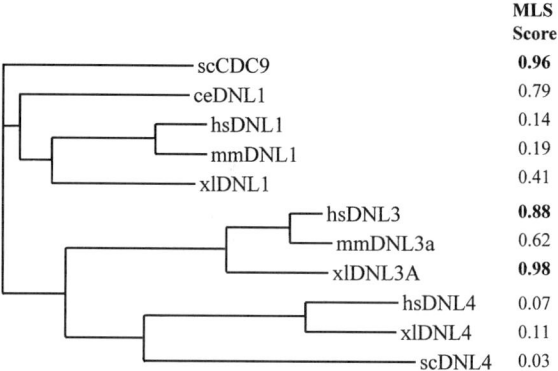

FIG. 4. Sequence relationships among potential mtDNA ligases. Sequences for DNA ligases from *S. cerevisiae* (scDNLI, NC 001136.1; scDNL4, NC 001147.1), *C. elegans* (Z73970.1), human (hsDNLI, NP 000225; hsDNL3, NM 002311.1; hsDNL4, NM 002312.1) mouse (mmDNL3A, U66058.1), and *X. laevis* (xlDNL1, P51892; xl DNL3A and XL DNL4; Perez-Jannotti and Bogenhagen, unpublished) were aligned using Vector NTI to produce the dendrogram shown. The N-terminal protein sequences were evaluated for the likelihood that they might encode a mitochondrial localization signal (MLS) using the MITOPROT program cited in the text.

codons resulting in polypeptides that either contain or lack an extra amino-terminal extension. In several cases, the extra N-terminal sequences provided by the upstream start sites are recognized by the MITOPROT program (15) as potential mitochondrial localization signals. Recently, the yeast DNA ligase I gene, *cdc9*, was shown to encode both a nuclear and a mitochondrial form of DNA ligase I by alternative translation initiation. Cells expressing only the nuclear isoform of CDC9 were defective in maintenance of the mitochondrial genome (35). It appears that while a form of DNA ligase I is directed to mitochondria in *S. cerevisiae,* the role of mtDNA ligase has been assumed by a form of DNA ligase III in vertebrates. To date, efforts to identify a role for DNA ligase IV in mtDNA maintenance have not been successful (34; Perez-Jannotti, unpublished).

2. DNA Ligase Provides an Alternative Source of AP Lyase Activity

As noted in Section III,C, both DNA pol γ and mtDNA ligase were found to be active in borohydride trapping assays for AP lyase (10). These results stimulated further investigation of the apparent dRP lyase activity in DNA ligases, in addition to the work summarized earlier on the dRP lyase activity in DNA pol γ and other family-A DNA polymerases. Bogenhagen and Pinz (37) showed that

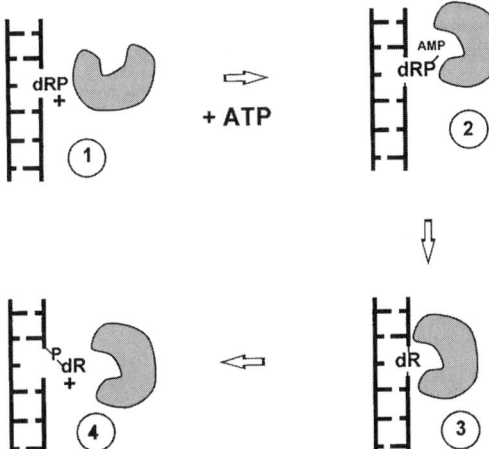

FIG. 5. Editing action of AP lyase in ATP-dependent DNA ligase. T7 DNA ligase is shown as a C-shaped structure with a groove containing the active site lysine residue adenylated during the ligase reaction. Binding of the adenylated DNA ligase to a 5'-dRP group (panel 1) results in transfer of the AMP to the DNA (panel 2) and ligation (panel 3). The deadenylated enzyme is able to use its AP lyase activity to incise the sealed AP site to produce a 3'-dRP residue.

other ATP-dependent DNA ligases, such as T4 and T7 DNA ligase, also contain AP lyase activity. This activity is referred to as an AP lyase, rather than as a dRP lyase since it was shown that T4 DNA ligase is capable of using this activity to incise the DNA strand at an AP site.

DNA ligases are adapted to bind avidly to nicks in DNA and to reseal them. Surprisingly, they are capable of resealing the nick created by AP endonuclease at an AP site, disregarding the absence of a base on the 5'-dRP residue (36, 37). This may be considered as an error, since the religation reverses the action of AP endonuclease. Bogenhagen and Pinz (37) suggested that the AP lyase activity in ATP-dependent DNA ligase has been adapted to correct these misligation events following the model diagrammed in Fig. 5. In a teleological sense, when DNA ligase makes the mistake of resealing a phosphodiester bond on the 5' side of an incised AP site, it recognizes this error and reopens the DNA backbone on the 3' side of the AP site using its intrinsic lyase activity. AP endonuclease is able to remove the resulting 3'-dRP residue, generating a clean one-nucleotide gap.

E. Is That All There Is?

The mtDNA repair field was ignored for a long time owing to the common misconception that damaged mtDNA was not repaired. Not surprisingly, the effort to identify enzymes that may act in BER of mtDNA has only just begun.

The search for mtDNA repair factors is complicated by the fact that mtDNA comprises less than 1% of total DNA in most cells, so that mitochondrial repair proteins may be particularly rare. A second difficulty is raised by the emerging theme that many mtDNA repair proteins appear to be alternative products of genes whose nuclear functions have already been well established. This is illustrated by the example cited above that mtDNA ligase appears to be encoded by the DNA ligase I gene in yeast but by the DNA ligase III gene in humans. This example also provides a sobering reminder that identification of a potential mtDNA repair protein in a lower eukaryote may have limited benefit for understanding the basis of mtDNA mutations in human disease.

There would be no reason to hold a meeting dedicated to BER if it were not for the fact that the simple short-patch repair scheme shown in Fig. 1 greatly oversimplifies the process of BER. Many aspects of BER of mtDNA require further study. For example, although we have succeeded in identifying DNA pol γ and mtDNA ligase as possible sources of dRP lyase activity, these enzymes are not as robust as one would expect. Thus, additional work is required to identify other dRP lyase or AP lyase activities in mitochondria. Repair events initiated by the mitochondrial isoforms of OGG1 or NTH may employ the AP lyase activities intrinsic to these enzymes. However, it is not known whether these enzymes may also function in general repair of AP sites produced by spontaneous base loss or by other glycosylases that lack AP lyase activity. A second issue is raised by the finding that some mitochondrial repair proteins are isoforms of nuclear gene products that require accessory factors for optimal stability or activity. It is appropriate to ask if the mitochondrial isoform is associated with the same or other binding partners. For example, the prospect that an isoform of hNTH1 is involved in repair of thymine glycol in mtDNA raises the possibility that mitochondria may contain a factor to facilitate the action of this enzyme in the way that XPG is thought to function in the nucleus (38). We have already seen how genomic and proteomic approaches can be applied to these questions. The identification of DNA ligase I as the mtDNA ligase in yeast was facilitated by the knowledge that yeast contains only two DNA ligase genes. We anticipate that the completion of the human genome project and efforts to identify mitochondrial proteins by direct sequencing will provide new tools to study the enzymology of mtDNA repair.

Acknowledgment

This work was supported by a grant from the NIEHS.

References

1. N. Gross, G. Getz, and M. Rabinowitz, *J. Biol. Chem.* **244,** 1552–1562 (1969).
2. D. Wallace, *Science* **283,** 1482–1488 (1999).

3. B. Chance, H. Sies, and A. Boveris, *Phys. Rev.* **59,** 527–605 (1979).
4. C. Richter, J.-W. Park, and B. N. Ames, *Proc. Natl. Acad. Sci. U.S.A.* **85,** 6465–6467 (1988).
5. R. Anson, S. Senturker, M. Dizdaroglu, and V. Bohr, *Free Radical Biol. Med.* **27,** 456–462 (1999).
6. D. Bogenhagen, *Am. J. Human Genet.* **64,** 1276–1281 (1999).
7. S. LeDoux, W. Driggers, B. Hollensworth, and G. Wilson, *Mutat. Res.* **434,** 149–159 (1999).
8. D. Sawyer and B. Van Houten, *Mutat. Res.* **434,** 161–176 (1999).
9. D. Crouteau, R. Stierum, and V. Bohr, *Mutat. Res.* **434,** 137–148 (1999).
10. K. Pinz and D. Bogenhagen, *Mol. Cell Biol.* **18,** 1257–1265 (1998).
11. R. Stierum, G. Dianov, and V. Bohr, *Nucleic Acids Res.* **27,** 3712–3719 (1999).
12. Y. Matsumoto and K. Kim, *Science* **269,** 699–702 (1995).
13. D. Srivastava, B. J. Berg, R. Prasad, J. T. Molina, W. A. Beard, A. E. Tomkinson, and S. H. Wilson, *J. Biol. Chem.* **273,** 21203–21209 (1998).
14. P. Fortini, E. Parlanti, O. Sidorkina, J. Laval, and E. Dogliotti, *J. Biol. Chem.* **274,** 15230–15236 (1999).
15. M. Claros and P. Vincens, *Eur. J. Biochem.* **241,** 779–786 (1996).
16. M. Takao, H. Aburatani, K. Kobayashi, and A. Yasui, *Nucleic Acids Res.* **26,** 2917–2922 (1998).
17. J. D. Domena, R. T. Timmer, S. A. Dicharry, and D. W. Mosbaugh, *Biochemistry* **27,** 6742–6751 (1988).
18. G. Slupphaug, F. H. Markussen, L. C. Olsen, R. Aasland, N. Aarsaether, O. Bakke, H. E. Krokan, and D. E. Helland, *Nucleic Acids Res.* **21,** 2579–2584 (1993).
19. K. Kim, S. Biade, and Y. Matsumoto, *J. Biol. Chem.* **273,** 8842–8848 (1998).
20. Y. Matsumoto, K. Kim, and D. F. Bogenhagen, *Mol. Cell Biol.* **14,** 6187–6197 (1994).
21. A. E. Tomkinson, R. T. Bonk, and S. Linn, *J. Biol. Chem.* **263,** 12532–12537 (1988).
22. Y. Matsumoto, K. Kim, D. Katz, and J.-A. Feng, *Biochemistry* **37,** 6456–6464 (1998).
23. J. Carrodeguas, R. Kobayashi, S. Lim, W. Copeland, and D. Bogenhagen, *Mol. Cell Biol.* **19,** 4039–4046 (1999).
24. M. Longley, R. Prasad, D. Srivastava, S. Wilson, and W. Copeland, *Proc. Natl. Acad. Sci. U.S.A.* **95,** 12244–12248 (1998).
25. A. Torri and P. Englund, *J. Biol. Chem.* **270,** 3495–3497 (1995).
26. S. Nielsen-Preiss and R. Low, *Arch. Biochem. Biophys.* **374,** 229–240 (2000).
27. K. Pinz and D. Bogenhagen, *J. Biol. Chem.* **275,** 12509–12514 (2000).
28. G. Dianov, B. Sedgwick, G. Daly, M. Olsson, S. Lovett, and T. Lindahl, *Nucleic Acids Res.* **22,** 993–998 (1994).
29. C. J. Levin and S. B. Zimmerman, *Biochem. Biophys. Res. Commun.* **69,** 514–520 (1976).
30. Y. F. Wei, P. Robins, K. Carter, K. Caldecott, D. J. Pappin, G. L. Yu, R. P. Wang, B. K. Shell, R. A. Nash, P. Schär, D. E. Barnes, W. A. Haseltine, and T. Lindahl, *Mol. Cell Biol.* **15,** 3206–3216 (1995).
31. J. Chen, A. E. Tomkinson, W. Ramos, Z. B. Mackey, S. Danehower, C. A. Walter, R. A. Schultz, J. M. Besterman, and I. Husain, *Mol. Cell Biol.* **15,** 5412–5422 (1995).
32. K. Frank, J. M. Sekiguchi, K. J. Seidl, W. Swat, G. A. Rathbun, H.-L. Cheng, L. Davidson, L. Kangaloo, and F. W. Att, *Nature* **396,** 173–177 (1998).
33. D. Lepetit, P. Thiebaud, S. Aoufouchi, C. Prigent, R. Guesne, and N. Theze, *Gene* **172,** 273–277 (1996).
34. U. Lakshmipathy and C. Campbell, *Mol. Cell Biol.* **19,** 3869–3876 (1999).
35. M. Willer, M. Rainey, T. Pullen, and C. Stirling, *Curr. Biol.* **9,** 1085–1094 (1999).
36. C. Goffin, V. Bailly, and W. G. Verly, *Nucleic Acids Res.* **15,** 8755–8771 (1987).
37. D. Bogenhagen and K. Pinz, *J. Biol. Chem.* **273,** 7888–7893 (1998).
38. A. Klungland, M. Hoss, D. Gunz, A. Constantinou, S. G. Clarkson, P. W. Doetsch, P. H. Bolton, R. D. Wood, and T. Lindahl, *Mol. Cell* **3,** 33–42 (1999).

Base Excision Repair of Mitochondrial DNA Damage in Mammalian Cells

S. P. LeDoux and G. L. Wilson

Department of Cell Biology and
Neuroscience
University of South Alabama
Mobile, Alabama 36688

I. Introduction 273
II. Evidence for BER in Mammalian Mitochondrial DNA 275
III. Repair of Oxidative Damage to mtDNA 276
IV. Repair of NO-Induced Damage in mtDNA 276
V. Evaluation of Damage and Repair to mtDNA at the Level of
  Individual Nucleotides 277
VI. Mechanisms Involved in the Repair of mtDNA 279
VII. Cell-Specific Differences in mtDNA Repair 279
VIII. Conclusions and Future Questions 282
  References 283

This review of the work from our laboratory describes initial studies in which base excision repair in mtDNA was first seen. It considers the results of experiments in which the substrates for mtDNA repair were identified. The discussion then focuses on studies during which the sequence context for mtDNA damage and repair were explored. Next, it addresses factors that have been identified that influence mtDNA repair. Finally, it summarizes the results of studies that evaluated cell-specific differences in the repair of mtDNA and explored some of the biological consequences of these differences. © 2001 Academic Press.

I. Introduction

Over the past 10 years, interest in mtDNA damage has risen with the discovery that defects in the mitochondrial genome are associated with several

Abbreviations: BER, base excision repair; mtDNA, mitochondrial DNA; CNS, central nervous system; PCR, polymerase chain reaction; ROS, reactive oxygen species; RNS, reactive nitrogen species; SZ streptozotocin; MNU, methylnitrosourea; DMS, dimethyl sulfate; CHO, Chinese hamster ovary; LM-PCR, ligation-mediated polymerase chain reaction; XO, xanthine oxidase; NO, nitric oxide; XP, Xeroderma pigmentosum; CuZn SOD, copper–zinc superoxide dismutase; MnSOD, manganese superoxide dismutase.

human hereditary diseases such as Kearns–Sayre syndrome, Leber's hereditary optic neuropathy, Pearson's syndrome, and some cases of chronic progressive external ophthalmoplegia (1, 2). Additionally, accumulations of mutations and deletions in mtDNA with their associated defects in oxidative phosphorylation have been implicated in diabetes, ischemic heart disease, Parkinson's disease, demyelinating polyneuropathy, cancer, and aging (1–4). A key question to be answered is how alterations in the mitochondrial genome, which basically affect only electron transport, can cause different diseases.

One mechanism that could lead to altered cellular function is related to differences in energy requirements in different cellular compartments. There is evidence that mitochondria are required to provide compartmentalized ATP to specific areas of the cell (5–8). For instance, the cell membrane requires ATP to energize specialized processes such as ion pumping, electrical transmission across the membrane as in Na^+–K^+ exchange, and neurotransmitter release (9). Since different cells have unique energy requirements for these processes, it is likely that they would be functionally affected in different ways by mutations in mtDNA. Another way that mtDNA damage could differentially affect cells is through progressive cell death. Interest in the initiation and regulation of apoptosis has resulted in an exponential growth in research in this area over the last few years. Heterogeneous death signals precede a common effector phase during which cells pass a threshold of "no return," and are engaged in a degradation phase that results in the disassembly of the cellular scaffolding. There have been numerous hypotheses which postulate that specific mediators of pathways are responsible for apoptosis. One hypothesis is that the mitochondrion plays a key role in the regulation of apoptosis (10).

Disruption of the mitochondrial membrane potential is associated with the induction of apoptosis following certain stimuli. The upstream factors that precede the disruption of the mitochondrial membrane potential have not been fully elucidated. Previously, the involvement of oxidative damage has been suggested. Because cells are continuously bombarded by reactive oxygen species (ROS), their survival depends upon a fine balance between radical production, damage repair, and antioxidant activity. The hypothesis that oxidative damage plays a role in apoptosis is supported by the observation that the addition of ROS or the removal of endogenous antioxidants results, in some cases, in apoptosis (11). Additionally, tumor necrosis factor-induced apoptosis is associated with increases in intracellular ROS levels (12). In attempting to provide a mechanism by which oxidative damage might induce apoptosis, a recent report presented data consistent with a pathway by which damage to mitochondrial DNA led to a bioenergetic crisis, disruption of mitochondrial membrane potential, and induction of apoptosis (13). Thus, an initial elevation in unrepaired oxidative damage to mtDNA could lead to defects in oxidative phosphorylation. This would cause additional ROS production which would heighten the stress and eventually push the cell into apoptosis.

Therefore, mtDNA repair may play a pivotal role in normal cellular defense mechanisms. While complex enzymatic mechanisms for recognition and repair of nuclear DNA damage have been demonstrated, the study of mitochondrial DNA repair processes was impaired greatly by the difficulty in isolating sufficient quantities of mitochondrial DNA free of nuclear DNA contamination. It was originally thought that the mitochondrion did not repair damage to its DNA. Rather, when mtDNA was damaged, it was believed that it was degraded and new mtDNA was synthesized from undamaged templates. However, using sequence-specific repair analysis, it has been demonstrated that mitochondria possess efficient base excision repair (BER) capacity.

II. Evidence for BER in Mammalian Mitochondrial DNA

In the initial studies that identified BER in mtDNA within mammalian cells (14), mtDNA was distinguished in Southern transfers of total cellular DNA, using a probe that contained the entire 16.5-kb mouse mitochondrial genome. The formation and repair of N-methylpurines, which are alkali-labile, was measured in an insulinoma cell line (RINr 38) after exposure to the naturally occurring nitrosoamide streptozotocin (SZ). Alkali-labile sites were formed in mtDNA in a dose-dependent fashion. Eight hours after exposure to the toxin, 55% of the lesions were removed. The amount of repair increased to 70% after 24 h. In comparison, only 46% of N^7-methylguanines were removed from the entire cellular genome at this time. These studies demonstrated that SZ causes appreciable mtDNA damage in a dose-dependent manner, and provided the first evidence that there is a repair mechanism in the mitochondrion for removal of these alkali-labile lesions.

Following the demonstration of an apparent BER mechanism in mitochondria, it remained to be determined whether mitochondria could repair only lesions produced by simple alkylating toxins, or whether they have the capability for correcting damage caused by other agents. Accordingly, studies were undertaken to investigate the repair of DNA lesions in mtDNA from CHO B11 cells following exposure to different agents (15), namely methylnitrosourea (MNU), dimethyl sulfate (DMS), nitrogen mustard, ultraviolet (UV) irradiation, and cisplatin. CHO cells were used because a wealth of information is already available concerning repair of different types of lesions in nuclear DNA from these cells. The results revealed that repair of mtDNA damage depends upon the type of lesion produced by the damaging agent. There was efficient repair of methylation damage following exposure to MNU or DMS, with approximately 70% of the lesions being removed by 24 h. However, more complex alkylation damage, such as that resulting from exposure to nitrogen mustard, was not repaired. Additionally, no repair of pyrimidine dimers after exposure to UV light was detected. Cisplatin intrastrand crosslinks also were not repaired; however, interstrand crosslinks resulting from this toxin were repaired to a significant degree. More than 70% of these lesions were removed from mtDNA by 24 h. Therefore, these studies showed that, while the nuclei of CHO cells possess mechanisms to repair DNA damage by all the agents used, the mitochondria were able to repair only specific types of injury.

III. Repair of Oxidative Damage to mtDNA

Since mitochondria are constantly exposed to high levels of reactive oxygen species, it is likely that oxidative damage to mtDNA may be responsible for some of the maladies associated with aging. To determine whether mitochondria are able to repair this type of damage, a modification of the same Southern blot technique utilized to study repair of alkylation damage was employed (16). Alloxan was used as an oxygen-radical generator. Insulinoma cells were exposed to this toxin for 1 h and the DNA was isolated immediately, or after repair intervals of up to 8 h. Alkali treatment was used to identify abasic (AP) sites and sugar lesions, endonuclease III was used to identify a variety of lesions associated with thymine and cytosine, and FAPY glycosylase was employed to recognize formamidopyrimidines and 8-oxoguanines in the restricted DNA. The results showed that all the forms of damage studied were repaired completely by 4 h, indicating that mitochondria are able to efficiently repair injury to their DNA caused by ROS. This was the first report directly showing repair of oxidative damage in mtDNA. We subsequently expanded these studies to show that damage induced by the radiomimetic drug bleomycin was also repaired rapidly (17).

IV. Repair of NO-Induced Damage in mtDNA

Most recently, we have directed our attention to the other reactive species to which mtDNA is frequently exposed, nitric oxide (NO). Initially, we showed that mtDNA from primary cultures of insulin-secreting β-cells is a more vulnerable target than nuclear DNA for damage caused by NO produced endogenously and exogenously (18). Therefore, whether NO damage is repaired became a crucial question that needed to be answered. To address this question, experiments were initiated in which normal human fibroblasts were exposed to NO generated by PAPA NONOate (PAPA/NO) (19). Cells were subjected acutely to different concentrations of the NO generator, total cellular DNA was isolated, and a Southern blot procedure was performed to determine damage to mtDNA. Extensive damage to mtDNA was revealed which was blocked by the NO scavenger carboxy-PTIO. Thus, the damage to the mtDNA was likely the result of deamination reactions. In addition to the deamination of guanine to xanthine and adenine to hypoxanthine, both of which are alkali-labile, it was possible that cytosine was deaminated to uracil. However, treatment of mtDNA that had been exposed to the NO generator with uracil DNA glycosylase did not reveal any additional damage. To assess repair, cells were treated with NO and allowed to repair the damage before the DNA was isolated. Most of the damage was repaired by 4 h after exposure. Therefore, these were the first studies to show that NO selectively damages mtDNA, and that this damage is of sufficient consequence that the mitochondrion has had to evolve efficient mechanisms to rapidly repair it (18).

V. Evaluation of Damage and Repair to mtDNA at the Level of Individual Nucleotides

In order to address mechanistic questions concerning mtDNA mutations, it was necessary to analyze damage and repair of mtDNA at the nucleotide level. Therefore, the technique of ligation-mediated PCR (LM-PCR), which had been used by Dr. Holmquist and colleagues to study nuclear repair (20), was adapted for the evaluation of mtDNA (21). Initially, the frequencies of single-strand breaks and oxidative base damage in mtDNA from insulinoma cells were measured. Addition of 5 mM alloxan to the cells increased the rate of oxidative base damage and the lesion frequency in mtDNA severalfold. Guanine positions showed the highest endogenous lesion frequencies and were the most responsive bases to alloxan-induced oxidative stress. Although specific bases were consistently hotspots for damage, there was no evidence that removal of these lesions occurred in a strand-specific manner. These data revealed nonrandom oxidative damage in several nucleotides which correlated with one of the break sites of the 5-kb "common deletion" seen with aging. Additionally, there was an apparent adaptive, non–strand-selective response for removal of such damage.

Following this initial observation, the technique was used to look at other types of damage (19). When a 200-bp sequence of mtDNA from cells exposed to NO produced by the NO generator PAPA/NO was analyzed, guanine was the predominant base that was damaged. However, there also was damage to specific adenines. No lesions were observed at pyrimidine sites, indicating that the predominant lesion is the deamination of guanine to xanthine. Work with the ROS-generating system xanthine oxidase/hypoxanthine showed that many of the same guanines were vulnerable to attack. However, other base damage is also seen. As expected, the methylating agent MNU also selectively alkylated guanines. It is intriguing that the pattern of damaged guanines was not identical to that damaged by NO. The studies using LM-PCR have shed new light on the damage caused by NO in mtDNA. They revealed that guanine is the most frequently damaged base, although there were some damaged adenines found when alkaline conditions were employed. No damage was detected at any pyrimidines in the sequence evaluated. For comparison, studies with the ROS generator XO showed that we were able to detect oxidative lesions to both specific purines and pyrimidines. The pattern and frequency of base damage was similar to that observed previously using the ROS generator alloxan (21). Therefore, it appears likely that the inability to detect pyrimidine lesions following exposure to NO was because few, if any, lesions are formed at these bases. These findings support the notion that the damage that occurred to mtDNA is through the formation of N_2O_3, which causes deamination of guanine to xanthine and adenine to hypoxanthine. In both cases, this lesion preceded depurination to produce an AP site which was converted to a strand break with an appropriate end for ligation by alkali treatment. That AP sites predominantly are formed in DNA is in agreement with work by Tamir et al. (22) studying plasmid DNAs which were electroporated into CHO cells. When these cells were treated with NO, a significant number of AP sites were produced in the DNA. Additional work by these investigators, using both DNA exposed to NO *in vivo* and *in vitro* has revealed that xanthine followed by hypoxanthine are the predominant base alterations (23, 24).

By comparing the pattern of damage produced by NO to that generated by the alkylating agent MNU or the ROS generator XO, it can be seen that, although all three of these agents damaged many of the same guanines, there were certain guanines that were only vulnerable to PAPA/NO. Therefore, it is possible to determine a signature damage pattern for reactive nitrogen species (RNS) that is different from that produced by ROS or methylating toxins. This finding may prove useful for future studies of mtDNA in which the identity of the damaging agent is unknown. Additionally, it will be important to compare damage produced in mtDNA by different agents with patterns of known mtDNA mutations. The fact that NO produced the same pattern of damage when exposed to a PCR-generated mtDNA sequence establishes that the pattern of damage produced is due to the chemical properties of NO interacting within the sequence context of the DNA, rather than being influenced by the association of the DNA with the mitochondrial matrix proteins or other DNA-binding proteins. When considering lesions in mtDNA, it is important to mention that in cases where a single mutation has been shown to cause a disease, the degree of heteroplasmy is usually between 60 and 90% (mutated genomes to nonmutated). However, in major neurodegenerative diseases such as Alzheimer's disease, Parkinson's disease, and amyotrophic lateral sclerosis (ALS), this is not the case. While increased mtDNA mutations are seen, none rises to a level sufficient by itself to cause disease. Thus, it is believed that it is a collection of mutations and heightened lesions in mtDNA that inactivates a critical number of mitochondrial genomes in a specific way to cause disease. LM-PCR allows one to determine what mutations might be expected following exposure to specific types of DNA damaging agents.

VI. Mechanisms Involved in the Repair of mtDNA

One approach to ascertain whether factors that affect repair in the nucleus also affect repair in the mitochondrion is to study repair in cells from individuals with known repair-deficient phenotypes such as xeroderma pigmentosum (XP). XP complementation group D (XP-D) is known to be defective in the repair of N-methylpurines in their nuclear DNA. To determine whether this defect also is seen in mtDNA from these cells, repair of MNU-induced N-methylpurines in XP-D cells was compared to that seen in normal WI-38 cells (25). At both 8 and 24 h after exposure to MNU, XP-D cells were found to normally repair N-methylpurine lesions in mtDNA, although the repair of the same lesions in the nuclear DNA from XP-D cells was significantly attenuated. These data indicated that the factors that retard repair of nuclear DNA in XP-D cells do not affect the repair of mtDNA in these cells.

Extracts from cells of individuals with XP complementation group A (XP-A) have been reported to be defective in the repair of some types of oxidative damage. Therefore, studies were performed to determine whether there was a correlation between the inadequate repair of oxidatively damaged nuclear DNA in XP-A cells and the capacity to repair similar damage to their mtDNA (26). The ability of karyotypically normal human fibroblasts and XP-A fibroblasts to repair alloxan-generated oxidative damage to nuclear and mtDNA was assessed. These data indicate that both nuclear and mitochondrial repair of DNA damage are appreciably more efficient in normal human fibroblasts. These findings suggest a similarity between the process(es) used to repair oxidative damage to nuclear and mtDNA in that both are inefficient in XP-A cells.

VII. Cell-Specific Differences in mtDNA Repair

Based on the studies described in the preceding sections, it is known that there are mechanisms for repair of endogenous damage in mtDNA. However, what is not as clear is how important mtDNA repair is to the cellular defenses of normal cells. The first indication of the importance of mtDNA repair to cellular defenses occurred when the simple question, "Are there cell-specific differences in repair of mtDNA?" was asked. Within the central nervous system (CNS) there are two predominant types of cells: neurons, which are the information processing cells of the CNS; and glial cells, which provide supportive functions for the neurons. Glial cells are composed mainly of three distinct populations of cells: astrocytes, oligodendrocytes, and microglia. While it was known that these cells play an important role in some diseases of the CNS, little was known about the repair mechanisms utilized by these cells in response to injury. Therefore, a well-characterized culture system (27) to generate pure primary cultures of astrocytes, oligodendrocytes, and microglia was used to evaluate mtDNA repair following both alkylation and oxidative damage.

For the studies of simple alkylation damage, MNU was used as the alkylating agent. Quantitative determinations of mtDNA initial break frequencies and repair efficiencies showed no cell type-specific differences in initial mtDNA damage. However, mtDNA repair of N-methylpurines in oligodendrocytes and microglia was significantly reduced compared to that of astrocytes. In astrocytes, and all other cell types previously evaluated in our laboratory, greater than 60% of N-methylpurines were removed from the mtDNA by 24 h. In contrast, only 35% of lesions were removed from mtDNA of oligodendrocytes and microglia during the same time period. Since mitochondrial perturbations by a variety of xenobiotics have been linked to apoptosis, analyses using DNA laddering and ultrastructural examination were performed. DNA fragmentation and morphological changes consistent with apoptosis were apparent following MNU treatment of cultured oligodendrocyte progenitors and microglia, but not astroglia. These data were the first to demonstrate a correlation between diminished mtDNA repair capacity and the induction of apoptosis (28).

Subsequently, oxidative mtDNA damage and repair also were assessed. Menadione, which undergoes redox cycling within the mitochondria, was used to generate oxygen radicals. The results from these studies showed that exposure to equimolar concentrations of menadione resulted in more initial mtDNA damage in oligodendrocytes and microglia as compared to astrocytes. Repair experiments then were performed using both equimolar concentrations and concentrations that resulted in comparable strand breaks. Under both conditions, astrocytes repaired the damage efficiently with all of the lesions being removed by 6 h in mtDNA of astrocytes as compared to approximately 60% repair in oligodendrocytes or microglia. Our previous findings of a correlation between lack of mtDNA repair of N-methylpurines and induction of apoptosis led us to ask the question whether oligodendrocytes and microglia undergo apoptosis in response to an oxidative insult. ApoTag and annexin V staining, DNA laddering and electron microscopy were used to evaluate apoptosis. There were no indications of apoptosis in any of the experiments with astrocytes, while oligodendrocyte and microglial cultures were positive in all cases. We also questioned whether the increased sensitivity could be due to decreases in antioxidants within the oligodendrocytes and microglia; so, in collaboration with Dr. Doug Spitz at Washington University, we measured enzyme activities for glutathione peroxidase, catalase, CuZn, and MnSOD, along with total, reduced, and oxidized glutathione levels.

For these studies, cells were prepared exactly as for the repair studies. The results demonstrated that under our culture conditions, there were no significant differences in catalase, glutathione peroxidase, or CuZnSOD between any of the cell types. Concerning glutathione, astrocytes had significantly lower levels than oligodendrocytes or microglia. The same was true for MnSOD (29). Thus, while antioxidant and repair capacity are both involved in protecting cells from oxidant insults, it appears that cells with efficient repair capacity may be spared, even in the presence of very low antioxidant levels, and that cells with less efficient repair are susceptible, even in the presence of higher levels of antioxidants. This observation of cell-specific differences in repair of oxidative damage among glial cells leads to the obvious question "Is there repair of oxidative mtDNA damage in neurons?" To address this question, primary cultures of cerebellar granule cells were exposed to comparable concentrations of menadione. More initial damage was observed in the neuronal cultures as compared to the glial cells. Because of the increased sensitivity, a 50-μM concentration was used for subsequent repair experiments. The results of these experiments demonstrated that the repair kinetics were slower in neurons as compared to glial cells, but by 48 h the lesions had been removed. As in the glial cultures with decreased repair capacity, apoptosis was induced in cerebellar granule cells by menadione exposure using quantitative electron microscopic evaluation, annexin V positive staining, or Apotag positive staining as the marker for apoptosis.

Since menadione redox cycles with complex I of the electron transport chain to produce superoxide, one could hypothesize that mitochondria may play a substantial role through activation of caspases. Of the caspases, caspase 9 has been associated with mitochondrial changes. The release of cytochrome c from the mitochondrial intermembrane space to the cytosol is necessary for the activation of this caspase. Western blots were performed using mitochondrial and cytosolic proteins and a monoclonal antibody to cytochrome c. No cytochrome c protein was detected in the cytosolic fraction prior to menadione treatment, indicating that the cell fractionation procedure had not disrupted the outer mitochondrial membrane. However, 2 h after exposure to 100 μM menadione, there was a decrease in the intensity of the mitochondrial cytochrome c band with a concomitant increase in the cytosolic band of oligodendrocytes and microglia. As expected, astrocytes, which show no evidence of cell death in response to menadione exposure, showed no increase in the cytosolic cytochrome c band. Therefore, release of cytochrome c from the mitochondria into the cytosol correlates with induction of apoptosis in CNS glial cells (29).

To determine if this release of cytochrome c resulted in the activation of caspase 9, a colorimetric activity assay based on cleavage of a caspase 9-specific substrate was performed. For a positive control, staurosporine, which has been shown to cause cytochrome c release in cerebellar granule neurons, was employed. After 3 h, caspase 9 activity in oligodendrocytes, microglia, and Jurkat cells exposed to menadione was elevated compared to control cells. No detectable increase in astrocyte caspase 9 activity was found. Subsequent experiments were performed using xanthine oxidase and hypoxanthine to generate the ROS. Again, activation of caspase 9 was observed (29).

Since it has been suggested that caspase 8 may play a role in the mitochondrial pathway of apoptosis, its activity following menadione exposure was tested in each of the cell types. Activation of caspase 8 has been well documented in the Fas-pathway of apoptosis; thus, Jurkat cells treated with an anti-Fas antibody were used for a positive control. No caspase 8 activity was detected in any of the cell types following exposure to menadione. Following treatment with the anti-Fas antibody, however, oligodendrocytes and microglia demonstrated caspase 8 activity. Conversely, astrocytes remained unresponsive to induction of apoptosis, showing no caspase 8 activity.

From these experiments, it appears that mtDNA repair plays a pivotal role in cellular defense mechanisms. This statement is based on the evidence that there are cell-specific differences in the repair of mtDNA damage. The importance of these differences is exemplified in the observation that decreases in mtDNA repair capacity correlates with increased cell killing. Cell death occurs through the induction of apoptosis involving a mitochondrial pathway that includes the release of cytochrome *c* and the activation of caspase 9.

VIII. Conclusions and Future Questions

It is becoming increasingly clear that lesions in mtDNA play an important role in a number of human diseases. Owing to the crucial role that mtDNA integrity plays in cellular processes, a new area for investigation into the pathogenesis of a number of age-related diseases has been identified. The premise upon which these investigations are based is that an acceptable lesion equilibrium must be maintained in mtDNA for normal mitochondrial function. This lesion equilibrium is controlled by a balance in the rate at which mtDNA is damaged and the rate at which these lesions are removed. If this critical balance is disrupted either by an increase in the rate of mtDNA damage or a decrease in the rate of repair, fewer functioning mitochondrial genomes will be available for transcription and cellular bioenergetics will decrease. This will cause cellular functions to diminish due to energy depletion. Additionally, there will be an increase in oxidative stress in the cell due to defects in electron transport caused by the heightened damage to mtDNA. If this deleterious process is allowed to continue, the cell will ultimately die via apoptotic or necrotic mechanisms. There are several ways in which intervention could alter this catastrophic cascade of events. However, before such intervention can be initiated, a more thorough understanding of the mechanisms involved in the alteration of the lesion equilibrium in mtDNA needs to be obtained. Therefore, many questions remain to be answered. The problems remaining to be addressed include a more precise definition of the components involved in mtDNA repair, a better comprehension of how they are regulated, a more thorough understanding of how they can malfunction to precipitate disease states, and the development of gene therapy protocols to reverse repair defects and prevent or delay the onset of disease.

Acknowledgments

This work was supported in part by the United States Public Health Service Grants ESO5865, ESO0313, and ESO3456 from the National Institute of Environmental Health Sciences, and AG12442 from the Institute of Aging.

References

1. D. C. Wallace, *Science* **256,** 628 (1992).
2. R. Taylor, *J. NIH Res.* **4,** 62 (1992).
3. S. Ikebe, M. Tanaka, K. Ohno, W. Sata, K. Hattori, T. Kondo, Y. Mizuno, and T. Ezawa, *Biochem. Biophys. Res. Commun.* **170,** 1044 (1990).
4. S. W. Ballinger, J. M. Shoffner, E. V. Hedaya, I. Trounce, M. A. Polak, D. A. Koontz, and D. C. Wallace, *Nature Genet.* **1,** 11 (1992).
5. C. A. Lindgren and D. O. Smith, *J. Neurosci.* **6,** 2644–2652 (1986).
6. R. F. Hevner, R. S. Duff, and M. T. Wong-Riley, *Neurosci. Lett.* **138,** 188–192 (1992).
7. R. L. Morris and P. J. Hollenbeck, *J. Cell Sci.* **104,** 917–927 (1993).
8. K. Stürmer, O. Baumann, and B. Walz, *J. Cell Sci.* **108,** 2273–2283 (1995).
9. E. A. Schon, E. Bonilla, and S. Dimauro, *J. Bioenerg. Biomemb.* **29,** 131–149 (1997).
10. V. Skulachev, *Quart. Rev. Biophys.* **29,** 169–202 (1996).
11. T. M. Buttke and P. A. Sandertrom, *Immunol. Today* **15,** 7–10 (1994).
12. V. Goossens, J. Grooten, K. De Vos, and W. Fiers, *Proc. Natl. Acad. Sci. U.S.A.* **92,** 8115–8119 (1995).
13. T. Ozawa, *Physiol. Rev.* **77,** 425–464 (1997).
14. C. C. Pettepher, S. P. LeDoux, V. A. Bohr, and G. L. Wilson, *J. Biol. Chem.* **266,** 3113–3117 (1991).
15. S. P. LeDoux, G. L. Wilson, E. J. Beecham, T. Stevnsner, K. Wassermann, and V. A. Bohr, *Carcinogenesis* **13,** 1967–1973 (1992).
16. W. J. Driggers, S. P. LeDoux, and G. L. Wilson, *J. Biol. Chem.* **268,** 22042–22045 (1993).
17. C. Shen, W. Wertelecki, W. J. Driggers, S. P. LeDoux, and G. L. Wilson, *Mutat. Res.* **337,** 19–23 (1995).
18. G. L. Wilson, N. J. Patton, and S. P. LeDoux, *Diabetes* **46,** 1291–1295 (1997).
19. V. I. Grishko, N. Druzhyna, S. P. LeDoux, and G. L. Wilson, *Nucleic Acids Res.* **27,** 4510–4515 (1999).
20. G. P. Pfeifer, R. Drouin, and G. P. Holmquist, *Mutat. Res.* **288,** 39–46 (1993).
21. W. J. Driggers, G. P. Holmquist, S. P. LeDoux, and G. L. Wilson, *Nucleic Acids Res.* **25,** 4362–4369 (1997).
22. S. Tamir, S. Burney, and S. R. Tannenbaum, *Chem. Res. Toxicol.* **9,** 821–827 (1996).

23. J. L. Caulfield, J. S. Wishnok, and S. R. Tannenbaum, *J. Biol. Chem.* **273,** 12689–12695 (1998).
24. S. Tamir, J. S. Wishnok, and S. R. Tannenbaum, *Methods Enzymol.* **269,** 230–243 (1996).
25. S. P. LeDoux, N. J. Patton, L. J. Avery, and G. L. Wilson, *Carcinogenesis* **14,** 913–917 (1993).
26. W. J. Driggers, V. I. Grishko, S. P. LeDoux, and G. L. Wilson, *Cancer Res.* **56,** 1262–1266 (1996).
27. K. D. McCarthy and J. deVellis, *J. Cell Biol.* **85,** 890–902 (1980).
28. S. P. LeDoux, C. Shen, V. I. Grishko, P. A. Fields, A. L. Gard, and G. L. Wilson, *Glia* **24,** 304–312 (1998).
29. B. S. Hollensworth, C. Shen, J. E. Sim, D. R. Spitz, G. L. Wilson, and S. P. LeDoux, *Free Radical Biol. Med.* **28,** 1161–1174 (2000).

Base Excision Repair in Nuclear and Mitochondrial DNA

> GRIGORY L. DIANOV, NADJA
> SOUZA-PINTO, SIMON G. NYAGA,
> TANJA THYBO,
> TINNA STEVNSNER, AND
> VILHELM A. BOHR
>
> *Laboratory of Molecular Genetics*
> *National Institute on Aging, NIH*
> *Baltimore, Maryland 21224*

I. Oxidative DNA Damage Processing in Nuclear DNA 286
   A. Roles of BER and NER in the Repair of Oxidative DNA
      Base Lesions 286
   B. Long- and Short-Patch BER 287
   C. Pol β Is Involved in Both Short- and Long-Patch BER 289
II. Mitochondrial DNA Repair in Mammalian Cells 291
   A. Identification of BER Enzymes in Mitochondria................. 292
   B. Oxidative DNA Damage Repair in Mitochondria
      of Human Lymphoblastoid Cell Lines 293
   C. *In Vitro* Repair of Hypoxanthine and Ethenoadenine in Rat and
      Human Mitochondrial Extracts. 293
III. Mitochondrial DNA Repair and Aging............................ 295
   References.. 296

Base excision repair mechanisms have been analyzed in nuclear and mitochondrial DNA. We measured the size and position of the newly incorporated DNA repair patch in various DNA substrates containing single oxidative lesions. Repair of 8-oxoguanine and of thymine glycol is almost exclusively via the base excision repair (BER) pathway with little or no involvement of nucleotide excision repair (NER). The repair mode is generally via the single-nucleotide replacement pathway with little incorporation into longer patches. Extension of these studies suggests that DNA polymerase β plays a critical role not only in the short-patch repair process but also in the long-patch, PCNA-dependent pathway. Mitochondria

---

Abbreviations: 8-oxoG, 8-oxoguanine; TG, thymine glycol; BER, base excision repair, NER, nucleotide excision repair; mt, mitochondrial; APE, AP endonuclease; Pol β/δ, DNA polymerase β/δ; PCNA, proliferating cell nuclear antigen; FEN1, flap endonuclease; εA, ethenoadenine; hX, hypoxanthine; mtODE, mitochondrial oxidative damage endonuclease, which recognizes 8-oxoG in rat. Fpg, fapy DNA-glycosylase; mtUDG, mitochondrial uracil DNA-glycosylase; endoG, mitochondrial endonuclease G.

are targets for a heavy load of oxidative DNA damage. They have efficient BER repair capacity, but cannot repair most bulky lesions normally repaired by NER. *In vitro* experiments performed using rat and human mitochondrial extracts suggest that the repair incorporation during the removal of uracil in DNA occurs via the short-patch repair BER pathway. Oxidative DNA damage accumulates with age in mitochondrial DNA, but this cannot be explained by an attenuation of DNA repair. In contrast, we observe that mitochondrial incision of 8-oxoG increases with age in rodents. © 2001 Academic Press.

I. Oxidative DNA Damage Processing in Nuclear DNA

A. Roles of BER and NER in the Repair of Oxidative DNA Base Lesions

Reactive oxygen species are formed continuously as a consequence of normal cellular metabolism and are also generated by a number of external factors. The reaction of active oxygen species with DNA results in numerous forms of base damage, and 8-oxoguanine (8-oxoG) and thymine glycol (TG) are some of the most abundant lesions generated. Increased levels of 8-oxoG and TG have been observed after treatment of cells with ultraviolet (UV), ionizing irradiation, or chemical mutagens that generate oxygen radicals (reviewed in Refs. *1* and *2*). 8-oxoG in a *syn* conformation can base-pair with adenine and induce transversion mutations (*3*). This mutation is thought to play a causal role in cancer and aging (*4, 5*). In contrast, TG constitutes a strong block for DNA replication (*6, 7*) and transcription (*8, 9*); therefore, it must be efficiently removed from DNA to maintain genome integrity. The base excision repair (BER) and nucleotide excision repair (NER) pathways are the two main excision repair systems processing TG and 8-oxoG in DNA. *In vitro* studies have shown that human NER enzymes can recognize both TG and 8-oxoG in DNA and excise them as a part of a 25–30 base-long oligonucleotide (*10, 11*). Recently, we have employed an *in vitro* repair system that is optimal for both BER and NER and examined the relative roles of these two pathways in the repair of oxidative damage in DNA (*12, 13*). Closed circular DNA constructs containing a single 8-oxoG or TG at a defined site were used as substrates to determine the amount and directionality of repair patch incorporation generated after *in vitro* repair by mammalian cell extracts (Fig. 1). Restriction analysis of the repair incorporation in the vicinity of the lesion indicated that most of the repair incorporation was localized 3′ to the lesion. Very little or no repair incorporation 5′ to the lesion, which is characteristic for the NER pathway, was found (Fig. 2). We thus conclude that BER is the primary repair system for removal of these oxidative DNA base lesions in human cell extracts.

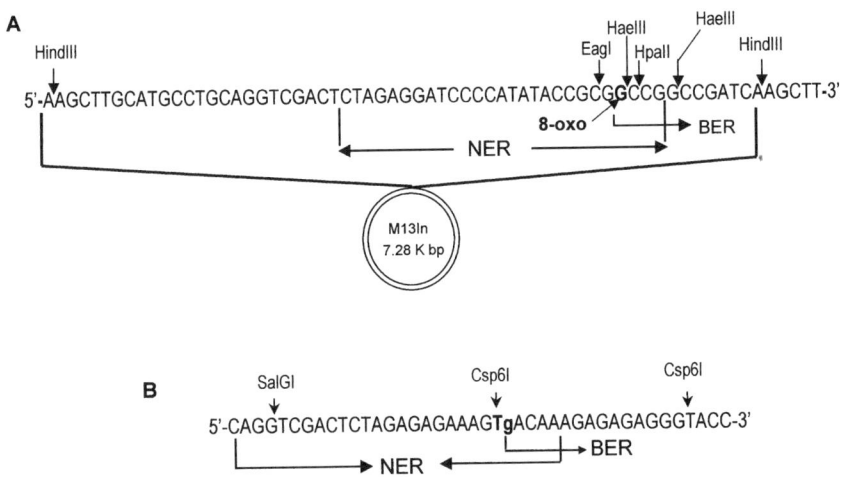

FIG. 1. DNA substrates containing a single 8-oxoG (A) or TG (B) residue. Nucleotide sequence and restriction sites of the lesion (8-oxoG or TG)-containing strand of the closed circular double-stranded DNA construct and the restriction sites used to analyze repair incorporation are shown. The direction of NER and BER incorporation is indicated.

B. Long- and Short-Patch BER

There are two pathways for BER involving different subsets of proteins and operating independently (14–17). Both pathways are initiated by DNA glycosylases that recognize and remove the damaged base leaving an abasic site (AP site). The AP site is further processed by AP endonuclease (APE1), which introduces a DNA strand break 5′ to the baseless sugar and then DNA polymerase β (pol β) catalyzes β-elimination of the 5′-sugar phosphate residue and fills the one-nucleotide gap. Finally, the nick is sealed by DNA ligase. These repair events result in a single-nucleotide repair patch (Fig. 3). The long-patch repair pathway, in addition to DNA glycosylase and AP endonuclease, also involves flap endonuclease (FEN1), proliferating cell nuclear antigen (PCNA), DNA polymerase δ (Pol δ), and DNA ligase (17, 18). To initiate this pathway, DNA polymerase first adds several nucleotides to the 3′-end of the nick and exposes the 5′-sugar phosphate as part of a single-stranded flap structure. This flap structure is recognized and excised by FEN1 and the DNA is finally ligated by DNA ligase. These repair events result in a 2–6-nucleotide-long repair patch (14, 19). The number and directionality of nucleotides replaced during *in vitro* repair of single lesion DNA substrates were assessed in order to determine whether the repair occurs via the single-nucleotide pathway or via the long-patch pathway. The substrates containing a single TG or 8-oxoG at a defined site were repaired

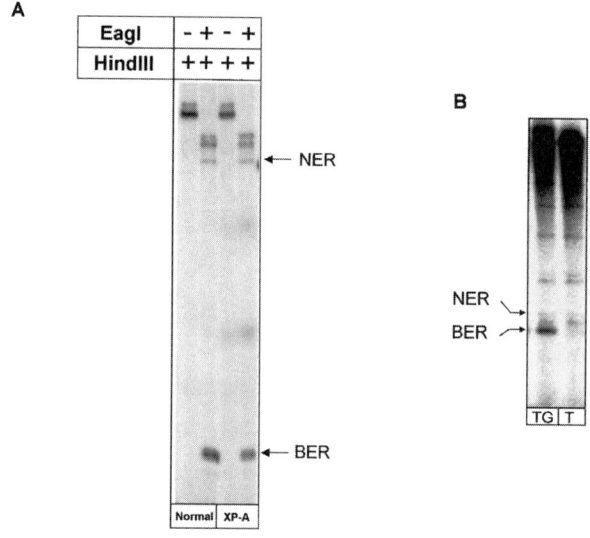

FIG. 2. Nucleotide excision repair is not involved in *in vitro* repair of 8-oxoG and TG. (A) Analysis of dGMP incorporation into the *Hin*dIII-*Eag*I fragment located 5' to the 8-oxoG. Substrate DNA (Fig. 1A) was incubated for 3 h at 30°C with 100 μg of whole-cell extract protein prepared from the normal human lymphoblastoid cell line AG9387 (lanes 1, 2), and xeroderma pigmentosum group A lymphoblastoid cell line GM2250 (lanes 3, 4). The DNA was recovered from the reaction mixture and cut with *Hin*dIII and *Eag*I restriction endonucleases. Repair incorporation into individual fragments was analyzed after electrophoresis on 15% polyacrylamide gel. (B) 200 ng of single TG-containing substrate DNA (Fig. 1B) was incubated with 100 μg of whole-cell extract protein from the human lymphoblastoid cell line AG9387 in the presence of [α-^{32}P]dATP for 3 h at 37°C. The DNA was recovered from the reaction mixture, restricted with *Sal*GI and *Csp*6I endonucleases, and repair incorporation was analyzed after electrophoresis on a 20% polyacrylamide gel.

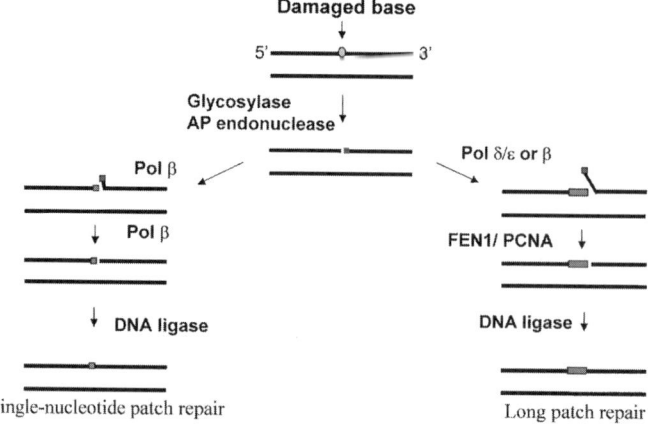

FIG. 3. Base excision repair pathways.

TABLE I

RESULTS FROM DIFFERENT LABORATORIES ON THE ROLE
OF SINGLE-NUCLEOTIDE AND LONG-PATCH BER INITIATED BY URACIL-DNA
GLYCOSYLASE (DNA GLYCOSYLASE WITHOUT AP LYASE ACTIVITY)
OR 8-OXOG-DNA GLYCOSYLASE (GLYCOSYLASE WITH AP LYASE ACTIVITY)

Extract	Damaged base	Single-nucleotide patch repair (%)	Long-patch repair (%)	References
Human	8-oxoG	85–90	10–15	20, 46, 47
	Uracil	75–80	20–25	
Mouse	8-oxogG	85–90	10–15	20, 46
	Uracil	70–80	20–30	47, 48
E. coli	Uracil	75–80	20–25	47

in human whole-cell extracts and the repair synthesis incorporation was measured in the vicinity of the lesion. Restriction analysis of the repair incorporation in the vicinity of the lesion indicated that the majority of 8-oxoG and TG (up to 90%) was repaired through the single-nucleotide repair mechanism and the remaining lesions were removed by the long-patch BER pathway. These data show that single-nucleotide patch BER is the primary pathway for repair of TG and 8-oxoG DNA lesions. In comparison to repair patch formation during the repair of uracil in DNA, the removal of 8-oxoG generally involves shorter repair patches (Table I). It has been speculated that the regulation of the patch size may be determined by the type of glycosylase that initiates incision (20), but that may not be the general situation. The type of DNA polymerase involved in different pathways and their ability to promote strand-displacement may be an important factor. Another possibility is that the size of the displacement is an important factor. The size of the repair patch is an operational parameter and more work is needed to investigate how this is regulated. The cell type, organelle, or tissue type may also be an important factor.

C. Pol β Is Involved in Both Short- and Long-Patch BER

Previous studies had shown that Pol β is the major DNA polymerase involved in the single-nucleotide BER pathway (21), while Pol δ/ε are thought to be involved in PCNA-dependent long-patch BER (15, 16). Recent findings, however, suggest that Pol β and Pol δ can substitute for each other in long-patch BER reconstituted with purified proteins (17). The substitution of different polymerases in BER was also confirmed by the competence of Pol β-null cell extracts in the *in vitro* repair of both natural and reduced abasic sites in closed circular DNA (22). Thus, the biochemical proficiency of Pol β and other polymerases in both subpathways for base excision repair has been documented. Yet, the question remained as to the DNA polymerase of choice for the long-patch BER

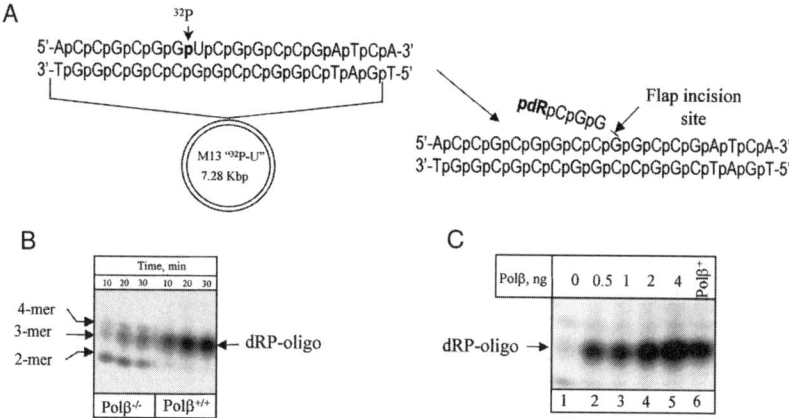

FIG. 4. Monitoring of the long-patch excision products generated during repair of uracil-containing DNA by human cell extract. (A) DNA substrate containing a single uracil residue and labeled phosphate group 5′ next to uracil (shown in bold). Nucleotide sequence in vicinity of uracil is shown. During long-patch repair FEN1 will release labeled dRP containing flap oligonucleotide (shown on the right). (B) Deficient dRP-oligo release in Pol β-null cell extracts. 100 ng of internally labeled uracil-containing DNA were incubated with 20 μg of whole-cell extract at 37°C for the indicated time. Reaction products were reduced with sodium borohydride and analyzed by electrophoresis on 20% polyacrylamide gel. (C) Pol β-dependence of dRP-oligo release. Cell extract (20 μg protein) derived from Pol β-null cells was preincubated on ice for 20 min with indicated amount of Pol β protein before the substrate U-DNA was added. Reactions were further incubated for 20 min at 37°C and processed as described above. dRP-oligo excision by normal (Pol β⁺) extract is presented in lane 6.

subpathway and its precise role(s) in the sequential mechanism. By analyzing products of long-patch excision generated during BER of a uracil-containing DNA substrate in mammalian cell extracts, we find that long-patch excision also depends on Pol β and that Pol β plays an essential role in long-patch BER by conducting strand-displacement synthesis and controlling the size of the excised flap (Fig. 4). Several experimental approaches were used to support this conclusion (18). First, we have shown that the excision step of long-patch BER is dependent upon the Pol β status of the cell extract. The release of a specific excision product (dRP-oligo) is reduced in Pol β-deficient cells, but can be reconstituted by addition of purified Pol β. In addition, Pol β-neutralizing antibody inhibits long-patch BER excision and especially the release of the dRP-oligo product. These results were unexpected since Pol β had not been thought to participate in the long-patch BER reaction. The striking homogeneity of the excision product size (i.e., the dRP-oligo) (18) indicates that the proteins involved in the excision step may predetermine the length of the excised oligonucleotides. Pol β is a good candidate for this function-displacement synthesis and may be controlling the size of the excised flap.

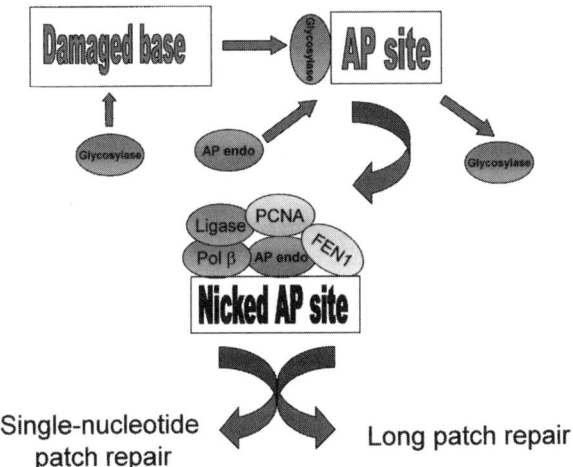

FIG. 5. A central role for Pol β in the coordination of BER pathways. A damaged base is recognized and removed by a DNA glycosylase that remains bound to the AP site and is later displaced by AP endonuclease that generates a strand break 5′ to the AP site. AP endonuclease interacts with Pol β and from this point Pol β plays a central role in determining the subpathway choice. Removal of the 5′-sugar phosphate by Pol β and filling the gap results in a single-nucleotide repair patch. Alternatively, a Pol β-dependent strand displacement and flap excision by FEN1 switches repair to the long-patch BER subpathway. In both pathways, DNA ligase interacts with Pol β and accomplishes DNA repair by sealing the strand break. PCNA interacts with both FEN1 and DNA ligase I and may stimulate both flap excision and ligation.

It appears that different steps of the BER reaction are coordinated and directed by multiple interactions between the participating proteins (Fig. 5). For example, after removal of the T base, from the G·T mismatch, G·T-DNA glycosylase remains bound to the AP site and must be displaced by APE (23). Further, APE was shown to interact with Pol β and this interaction stimulates the Pol β lyase activity (24). After this step in the repair process, Pol β appears to play a central role in determining the subpathway: removal of 5′-sugar phosphate by Pol β and interaction with DNA ligase I results in single-nucleotide BER. Alternatively, a Pol β-dependent excision reaction may switch repair to the long-patch BER subpathway (Fig. 5).

II. Mitochondrial DNA Repair in Mammalian Cells

Oxidative phosphorylation in mitochondria results in the production of reactive oxygen species (ROS). Several other processes—including peroxisomal metabolism, enzymatic synthesis of nitric oxide, metabolism of phagocytic leukocytes, heat, ultraviolet light, various drugs such as those used in the treatment of HIV, ionizing radiation, and redox cycling compounds—also contribute significantly to the pool of ROS. Hydrogen peroxide, singlet oxygen, and hydroxyl radicals are among the ROS produced. Mitochondrial DNA (mtDNA), being in close proximity to the electron transport chain, is a good target of ROS. This interaction results in oxidation of specific mtDNA bases. The most common oxidation products of mtDNA include 8-oxoG and TG, and these base modifications have been detected in human cells (4, 25). In addition, deamination of cytosine and incorporation of dUTP results in the formation of uracil in mtDNA. If unrepaired, this uracil can cause GC–AT transition mutations. Thus, insufficient mtDNA repair may result in mitochondrial dysfunction and thereby cause degenerative diseases, loss of energy formation, and pathophysiological processes leading to aging and cancer.

A. Identification of BER Enzymes in Mitochondria

Early indications for a BER mechanism in mitochondria came with the isolation of a mammalian mitochondrial endonuclease that specifically recognizes AP sites and cleaves the DNA strand (26). Later, it was demonstrated that a combination of enzymes purified from *Xenopus laevis* mitochondria efficiently repairs abasic sites in DNA (27). In the same study a mtDNA ligase was purified that seemed to be related to nuclear DNA ligase III. Recently, it was confirmed that the human DNA ligase III gene encodes both a nuclear and a mitochondrial enzyme (28). The mitochondrial DNA polymerase γ (Pol γ) has been shown to possess a 5′-deoxyribose phosphate lyase activity, and, like Pol β, it can remove dRP moieties through a β-elimination process, suggesting a role for Pol γ in mitochondrial BER (29). PCNA has been shown to stimulate Pol γ-mediated DNA synthesis, suggesting a role for PCNA as auxiliary factor in mitochondrial DNA replication and repair (30).

The isolation of mitochondrial glycosylases has provided further evidence for a BER mechanism in mitochondria. An endonuclease specific for oxidative damage (mtODE: mitochondrial oxidative damage endonuclease) has been purified from rat liver mitochondria (31). This enzyme introduces single-strand breaks at 8-oxoG sites and is a putative glycosylase/AP-lyase. mtODE might be a mitochondrial form of OGG1 since three forms of hOGG1 contain a putative mitochondrial localization sequence and have been shown to be targeted to mitochondria (32). A human 8-oxo-dGTPase that prevents misincorporation of 8-oxodG in DNA by hydrolyzing 8-oxo-dGTP to 8-oxo-dGMP also has been found to localize to the mitochondria (33).

Another rat mitochondrial endonuclease (mtTGendo), which specifically recognizes TG, has been purified and characterized. Like the nuclear Nth1 protein, it has an associated glycosylase activity (34), which suggests that it might be a functional mitochondrial homolog of Nth1. In fact, hNth1 has been shown to be transported both to the nucleus and to mitochondria (32). In yeast, two functional homologs, Ntg1 and Ntg2, exist, and it was recently shown that Ntg1 localizes primarily to the mitochondria and Ntg2 exclusively localizes to the nucleus (35). Finally, the UNG gene has been shown to generate both a mitochondrial and a nuclear uracil DNA glycosylase by alternative splicing (36). The pathway for repair of uracil in mitochondria has been investigated, and it was demonstrated that only one nucleotide is incorporated during the repair process, suggesting a repair pathway similar to nuclear short-patch BER (37).

B. Oxidative DNA Damage Repair in Mitochondria of Human Lymphoblastoid Cell Lines

Whereas *in vitro* assays have been established for nuclear DNA repair studies and yielded tremendous insight, such assays have not been available for mitochondrial studies. We first succeeded in establishing an *in vitro* repair assay using mitochondrial extracts from rat liver (37). More recently, we have extended this work to utilize human cells in culture, thereby allowing much easier access to material and the possibility to use various mutant cell lines. We chose human lymphoblasts for our studies and have examined the removal of 8-oxoG, TG, and uracil using *in vitro* assays.

Mitochondrial extracts were prepared as previously described (31) and screened for incision of oligonucleotides containing single lesions at specific positions. The results show that mitochondrial extracts from human lymphoblasts incised all three lesions—TG, 8-oxoG, and uracil—efficiently. There was no incision of an undamaged oligonucleotide, demonstrating that this was damage-specific incision. These results indicate that mitochondria from human lymphoblasts harbor enzymatic activity specific for the repair of these lesions. The incision of 8-oxoG containing oligonucleotides is shown in Fig. 6. This experiment revealed that a 10-mer product accumulated over time (Fig. 6, lanes 2-6). This product was consistent with the one produced by formamidopyrimidine DNA glycosylase (Fpg) (Fig. 6, lane 8). Investigations are currently under way to determine whether repair of these lesions occur via the long-, short-, or both BER pathways in human mitochondria.

C. *In Vitro* Repair of Hypoxanthine and Ethenoadenine in Rat and Human Mitochondrial Extracts

Hypoxanthine (hX) and ethenoadenine (εA) are both efficiently removed from nuclear DNA by the 3-methyladenine DNA glycosylase (38), but so far no study has addressed the question of whether these lesions are repaired in mitochondria. Mitochondria were isolated from fresh rat liver and from human lymphoblasts by a method that previously has been shown to involve no

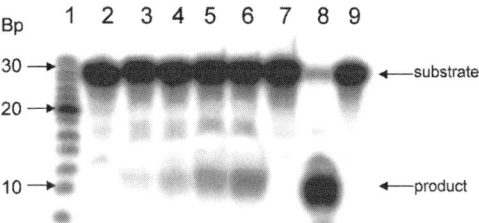

FIG. 6. Kinetics of incision of substrate containing a single 8-oxoG lesion in mitochondrial extracts from human lymphoblastoid cells. 1.5 ng of a 5′-end-labeled 30-mer containing 8-oxoG was incubated at 37°C with 50 μg of mitochondrial extracts from human lymphoblastoid cell lines. Aliquots were taken at various time points, the reactions were stopped, and the products were analyzed by electrophoresis through a 20% polyacrylamide gel containing 7 M urea. The lane assignments are: (1) oligonucleotide sizing markers; (2) 0 h; (3) 1 h; (4) 2 h; (5) 3 h; (6) 4.5 h; (7) 5.5 h; (8) Fpg + oligonucleotide; and (9) control oligonucleotide + extracts.

significant contamination with nuclear enzymes (31). Western blots with rat mitochondria and crude rat liver and human cell extract confirmed that these mitochondrial extracts were free of nuclear contamination.

By incubating the mitochondrial extracts with 5′-labeled single lesion containing oligonucleotides, we have for the first time shown that hX and εA are excised from DNA in both rat and human mitochondria (unpublished). The data suggest that either nuclear 3-methyladenine DNA glycosylase can localize to mitochondria, or that a novel mitochondrial enzyme excises the lesions. We find that hX is removed more efficiently than εA from mtDNA. It has been reported that 3-methyladenine DNA glycosylase recognizes εA more efficiently than hX (38, 39), and this may suggest that 3-methyladenine DNA glycosylase is not responsible for the observed mitochondrial repair. Several alternatively spliced glycosylases containing a nuclear and mitochondria localization signal have been reported (32, 36). Two forms of 3-methyladenine glycosylase are expressed simultaneously in human cells, suggesting that one form may be mitochondrial (40). The alternative splicing may alter the substrate specificity, and this could explain the observed discrepancy between substrate preferences observed in the present study and previous observations for 3-methyladenine DNA glycosylase. Alternatively, hX and εA may be excised by a previously uncharacterized mitochondrial enzyme.

Uracil in DNA has been shown to be repaired by the short-patch BER when a uracil-containing oligonucleotide was incubated with rat mitochondrial extract (37). No long-patch repair has so far been shown to exist in mitochondria and hX is known to be repaired mainly by the long-patch PCNA-dependent pathway in nucleus (20). We are currently attempting to measure the patch size of the repair incorporation of hX and εA lesions in human mitochondria.

III. Mitochondrial DNA Repair and Aging

The molecular mechanisms that lead to aging in multicellular organisms are still subject to strong controversy. Many theories have arisen to explain the aging process, and among them, the mitochondrial theory of aging has received much attention. This theory postulates that accumulation of DNA damage and mutations in the mitochondrial genome leads to mitochondrial dysfunction and consequent cell loss (41). In postmitotic tissues, such as heart and brain, or in tissues with high energy demand, such as muscle, loss of mitochondrial function would have greater impact on their normal physiology.

Several lines of evidence support the involvement of mitochondria in the aging process. Changes in mitochondrial function with age have been observed in several organisms. Experimental data from this and many other laboratories suggest that indeed the mitochondrial genome accumulates DNA damage with age (42). The levels of 8oxoG increase in the mtDNA with age. We found 4 times more 8-oxoG in mtDNA isolated from 23-month-old rat liver mitochondria than in mtDNA isolated from 6-month-old animals (43). In contrast, no significant change was found in nuclear DNA isolated from those same animals (43).

Another aspect that has received much attention in correlation with age is alterations in DNA repair capacity. Attempts to correlate repair capacity of different organisms with maximum life span have been made. Hart and Setlow (44) demonstrated a linear correlation between the logarithm of life span and the DNA repair capacity in cells from different mammalian species, suggesting that higher DNA repair activity is associated with longer life span. Further studies showed limitations in this correlation, and there have since been numerous studies on the changes in repair with age (reviewed in Ref. 42). The results have varied and there is no consensus on this matter.

Since oxidative DNA damage accumulates in mitochondria, we have studied changes in mitochondrial DNA repair with age. Initially, we measured the activity of the rat mtODE (31). We compared mtODE activity in mitochondrial extracts obtained from livers and hearts of 6-, 12-, and 23-month-old rats. In contrast to the common notion that DNA repair decreases with age, we found an increase in mtODE activity with increasing age. In both organs, 12- and 23-months activities were significantly higher than 6 months ($p < 0.01$) (45). We also measured the activities of the mitochondrial uracil DNA glycosylase (mtUDG) and mitochondrial endonuclease G (Endo G). In both cases we found no difference with age in either heart or liver mitochondria (45). Together, these results suggested that the changes observed in mtODE activity reflected a specific upregulation of the oxidative DNA damage repair mechanisms. Similar results were obtained when we investigated mtODE activity in extracts from mouse liver mitochondria. The activity increased from 6 to 14 months of age while there was no change in the activity of mitochondrial uracil glycosylase or mtEndonuclease G.

In conclusion, our results show that mtDNA repair of 8-oxoG does not decline with age, but rather increases. Even with this induction and increase in activity of the 8-oxoG DNA glycosylase, there is still an age-dependent accumulation of this DNA base lesion. Maybe the repair activity reaches a maximum repair capacity which is still not enough to maintain a status quo in lesion frequency. This suggests that the frequency of ROS formation increases in mitochondria with age.

REFERENCES

1. B. Demple and L. Harrison, *Annu. Rev. Biochem.* **63,** 915–948 (1994).
2. M. Dizdaroglu, *Mutat. Res.* **275,** 331–342 (1992).
3. S. Shibutani, M. Takeshita, and A. P. Grollman, *Nature (London)* **349,** 431–443 (1991).
4. B. N. Ames, *Free Radical Res. Comm.* **7,** 121–128 (1989).
5. T. Lindahl, *Nature (London)* **362,** 709–715 (1993).
6. H. Ide, Y. W. Kow, and S. S. Wallace, *Nucleic Acids Res.* **13,** 8035–8052 (1985).
7. J. M. Clark and G. P. Beardsley, *Biochemistry* **26,** 5398–5403 (1987).
8. Z. Hatahet, A. A. Purmal, and S. S. Wallace, *Ann. N.Y. Acad. Sci.* **726,** 346–348 (1994).
9. H. Htun and B. H. Johnston, *Methods Enzymol.* **212,** 272–294 (1992).
10. Y. W. Kow, S. S. Wallace, and B. Van Houten, *Mutat. Res.* **235,** 147–156 (1990).
11. J. T. Reardon, T. Bessho, H. C. Kung, P. H. Bolton, and A. Sancar, *Proc. Natl. Acad. Sci. U.S.A.* **94,** 9463–9468 (1997).
12. G. Dianov, C. Bischoff, M. Sunesen, and V. A. Bohr, *Nucleic Acids Res.* **27,** 1365–1368 (1999).
13. G. L. Dianov, T. Thybo, I. Dianova, L. J. Lipinski, and V. A. Bohr, *J. Biol. Chem.* **275,** 11809–11813 (2000).
14. G. Dianov and T. Lindahl, *Curr. Biol.* **4,** 1069–1076 (1994).
15. Y. Matsumoto, K. Kim, and D. F. Bogenhagen, *Mol. Cell Biol.* **14,** 6187–6197 (1994).
16. G. Frosina, P. Fortini, O. Rossi, F. Carrozzino, G. Raspaglio, L. S. Cox, D. P. Lane, A. Abbondandolo, and E. Dogliotti, *J. Biol. Chem.* **271,** 9573–9578 (1996).
17. A. Klungland and T. Lindahl, *EMBO J.* **16,** 3341–3348 (1997).
18. G. L. Dianov, R. Prasad, S. H. Wilson, and V. A. Bohr, *J. Biol. Chem.* **274,** 13741–13743 (1999).
19. Y. Matsumoto, K. Kim, J. Hurwitz, R. Gary, D. S. Levin, A. E. Tomkinson, and M. S. Park, *J. Biol. Chem.* **274,** 33703–33708 (1999).
20. P. Fortini, E. Parlanti, O. M. Sidorkina, J. Laval, and E. Dogliotti, *J. Biol. Chem.* **274,** 15230–15236 (1999).
21. R. W. Sobol, J. K. Horton, R. Kuhn, H. Gu, R. K. Singhal, R. Prasad, K. Rajewsky, and S. H. Wilson, [published erratum appears in *Nature* **383**(6599), 457 (1996)]. *Nature (London)* **379,** 183–186 (1996).
22. S. Biade, R. W. Sobol, S. H. Wilson, and Y. Matsumoto, *J. Biol. Chem.* **273,** 898–902 (1998).
23. T. R. Waters, P. Gallinari, J. Jiricny, and P. F. Swann, *J. Biol. Chem.* **274,** 67–74 (1999).
24. R. O. Bennett, D. M. Wilson, D. Wong, and B. Demple, *Proc. Natl. Acad. Sci. U.S.A.* **94,** 7166–7169 (1997).
25. M. Dizdaroglu, "Chemistry of Free Radical Damage to DNA and Nucleoproteins." Ellis Horwood, Ltd., London, 1993.
26. A. E. Tomkinson, R. T. Bonk, and S. Linn, *J. Biol. Chem.* **263,** 12532–12537 (1998).
27. K. G. Pinz and D. F. Bogenhagen, *Mol. Cell Biol.* **18,** 1257–1265 (1998).

28. U. Lakshmipathy and C. Campbell, *Mol. Cell Biol.* **19,** 3869–3876 (1999).
29. M. J. Longley, R. Prasad, D. K. Srivastava, S. H. Wilson, and W. C. Copeland, *Proc. Natl. Acad. Sci. U.S.A.* **95,** 12244–12248 (1998).
30. S. Jinno, K. Kida, and T. Taguchi, *Mech. Age. Dev.* **85,** 95–107 (1995).
31. D. L. Croteau, C. M. J. ap Rhys, E. K. Hudson, G. L. Dianov, R. G. Hansford, and V. A. Bohr, *J. Biol. Chem.* **272,** 27338–27344 (1997).
32. M. Takao, H. Aburatani, K. Kobayashi, and A. Yasui, *Nucleic Acids Res.* **26,** 2917–2922 (1998).
33. D. Kang, J. Nishida, A. Iyama, Y. Nakabeppu, M. Furuichi, T. Fujiwara, M. Sekiguchi, and K. Takeshige, *J. Biol. Chem.* **270,** 14659–14665 (1995).
34. R. H. Stierum, D. L. Croteau, and V. A. Bohr, *J. Biol. Chem.* **274,** 7128–7136 (1999).
35. H. J. You, R. L. Swanson, C. Harrington, A. H. Corbett, S. Jinks-Robertson, S. Senturker, S. S. Wallace, S. Boiteux, M. Dizdaroglu, and P. W. Doetsch, *Biochemistry* **38,** 11298–11306 (1999).
36. H. Nilsen, M. Otterlei, T. Haug, K. Solum, T. A. Nagelhus, F. Skorpen, and H. E. Krokan, *Nucleic Acids Res.* **21,** 2579–2584 (1997).
37. R. H. Stierum, G. L. Dianov, and V. A. Bohr, *Nucleic Acids Res.* **27,** 3712–3719 (1999).
38. M. Saparbaev and J. Laval, *Proc. Natl. Acad. Sci. U.S.A.* **91,** 5873–5877 (1994).
39. M. Saparbaev, K. Kleibl, and J. Laval, *Nucleic Acids Res.* **23,** 3750–3755 (1995).
40. A. Pendlebury, I. M. Frayling, M. F. Santibanez Koref, G. P. Margison, and J. A. Rafferty, *Carcinogenesis* **15,** 2957–2960 (1994).
41. D. Harman, *J. Anti-Aging Med.* **2,** 15–36 (1999).
42. V. A. Bohr and R. M. Anson, *Mutat. Res.* **338,** 25–34 (1995).
43. E. K. Hudson, N. Tsuchiya, and R. G. Hansford, *Mech. Age Dev.* **103,** 179–193 (1998).
44. R. W. Hart and R. B. Setlow, *Proc. Natl. Acad. Sci. U.S.A.* **71,** 2169–2173 (1974).
45. N. C. Souza-Pinto, D. L. Croteau, E. K. Hudson, R. G. Hansford, and V. A. Bohr, *Nucleic Acids Res.* **27,** 1935–1942 (1999).
46. G. Dianov, C. Bischoff, J. Piotrowski, and V. A. Bohr, *J. Biol. Chem.* **273,** 33811–33816 (1998).
47. G. Dianov, A. Price, and T. Lindahl, *Mol. Cell Biol.* **12,** 1605–1612 (1992).
48. J. K. Horton, R. Prasad, E. Hou, and S. H. Wilson, *J. Biol. Chem.* **275,** 2211–2218 (2000).

Session 6
Structural Implications of BER Enzymes: Dragons Dancing—The Structural Biology of DNA Base Excision Repair

JOHN A. TAINER

The Scripps Research Institute

The combination of mythical power and potential for devastating damage implicit in the idea of dragons is somehow appropriate for DNA repair enzymes. These enzymes possess almost magical abilities for DNA damage detection, excisions, and couplings that thereby protect the treasure of our genetic integrity. Yet DNA repair enzymes implicitly have the potential to inflict terrible damage should the structural forces controlling their behavior be compromised by mutations. These evident mysteries and paradoxes are addressed by the papers in this section establishing the structural biochemistry and likely macromolecular partnerships that drive structurally coupled conformational changes, handoffs, and timing of subsequent enzymatic steps. Such keystone interactions thereby regulate and connect both repair steps and pathways. The themes that echo throughout these reports include DNA nucleotide flipping for specific damage recognition that underlie protein interactions controlling pathway and network connections. The structural biology of macromolecular complexes acting in DNA base excision repair (BER) is consequently pointing toward an emerging understanding not only of the spatial geometry linking individual repair steps, but also of a structurally encoded regulation of enzyme partnerships that ensure efficient repair of DNA damage. So the dragon dilemmas we wish to envision and resolve are not standing still, but are engaged in a choreographed dance with specific movements, steps, and partner exchanges. Although we are only beginning to see this dance at the molecular level, the combination of structural results with biochemical and genomic data is providing a rich context for the analysis of these interactions at the atomic level. The articles in this session provide detailed structural visions and insights into BER systems that we view as crucial to the field. These structures are at the heart of an emerging view of a coherent, but fluid mosaic of DNA repair networks within the cell.

The goal of understanding DNA BER at the molecular level will ultimately require addressing the full range of processes that interconnect BER steps of

DNA base damage recognition, damage removal, and repair synthesis that coordinate BER with replication, transcription, and other cell cycle processes. Yet, as more results and details of DNA BER have emerged, it has become more difficult, but also more important, to establish clear-cut general principles and concepts. The papers in this section provide a significant advance in this direction by identifying common and fundamental structural principles for enzymes initiating BER and by providing the structural framework for a unified understanding of the biochemistry, genetics, and biology needed for this field. Hypotheses that made sense in two-dimensional DNA models and in consideration of individual steps became paradoxical upon consideration of the three-dimensional DNA double-helical structure and the complex multistep chemistry required for BER. Fortunately, the experimental structural biochemistry of BER enzymes and their damaged DNA substrates, as discussed in the following papers, has proven powerful for resolving the apparent paradoxes and for addressing paramount issues in BER biology. In this overview, key conceptual issues are outlined that emerge from the structural biology of BER. Taken together, these results argue against the idea of a periodic table of discrete structural domains. Rather, the reported structures reveal a continuum of variations and elaborations upon functional and structural classes. These structures furthermore show that distinct folds and assemblies share common functional mechanisms and activities to reveal the frequent disconnection between domain fold and function. As an integrated whole, the structural biology of BER enzymes supports two unifying hypotheses: (1) DNA repair proteins function as chemomechanical devices in the detection, removal, and repair of DNA damage via protein and DNA conformational switching; (2) BER proteins interact transiently to couple repair steps and utilize cooperativity, allostery, and conformational change to accomplish self-regulation and functional coordination.

Prior to detailed three-dimensional structural results on BER enzymes, hypotheses for damaged-base recognition and excision of bases from the DNA duplex, and for coordination of the multistep repair chemistries, recalled two of Escher's depictions of geometrically impossible dragons. One such Escher dragon, which places its head and tail impossibly through its body, dramatically showed how the two-dimensional conceptual models for DNA damage detection obscured spatial and geometric impossibilities in three-dimensional structures. This first Escher-dragon dilemma thus concerned the apparent paradox that the three-dimensional DNA double helix structure sterically precludes the specific, sequence-independent, *in situ* recognition and excision of damaged bases, which was evidently required for the biological function of DNA glycosylases to initiate BER. This critical spatial paradox was only resolved satisfactorily by crystallographic structures establishing nucleotide flipping by DNA repair enzymes, as elucidated and outlined by the authors and articles within this section. The second Escher-dragon dilemma, which is depicted by a cycle of dragons emerging from a tightly knit mosaic pattern to emit a dangerous smoky blast and then reentering the mosaic of interactions, concerns enigmas in the coordination of separate steps and distinct pathways for the complex, multistep chemistry of the BER cycle. First, what protects cells from the equivalent of the fire-emitting dragons of the cycle—the dangerously cytotoxic and mutagenic abasic DNA sites created by DNA glycosylases and the subsequent single-strand DNA breaks created by the abasic site or AP endonucleases? Second, how are individual steps integrated into a coordinated repair cycle, how are the various alternative BER pathways selected, and how are these separate pieces integrated into the coherent mosaic pattern with other DNA repair pathways. The simple solution for the second Escher-dragon dilemma of assuming a multiprotein complex to regulate the conformational and temporal progression of the multistep chemistry does not agree with the biochemical data. Yet the experimental three-dimensional structures of target DNA complexes with damage excision enzymes are not only revealing how conformational changes accomplish damage recognition and excision, but are also suggesting testable hypotheses for how conformational switching couples individual enzymes and steps to coordinate and integrate pathway progression into a mosaic that avoids the release of DNA repair intermediates. Thus, the structural biology presented in this section supports the emerging view that connections are as important as chemistry for BER enzymes.

Three-dimensional structures are therefore telling us fundamental characteristics about BER enzyme biology. Compelling results suggest that evolutionary selection has optimized these enzyme structures for tight control of pathway coordination. When BER enzymes bind DNA, they funnel part of their binding energy into strained DNA conformations. The energetic cost of this strain need be paid only once and then is likely reduced by diffusion along DNA toward damaged sites that require less energy to distort, and by flipping and excising the damaged nucleotides the strain from DNA backbone compression is reduced. Strain in the BER enzyme:DNA substrate complex favors excision chemistry and product recognition that in turn avoids the release of toxic products by forming stable product complexes. Thus, BER enzyme structures suggest how their chemistry is coupled to DNA repair biology. Although we are only beginning to glimpse these connections, it is already clear from the papers in this section that the pathway context is important for understanding structure–function information. Viewed as a language translation problem, this importance becomes evident in considering a context-independent computer translation from English to Japanese and back into English. Thus for example, the meaning of the statement "out of sight, out of mind" is altered by out-of-context translation to "invisible idiot." Our understanding of the structural chemistry for DNA damage recognition and removal is most developed and detailed for BER, and we are fortunate in having four articles highlighting structural insights on this pathway.

Tom Ellenberger and colleagues review the important structural biochemistry of glycosylases that recognize and repair a variety of alkylated bases. Their results establish damage recognition and excision mechanisms for these enzymes and suggest that the DNA bending and nucleotide flipping features first identified for UDG are common themes for DNA glycosylases. Their crystallographic structures of AlkA in complex with product-mimicking azaribose-containing DNA reveal the role of the helix–hairpin–helix (HhH) motif for DNA backbone binding. These results provide a framework for understanding the recognition of alkylation damage and for the HhH motif, and strongly suggest that catalysis proceeds via a dissociative S_N1-type mechanism rather than activating a water molecule for nucleophilic attack. The flip translation model, proposed for human 3-methyladenine glycosylase, whereby DNA binding induces nucleotide flipping that is then sequentially translated along the DNA, provides an intriguing explanation for base damage detection and needs to be experimentally tested.

John Tainer and colleagues present structural results on the direct damage reversal protein O^6-alkylguanine alkyltransferase (AGT) as compared to BER enzymes UDG, APE1, Fen-1, and endonuclease IV. The stoichiometric repair of alkylated DNA by AGT is now defined by the structures of the human, bacterial, and archael proteins. These structures elucidate motifs that govern recognition and direct one-step repair of O^6-alkylguanine lesions, proving a striking counterpoint to the complex base-excision, mismatch, and nucleotide-excision repair pathways. The results of substrate destabilization by UDG support the idea that DNA glycosylases catalyze base removal by destabilization of the base–sugar bond. All DNA glycosylases seem to bind primarily one DNA strand as first established for UDG, which has prompted Tainer and colleagues to propose a backbone pinch hypothesis for damage detection as an alternative to the flip translation hypothesis. Comparisons of UDG in unbound, enzyme:substrate, and enzyme:product complexes suggests these enzymes act as chemomechanical devices so that focused strain, as well as active site functional group chemistry, acts in bond cleavage. This paper furthermore considers the structural results for the apurinic/apyrimidinic endonuclease (AP-endonuclease) families, their relationships to the DNA glycosylases, and the implications for pathway coordination at the junction between the damage-specific and damage-general steps of BER. The increased interface for the APE1:DNA complex and the significant activity enhancement of UDG by the APE1 enzyme provides a biological explanation for nature's usage of the nucleotide flipping mechanism beyond the obvious structural one. The beauty of backbone compression and nucleotide flipping is that it links specific damaged-base recognition with general AP site recognition, thus providing an elegant means of connecting the initial damage-specific enzyme with the subsequent damage-general BER enzymes. Surprisingly, these enzyme–DNA complexes provide a basis for understanding not only individual reaction steps, but also the handoff that connects steps to coordinate the BER pathway. These high-resolution structures of complexes that define activities and conformational changes along reaction pathways, such as now achieved for the human repair enzymes UDG, APE1, and endonuclease IV, show the power of multiple structures to integrate results and reveal functional relationships.

R. Stephen Lloyd and colleagues provide an in-depth understanding of the adenine-specific DNA glycosylase MutY, including a detailed structure-based mechanism for base excision of adenine mispaired with 8-oxoguanine (8-oxoG). Together, crystallographic and NMR results on MutY imply that double nucleotide flipping may explain the specificity of MutY for 8-oxoG paired to A. Notably, these studies extend the existing understanding for damage specificity to include information from the partner strand. Modeling and NMR results show that the C-terminal domain of MutY resembles the 8-oxoGTPase MutT, suggesting that this domain specifically recognizes and binds 8-oxoG in an extrahelical conformation. This cross-disciplinary work links MutY chemistry, structure, and biology and provides the prototype for successful analyses of DNA repair enzymes.

Hans Krokan and his colleagues have helped establish UDG as the archetypal BER enzyme. Structural and mutational analyses provided the first experimental evidence that damage specificity is controlled by an enzyme pocket for binding extrahelical DNA bases. The reported discoveries of alternative splicing of 5′ exons to produce UDG enzyme variants provide a basis for understanding differential behavior of BER enzymes in response to the cell's internal and external environment. Single-site mutations in UDG alter specificity from uracil to thymine and cytosine, resulting in a mutator phenotype. These informative results suggest how polymorphisms in BER enzymes may have profound public health consequences in terms of environmental risks. The relatively low impacts of knockout mice support other results that suggest multiple enzymes are capable of removing uracil from DNA and that redundancy in removal of uracil is biologically important for maintaining genome integrity. Importantly, UDG binds to both PCNA and RPA, and results from the Krokan laboratory suggest that UDG functions primarily *in vivo* to remove misincorporated uracils from A:U base pairs arising from DNA replication errors.

In terms of understanding genomic information, protein structures help provide a basis for successfully applying the comparative knowledge of the completely sequenced bacterial, archael, and yeast genomes to an understanding of the structural cell biology of DNA repair in humans. The DNA repair protein structures and their complexes are providing a critical Rosetta stone for the rational interpretation of the biological implications of DNA repair protein sequence polymorphisms. The elegance and general principles of BER structural biology originally established from the *E. coli* system have thus provided a basis to understand new results on the human enzymes. These new structural insights include conformational changes in enzyme and DNA, funneling enzyme-damaged DNA partnership interactions into chemomechanical strain for damage excision, and strain release in the cleaved product to favor abasic DNA product binding that links the enzyme chemistry to its ability to coordinate handoff within the BER pathway.

These structural biology results, which indicate that connections are as important as chemistry, prompt some key biological questions. First, do some cancers initiate or perpetuate via loss of BER enzyme specificities? The deliberate generation of mutators that reduce damage specificity by UDG suggests this is a likely possibility. Second, is the overproduction of DNA enzymes protective or biologically negative due to pathway disruption? Current genetic and biochemical results suggest that it is biologically easier to compensate for an absence by redundancy than to nullify excess or altered activities; i.e., deregulation is more destructive than deactivation for DNA repair enzymes. Third, can we use specific damaging agents and/or knockout mice to cleanly dissect the roles of individual repair enzymes and steps? The current structural views would suggest that environmental agents and genetic knockout animals represent a reprogramming of the genetic systems that can alter existing mosaics of enzyme interactions rather than truly defining the results from a single damage repair step or single enzyme change.

In general, the emerging understanding of DNA and BER enzyme structures and complexes, as described in the following papers, is taking our understanding of BER to a higher level that may prove relevant to many cellular processes requiring the coordination of complex multistep chemistry. Structural biology of BER is thus an important prototype for understanding protein interactions controlling systems in the post-genomic area. Taken together, these papers provide a fundamental understanding of DNA repair events underlying the function of DNA in the cell biology of life, especially for processes controlling genetic integrity.

Crystallizing Thoughts about DNA Base Excision Repair

THOMAS HOLLIS, ALBERT LAU, AND TOM ELLENBERGER

Department of Biological Chemistry and Molecular Pharmacology
Harvard Medical School
Boston, Massachusetts 02115

I. Introduction.. 305
II. Structures of the Human Alkyladenine
   DNA Glycosylase 307
III. Structures of *E. coli* AlkA.. 309
IV. Recognition of Alkylated Bases 311
V. Catalysis of Glycosylic Bond Cleavage 312
VI. Future Directions.. 313
   References... 313

Chemically damaged bases are removed from DNA by glycosylases that locate the damage and cleave the bond between the modified base and the deoxyribose sugar of the DNA backbone. The detection of damaged bases in DNA poses two problems: (1) The aberrant bases are mostly buried within the double helix, and (2) a wide variety of chemically different modifications must be efficiently recognized and removed. The human alkyladenine glycosylase (AAG) and *Escherichia coli* AlkA DNA glycosylases excise many different types of alkylated bases from DNA. Crystal structures of these enzymes show how substrate bases are exposed to the enzyme active site and they suggest mechanisms of catalytic specificity. Both enzymes bend DNA and flip substrate bases out of the double helix and into the enzyme active site for cleavage. Although AAG and AlkA have very different overall folds, some common features of their substrate-binding sites suggest related strategies for the selective recognition of a chemically diverse group of alkylated substrates. © 2001 Academic Press.

I. Introduction

The maintenance of DNA integrity is essential for normal cellular function and for the propagation of the genetic code to successive generations. The bases of DNA are chemically reactive and subject to a variety of modifications, so cells rely on enzymes that can accurately detect, remove, and replace damaged bases in DNA. The primary method for removing single aberrant bases from DNA is the base excision repair (BER) pathway. BER begins with a DNA N-glycosylase that locates a damaged base within a vast excess of normal DNA, then exposes the target nucleotide and cleaves the glycosylic bond. The resulting abasic nucleotide is removed by endonucleolytic cleavage and the original sequence is restored by repair DNA synthesis. Although the crystal structures of several BER glycosylases are known, we know little about how these enzyme locate damaged bases in DNA or what sequence of DNA distortions leads to capture of the damaged base in the enzyme active site.

Most BER glycosylases are highly specific and excise only one type of modified base or a few closely related modifications. The alkylation damage repair glycosylases, *E. coli* AlkA and human alkyladenine glycosylase (AAG) are unusual in catalyzing the removal of a structurally diverse group of modified bases. AlkA and AAG efficiently excise 3-methyladenine, 7-methyladenine, and 7-methylguanine from DNA (*1*), and AAG also efficiently removes 1,N^6-ethenoadenine and hypoxanthine (*2, 3*). Although most of the alkylation damage-specific glycosylases from various organisms have broadly overlapping functions, they belong to several different structural families. The *E. coli* AlkA protein is a member of the helix–hairpin–helix (HhH) family of DNA glycosylases that includes endonuclease III, MutY, and 8-oxoguanine glycosylase (OGG1) (*4*). Aside from the HhH motif and a neighboring catalytic aspartic acid, these proteins share little amino acid sequence similarity, yet their three-dimensional structures are remarkably similar (*5–8*). The *Saccharomyces cerevisiae* 3-methyladenine DNA glycosylase, Mag1, shares the conserved HhH region and catalytic aspartate and is likely to be mostly α-helical with a protein fold that is similar to AlkA's. The human functional analog of AlkA, the AAG glycosylase, belongs to a separate family consisting of the mammalian glycosylases that have a mixed α/β fold without a HhH motif (*9*) (Fig. 1). The distinctive structures of AAG and AlkA suggest that these enzymes could have very different catalytic mechanisms as well.

AAG and AlkA act on many of the same lesioned bases by a base-flipping mechanism that exposes the substrate base into the enzyme active site. First described for cytosine methyltransferases (*10, 11*), base flipping has since been reported for many different enzymes catalyzing nucleotide modifications within duplex DNA (*12*). This scheme of rotating a base out of the DNA helix and into an active site pocket exposes the substrate and allows control of substrate solvation and reaction stereochemistry. It remains to be determined if base flipping is induced by the enzyme binding to DNA or if the enzyme passively captures nucleotides that are spontaneously exposed to solvent by the transient breathing of the DNA duplex. The recent proliferation of structures and biochemical studies of DNA glycosylases and other base-flipping enzymes has provided new insights into the mechanics of DNA repair. Here we describe the structural progress made for the alkylation repair glycosylases AAG and AlkA and outline some of their mechanistic implications for this essential process.

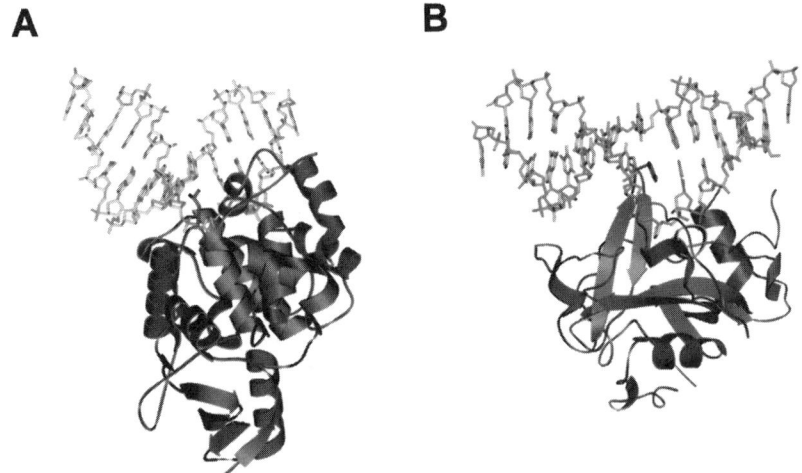

FIG. 1. Crystal structures of (A) the *E. coli* AlkA–DNA complex (17) and (B) the human AAG–DNA complex (9). Both enzymes insert residues into the minor groove of DNA and flip the nucleotide targeted for cleavage out of DNA and into the enzyme active site. However, these enzymes cause different DNA bends that expose substrate nucleotides, and the active site structures suggest different mechanisms for catalysis of glycosylic bond cleavage.

II. Structures of the Human Alkyladenine DNA Glycosylase

The crystal structure of AAG was initially determined in complex with DNA containing the pyrrolidine abasic nucleotide (*pyr*), which is a transition-state mimic that potently inhibits glycosylase activity (9, 13) (Fig. 1). When complexed to AAG, the *pyr* is flipped out of the DNA duplex and into a pocket on the enzyme surface. Tyr162 inserts into the minor groove of the DNA where its side chain fills the space vacated by the flipped-out nucleotide and acts as a surrogate base that presumably stabilizes the nucleotide targeted for repair in an extrahelical conformation. A water molecule bound in the enzyme active site is positioned for an in-line attack on the flipped-out nucleotide, suggesting that the water directly displaces the damaged base.

The AAG fragment that was crystallized lacks residues 1–79, yet it retains full enzymatic activity and specificity for alkylated bases. Correspondingly, the analogous truncation mutant of the homologous mouse protein, methylpurine glycosylase (MPG), has enzymatic properties that are similar to those of full-length MPG (14). The AAG protein consists of a single domain with a mixed α/β structure that does not closely resemble any other protein in the Protein Data Bank. Clusters of basic residues on the protein surface flank the flipped-out *pyr* and anchor the DNA across AAG's substrate-binding pocket. A β-hairpin jutting out from this relatively flat DNA-binding surface inserts the Tyr162 side chain into the minor groove of the DNA. AAG acts only on duplex DNA, and both DNA strands receive a similar number of contacts from the protein. On the whole, the bound DNA is a B-form helix that bends away from the protein by an angle of about 22°. This limits the surface area of the protein–DNA interaction, which together with the absence of sequence-specific interactions could facilitate sliding of the protein along DNA in its search for substrate (15).

The pyrrolidine abasic nucleotide is rotated out of the DNA double helix into the protein active site. Here a bound water molecule is inserted between the *pyr* N4' position and the side chains of Glu125 and Arg182 and the main-chain carbonyl of Val262 through hydrogen-bonding interactions. This hydrogen-bonding network is likely to be important for tight binding of the *pyr*-containing DNA to AAG (13). The position of the water with respect to the flipped-out *pyr* and its interaction with Glu125 strongly imply that Glu125 is the general base that deprotonates water for attack of the *N*-glycosylic bond. The hydrogen bonding between Arg182 and the water might assist nucleophilic attack of the nucleotide by stabilizing an incipient hydroxide ion. The recently determined crystal structure of AAG complexed to the DNA containing the bulky alkylation adduct

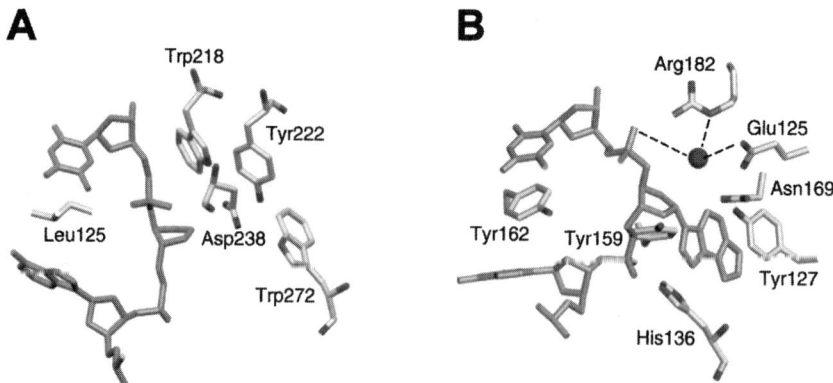

FIG. 2. A comparison of the active sites of AlkA and AAG. (A) In the AlkA–DNA complex, the 1-azaribose is flipped out of the DNA helix into the active site such that the positively charged nitrogen of 1-azaribose contacts the carboxylate of the Asp238, a catalytically essential residue. Trp218 shields the substrate from solvent and it leaves no room on the backside of the flipped-out ribose for a nucleophilic water. (B) In the AAG–εA complex, the εA is flipped into the protein active site, where it stacks snugly between Tyr127 on one side and His136 and Tyr159 on the other. Glu125 is well positioned to deprotonate a bound water (red sphere) for a backside attack of the *N*-glycosylic bond.

1,N^6-ethenoadenine (εA) (*15a*) shows a similarly bound water and interactions of the εA base with the active site. The εA base stacks between Tyr127 on one side and His136 and Tyr159 on the other (Fig. 2). A combination of hydrogen-bonding interactions and the steric constraints of the base binding pocket could account for AAG's low catalytic activity toward unmodified purines and its preference for some alkylated substrates. However, this initial structure does not readily explain the efficient cleavage of inosine and methylated purines by AAG. Additional structural and biochemical characterizations with other preferred substrates might reveal alternative binding conformations that accommodate different substrates in the active site.

III. Structures of *E. coli* AlkA

Crystal structures of *E. coli* AlkA (*5, 16*) reveal a compact globular protein consisting of three subdomains (Fig. 1A). An amino-terminal subdomain (residues 1–88) forms a mixed α–β structure that is similar to that of the TATA-binding protein, but it plays no direct role in AlkA binding to DNA (*17*). The middle (residues 113–230) and carboxyl-terminal (residues 231–282) subdomains of AlkA are α-helical bundles that are structurally homologous to MutY (*6*; see also House *et al.*, this volume) and Endo III (*18*), despite very limited sequence homology. The conserved HhH motif of all three proteins is located at the interface of these subdomains, adjacent to a catalytically essential aspartic acid (Asp238 in AlkA) in the enzyme active site. The residues lining the interdomain cleft are different in all three enzymes, probably reflecting their distinctive substrate preferences. AlkA lacks the iron–sulfur cluster located near the carboxyl termini of Endo III and MutY, where it presumably contacts the DNA.

The recent crystal structure of an AlkA–DNA complex has provided the first glimpse of the mode of DNA binding by a member of the HhH superfamily, as well as some insight into AlkA's base recognition and catalytic mechanism (*17*). AlkA induces a sharp bend (66°) in DNA centered around the flipped-out 1-azaribose abasic nucleotide, markedly widening the minor groove (Fig. 1). A combination of polar and nonpolar interactions involving several loops near the enzyme active site stabilizes the DNA in a distorted conformation. These loops form a "wedge" that is inserted into the minor groove surface of the DNA (Fig. 2). At the tip of the wedge is the side chain of Leu125, which occupies the position of the flipped-out nucleotide like the analogous residue of AAG, Tyr162. Most of AlkA's DNA interaction surface is nonpolar, except for a few polar contacts made by the HhH motif that anchors the DNA to the protein on the 3' side of the flipped-out nucleotide. Residues 202–227, which form the HhH motif of AlkA, provide several hydrogen bonds and a metal-mediated interaction with the DNA. The metal ion, modeled as sodium, is coordinated by the main-chain carbonyl oxygens of residues 210, 212, and 215 in the hairpin of the HhH, a DNA phosphate oxygen and a water molecule (17). The metal is not stably bound in the absence of DNA, as evidenced by the fact that no metal is detected in the high-resolution crystal structures of unliganded AlkA (5, 16), Endo III (8, 18), or MutY (6). Similar DNA binding motifs have been seen in enzymes such as DNA polymerase β (19) and more recently in interferon regulatory factor (IRF) (20). Polymerase β has two HhH domains that are instrumental in binding to DNA. These HhH motifs ligate a metal ion only while DNA is bound and prefer monovalent sodium and potassium over divalent calcium and magnesium. In the IRF crystal structure, a related helix–hairpin–strand motif was found to ligate a potassium ion in its interaction with DNA. A number of other DNA binding proteins are predicted to contain this motif (19). This method of DNA interaction has probably evolved as an efficient platform for tight binding, nonspecific interaction with duplex DNA. Furthermore, the AlkA–DNA structure shows that the HhH motif serves as a stable support for the protein-induced DNA distortions, but it does not directly participate in the flipping of a nucleotide substrate.

In the structure of the AlkA–DNA complex (17), the 1-azaribose is flipped-out of the DNA helix and into the enzyme active site. The positively charged nitrogen of the 1-azaribose is situated about 3.2 Å from the carboxylate of Asp238, the catalytically essential residue. Leu125 is pushed into the minor groove where it occupies the position of the flipped-out nucleotide. The geometry of the DNA distortion is quite different from that seen in the human uracil DNA glycosylase (UDG) (21, 22). UDG creates a "pinch" in the DNA backbone, compressing the inter-phosphate distance around the flipped-out base, thereby producing a zig-zag conformation in the phosphodiester backbone (21). In the AlkA–DNA complex, the phosphate backbone around the 1-azaribose is fully extended and unlike the DNA distortion induced by UDG.

AlkA, Endo III, and MutY have remarkably similar protein folds despite their minimal sequence similarity (5, 6, 8, 16, 18). Conservation of the HhH motif, as well as overall structure, implies that these proteins may interact with DNA in a similar fashion. Superposition of the DNA from the AlkA–DNA complex onto the unliganded structures of Endo III and MutY reveals that the proposed DNA-binding surfaces interact with the DNA without any major steric clashes and contain a similar nonpolar chemistry as the AlkA-binding surface (17). The residues of MutY (Gln42) and Endo III (Gln41) that have been proposed to be involved in base flipping (6) are properly positioned to intercalate into the DNA in these modeled complexes.

IV. Recognition of Alkylated Bases

Whereas most BER glycosylases are specific for a particular type of abnormal base, AlkA and AAG recognize and excise chemically and structurally diverse group of alkylated bases from DNA. The basis for this broad, yet well defined substrate specificity is not completely understood. Electron-deficient alkylated bases might be recognized and targeted for excision from DNA by virtue of their positive charge (5, 9, 23, 24). Positively charged bases might stack more tightly against the aromatic residues of the active site than uncharged normal bases, thereby allowing the enzyme to distinguish alkylated bases from normal bases. Superposition of a 3-methyladenine nucleoside substrate on the flipped-out 1-azaribose abasic nucleotide in the active site of AlkA shows that the 3-methyladenine base stacks against Trp272 and that C1′ of the ribose is adjacent to Asp238. This stacking arrangement of a positively charged substrate against the aromatic tryptophan constitutes a cation–π interaction that could stabilize the extrahelical conformation of the substrate base. The open architecture of the binding site, identified by modeling, allows ample room to accommodate all alkylated substrate bases. Interestingly, the modeled base makes no detectable hydrogen-bonding interactions with the AlkA active site, suggesting that the enzyme's activity toward alkylated substrates may result from factors other than binding to a modified base. Additionally, AlkA's indifference to the base opposite of the flipped-out nucleotide is readily explained by the lack of any direct protein contacts to that base.

In contrast to AlkA, some of the preferred substrates of the human AAG glycosylase are not positively charged, e.g., εA and inosine, suggesting that catalytic specificity is achieved by a different mechanism. The AAG active site accommodates the εA very snugly between Tyr127 on one side and His136 and Tyr159 on the other (15a). This complementarity does not result from an induced fit because the conformation of the base binding pocket is the same when complexed to DNAs containing the εA substrate base or the abasic *pyr* inhibitor. An additional 17 Å^2 of surface area is buried by the εA substrate in comparison to an adenine modeled in the active site. This additional van der Waals contact surface might favor binding of the alkylated substrate in preference to normal bases. Also, a hydrogen bond between the main-chain amide of His136 and N^6 of εA would not form with unmodified adenine. The N^6 of a normal adenine could not accept this bond and would instead be repelled. AAG might therefore preferentially bind εA over an unmodified adenine by selective hydrogen bond formation and additional van der Waals interactions. How is a modified guanine recognized as being different from a normal guanine? The O^6 of a normal guanine, when superimposed in the AAG active site, could accept the hydrogen bond from the main-chain amide of His136 like the N^6 of εA, but the guanine's

N^2 would create a steric clash with Asn169, resulting in a repulsive interaction that might cause the guanine to be expelled from the active site. The positive charge, however, on a substrate base such as 7-methylguanine might cause the base to be pulled into the active site to satisfy cation–π interactons with Tyr127 and His136 strongly enough so as to push aside Asn169 (23, 24). This hypothesis has yet to be tested. Like guanine, the O^6 of an inosine substrate could accept the hydrogen bond from the main-chain amide of His136, but inosine lacks an N^2, allowing the base to fit comfortably in the active site.

V. Catalysis of Glycosylic Bond Cleavage

The proposed transition state for hydrolysis of N-glycosidic bonds by the well studied nucleoside hydrolases involves a positively charged intermediate localized to the C1′–O4′ region of the ribose sugar (25). The pyrrolidine and 1-azaribose abasic nucleotides, which have a positively charged N4′ or N1′, respectively, were designed as transition state mimics of the analogous intermediate for the DNA N-glycosylase reaction. DNAs containing these positively charged abasic nucleotides bind very tightly to AAG and to AlkA (13, 26). The observed interactions between the anomeric position of these designed inhibitors and the glycosylase active site are suggestive of a possible mechanism of glycosylic bond cleavage.

In the AAG–εA complex, a bound water molecule is positioned in the active site by interaction with Glu125 about 3.1 Å from the N4′ of the pyrrolidine (15a). This binding geometry is nearly ideal for an in-line displacement of the glycosylic bond via backside attack of the ribose (9) (Fig. 2). Glu125 is likely to deprotonate the water to create the hydroxyl nucleophile that attacks the C1′ of a substrate nucleotide to release the base.

It was assumed that AlkA's catalytic mechanism would involve a similar direct displacement mechanism, with activation of a water nucleophile by the conserved residue Asp238 (5, 13, 27). However, the crystal structure of the AlkA–DNA complex reveals there is no room on the backside of the flipped-out ribose for a water nucleophile. Instead, the anomeric N1′ position of the 1-azaribose directly contacts the carboxylate oxygen of Asp238 and it is further shielded from solvent by Trp218, a geometry that is inconsistent with an S_N2 mechanism of glycosylic bond cleavage (17). The position of the flipped-out 1-azaribose is constrained by numerous contacts to the flanking phosphates and van der Waals interactions with the abasic sugar, suggesting any change in orientation would require a significant adjustment of the protein's DNA-binding surface. An alternative, S_N1, mechanism is consistent with the crystal structure of AlkA (17). AlkA's positively charged substrate bases are favorable leaving groups because they have destabilized glycosylic bonds. The close juxtaposition of the

Asp238 side chain to C1′ of the substrate deoxyribose might stabilize a carbocation sugar intermediate formed by loss of the alkylated base, either through a charge–charge interaction or by the transient formation of a covalent intermediate. AlkA's catalytic selectivity might result from the favorable binding of alkylated substrates to the enzyme active site and the inherent instability of the glycosylic bond of alkylated nucleotides (28).

VI. Future Directions

Recent crystal structures of several DNA glycosylases have offered clues about how abnormal bases in DNA are recognized and possible mechanisms for catalytic selectivity of glycosylic bond cleavage. Lingering questions remain about the initial steps of locating damaged bases in a vast excess of normal DNA and the mechanism of extruding bases into the enzyme active site by base flipping. The highly distorted structures of DNAs complexed to several different repair glycosylases support the notion that base flipping is an active process, induced by binding to the DNA glycosylase. Several models for the process of locating DNA damage and base flipping have been extrapolated from the crystal structures (9, 15, 21, 22, 29). However, more experimental tests of this dynamic process are needed to identify the general principles and the enzyme-specific mechanisms for cleansing chemical damage from the genome.

Acknowledgments

We thank our many colleagues and collaborators who have taught us about the intricacies of DNA damage and repair. A special thanks to Yoshi Ichikawa, Leona Samson, Orlando Schärer, and Greg Verdine for championing our efforts to grow crystals. Our work on DNA repair enzymes has been supported by grants from the National Institutes of Health, the Lucille P. Markey Foundation, and the Armenise-Harvard Foundation for Advanced Biomedical Research.

References

1. B. Singer and B. Hang, *Chem. Res. Toxicol.* **10**, 713–732 (1997).
2. M. Saparbaev, K. Kleibl, and J. Laval, *Nucleic Acids Res.* **23**, 3750–3755 (1995).
3. M. Saparbaev and J. Laval, *Proc. Natl. Acad. Sci. U.S.A.* **91**, 5873–5877 (1994).
4. H. M. Nash, S. D. Bruner, O. D. Schärer, T. Kawate, T. A. Addona, E. Spooner, W. S. Lane, and G. L. Verdine, *Curr. Biol.* **6**, 968–980 (1996).
5. J. Labahn, O. D. Schärer, A. Long, K. Ezaz-Nikpay, G. L. Verdine, and T. E. Ellenberger, *Cell (Cambridge, Mass.)* **86**, 321–329 (1996).

6. Y. Guan, R. C. Manuel, A. S. Arvai, S. S. Parikh, C. D. Mol, J. H. Miller, S. Lloyd, and J. A. Tainer, *Nat. Struct. Biol.* **5,** 1058–1064 (1998).
7. S. D. Bruner, D. P. Norman, and G. L. Verdine, *Nature (London)* **403,** 859–866 (2000).
8. M. M. Thayer, H. Ahern, D. Xing, R. P. Cunningham, and J. A. Tainer, *EMBO J.* **14,** 4108–4120 (1995).
9. A. Y. Lau, O. D. Schärer, L. Samson, G. L. Verdine, and T. Ellenberger, *Cell (Cambridge, Mass.)* **95,** 249–258 (1998).
10. S. Klimasauskas, S. Kumar, R. J. Roberts, and X. Cheng, *Cell (Cambridge, Mass.)* **76,** 357–369 (1994).
11. K. M. Reinisch, L. Chen, G. L. Verdine, and W. N. Lipscomb, *Cell (Cambridge, Mass.)* **82,** 143–153 (1995).
12. R. J. Roberts and X. Cheng, *Annu. Rev. Biochem.* **67,** 181–198 (1998).
13. O. D. Schärer, H. M. Nash, J. Jiricny, J. Laval, and G. L. Verdine, *J. Biol. Chem.* **273,** 8592–8597 (1998).
14. R. Roy, T. Biswas, T. K. Hazra, G. Roy, D. T. Grabowski, T. Izumi, G. Srinivasan, and S. Mitra, *Biochemistry* **37,** 580–589 (1998).
15. G. L. Verdine and S. D. Bruner, *Chem. Biol.* **4,** 329–334 (1997).
15a. A. Y. Lau, M. D. Wyatt, B. J. Glassner, L. D. Samson, and T. E. Ellenberger, *Proc. Natl. Acad. Sci. U.S.A.* **97,** 13573–13578 (2000).
16. Y. Yamagata, M. Kato, K. Odawara, Y. Tokuno, Y. Nakashima, N. Matsushima, K. Yasumura, K. Tomita, K. Ihara, Y. Fujii, Y. Nakabeppu, M. Sekiguchi, and S. Fujii, *Cell (Cambridge, Mass.)* **86,** 311–319 (1996).
17. T. Hollis, Y. Ichikawa, and T. Ellenberger, *EMBO J.* **19,** 758–766 (2000).
18. C. F. Kuo, D. E. McRee, C. L. Fisher, S. F. O'Handley, R. P. Cunningham, and J. A. Tainer, *Science* **258,** 434–440 (1994).
19. H. Pelletier and M. R. Sawaya, *Biochemistry* **35,** 12778–12787 (1996).
20. Y. Fujii, T. Shimizu, M. Kusumoto, Y. Kyogoku, T. Taniguchi, and T. Hakoshima, *EMBO J.* **18,** 5028–5041 (1999).
21. S. S. Parikh, C. D. Mol, G. Slupphaug, S. Bharati, H. E. Krokan, and J. A. Tainer, *EMBO J.* **17,** 5214–5226 (1998).
22. G. Slupphaug, C. D. Mol, B. Kavli, A. S. Arvai, H. E. Krokan, and J. A. Tainer, *Nature (London)* **384,** 87–92 (1996).
23. J. P. Gallivan and D. A. Dougherty, *Proc. Natl. Acad. Sci. U.S.A.* **96,** 9459–9464 (1999).
24. G. Hu, P. D. Gershon, A. E. Hodel, and F. A. Quiocho, *Proc. Natl. Acad. Sci. U.S.A.* **96,** 7149–7154 (1999).
25. V. L. Schramm, B. A. Horenstein, and P. C. Kline, *J. Biol. Chem.* **269,** 18259–18262 (1994).
26. K. Makino and Y. Ichikawa, *Tetrahedron Lett.* **39,** 8245–8248 (1998).
27. B. Sun, K. A. Latham, M. L. Dodson, and R. S. Lloyd, *J. Biol. Chem.* **270,** 19501–19508 (1995).
28. K. G. Berdal, R. F. Johansen, and E. Seeberg, *EMBO J.* **17,** 363–367 (1998).
29. T. E. Barrett, R. Savva, G. Panayotou, T. Barlow, T. Brown, J. Jiricny, and L. H. Pearl, *Cell (Cambridge, Mass.)* **92,** 117–129 (1998).

DNA Damage Recognition and Repair Pathway Coordination Revealed by the Structural Biochemistry of DNA Repair Enzymes

DAVID J. HOSFIELD, DOUGLAS
S. DANIELS, CLIFFORD D. MOL,
CHRISTOPHER D. PUTNAM,
SUDIP S. PARIKH, AND
JOHN A. TAINER

Department of Molecular Biology
Skaggs Institute for Chemical Biology
The Scripps Research Institute
La Jolla, California 92037

I. DNA Damage Reversal and Avoidance	316
A. O^6-Alkylguanine-DNA Alkyltransferase	316
B. Deoxyuridine Triphosphate Pyrophosphatase	319
II. DNA Base Excision Repair Glycosylases	323
A. Helix–Hairpin–Helix Glycosylases/AP Lyases	324
B. Uracil-DNA Glycosylase	324
III. 5′ Apurinic/Apyrimidinic (AP) Endonucleases	328
A. APE1	329
B. Endonuclease IV	335
IV. FEN-1 and Long-Patch Base Excision Repair	338
V. Perspectives and Implications	342
References	342

Cells have evolved distinct mechanisms for both preventing and removing mutagenic and lethal DNA damage. Structural and biochemical characterization of key enzymes that function in DNA repair pathways are illuminating the biological and chemical mechanisms that govern initial lesion detection, recognition, and excision repair of damaged DNA. These results are beginning to reveal a higher level of DNA repair coordination that ensures the faithful repair of damaged DNA. Enzyme-induced DNA distortions allow for the specific recognition of distinct extrahelical lesions, as well as tight binding to cleaved products, which has implications for the ordered transfer of unstable DNA repair intermediates between enzymes during base excision repair. © 2001 Academic Press.

The viability and genomic stability of all cells are jeopardized by DNA base damage. Oxidation, deamination, alkylation, depurination, depyrimidination, and even normal cellular metabolic processes, generate ~10,000 mutagenic DNA base lesions per day in each human cell (1–3). These damaged bases can block DNA polymerase progression and halt DNA replication (4, 5), thus interfering with cell-cycle progression. When the amount of DNA damage is exceedingly high, cells induce error-prone DNA polymerases that can synthesize DNA past blocking lesions (6, 7). Such translesion synthesis is inherently mutagenic as the identity of the inserted base cannot be derived without correct base-pairing interactions with template nucleotides. This strategy to continue replication in the presence of DNA damage predisposes the cell to an increased mutation rate, and underscores the need for efficient DNA repair processes to circumvent the need for error-prone polymerases.

In recent years, our understanding of the cellular mechanisms underlying different DNA repair pathways has benefited from combined biochemical, mutagenesis, and structural studies on enzymes involved in the individual processes. Enzyme networks ensure the maintenance of cellular nucleotide pools to safeguard pathways for short-patch damaged base excision repair (BER) and long-patch nucleotide excision repair (NER), while cells also employ strategies for the highly efficient transcription-coupled repair (TCR) of oxidized bases as well as the direct repair of damaged DNA bases. New results are providing an integrated view for how DNA repair is coordinated within cells.

I. DNA Damage Reversal and Avoidance

Conceptually, the simplest form of DNA repair is the direct removal of the damaged moiety from the DNA base. Although in bacterial systems enzymes such as photolyase can directly reverse thymine dimers in DNA to restore two adjacent thymine bases (8), the only known human protein that directly reverses DNA damage is O^6-alkylguanine-DNA alkyltransferase (AGT; also called O^6-methylguanine-DNA methyltransferase, or MGMT). AGT suicidally repairs O^6-alkylguanine lesions in DNA, which result from endogenous sources such as S-adenosylmethionine (9) and environmental toxins (10). These lesions are mutagenic, causing GC→AT transition mutations (11), and cytotoxic, forming the basis of anticancer chemotherapies that involve DNA methylation or chloroethylation. Hence, increased AGT levels confer alkylation resistance in human tumors, (12) and inhibitors of AGT, such as O^6-benzylguanine, are potential adjuvants for these antineoplastic therapies (13).

A. O^6-Alkylguanine-DNA Alkyltransferase

The structure of human AGT (14) revealed a two-domain α/β protein (Fig. 1A) with an N-terminal α/β roll containing a tetrahedral Cys_2His_2

Zn-binding site that stabilizes a short helix bridging the two domains. The C-terminal domain contains a conserved active-site cysteine motif that is coupled to a nonspecific, helix–turn–helix (HTH) DNA-binding motif via two Asn-stabilized tight turns dubbed the "Asn-hinge" (14). This overall fold is highly conserved in Zn-bound (14) and apo (15) human forms, as well as in bacterial (16) and archaebacterial (17) AGTs (18).

Structures of methylated and benzylated AGT, obtained by the reaction of AGT with the biologically relevant O^6-methylguanine and therapeutically relevant O^6-benzylguanine substrates, definitively establish the alkyl-binding pocket and reveal the structural basis for O^6-alkylguanine recognition (14). The benzyl group, which is covalently bound to the nucleophilic Cys, stacks between conserved amino acids (Fig. 1B), and has an edge-on hydrophobic interaction with an aromatic Tyr side chain. This product complex suggests a model for guanine base binding that brings the guanine N3 within hydrogen-bond distance of a key Tyr hydroxyl, and positions the exocyclic amino and O^6 groups for hydrogen bonding with backbone carbonyl and nitrogen atoms.

DNA binding to AGT can be modeled with the recognition helix of the HTH motif inserted into the DNA major groove (Fig. 1C). Several non–hydrogen-bonding residues on the DNA interacting face of this helix mediate sequence-independent DNA recognition while the N-terminal residues of two adjacent α helices are poised to interact with the phosphate backbone. Additionally, an AGT β-turn, which is similar in structure to a winged-HTH DNA-binding motif (19), likely interacts with the minor groove. Like other DNA repair enzymes, the DNA binding surface is positively charged to provide electrostatic complementarity for the negatively charged DNA phosphodiester backbone (14). Importantly, this inferred DNA-binding mode places a key recognition-helix Arg within the DNA base stack (Fig. 1C) and suggests that AGT employs an "arginine finger" to extrude target O^6-alkylguanine nucleotides from duplex DNA. As O^6-alkyl lesions occupy the major groove, Arg insertion into the major groove could promote the extrusion of O^6-alkylguanines through the minor groove and into the enzyme active site. The gap left by the resultant extrahelical nucleotide would be stabilized by the intercalated Arg side chain. Structural and mutational data (14) support an active role of the Arg finger in catalysis and are not consistent with its having a role in general phosphate recognition.

The AGT active site displays an extensive hydrogen-bond network that is not only critical for protein stability (20), but is also ideally poised to drive the dealkylation reaction (Fig. 1B). His146, acting as a water-mediated general base, deprotonates Cys145 which directly attacks the alkylated base in an irreversible and stoichiometric reaction. The resulting dealkylated base is likely protonated by Tyr114 at the N3 position, and this is consistent with both mutagenesis data (21) and biochemical results that show a decreased reaction rate when N3 is substituted with carbon (22).

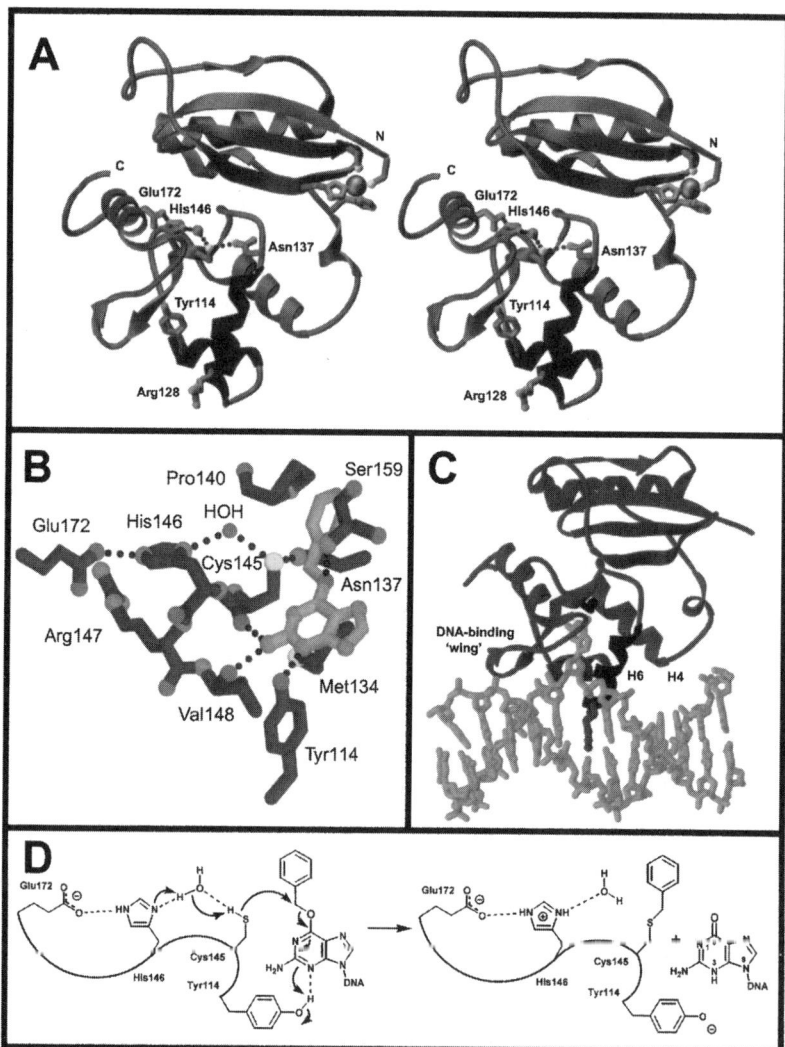

FIG. 1. Human AGT secondary structure, active site, inferred DNA-binding mode and mechanism. (A) Overall AGT fold, displaying the active-site cysteine and its surrounding hydrogen-bond network, Zn-binding site, and Arg128 "arginine finger." A central interdomain cleft separates the N-terminal α/β roll from the C-terminal domain, which contains the active site and helix–turn–helix (HTH) DNA-binding motif. (B) AGT active site (dark gray) and O^6-benzylguanine (light gray) binding model, inferred from the O^6-benzylguanine product complex. Guanine inserts into the active site channel, stacking against Met134. Guanine-specific hydrogen bonds occur between the Watson–Crick base-pairing atoms of guanine and protein main-chain atoms. The benzyl lesion packs against Pro140 and is oriented for optimal nucleophilic attack by Cys145. (C) Inferred DNA-binding mode. By analogy to homologous HTH DNA-binding proteins, AGT inserts the recognition helix (H6) into the major groove of DNA, with the adjacent DNA-binding "wing" contacting the minor groove.

After completing the dealkyation reaction, AGT is inactivated and shows an increased Stokes radius (23), an increased susceptibility to proteolysis (24), and is rapidly degraded in HT29 cells and cell-free extracts (25, 26). These results, and comparison of the AGT structures before and after alkylation (14), suggest that disruption of the active-site hydrogen-bond network and a sterically driven helix displacement result in a conformational change that promotes release of repaired DNA and *in vivo*, protein degradation. The methyl and benzyl adducts of the two product complexes lie in close van der Waals contact (3.2 and 3.0 Å, respectively) with the H6 recognition helix, and movement of this helix away from the active site relative to the native structure partially relieves the strain from this close contact. Since the recognition helix is expected to bind the DNA major groove, this movement may facilitate release of repaired DNA. Additionally, as alkylation of AGT disrupts the active-site hydrogen-bonding network which is crucial for protein stability (20), the dealkylation reaction also initiates the cellular turnover of inactivated enzyme.

B. Deoxyuridine Triphosphate Pyrophosphatase

A key enzyme that protects cells from undergoing destructive cycles of futile uracil BER is deoxyuridine triphosphate pyrophosphatase (dUTPase; also called deoxyuridine triphosphatase, or dUTP nucleotidohydrolase). Although the enzyme is not classically defined as a DNA repair enzyme, dUTPase converts dUTP, which arises from normal cellular RNA metabolism, to dUMP, which is the precursor for *de novo* dTTP biosynthesis (Fig. 2). This essential enzyme (27, 28) maintains DNA replication fidelity since DNA polymerase cannot discriminate between dTTP and dUTP when base-pairing with template adenine (29). When the concentration of dUTP in cells is high, DNA polymerase misincorporation creates A·U base pairs that disrupt DNA transcription sequences (30–32) and poison the uracil BER pathway. Uracil excision from A·U base pairs by uracil-DNA glycosylase (UDG) leads to destructive cycles of uracil misincorporation by DNA polymerase and futile cycles of replicative repair, which ultimately cause cell death (33–36). Thus, the cellular localization and gene expression of dUTPase and UDG are closely linked. As with human UDG (37, 38), human dUTPase is localized to the nucleus and mitochondria (39), with both isoforms generated from the same gene through splicing of alternate 5′ exons (40), and dUTPase gene expression is tied to DNA replication (40).

The biological function of dUTPase depends upon its exquisite specificity for dUTP, with no other deoxy- and ribonucleotide triphosphates acting as

---

Arg128, extending from the recognition helix, penetrates the DNA base stack and is ideally positioned to facilitate flipping of target O^6-alkylguanine nucleotides. (D) Proposed reaction mechanism. His146, acting as a water-mediated general base, deprotonates Cys145 to facilitate attack at the O^6-alkyl carbon atom. Repaired guanine is then released following protonation by the hydrogen-bond donor Tyr114.

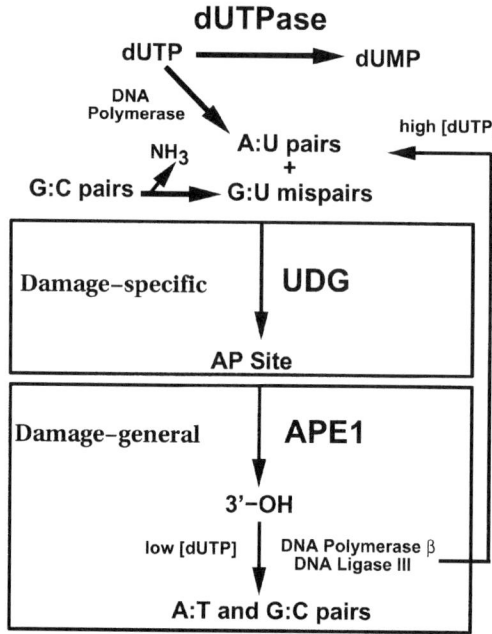

FIG. 2. dUTPase safeguards uracil base excision repair. The normal RNA base uracil arises in DNA through misincorporation by DNA polymerase, giving A:U pairs, or via cytosine deamination, to create promutagenic G:U wobble mispairs. Uracil-DNA glycosylase (UDG) cleaves the uracil base from both A:U pairs and G:U mispairs to create an abasic, or AP, site. AP sites are recognized by 5′ AP-endonucleases, and the DNA phosphodiester backbone is cleaved 5′ to create a suitable 3′-OH template for repair synthesis by DNA polymerase. dUTPase safeguards this DNA repair pathway by maintaining the cellular dUTP/dTTP ratio at a low level by hydrolyzing dUTP to dUMP and pyrophosphate. As dUTP is a normal by-product of cellular RNA metabolism, cells without dUTPase have high dUTP concentrations and experience repetitive cycles of futile DNA repair with uracil misincorporation followed by excision, leading to multiple DNA strand breaks that ultimately lead to cell death.

substrates. The structure of dUTPase is conserved from bacteria (41, 42) to viruses (43, 44) to man (45), studies of which reveal that dUTPase is a homotrimer with each subunit ~150 residues long and with conserved amino acid residues clustered at the domain interfaces. Each dUTPase subunit is a β-barrel with the carboxy-terminal β-strand interchanged, or domain-swapped (46), among the three subunits to form an interlocked, propeller-shaped trimer (Fig. 3A). Nucleotide binding (44, 45, 47) confirms that the biological dUTPase trimer has three active sites, each composed of amino acids contributed by all three different subunits, with the exchanged glycine-rich carboxy termini forming flexible tails that cap the substrate-bound active sites (Fig. 3A).

A distorted β-hairpin within each dUTPase subunit forms a pocket that specifically recognizes the uracil base and discriminates against ribonucleotides (Fig. 3B). In human dUTPase, the uracil Watson–Crick base-pairing atoms form hydrogen bonds to enzyme polypeptide backbone atoms and a bound water molecule, effectively excluding cytosine and thymine binding (Fig. 3C). Ribonucleotide binding is precluded by the close approach of the deoxyribose 2′ position to the aromatic ring of a conserved tyrosine residue in the uracil-recognition β-hairpin. Truncation of this tyrosine to alanine by site-directed mutagenesis abolishes dUTPase specificity for uracil deoxyribonucleoside (48, 49).

While uracil recognition is mediated by the distorted β-hairpin of one subunit, the carboxy-terminal tail from another subunit forms a "phenylalanine lid" that caps the substrate-bound active sites (45). In these tails, a conserved phenylalanine residue stacks with the bound uracil, an arginine contacts the dUTP α-phosphate, and conserved glycine residues interact with the phosphates through their backbone nitrogens (Fig. 3C). The flexible tails are important for dUTPase activity, with carboxy-terminal truncated dUTPase enzymes strongly impaired, showing a 40-fold decrease in k_{cat} and a doubling of K_M (50).

dUTPase interactions with the phosphate groups and metal ions are via conserved residues of the third subunit of the trimer (Fig. 3C). Oxygen-isotope experiments reveal that dUTPase catalysis takes place by a nucleophilic attack of a water molecule on the dUTP α-phosphate, with the oxygen bound to the dUMP reaction product (47). The position of a conserved aspartic acid in each of the dUTPase active sites in the trimer suggests its role in catalysis. When the active sites are capped by the phenylalanine lids, the negatively charged aspartic acid carboxylate is buried from bulk solvent. Protonation of this aspartic acid creates the hydroxyl nucleophile that attacks the dUTP α-phosphate, with the pyrophosphate leaving group stabilized by interactions with the metal ions and conserved arginine residues (Fig. 3D).

The purpose of evolving a complex interlocked trimer for dUTP hydrolysis may derive from the need to overcome product inhibition, highlighting the biological necessity of maintaining low cellular dUTP concentrations. As substrate dUTP and Mg^{2+} ions bind to any particular active site, there is a slight closure of the adjacent subunits and ordering of the C-terminal tail of the third subunit, which likely opens the other active sites of the trimer. At a dUMP-bound active site after bond cleavage, the pyrophosphate–metal ion moiety likely dissociates (51, 52), thereby facilitating the uncapping of the active site by the C-terminal tail. The flexible tails make little specific contact with the enzyme; thus, by uncapping the active site, they can pluck the bound dUMP out of the pocket through the interactions of the phenylalanine and arginine with the uracil base and dUMP phosphate, respectively. In this coordinated manner, the dUTPase biological trimer may have evolved an effective means of not only overcoming product inhibition by dUMP, but also ensuring that

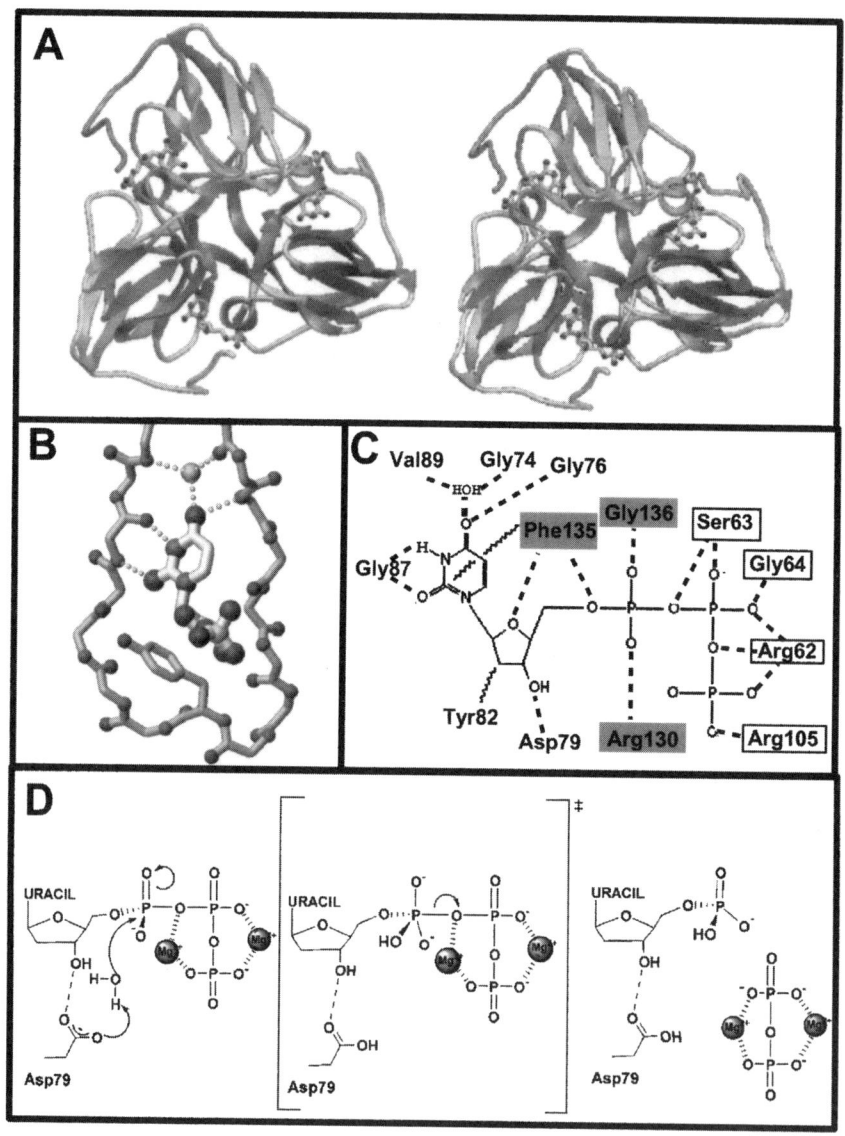

FIG. 3. The interlocked dUTPase biological trimer, dUTP recognition, and catalysis. (A) The dUTPase trimer and three active-site regions shown as a stereo ribbon diagram. The individual subunits of the trimer are colored white, light gray, and dark gray. The carboxy-terminal β-strand of each subunit is swapped with the other subunits to interlock the trimer and extends toward the active sites at the subunit interfaces to cap the bound nucleotides. The positions of the three enzyme active sites are shown by the bound dUMP molecules. (B) The novel β-hairpin recognition motif for uracil in dUTPase with a tyrosine corner motif providing the discrimination against ribose. The bound dUMP is shown from the enzyme–dUMP complex with uracil recognized entirely by

at least one active site of the trimer is always available to bind a fresh dUTP substrate.

II. DNA Base Excision Repair Glycosylases

DNA glycosylases recognize base damage and are key enzymes for the initiation of BER. Pure DNA glycosylases cleave the N–C1′-glycosylic bond between a target damaged DNA base and the deoxyribose sugar. Combined DNA glycosylase/AP lyases remove damaged bases and, as a result of their β-elimination chemistry, also cleave the DNA backbone 3′ of the damage. This damage-specific stage of BER precedes a damage-general stage in which 5′-AP endonucleases cleave the DNA phosphodiester backbone 5′ of the products created by both DNA glycosylases and DNA glycosylase/lyases to provide a 3′-OH end needed to prime DNA repair synthesis.

Individual glycosylases can have distinct folds and domain structures. Structural and biophysical analyses of two structural families of glycosylases, typified by uracil-DNA glycosylase (UDG) and the endonuclease III helix–hairpin–helix (HhH) enzyme family of combined glycosylase/AP lyases, reveal common mechanisms for DNA damage detection, recognition, and catalysis. Both enzyme

---

main-chain hydrogen bonds (dotted lines) from the distorted β-hairpin motif. A tightly bound water provides water-bridged hydrogen bonds from uracil O4 to the Gly74 carbonyl and Val89 amide. The interrupted β-strand hydrogen bonds at the bulged conformation of Ala75 favor the hydrogen bond to uracil O4 from the Gly76 amide, and additional hydrogen bonds to uracil N3 and O2 with dUTPase main-chain atoms discriminate against cytosine binding. Thymine binding is precluded by steric clash of a C5-methyl group with dUTPase backbone atoms. The Tyr82 side chain, which provides steric discrimination at the C2′ position against ribonucleotides, is positioned by a hydrogen bond from the Tyr hydroxyl to the Asn85 carbonyl. (C) Schematic of dUTP contacts by the three separate dUTPase subunits. Human dUTPase interactions with bound uracil deoxyribonucleotides at each of the three active sites of the trimer are shown with hydrogen bonds (dashed lines, <3.4 Å) and van der Waals contacts (squiggly lines) of the Tyr82 side-chain packing against the deoxyribose and the side chain of Phe135 stacking above the uracil base. One subunit recognizes the uracil via insertion into a distorted β-hairpin, whereas contacts by the second and third subunits of the trimer are illustrated in open and shaded boxes, respectively. (D) Structure-based reaction mechanism for dUTPase catalysis. Upon binding dUTP and Mg^{2+} ions at the interface between two subunits of the trimer, the C-terminal capping tail of the third subunit covers the substrate-bound active site, excluding it from bulk solvent and prompting the deprotonation of a water molecule by the buried Asp79 side-chain carboxylate (left panel). The hydroxyl ion performs an in-line attack on the α-phosphate of the dUTP, proceeding through a pentacovalent transition state (middle panel). Collapse of the transition state yields the reaction products dUMP and pyrophosphate (right panel). Interactions of the C-terminal capping tail with the uracil and phosphate of the dUMP product may facilitate removal of the dUMP from the active site, providing a method for overcoming product inhibition to ensure that a dUTPase active site is always available to bind a fresh $dUTP:Mg^{2+}$ substrate.

families employ nucleotide flipping to remove the damaged nucleotide from the DNA base stack and sequester it an enzyme active site pocket. Nucleotide flipping is a general mechanism for recognizing potentially mutagenic lesions in DNA. DNA damage-specific BER enzymes must detect and remove various distinct base lesions within a large pool of undamaged DNA. The elegance of nucleotide flipping for DNA BER is that efficient initial lesion detection, ease of damaged DNA base flipping, complementarity of interactions with damaged base partners, and stabilization of the extrahelical nucleotide, all can improve specificity for particular DNA base damage. Once a damaged DNA nucleotide is detected and sequestered in an enzyme active site, however, the precise interaction chemistry at the target scissile bond dictates each individual enzyme's catalytic efficiency. Thus, UDG is an extremely efficient enzyme with a narrow specificity for a particular damaged base (uracil), while broad-specificity enzymes, such as the HhH glycosylase AlkA, excise damaged bases at decidedly slower rates.

A. Helix–Hairpin–Helix Glycosylases/AP Lyases

The structure of endonuclease III (63) revealed a novel helix–hairpin–helix (HhH) DNA-binding motif (64) that has since been seen in a superfamily of DNA repair enzymes with distinct activities (65, 66). The HhH glycosylase/AP lyase fold consists of two domains: a 6 α-helix barrel domain and a 4 α-helix iron–sulfur cluster domain. Conserved catalytic residues bracket the interdomain cleft, as does the hairpin loop between two α-helices that comprise the HhH motif. The HhH motif is found in a number of different homologous enzyme familes, including DNA polymerase I and mammalian DNA polymerase β (67), and was originally proposed to interact nonspecifically with the DNA sugar phosphate backbone (64); and this function has been directly observed in structures of human DNA polymerase β bound to DNA (68).

The signature HhH motif has also been observed in crystal structures of the mismatch repair enzyme MutY (66), which cleaves mispaired adenines (69); the 3-methyladenine DNA glycosylase II (AlkA) (70, 71), which removes alkylated purine bases (72); and in human Ogg1, which cleaves 8-oxoguanine paired with cytosine (65). Details of how the HhH glycosylases/AP lyases function in DNA damage recognition, specificity, and catalysis have benefited from combined biochemical and structural results on MutY (66, 73), and have recently been clarified with NMR solution and co-crystal structures of MutY and AlkA DNA (74, 75). These results are presented in other chapters of this volume contributed by the research groups of R. S. Lloyd and T. Ellenberger, respectively.

B. Uracil-DNA Glycosylase

UDG was the first DNA glycosylase to be discovered (76) and is among the best characterized biochemically (5, 77, 78). UDG removes uracil from double- and single-stranded DNA (77, 78), but is inactive on uracil in RNA. The structure of UDG is highly conserved among viral (79), human (80), and E. coli (81–83) homologs, with a central, four-stranded parallel β sheet surrounded by α-helices. Interestingly, this protein architecture is also seen in the related G·T/U mismatch-specific glycosylase (84). A conserved feature of these structures is a loop above the active site containing a solvent-exposed leucine residue proposed to facilitate nucleotide flipping (80). The biological importance of this exposed UDG leucine loop is underscored by structures of UDG in complex with the irreversible protein inhibitor Ugi from the PBS1/2 *Bacillus subtilis* bacteriophages (81, 83, 85, 86). Ugi is a protein mimic of DNA in that it targets the DNA-binding face of UDG, inserting the exposed UDG leucine residue into a hydrophobic Ugi pocket. Structures of UDG:DNA complexes (87, 88) reveal an enzyme conformational change from an open unbound state to a closed DNA-bound state, corresponding to UDG engulfing the deoxyuridine that is flipped out of the DNA helix and into the UDG active center. These results showed that Ugi locks UDG in its open unbound conformation.

Structural analyses of wild-type UDG bound to both U·A and U·G containing DNA confirm that the exposed UDG leucine residue penetrates the DNA base stack from the DNA minor groove (87). Three rigid enzyme loops compress the distance betweeen the DNA phosphate groups on either side of the uracil nucleotide by ~4 Å, bending the DNA by ~45° and displacing the DNA helix axis by ~2 Å. These loops would clash with, and slightly bend, the DNA phosphodiester backbone in an initial UDG–DNA complex, providing a mechanism for initial lesion detection. However, processive scanning by UDG (89) and binding to extrahelical uracil nucleotides occur in a two-step process (90), with rapid scanning by DNA backbone compression (pinch), followed by full DNA bending and kinking by penetration of the leucine minor-groove intercalation loop into the DNA base stack (push), and the attraction of the uracil-specific recognition pocket (pull).

This pinch–push–pull model for lesion detection and recognition explains UDG's preference for cleaving uracil from U·G mispairs over U·A pairs (77). The U·G wobble displacement shifts the guanine into the DNA minor groove. The co-crystal structure of a UDG:U·G–DNA complex suggests that the branched leucine side chain pushes the guanine slightly into, and the uracil completely through, the DNA major groove (87, 91). Thus, the preference of human UDG for U·G mispairs is likely dictated by the initial UDG–DNA complex that forms before the target uridine is flipped out of the DNA base stack. Other damage-specific DNA repair enzymes likely also utilize aspects of this UDG pinch–push–pull method for lesion detection and recognition.

When a uracil in DNA is detected and flipped into the UDG active site, concerted loop movements clamp UDG around the bound, flipped-out nucleotide. This discrete structural change is centered on a β-zipper conformational motif (81) that allows the entire UDG fold to stabilize the global enzyme state and brings key UDG residues into their functional positions.

The recently determined structure of wild-type human UDG bound to DNA containing a noncleavable deoxypseudouridine nucleotide reveals that the UDG active center severely distorts the target uracil base and deoxyribose to facilitate catalysis (92, 93). In the uncleaved complex, the uracil ring is rotated ~90° on its N1–C4 axis from its normal position in DNA to a conformation about

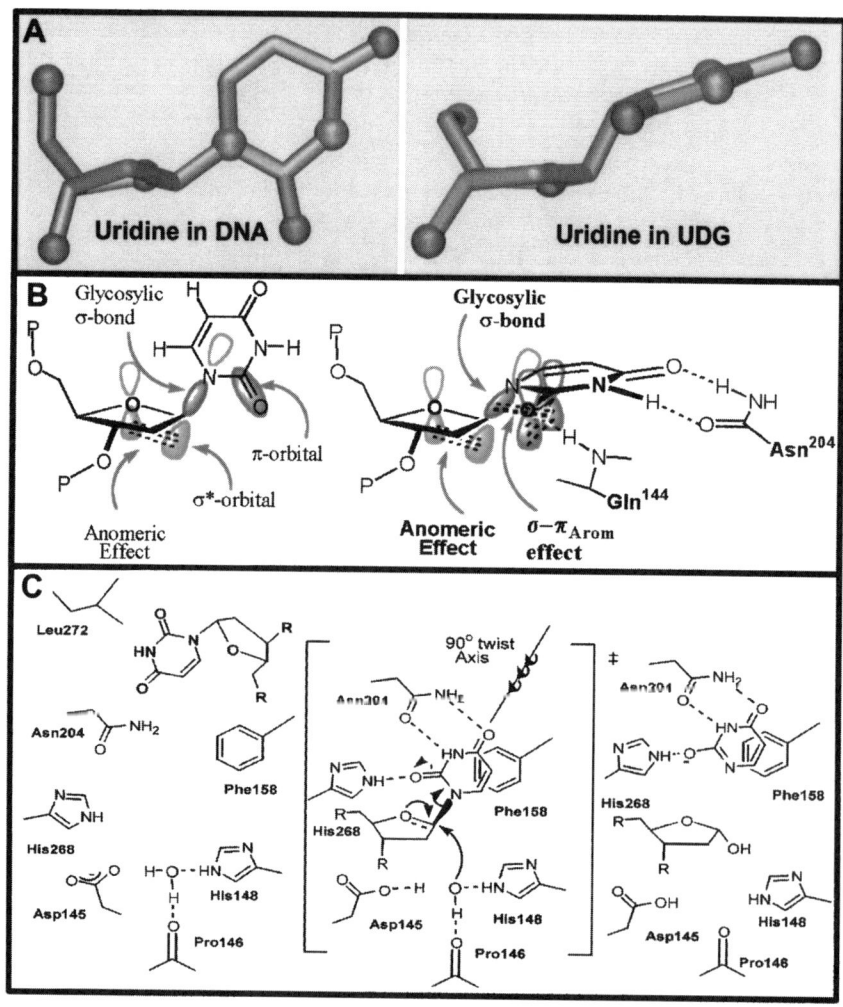

FIG. 4. Enzyme-induced distortions to maximize orbital overlap for efficient catalysis at the UDG active site. (A) Structural comparison illustrating the conformational changes between a deoxyuridine nucleotide in DNA and the uncleaved deoxypseudouridine bound in the UDG active site.

halfway between *anti* and *syn* (Fig. 4A), causing a steric clash between the uracil C6 hydrogen and the deoxyribose O4′ atom that significantly stretches the glycosylic bond (94). This UDG-induced twisting of the uracil base distorts the normally trigonal-planar uracil N1 atom toward a tetrahedral geometry and flattens the sugar pucker of the deoxyribose such that the N–C1′-glycosylic bond is raised to a semi-axial position, creating a p–σ^* overlap with O4′ (Fig. 4A,B). This anomeric effect is coupled to the π-electron system of the uracil ring and this orbital overlap significantly aids bond cleavage. Thus, UDG couples two distinct stereoelectronic effects by distorting the structure of the substrate to increase sequential linear orbital overlap from O4′ to O2. Stabilization of the developing negative charge at the uracil O2 position is provided by hydrogen bonds from Gln144 and His268. This channeling of binding energy into physical distortion of the target nucleotide allows UDG to couple two distinct and complementary stereoelectronic effects to promote efficient catalysis (Fig. 4C).

The UDG active center provides a special environment for the uracil and deoxyribose, which suggests a mechanism that is a substantially dissociative (S_N1-like) nucleophilic substitution process involving formation of a stable uracil enolate anion coupled to enzyme-induced substrate strain (Fig. 4C). This dissociation reaction requires that the normally orthogonal three transposing

---

Relative to its structure within B-DNA, the uracil ring is rotated by ~90° about the N–C1′-glycosidic bond. Steric constraints help to flatten out the sugar ring and force the pseudouracil C1 (uracil N1) position to become more *sp*3-hybridized. (B) Electron orbital diagrams of the deoxyuracil in DNA and in the UDG active site illustrate the importance of the conformational changes induced by UDG for the dissociation of uracil. In DNA, the three orbitals involved in glycosylic bond cleavage are orthogonal to each other. The lone-pair orbitals of O4′, the σ^* orbital of the glycosylic bond, and the π orbital of the C2=O carbonyl are aligned only after binding UDG (right side of arrow). This orbital alignment involves both the anomeric effect between the O4′ lone pair and the glycosylic bond σ^* orbital and a σ–π_{Arom} effect due to overlap of the σ^* orbital and the C2=O carbonyl. This conformational change is independent of the functional group chemistry of the active site and explains the substantial rate enhancements observed even when the residues that help stabilize the reaction are mutagenized. (C) UDG reaction mechanism. The empty UDG active site prior to flipping of the deoxyuridine out of the DNA (left panel). Once the deoxyuridine is flipped into the active site, conformational shifts cause UDG to close around the bound uracil base and deoxyribose sugar, bringing His268 into hydrogen-bonding distance of the uracil O2 atom and flipping the side chain of Asp145 (middle panel). Upon entering the active site pocket, the deoxyuridine is destabilized by an ~90° twist about the N–C1′-glycosylic bond and distorted toward a tetrahedral geometry at uracil N1 by collision with the aromatic ring of Phe158. These distortions allow delocalization of electrons from the O4′ of the deoxyribose through the glycosylic bond and into the aromatic ring (middle panel). The transient formation of an oxocarbenium cation at the C1′ of the deoxyribose, coupled with addition of hydroxyl ion to C1′, cleaves the scissile bond in a reaction with substantial S_N1 dissociative character. Thus, the uracil ring leaves as an enolate anion (right panel), which is stabilized by hydrogen bonds and is protonated to give uracil.

electron pairs have overlapping orbitals. The strained deoxyuridine conformation induced by UDG provides the required electronic overlap (Fig. 4A,B). As the strained N–C1'-glycosylic bond disintegrates, a water molecule positioned below the deoxyribose C1' atom can provide the C1'–α-OH to the incipient deoxyribose oxocarbenium ion (Fig. 4C). The released uracil base has improved hydrogen-bonding interactions and can stack more efficiently with the conserved phenylalanine ring. Moreover, steric strain is also reduced at the abasic deoxyribose, which relaxes into a C2'-endo sugar pucker and withdraws slightly from the enzyme.

Thus, UDG preferentially binds its unstrained cleaved products, in agreement with solution DNA-binding kinetics (87), suggesting that *in vivo* UDG remains bound to its cleaved products to avoid the release of toxic DNA repair intermediates. The fact that the next enzyme in the human BER pathway, the 5' AP endonuclease APE1, stimulates UDG activity (87), provided the first hint that DNA repair steps *in vivo* may be coordinated by enzyme partnerships mediated through the distorted DNA.

III. 5' Apurinic/Apyrimidinic (AP) Endonucleases

Specific removal of damaged bases by DNA glycosylases generates abasic sites in DNA, and these lesions are recognized by 5' AP-endonucleases. Abasic sites also arise spontaneously in the cell through depurination and depyrimidation events and thus constitute the damage-general central intermediate in DNA BER. Importantly, the efficient repair of AP sites is crucial for cell survival due to their cytotoxic (95–97), potentially mutagenic, and apoptotic effects (3, 98, 99).

Two conserved families of 5' AP-endonucleases are responsible for abasic site processing in cells. The first family is typified by APE1, which is also known by the names HAP1 (100), REF1 (101), and APEX (102), and is the major constitutive 5' AP-endonuclease in human cells. APE1 is a small, monomeric, Mg^{2+}-dependent enzyme, whose importance in human DNA repair is underscored by the fact that APE1 gene expression is activated by DNA-damaging agents and reactive oxygen species (103), and homozygous APE1 knockout mice are embryonic lethal (104). Homologs of APE1 exist in several organisms (105), including Exonuclease III (Exo III), which is the major constitutively expressed AP endonuclease in *E. coli* (61).

The second conserved family of 5' AP endonucleases is typified by the *E. coli* enzyme Endonuclease IV (Endo IV). Like APE1, Endo IV is a small, monomeric enzyme, but differs in its metal ion dependence by requiring Zn^{2+} rather than Mg^{2+}. In *E. coli*, Endo IV gene expression is induced by superoxide anion generators (106, 107); but in *Saccharomyces cerevisiae*, the Endo IV homolog APN1 is the predominant constitutively expressed AP endonuclease. In yeast, the importance of APN1 is underscored by the fact that gene deletion increases the AT to GC transversion rate ~60-fold (*108*) and renders the cells hypersensitive to alkylating agents and chemical oxidants (*109*). The combination of recent biochemical and genome sequencing efforts has identified Endo IV homologs in several additional organisms including *Caenorhabditis elegans, Candida albicans*, and *Plasmodium falciparum* (*110, 111*); however, the precise role these enzymes play in DNA repair processes in these organisms remains to be determined.

Structural results obtained for enzymes from both conserved families of 5′ AP endonucleases, in their free forms (*112–114*) and also in complex with abasic DNA (*114, 115*), have recently provided key insights into how these enzymes function and serve to illustrate the biological importance of efficient AP site processing. Notably, these results reveal the mechanisms employed by both APE1 and endo IV for AP site detection and recognition, and also the chemical mechanisms employed for phosphodiester bond cleavage.

A. APE1

Crystal structures of human APE1 bound to synthetic abasic site-containing DNA, both with and without the divalent metal ion, show how APE1 recognizes abasic sites and cleaves the target bond (*115*). In the co-crystal structures, APE1 binds a flipped-out abasic nucleotide with residues penetrating both the DNA minor and major grooves to engulf the extrahelical AP site in a sequestered active site pocket (Fig. 5A,B). The APE1–DNA interface is centered around the extrahelical abasic deoxyribose and its flanking phosphates, which contribute nearly one-third of the buried surface area of the complex. The interface encompasses both DNA strands, but most contacts are made with the AP-DNA strand (Fig. 5A). These results agree with solution data from footprinting experiments, which show that APE1 contacts two DNA phosphates 3′ of the AP site (*116*) and that ~3 base pairs 3′ and ~4 base-pairs 5′ of the AP site are needed for bond cleavage (*117*).

APE1 is preformed for recognizing and binding abasic DNA. The enzyme deviates only ~0.7 Å in Cα atom positions between the AP-DNA bound and free enzyme structures. The AP-DNA structure, however, is severely distorted from a standard B-DNA helix. The enzyme-bound AP-DNA is bent ~35° and there is a large ~5-Å displacement, or kinking, of the DNA helical axis, where the two inserted APE1 loops cap the extrahelical abasic site (Fig. 5A). The kinked AP-DNA structure is stabilized on the 5′ side by an α-helix, which braces the DNA backbone, and a loop, which spans and widens the DNA minor groove. These contacts anchor the DNA for the bending and extreme kinking centered on the extrahelical AP site. At the AP site, an APE1 loop buttresses the DNA

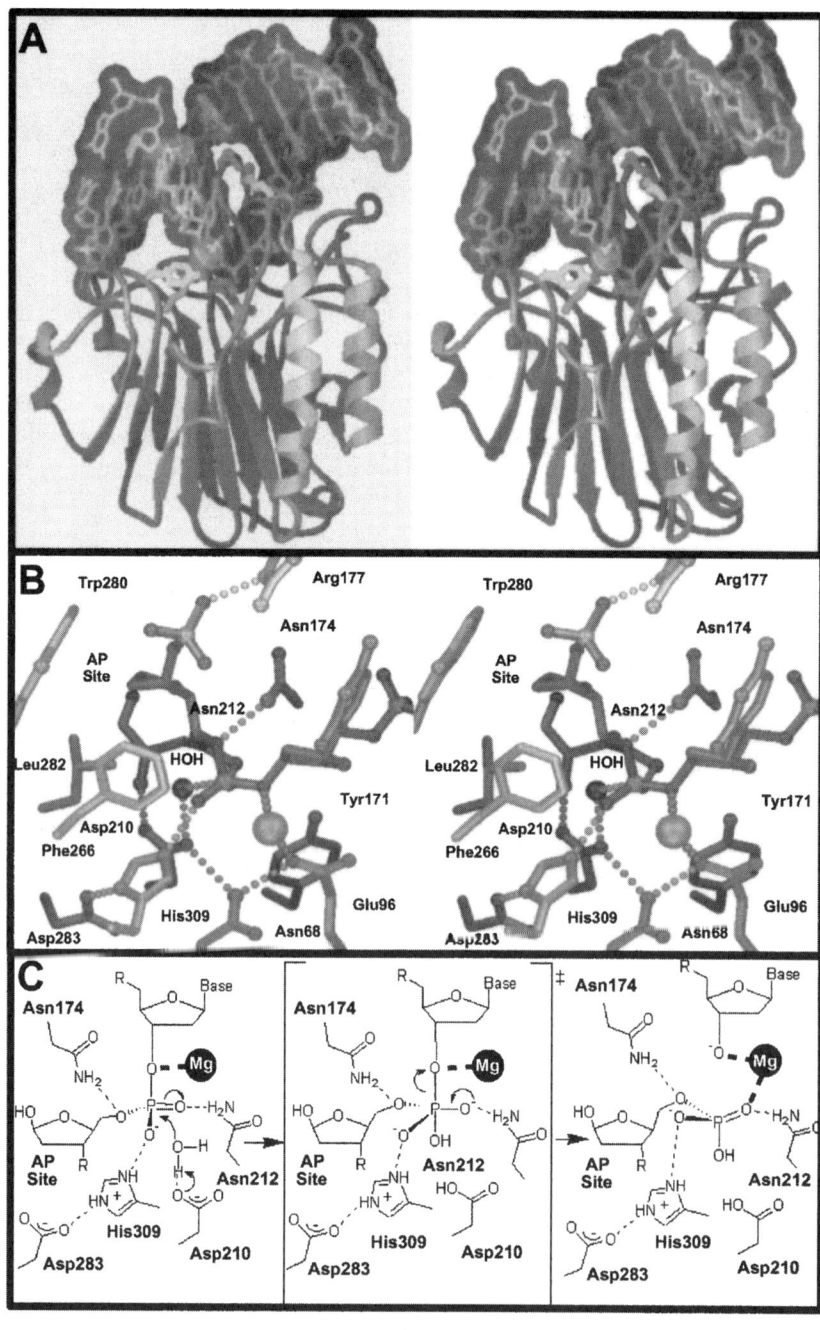

backbone for double-loop insertion of methionine residues into the DNA minor groove (Met270 and Met271), and an arginine residue into the DNA major groove (Arg177). The arginine side chain delivers a charged hydrogen bond to the nontarget DNA phosphate group 3' of the AP site (Fig. 5B), stabilizing the extrahelical conformation of the AP site and effectively locking APE1 onto the AP-DNA. At the APE1 active site, direct hydrogen bonds from conserved APE1 residues, as well as from the divalent metal ion, to all four oxygen atoms of the 3' AP site phosphate, orient and polarize the target scissile P–O3' bond for cleavage (Fig. 5B). APE1 specifically binds the extrahelical abasic site in a pocket bordered by aromatic and hydrophobic amino acid residues that pack with the hydrophobic side of the abasic deoxyribose (Fig. 5B).

The effects on enzyme activity of site-directed alanine mutants of the APE1 residues that penetrate the DNA minor groove (Met270, Met271), and DNA major groove (Arg177) showed, surprisingly, that none of these residues are required to flip AP sites out of the DNA helix (Fig. 6A). Both the Met270Ala and Met271Ala mutant enzymes retain wild-type activity; but for the Arg177Ala variant, enzyme activity is greater than for wild-type APE1 (Fig. 6A). Therefore, the Arg177 intercalation into the DNA base stack and DNA phosphate interaction apparently slows APE1 dissociation from the cleaved product (118). Enzyme kinetic analyses confirm that the Arg177Ala mutant is a more efficient enzyme than wild-type APE1 (115). The conformation and amino acid sequence of the Arg177

FIG. 5. APE1 binding to flipped-out AP-DNA via double-loop penetration of the DNA major and minor grooves, abasic site recognition, and 5' phosphodiester bond cleavage. (A) Stereo view of AP-DNA binding to APE1. The position of the divalent metal ion (large gray sphere) 5' of the AP site and the side chains for selected key residues are shown, including: Tyr128, inserting into the DNA minor groove 5' of the AP site; Arg177 (front) and Met270 (back) penetrating the DNA major and minor grooves, respectively, above the flipped-out AP site; Glu96, binding the metal ion; and His309 and Asp210 in the APE1 active site. (B) Stereo view of the APE1 active site illustrating interactions with the flipped-out abasic deoxyribose and target 5' phosphate. The hydrophobic face of the extrahelical AP site packs within a complementary APE1 pocket formed by the side chains of Phe166, Trp280, and Leu282. The side chain of Arg177 inserts through the DNA major groove to form a hydrogen bond to the nontarget AP site 3' phosphate. The target AP site 5' phosphate O5' and O2P atoms form hydrogen bonds with Asn174 and Asn212, respectively, while His309 forms a direct hydrogen bond to a third phosphate oxygen. Asp210, which is positioned to activate a water molecule (HOH) for the cleavage reaction, is oriented by hydrogen bonds from the Asn212 backbone amide and the Asn68 side chain, which also delivers a hydrogen bond to the metal ligand Glu96. The metal ion (large gray sphere) contacts the O3' leaving group. (C) Structure-based reaction mechanism for phosphodiester bond cleavage by APE1. Substrate AP-DNA is oriented by the bound divalent metal ion (black circle) and APE1 active-site residues for attack of a hydroxyl nucleophile activated by Asp210 (left panel). Collapse of the pentacovalent transition state (middle panel) leads to cleavage of the scissile P–O3' bond, with the transition state and O3' leaving group stabilized by the metal ion, and the inversion of the stereochemical configuration of the phosphate.

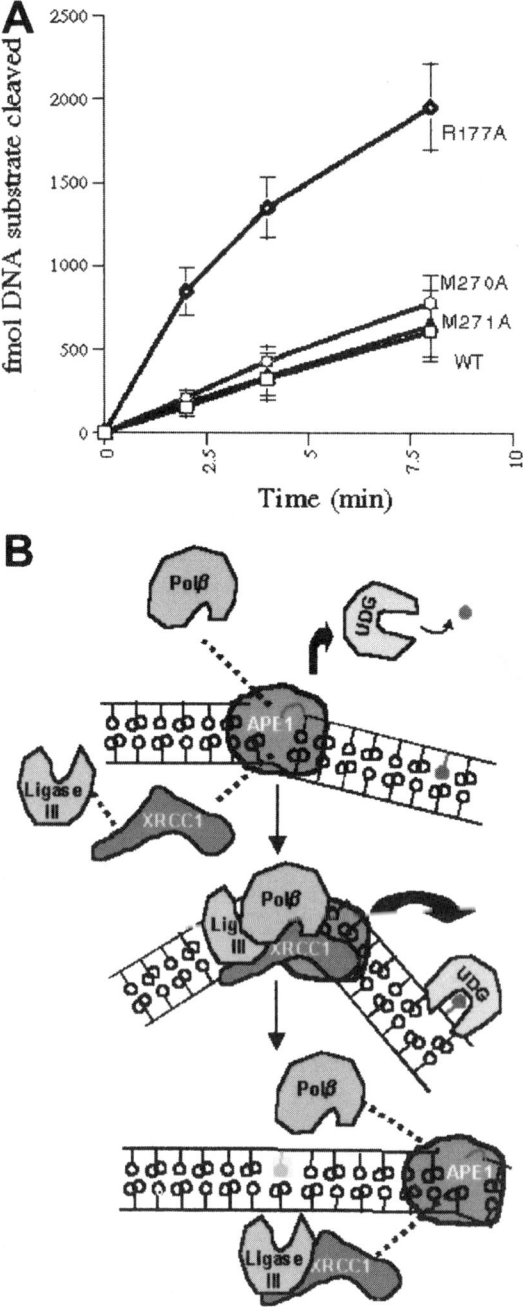

loop in APE1 is not conserved in Exo III or in other nonmammalian members of this AP endonuclease family, suggesting that the interactions of Arg177 reflect a biological function in human cells. As evolutionary selective pressure acts at the level of biological function, rather than single enzyme rates, these results support a biological role for tight product binding by APE1. The biological need to coordinate APE1 with downstream DNA BER enzymes (43, 119, 120) has evidently guided evolution of the APE1 structure, rather than maximizing catalytic efficiency and allowing rapid release of toxic, incised DNA products.

The recognition of DNA bases through nucleotide flipping is a common method for damage detection employed by DNA glycosylases and glycosylase/lyases (91). Abasic nucleotide flipping by APE1 allows for the sharp DNA kinking at the damage locus. Both the amount of helical displacement and the extent of the APE1:DNA interface are larger than that seen in DNA complexes with damage-specific DNA glycosylases (84, 87, 88, 121). Acting nonspecifically, APE1 may simply displace these enzymes from their inhibitory AP-site products and thereby enhance their DNA glycosylase activities (87, 122), suggesting that *in vivo* the AP site is passed directly to APE1 from the damage-specific enzymes. This partner exchange of DNA substrates between BER enzymes avoids the release of repair intermediates into solution, and suggests a model for the orderly transfer of DNA damage from DNA glycosylases to APE1 and from APE1 to Pol β (Fig. 6B). The APE1 adaptation of tight product binding allows for the recruitment of the DNA repair synthesis enzymes to areas of damaged DNA through the interaction of the APE1:DNA complex with Pol β. Functionally relevant APE1 interactions with Pol β likely depend largely upon the distorted

FIG. 6. APE1 penetrating loop mutants and model for DNA repair coordination. (A) AP endonuclease activities of wild-type APE1 and site-directed mutant enzymes of the residues that penetrate the DNA major groove (Arg177Ala), and DNA minor groove (Met270Ala and Met271Ala). None of these residues is required to flip the AP site out of the DNA helix, but the Arg177Ala is actually a more efficient enzyme (115, 157), suggesting that wild-type human APE1 has evolved to remain bound to its nicked DNA product and coordinate DNA repair with downstream synthesis enzymes. (B) Coordination of DNA repair steps by APE1. A damage-specific DNA repair enzyme, such as UDG, is nonspecifically displaced from its bound abasic DNA products by the more extensive DNA interaction surface and more pronounced DNA kinking induced by APE1. UDG releases the cleaved uracil base (small dot) and is free to detect another uracil-containing nucleotide (top panel). After phosphodiester bond cleavage, APE1 presents its nicked product DNA to DNA polymerase β and/or possibly XRCC1, with this product–substrate exchange partially mediated through the unique APE1 N-terminus interacting with the Polβ deoxyribophosphodiesterase domain. Through the known interactions of APE1:DNA with Polβ, and of XRCC1 with both Pol β and DNA ligase III, the DNA repair synthesis enzymes can assemble at the site of DNA damage (middle panel). Following repair and religation, the DNA can no longer accommodate bending and the component enzymes likely dissociate, but are now localized to a region of damaged DNA and can reassemble at the next lesion (bottom panel).

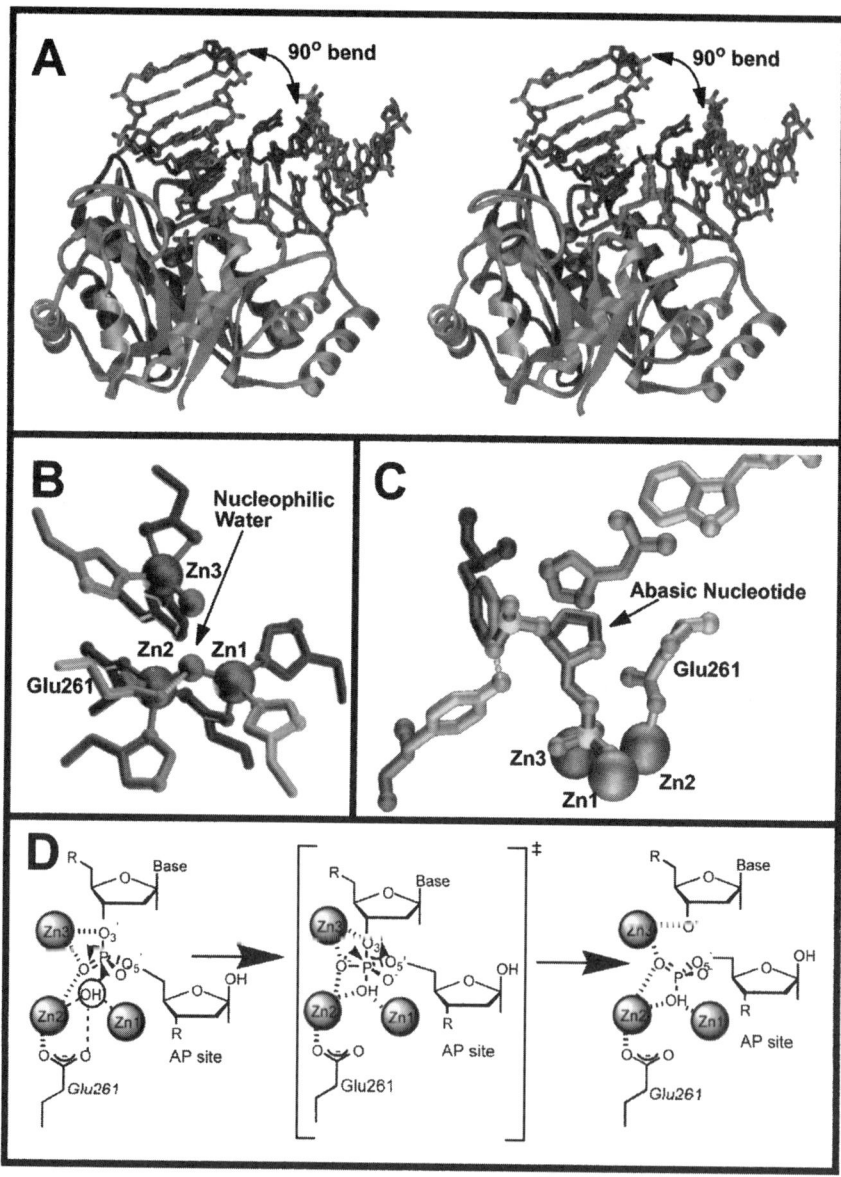

FIG. 7. Endo IV structure and active site. (A) Stereo view of the Endo IV:DNA product complex. In the complex, the DNA is kinked by ~90°, with the orphan guanine nucleotide displaced from the DNA helix and the abasic nucleotide flipped by 180° out of the DNA helix into the enzyme active site. The phosphate bond connecting C6 and the abasic nucleotide is cleaved and the free 5′ phosphate is tightly bound to the three Zn ions. Insertion of Arg37, Tyr72, and

APE1:DNA complex, with APE1 able to stage repair synthesis by presenting the bent and kinked nicked abasic DNA to Pol β, such that the smaller polymerase deoxyribophosphodiesterase domain contacts the APE1:DNA complex 3' of the AP site. The polymerase dislodges APE1 and increases the DNA bending to that needed for an active Pol β:DNA complex (Fig. 6B). The entire DNA repair synthesis machinery can thus be assembled at damaged loci through the known interactions of Pol β, XRCC1, and DNA ligase III (123–125); and, as most human BER synthesis replaces only the single damaged nucleotide (126), the religated and repaired DNA can no longer bend sufficiently to bind Pol β. Without an amenable DNA substrate, the component BER enzymes likely dissociate after repair synthesis. However, the detection and repair of a single damaged locus may indicate that other damaged DNA bases are nearby, and by recruiting enzymes to a single damaged site, DNA repair at adjacent sites can be more efficiently completed.

B. Endonuclease IV

The structure of Endo IV from *E. coli*, determined to ultra-high 1.0-Å resolution, provides the highest resolution atomic description yet obtained for any DNA repair or Zn-containing enzyme (114). The structure revealed that the enzyme is a single-domain α/β protein with secondary structure elements arranged as a β-barrel having eight parallel β-strands surrounded by eight peripheral α-helices (Fig. 7A). This common fold was first observed in triose phosphate isomerase (TIM) (127), and occurs in many enzyme families of diverse function (128, 129), but Endo IV is the first structure to demonstrate how the ubiquitous TIM barrel fold can be used for binding a large macromolecule such as DNA.

As seen in other TIM barrel enzymes, the active site is situated at the C-terminal end of the β-barrel, where the protruding enzyme loops that connect

---

Leu73 through the DNA minor groove and into the base stack stabilizes the bent DNA and assists in double-nucleotide flipping. (B) Illustration of the Endo IV active site and trinuclear zinc cluster in the uncomplexed enzyme. Conserved Endo IV His, Asp, and Glu residue side chains ligate the three Zn^{2+} ions. The proposed catalytic water molecule bridges Zn1 and Zn2, while a second well-ordered water molecule serves as the final ligand to Zn3. (C) The abasic sugar of the flipped-out abasic nucleotide is bound in a sequestered active-site pocket lined by conserved hydrophobic side chains. The pocket can accommodate a hydroxyl at the C1' atom of deoxyribose, yet normal DNA nucleotides are sterically prevented from binding in the pocket. The 5' phosphate of the flipped-out AP site is shown ligated to the three active-site Zn^{2+} ions. (D) Structure-based three-metal-ion–mediated reaction mechanism for phosphodiester bond cleavage for the Endo IV family of 5' AP endonucleases. Nucleophilic attack of the scissile P–O3' bond by the hydroxide bridging Zn1 and Zn2 (left panel) proceeds through a pentacoordinate transition state stabilized by all three Zn^{2+} ions (middle panel). The phosphate oxygen atom bridging Zn2 and Zn3 remains bound to these metal ions, and collapse of the transition state inverts the stereochemistry at the target phosphate with the developing negative charge at O3' stabilized by interaction with Zn3 (right panel).

individual α-helices and β-strands of the barrel form a deep crescent-shaped groove that complements DNA binding. Three active site Zn^{2+} ions are ligated by conserved residues that emanate from individual strands and loops of the barrel and are precisely situated within the deep groove to effect phosphodiester bond cleavage. Like other trinuclear Zn^{2+} cluster–containing enzymes, such as P1 nuclease from *Penicillium citrinum* (*130, 131*) and phospholipase C from *Bacillus cereus* (*132–134*), two of the Zn atoms are bridged by a tightly bound water molecule (Fig. 7B). This water molecule, which is likely deprotonated in the enzyme active site, serves as the nucleophile in the hydrolysis reaction. The third Zn atom is not involved in activation of the catalytic water, but plays a key role by stabilizing the 3′ hydroxyl leaving group liberated in the reaction (see below). The arrangement of these three Zn ions in the active site illustrates the utility of the trinuclear Zn cluster for catalyzing phosphodiester bond cleavage, and the structure of Endo IV bound to abasic DNA revealed the specific structural features of the enzyme that allow it to specifically cleave such a bond at AP sites in dsDNA.

The crystal structure of Endo IV bound to an abasic site-containing dsDNA revealed the structural features that underlie AP site recognition and cleavage by this conserved family of 5′ AP endonucleases (*114*). In the Endo IV:DNA complex structure, the AP-DNA is bent by ∼90° and both the abasic site and its partner nucleotide are flipped out of the DNA helix (Fig. 7A). DNA recognition is mediated exclusively through interactions with nucleotides in the minor groove and enzyme backbone and side-chain atoms from five DNA-recognition loops. Specific contacts are made with both DNA strands through direct and water-mediated interactions and each DNA strand is contacted almost equally. The Endo IV TIM-barrel fold orients the eight α-helices surrounding the β-barrel with their positive helix dipoles facing the DNA-binding surface; and in combination with positively charged Arg and Lys residues and the active-site Zn^{2+} ions, the DNA-recognition loops form a positively charged, crescent-shaped groove that provides both shape and electrostatic complementarity for the negatively charged phosphate backbone of double-stranded DNA. Notably, the enzyme loop contacts, which brace the DNA in the dramatically bent conformation, compress the DNA backbone ∼6.4 Å between the AP site-flanking phosphates, and precisely deliver the AP site 5′ phosphate into the active site where it is tightly anchored to all three active site Zn^{2+} ions (Fig. 7C).

While most of the Endo IV scaffold is preformed for binding damaged DNA, striking structural rearrangements of two minor-groove DNA recognition loops facilitate DNA binding. Importantly, these conformational changes permit insertion of key residues into the DNA base stack to occupy the gap left by the flipped-out AP site and its orphan nucleotide partner. Stacking interactions with both base pairs 3′ and 5′ of the AP site are preserved in the complex. Conserved Tyr and Ile side chains on one of the minor-groove DNA-recognition loops undergo backbone and side-chain shifts to stack with and stabilize the base pair 5′ of the AP site. A second minor-groove DNA-recognition loop also undergoes significant conformational changes, allowing a conserved Arg to stack with the 3′ base pair. Consistent with biochemical results showing that Endo IV has no preference for the base opposite an AP site (135), no specific contacts are made to the orphan base; yet the extrahelical conformation of this orphan nucleotide is stabilized by a solvent-exposed Trp, whose backbone amide and side-chain indole nitrogen atoms also move from their positions in the unbound enzyme structure to interact with the two phosphates flanking the orphan nucleotide. As a result of these structural rearrangements that facilitate side-chain penetration of the DNA helix, the minor groove is grossly deformed, becoming ~5.5 Å wider than normal B-DNA.

AP site detection by Endo IV is likely mediated by scanning of the DNA minor groove for regions of DNA that can be distorted to fit the enzyme surface. Two minor-groove DNA-binding loops insert residues into the minor groove, and these enzyme:DNA contacts would slightly expand the minor groove of normal DNA and cause initial DNA bending. Upon detecting an AP site, damage-specific conformational changes in both the enzyme and DNA lead to increased DNA bending and double-nucleotide flipping and ultimately the formation of a catalytically competent complex. Specificity for AP sites thus results from these concerted changes that are damage-specific, as only AP-DNA can be deformed in this manner consistent with the constraints of the abasic nucleotide-binding pocket.

Endo IV specifically recognizes AP sites in dsDNA by binding the flipped-out sugar in a sequestered pocket that sterically excludes binding of normal nucleotides (Fig. 7C). The Endo IV active-site pocket is lined with conserved residues that pack against the deoxyribose, and both the α and β anomers of abasic sites can be accommodated in this pocket. The active-site structure explains Endo IV's 3′ repair diesterase activity as the sequestered pocket can accommodate any base-free nucleotide, such as 3′ α, β-unsaturated aldehydes, 3′ phosphates, and 3′ phosphoglycolates. All these substrates would fit in the AP-site–binding pocket, while retaining the normal spacing of the DNA backbone atoms that permit the target phosphate to be intimately coordinated by the three Zn^{2+} ions; and these interactions are critical for catalysis.

Endo IV utilizes a three-metal-ion mechanism for phosphodiester bond hydrolysis where all three Zn^{2+} ions participate directly in catalysis (Fig. 7D). In the complex structure, the target AP site 5′ phosphate is intimately coordinated by enzyme residues and the bound metal ions with two of the target phosphate oxygen atoms bridging adjacent Zn^{2+} ions. This product complex thus suggests that catalysis proceeds through a pentacoordinate transition state arising from nucleophilic attack by the hydroxide bridging the two Zn^{2+} ions. A conserved active-site Glu, which is also a Zn ligand, assists in orienting the hydroxide for in-line attack on the phosphorus atom; and charge neutralization of the phosphate group by the three Zn^{2+} ions renders the phosphorus atom susceptible to nucleophilic substitution. Collapse of the transition state to products proceeds with an inversion of the configuration of the phosphate group, and the developing negative charge at O3′ is stabilized by interactions with a Zn^{2+} ion (Fig. 7D).

IV. FEN-1 and Long-Patch Base Excision Repair

In human cells, AP site adducts that are not substrates for the dRPase activity of Pol β can be repaired through an alternate long-patch BER pathway (Fig. 8). In long-patch BER a multiprotein complex, which includes a DNA

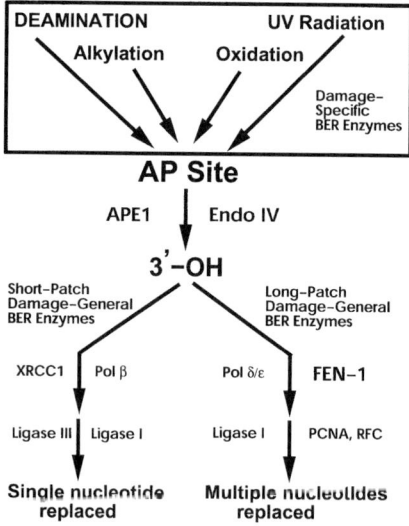

FIG. 8. DNA base excision repair (BER) consists of damage-specific DNA glycosylases that recognize and remove bases damaged by exogenous and endogenous agents. The AP site products of the glycosylases are recognized and cleaved by one of two damage-general 5′-AP endonucleases (APE1 or Endo IV) and the resulting nicked DNA, containing a free 3′-OH, is further processed by the short-patch or long-patch BER pathway. In the short-patch BER pathway, the combined polymerase and dRPase activities of polymerase β replace the incised AP site, and DNA ligase seals the nick to complete repair. XRCC1, the short-patch BER scaffolding protein, coordinates the activities of these enzymes to ensure that toxic AP sites are not left exposed in the cell. In long-patch repair, the incised AP site is replaced by the combined activities of several enzymes that also function in DNA replication. PCNA coordinates long-patch BER by localizing all the repair factors to the free 3′-OH, and displacement synthesis generates a 5′ flap strand that is recognized and removed by FEN-1. The resulting nick is recognized by DNA ligase, which restores the integrity of the DNA backbone to complete repair.

polymerase, proliferating cell nuclear antigen (PCNA), and replication factor C (RF-C), displaces the AP-site–containing strand to create a 5′ flap (126, 136). The damage-containing 5′ flap is recognized and removed by flap Endonuclease I (FEN-1), an essential long-patch BER nuclease that, along with DNA ligase I, interacts with PCNA to complete repair synthesis (126, 136, 137). Recent results from combined structure–function studies on FEN-1 are beginning to illuminate how this key enzyme in long-patch BER recognizes and excises damaged DNA.

FEN-1 is a multifunctional, structure-specific nuclease that cleaves branched DNA structures with an overhanging, single-stranded 5′ flap and also acts as a 5′→3′ exonuclease where it processively removes mononucleotides at a nick and with lower efficiency at a gap (138–140). This ∼40-kDa, Mg^{2+}-dependent metallonuclease cleaves either RNA or DNA (141), and its activity is independent of 5′ flap length (142). The enzyme is highly conserved in archaebacterial and eukaryotic organisms where it is integral to lagging-strand DNA replication by processing the 5′ ends of Okazaki fragments. FEN-1 interacts with PCNA through a conserved motif (143, 144), and its activity is modulated by this interaction (145). When binding to PCNA, the enzyme competes with the cell cycle regulatory protein p21WAF1=CIP1 (146), and this competition may provide a mechanism to coordinate the functions of PCNA in DNA replication and repair (143). FEN-1 also interacts with other proteins at the replication fork, including the DNA2 helicase, which may unwind Okazaki fragments preparing them for endonucleolytic cleavage by FEN-1 (147).

Crystal structures of FEN-1 enzymes from the thermophiles *Pyrococcus furiosus* (148) and *Methanococcus jannaschii* (149) reveal a saddle-shaped single-domain α/β protein with a deep groove for DNA binding (Fig. 9A). The groove is positively charged to interact with DNA; and strictly conserved acidic residues, which bind the divalent metal ions required for activity, cluster at the bottom of the groove (Fig. 9B). FEN-1 shares the same overall protein topology as the bacteriophage T4 (150) and T5 (151) 5′→3′ exonucleases and the 5′ exonuclease domain of *Thermus aquaticus* DNA polymerase (152), yet amino acid sequence identity between FEN-1 and these latter enzymes is restricted to the active site metal-liganding acidic residues (153). Notably, the relative positions of the two bound metal ions is variable in all the published 5′→3′ nuclease structures (148–152), suggesting that the active site geometry and metal ion positions may change upon substrate binding.

Binding to the dsDNA portion of a flap substrate is likely mediated by a highly conserved helix–3turn–helix (H3TH) motif that shares structural homology with the nonspecific backbone-binding helix–hairpin–helix motif (64, 67). The FEN-1 H3TH motif consists of two α-helices connected by a 13-residue structured loop formed by three consecutive four-residue turns (Fig. 9C). Conserved glycine-backbone amides on the third turn allow close approach of the DNA phosphate backbone, and two residues at the apex of the first turn protrude

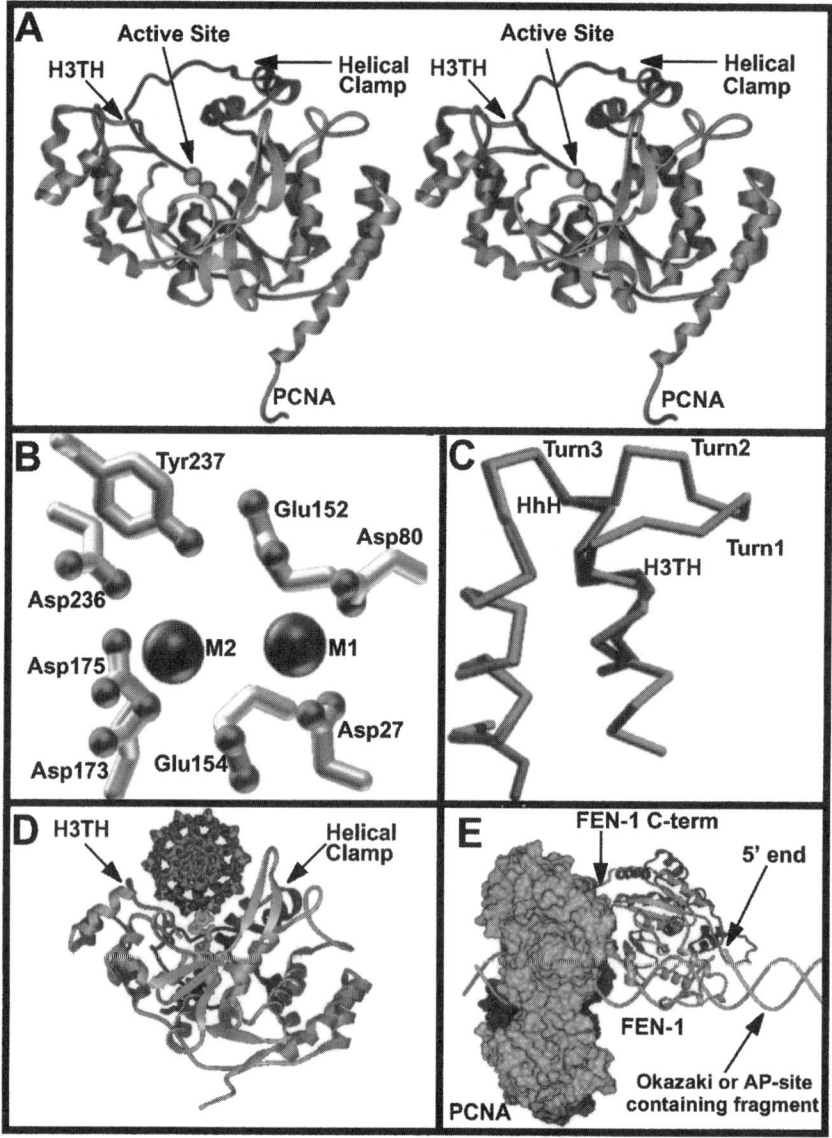

FIG. 9. FEN-1 structure, active-site, double-stranded DNA-binding helix–3turn–helix motif, and model for flap DNA recogniton and coordination with PCNA during DNA synthesis. (A) Stereo view of the FEN-1 structure illustrating the overall α/β fold and functionally important motifs. The enzyme active site is located by two divalent metal ion-binding sites that are critical for activity. (B) Two divalent metal ions are bound to the FEN-1 active site (M1 and M2) through interactions with negatively charged, conserved aspartate, glutamate, and tyrosine side chains.

into the active site. Active-site rearrangements initiated by dsDNA binding to the H3TH motif may alter the relative positions of these residues; thus, this conserved motif likely couples flap DNA binding to enzyme activity (Fig. 9D).

The FEN-1 and bacterial 5′ nuclease structures suggest a model for binding the 5′ overhanging flap strand by threading the ssDNA through a hole in the enzyme surface formed by a "helical arch" in the T5 enzyme (*151*), an extended loop in *M. jannaschii* FEN-1, and a "helical clamp" in *P. furiosus* FEN-1. In each enzyme the hole is lined by polar and charged residues and the space through which the single-stranded 5′ flap must be threaded is relatively small. Double-stranded DNA is not able to pass through the opening, yet FEN-1 must be capable of expanding this hole in order to slip past bulky adducts on the flap strand (*154*). It is likely that the conformational variability of this region is critical for enzyme function. Thus, a productive complex likely forms at the junction of a 5′ flap strand and dsDNA where the helical clamp and H3TH motifs, respectively, coalesce around the target phosphate to bring key catalytic residues and metal ions into position for phosphodiester bond cleavage (Fig. 9D).

The dual role of FEN-1 in DNA repair and replication is modulated by PCNA. The conserved FEN-1 C-terminus extends away from the enzyme surface opposite the DNA-interacting groove to provide a motif that is structurally independent of the FEN-1 catalytic domain. As PCNA coordinates long-patch BER by providing the scaffold that recruits the different repair factors to the site of damage, interaction with FEN-1 facilitates flap-strand incision. In replication, PCNA-bound FEN-1 is localized to the replication fork and positioned to thread the downstream Okazaki fragment through the active site, releasing mononucleotides or short oligonucleotides as DNA synthesis proceeds (Fig. 9E).

The positions of the metal-ion–binding sites likely change upon binding flap DNA. (C) Structural superposition of the pFEN-1 H3TH motif with a typical helix–hairpin–helix motif reveals the similarity of the two motifs. Glycine-rich turn3 of the H3TH is very similar to the HhH β-hairpin, indicating how this key motif likely interacts nonspecifically with dsDNA phosphate groups. Absolutely conserved Asp236 and Tyr237 are located at the apex of turn 1 and protrude into the active site of the enzyme where they interact with the divalent metal ions. (D) The FEN-1 fold and dsDNA-binding site viewed along the DNA-binding cleft. A flap DNA substrate was modeled into the cleft with the phosphodiester backbone poised to interact with the H3TH domain. The helical clamp caps the active site, and may serve to stabilize the bound substrate by interacting with single-stranded and/or double-stranded DNA. During repair, the 5′ AP-site–containing flap strand passes over the active site and through the helical clamp. (E) The FEN1 C terminus was docked onto the surface of the human PCNA trimer and double-stranded DNA containing a short 5′ flap was placed through the center of the PCNA trimer with the 5′ end of the Okazaki or AP-site–containing fragment poised to be threaded through the FEN-1 helical clamp and into the FEN-1 active site. With FEN-1 facing forward during DNA repair and replication, the proximity of the enzyme to the growing flap would prevent the generation of long Okazaki and AP-site–containing fragments that could form FEN-1–resistant secondary structures. PCNA coordinates long-patch BER as the additional binding sites on the PCNA trimer could be occupied by other repair factors.

FEN-1 mutations that alter PCNA binding would reduce enzyme activity during repair and replication and may explain the expansion of flap endonuclease resistant short DNA repeats in some cancers and genetic diseases (13, 155, 156).

V. Perspectives and Implications

Together, these structural results have revealed the elegant efficiency of damage recognition and chemical catalysis employed by these key DNA repair enzymes, and, when integrated with biochemical and genetic data, provide important insights into how both indirect and direct protein–protein and protein–DNA interactions influence the activities of the individual enzymes. The emerging theme that DNA repair pathways can be likened to an intricate dance where partner exchanges are critical for the *in vivo* function has clearly benefited from these structural results.

Acknowledgments

This work could not have been possible without the patient and expert collaboration with the laboratories of R. P. Cunningham at the State University of New York at Albany; S. Mitra and R. S. Lloyd at the Sealy Center for Molecular Science, University of Texas Medical Branch (UTMB) in Galveston, Texas; B. Shen at the City of Hope National Medical Center and Beckman Research Institute; A. E. Pegg at the Pennsylvania State University College of Medicine; H. E. Krokan and G. Slupphaug at the Norwegian University of Science and Technology; and G. M. Blackburn at the University of Sheffield. We thank the staff and facilities at the Cornell High Energy Synchrotron Source (CHESS), the Stanford Synchrotron Radiation Laboratory (SSRL), the Advanced Light Source (ALS), and the Advanced Photon Source (APS), which are supported by the National Science Foundation and Department of Energy. Work on DNA repair in the Tainer and Cunningham laboratories is supported by the National Institutes of Health grant GM46312, while work in the Tainer and Shen laboratories is supported by the National Institutes of Health grant CA57348. In addition, work on DNA repair is supported by a Special Fellowship (to C.D.M) from the Leukemia and Lymphoma Society (formerly the Leukemia Society of America), The Skaggs Institute for Chemical Biology (D.J.H., S.S.P.), and graduate research fellowships from the Howard Hughes Medical Institute (C.D.P.) and the National Science Foundation (D.S.D., S.S.P.).

References

1. T. Lindahl, *Nature* (*London*) **362**, 709–715 (1993).
2. T. Lindahl and B. Nyberg, *Biochemistry* **11**, 3610–3618 (1972).
3. T. Lindahl, *Mutat. Res.* **238**, 305–311 (1990).
4. B. T. Smith and G. Walker, *Genetics* **148**, 1599–1610 (1998).
5. E. C. Friedberg, G. C. Walker, and W. Siede, "DNA Repair and Mutagenesis." ASM Press, Washington, D. C., 1995.
6. O. J. Becherel and R. P. Fuchs, *J. Mol. Biol.* **294**, 299–306 (1999).

7. I. Baynton and R. P. Fuchs, *Trends Biochem Sci.* **25,** 74–79 (2000).
8. H.-W. Park, S.-T. Kim, A. Sancar, and J. Deisenhofer, *Science* **268,** 1866–1872 (1995).
9. B. Rydberg and T. Lindahl, *EMBO J.* **1,** 211–216 (1982).
10. D. T. Beranek, *Mutat. Res.* **231,** 11–30 (1990).
11. S. A. Kyrtopoulos, L. M. Anderson, S. K. Chhabra, V. L. Souliotis, V. Pletsa, C. Valavanis, and P. Georgiadis, *Cancer Detect. Prev.* **21,** 391–405 (1997).
12. J. Mattern, U. Eichhorn, B. Kaina, and M. Volm, *Int. J. Cancer* **77,** 919–922 (1998).
13. C. Spiro, P. Pelletier, M. L. Rolfsmeier, M. J. Dixon, R. S. Lahue, G. Gupta, M. S. Park, X. Chen, S. V. Mariappan, and C. T. McMurray, *Mol. Cell* **4,** 1079–1085 (1999).
14. D. S. Daniels, C. D. Mol, A. S. Arvai, S. Kanugula, A. E. Pegg, and J. A. Tainer, *EMBO J.* **19,** 1719–1730 (2000).
15. J. E. A. Wibley, A. E. Pegg, and P. C. E. Moody, *Nucleic Acids Res.* **28,** 393–401 (2000).
16. M. H. Moore, J. M. Gulbis, E. J. Dodson, B. Demple, and P. C. E. Moody, *EMBO J.* **13,** 1495–1501 (1994).
17. H. Hashimoto, T. Inoue, M. Nishioka, S. Fujiwara, M. Takagi, T. Imanaka, and Y. Kai, *J. Mol. Biol.* **292,** 707–716 (1999).
18. D. S. Daniels and J. A. Tainer, *Mutat. Res.* **460,** 151–163 (2000).
19. R. G. Brennan, *Cell (Cambridge, Mass.)* **74,** 773–776 (1993).
20. T. M. Crone, K. Goodtzova, and A. E. Pegg, *Mutat. Res.* **363,** 15–25 (1996).
21. K. Goodtzova, S. Kanugula, S. Edara, and A. E. Pegg, *Biochemistry* **37,** 12489–12495 1998.
22. T. E. Spratt, J. D. Wu, D. E. Levy, S. Kanugula, and A. E. Pegg, *Biochemistry* **38,** 6801–6806 (1999).
23. J. F. Hora, A. Eastman, and E. Bresnick, *Biochemistry* **22,** 3759–3767 (1983).
24. S. Kanugula, K. Goodtzova, and A. E. Pegg, *Biochem. J.* **329,** 545–550 (1998).
25. A. E. Pegg, L. Wiest, C. Mummert, L. Stine, R. C. Moschel, and M. E. Dolan, *Carcinogenesis* **12,** 1679–1683 (1991).
26. K. S. Srivenugopal, X.-H. Yuan, H. S. Friedman, and F. Ali-Osman, *Biochemistry* **35,** 1328–1334 (1996).
27. H. El-Hajj, H. Zhang, and B. Weiss, *J. Bacteriol.* **170,** 1069–1075 (1988).
28. M. H. Gadsden, E. M. McIntosh, J. C. Game, P. J. Wilson, and R. H. Haynes, *EMBO J.* **12,** 4425–4431 (1993).
29. B. K. Tye, P. O. Nyman, I. R. Lehman, S. Hochhauser, and B. Weiss, *Proc. Natl. Acad. Sci. U.S.A.* **74,** 154–157 (1977).
30. H. H. El-Hajj, L. Wang, and B. Weiss, *J. Bacteriol.* **174,** 4450–4456 (1992).
31. W. T. Pu and K. Struhl, *Nucleic Acids Res.* **20,** 771–775 (1992).
32. R. Ivarie, *Nucleic Acids Res.* **15,** 9975–9983 (1987).
33. H. A. Ingraham, L. Dickey, and M. Goulian, *Biochemistry* **25,** 3225–3230 (1986).
34. R. G. Richards, L. C. Sowers, J. Laszlo, and W. D. Sedwick, *Adv. Enzyme Reg.* **22,** 157–185 (1984).
35. B. J. Barclay, B. A. Kunz, J. G. Little, and R. H. Haynes, *Can. J. Biochem.* **60,** 172–194 (1982).
36. M. Goulian, B. Bleile, and B. Y. Tseng, *J. Biol. Chem.* **255,** 10630–10637 (1980).
37. H. Nilsen, M. Otterlei, T. Haug, K. Solum, T. A. Nagelus, F. Skorpen, and H. E. Krokan, *Nucleic Acids Res.* **25,** 750–755 (1997).
38. M. Otterlei, T. Haug, T. A. Nagelhus, G. Slupphaug, T. Lindmo, and H. E. Krokan, *Nucleic Acids Res.* **26,** 4611–4617 (1998).
39. R. D. Ladner, D. E. McNulty, S. A. Carr, G. D. Roberts, and S. J. Caradonna, *J. Biol. Chem.* **271,** 7745–7751 (1996).
40. R. D. Ladner and S. J. Caradonna, *J. Biol. Chem.* **272,** 19072–19080 (1997).
41. E. S. Cedergren-Zeppezauer, G. Larsson, P. O. Nyman, Z. Dauter, and K. S. Wilson, *Nature (London)* **355,** 740–743 (1992).

42. Z. Dauter, K. S. Wilson, G. Larsson, P. O. Nyman, and E. S. Cedergren-Zeppezauer, *Acta Cryst.* **D54**, 735–749 (1998).
43. R. Prasad, R. K. Singhal, D. K. Srivastava, J. T. Molina, A. E. Tomkinson, and S. H. Wilson, *J. Biol. Chem.* **271**, 16000–16007 (1996).
44. Z. Dauter, R. Persson, A. M. Rosengren, P. O. Nyman, K. S. Wilson, and E. S. Cedergren-Zeppezauer, *J. Mol. Biol.* **285**, 655–673 (1999).
45. C. D. Mol, J. M. Harris, E. M. McIntosh, and J. A. Tainer, *Structure* **4**, 1077–1092 (1996).
46. M. J. Bennett, S. Choe, and D. Eisenberg, *Proc. Natl. Acad. Sci. U.S.A.* **91**, 3127–3131 (1994).
47. G. Larsson, L. A. Svensson, and P. O. Nyman, *Nat. Struct. Biol.* **3**, 532–538 (1996).
48. B. G. Vertessy, P. Zalud, P. O. Nyman, and M. Zeppezauer, *Biochim. Biophys. Acta* **1205**, 146–150 (1994).
49. P. C. Wagaman, C. S. Hasselkus-Light, M. Henson, D. L. Lerner, T. R. Phillips, and J. H. Elder, *Virology* **196**, 451–457 (1993).
50. B. G. Vertessy, *Proteins* **28**, 568–579 (1997).
51. B. G. Vertessy, G. Larsson, T. Persson, A. C. Bergman, R. Persson, and P. O. Nyman, *FEBS Lett.* **421**, 83–88 (1998).
52. G. Larsson, P. O. Nyman, and J. O. Kvassman, *J. Biol. Chem.* **271**, 24010–24016 (1996).
53. B. Demple and L. Harrison, *Annu. Rev. Biochem.* **63**, 915–948 (1994).
54. B. A. Gilchrest and V. A. Bohr, *FASEB J.* **11**, 322–330 (1997).
55. M. Radman, *J. Biol. Chem.* **251**, 1438–1445 (1976).
56. B. Demple and S. Linn, *Nature (London)* **287**, 203–208 (1980).
57. R. J. Boorstein, T. P. Hilbert, J. Cadet, R. P. Cunningham, and W. G. Teebor, *Biochemistry* **28**, 6164–6170 (1989).
58. Z. Hatahet, Y. W. Kow, A. A. Purmal, R. P. Cunningham, and S. S. Wallace, *J. Biol. Chem.* **269**, 18814–18820 (1994).
59. G. F. Strniste and S. S. Wallace, *Proc. Natl. Acad. Sci. U.S.A.* **72**, 1997–2001 (1975).
60. F. T. Gates III and S. Linn, *J. Biol. Chem.* **252**, 2802–2807 (1977).
61. P. W. Doetsch and R. P. Cunningham, *Mutat. Res.* **236**, 173–201 (1990).
62. H. L. Katcher and S. S. Wallace, *Biochemistry* **22**, 4071–4081 (1983).
63. C. F. Kuo, D. E. McRee, C. L. Fisher, S. F. O'Handley, R. P. Cunningham, and J. A. Tainer, *Science* **258**, 434–440 (1994).
64. M. M. Thayer, H. Ahern, D. Xing, R. P. Cunningham, and J. A. Tainer, *EMBO J.* **14**, 4108–4120 (1995).
65. H. M. Nash, S. D. Bruner, O. D. Scharer, T. Kawate, T. A. Addona, E. Spooner, W. S. Lane, and G. L. Verdine, *Curr. Biol.* **6**, 968–980 (1996).
66. Y. Guan, R. C. Manuel, A. S. Arvai, S. S. Parikh, C. D. Mol, J. H. Miller, R. S. Lloyd, and J. A. Tainer, *Nat. Struct. Biol.* **5**, 1058–1064 (1998).
67. A. J. Doherty, L. C. Serpell, and C. P. Ponting, *Nucleic Acids Res.* **24**, 2488–2497 (1996).
68. H. Pelletier, M. R. Sawaya, W. Wolfe, S. H. Wilson, and J. Kraut, *Biochemistry* **35**, 12742–12761 (1996).
69. M. L. Michaels, L. Pham, Y. Nghiem, C. Cruz, and J. H. Miller, *Nucleic Acids Res.* **18**, 3841–3845 (1990).
70. J. Labahn, O. D. Scharer, A. Long, K. Ezaz-Nikpay, G. L. Verdine, and T. E. Ellenberger, *Cell (Cambridge, Mass.)* **86**, 321–329 (1996).
71. Y. Yamagata, M. Kato, K. Odawara, Y. Tokuno, Y. Nakashima, N. Matsushima, K. Yasumura, K. Tomita, K. Ihara, Y. Fujii, Y. Nakabeppu, M. Sekiguchi, and S. Fujii, *Cell (Cambridge, Mass.)* **86**, 311–319 (1996).
72. B. W. Mattes, C.-S. Lee, J. Laval, and T. R. O'Connor, *Carcinogenesis* **17**, 643–648 (1996).
73. R. C. Manuel and R. S. Lloyd, *Biochemistry* **36**, 11140–11152 (1997).
74. T. Hollis, Y. Ichikawa, and T. Ellenberger, *EMBO J.* **19**, 758–766 (2000).

75. S. D. Bruner, D. P. G. Norman, and G. L. Verdine, *Nature (London)* **403**, 859–866 (2000).
76. T. Lindahl, *Proc. Natl. Acad. Sci. U.S.A.* **71**, 3649–3653 (1974).
77. G. Slupphaug, I. Eftedal, B. Kavli, S. Bharati, N. M. Helle, T. Haug, D. W. Levine, and H. E. Krokan, *Biochemistry* **34**, 128–138 (1995).
78. A. Verri, P. Mazzarello, S. Spadari, and F. Focher, *Biochem. J.* **287**, 1007–1010 (1992).
79. R. Savva, K. McAuley-Hecht, T. Brown, and L. Pearl, *Nature (London)* **373**, 487–493 1995.
80. C. D. Mol, A. S. Arvai, R. J. Sanderson, G. Slupphaug, B. Kavli, H. E. Krokan, D. W. Mosbaugh, and J. A. Tainer, *Cell (Cambridge, Mass.)* **82**, 701–708 (1995).
81. C. D. Putnam, M. J. Shroyer, A. J. Lundqvist, C. D. Mol, A. S. Arvai, D. W. Mosbaugh, and J. A. Tainer, *J. Mol. Biol.* **287**, 331–346 (1999).
82. G. Xiao, M. Tordova, J. Jagadeesh, A. C. Drohat, J. T. Stivers, and G. L. Gilliland, *Proteins* **35**, 13–24 (1999).
83. R. Ravishankar, M. B. Sagar, S. Roy, K. Purnapatre, P. Handa, U. Varshney, and M. Vijayan, *Nucleic Acids Res.* **26**, 4880–4887 (1998).
84. T. E. Barrett, R. Savva, G. Panayotou, T. Barlow, T. Brown, and L. H. Pearl, *Cell (Cambridge, Mass.)* **92**, 117–129 (1998).
85. C. D. Mol, C.-F. Kuo, M. M. Thayer, R. P. Cunningham, and J. A. Tainer, *Nature (London)* **374**, 381–386 (1995).
86. R. Savva and L. H. Pearl, *Nat. Struct. Biol.* **2**, 752–757 (1995).
87. S. S. Parikh, C. D. Mol, G. Slupphaug, S. Bharati, H. E. Krokan, and J. A. Tainer, *EMBO J.* **17**, 5214–5226 (1998).
88. G. Slupphaug, C. D. Mol, B. Kavli, A. S. Arvai, H. E. Krokan, and J. A. Tainer, *Nature (London)* **384**, 87–92 (1996).
89. M. Higley and R. S. Lloyd, *Mutat. Res.* **294**, 109–116 (1993).
90. J. T. Stivers, K. W. Pankiewicz, and K. A. Watanabe, *Biochemistry* **38**, 952–963 (1999).
91. C. D. Mol, S. S. Parikh, C. D. Putnam, T. P. Lo, and J. A. Tainer, *Annu. Rev. Biophys. Biomol. Struct.* **28**, 101–128 (1999).
92. S. S. Parikh, G. Walcher, G. D. Jones, G. Slupphaug, H. E. Krokan, G. M. Blackburn, and J. A. Tainer, *Proc. Natl. Acad. Sci. U.S.A.* **97**, 5083–5088 (2000).
93. S. S. Parikh, C. D. Putnam, and J. A. Tainer, *Mutat. Res.* **460**, 183–199 (2000).
94. W. Saenger, "Principles of Nucleic Acid Structure." Springer-Verlag, New York, 1984.
95. P. Pourquier, L. M. Ueng, J. Fertala, D. Wang, H. J. Park, J. M. Essigman, M. A. Bjornsti, and Y. Pommier, *J. Biol. Chem.* **274**, 8516–8523 (1999).
96. P. Pourquier, A. A. Pilon, G. Kohlhagen, A. Mazumder, A. Sharma, and Y. Pommier, *J. Biol. Chem.* **272**, 26441–26447 (1997).
97. P. Pourquier, L.-M. Ueng, G. Kohlhagen, A. Mazumder, M. Gupta, K. W. Kohn, and Y. Pommier, *J. Biol. Chem.* **272**, 7792–7796 (1997).
98. B. Demple, A. Johnson, and D. Fung, *Proc. Natl. Acad. Sci. U.S.A.* **83**, 7731–7735 (1986).
99. L. A. Loeb, *Cell (Cambridge, Mass.)* **40**, 483–484 (1985).
100. C. N. Robson, D. Hochhauser, R. Craig, K. Rack, V. J. Buckle, and I. D. Hickson, *Nucleic Acids Res.* **20**, 4417–4421 (1992).
101. S. Xanthoudakis and T. Curran, *EMBO J.* **11**, 653–665 (1992).
102. S. Seki, S. Ikeda, S. Watanabe, M. Hatsushika, K. Tsutsui, and K. A. Zhang, *Biochim. Biophys. Acta* **1079**, 57–64 (1991).
103. C. V. Ramana, I. Boldogh, T. Izumi, and S. Mitra, *Proc. Natl. Acad. Sci. U.S.A.* **95**, 5061–5066 (1998).
104. D. M. Wilson III and L. H. Thompson, *Proc. Natl. Acad. Sci. U.S.A.* **94**, 12754–12757 (1997).
105. C. D. Mol, S. S. Parikh, T. P. Lo, and J. A. Tainer, *in* "Nucleic Acids and Molecular Biology." (F. Eckstein and D. M. J. Lilley, eds.), Vol. 12. Springer, Berlin, 1998.

106. E. Chan and B. Weiss, *Proc. Natl. Acad. Sci. U.S.A.* **84,** 3189–3193 (1987).
107. L. K. Walkup and T. Kogoma, *J. Bacteriol.* **171,** 1476–1484 (1989).
108. B. A. Kunz, E. S. Henson, H. Roche, D. Ramotar, T. Nunoshiba, and B. Demple, *Proc. Natl. Acad. Sci. U.S.A.* **91,** 8165–8169 (1994).
109. D. Ramotar, S. C. Popoff, E. B. Gralla, and B. Demple, *Mol. Cell Biol.* **11,** 4537–4544 (1991).
110. J. Y. Masson, S. Tremblay, and D. Ramotar, *Gene* **179,** 291–293 (1996).
111. B. M. Haltiwanger, N. O. Karpinich, and T. F. Taraschi, *Biochem. J.* **345,** 85–89 (2000).
112. C. D. Mol, C. F. Kuo, M. M. Thayer, R. P. Cunningham, and J. A. Tainer, *Nature (London)* **374,** 381–386 (1995).
113. M. A. Gorman, S. Morera, D. G. Rothwell, E. de La Fortelle, C. D. Mol, J. A. Tainer, I. D. Hickson, and P. S. Freemont, *EMBO J.* **16,** 6548–6558 (1997).
114. D. J. Hosfield, Y. Guan, B. J. Haas, R. P. Cunningham, and J. A. Tainer, *Cell (Cambridge, Mass.)* **98,** 397–408 (1999).
115. C. D. Mol, T. Izumi, S. Mitra, and J. A. Tainer, *Nature (London)* **403,** 451–456 (2000).
116. D. M. Wilson III, M. Takeshita, and B. Demple, *Nucleic Acids Res.* **25,** 933–939 (1997).
117. D. M. Wilson III, M. Takeshita, A. P. Grollman, and B. Demple, *J. Biol. Chem.* **270,** 16002–16007 (1995).
118. Y. Masuda, R. A. Bennett, and B. Demple, *J. Biol. Chem.* **273,** 30352–30359 (1998).
119. R. A. O. Bennett, D. M. Wilson III, D. Wong, and B. Demple, *Proc. Natl. Acad. Sci. U.S.A.* **94,** 7166–7169 (1997).
120. S. S. Parikh, C. D. Mol, D. J. Hosfield, and J. A. Tainer, *Curr. Opin. Struct. Biol.* **9,** 37–47 (1999).
121. A. Y. Lau, O. D. Schärer, L. Samson, G. L. Verdine, and T. Ellenberger, *Cell (Cambridge, Mass.)* **95,** 249–258 (1998).
122. T. R. Waters, P. Gallinari, J. Jiricny, and P. F. Swann, *J. Biol. Chem.* **274,** 67–74 (1999).
123. A. Marintchev, M. A. Mullen, M. W. Maciejewski, B. Pan, M. R. Gryk, and G. P. Mullen, *Nat. Struct. Biol.* **6,** 884–893 (1999).
124. Y. Kubota, R. A. Nash, A. Klungland, P. Schar, D. E. Barnes, and T. Lindahl, *EMBO J.* **15,** 6662–6670 (1996).
125. K. W. Caldecott, C. K. McKeown, J. D. Tucker, S. Ljungquist, and L. H. Thompson, *Mol. Cell. Biol.* **14,** 68–76 (1994).
126. A. Klungland and T. Lindahl, *EMBO J.* **16,** 3341–3348 (1997).
127. D. W. Banner, A. C. Bloomer, G. A. Petsko, D. C. Phillips, C. I. Pogson, I. A. Wilson, P. H. Coran, A. J. Furth, J. D. Milman, R. E. Offord, J. D. Priddle, and S. G. Waley, *Nature (London)* **255,** 609–614 (1975).
128. G. K. Farber and G. A. Petsko, *Trends Biochem. Sci.* **15,** 228–234 (1990).
129. D. Reardon and G. K. Farber, *FASEB J.* **9,** 497–503 (1995).
130. A. Volbeda, A. Lahm, F. Sakiyama, and D. Suck, *EMBO J.* **10,** 1607–1618 (1991).
131. C. Romier, R. Dominguez, A. Lahm, O. Dahl, and D. Suck, *Proteins: Struct. Funct. Genet.* **32,** 414–424 (1998).
132. E. Hough, L. K. Hansen, B. Birknes, K. Jynge, S. Hansen, A. Hodvik, C. Little, E. Dodson, and Z. Derewenda, *Nature (London)* **338,** 357–360 (1989).
133. S. Hansen, L. K. Hansen, and E. Hough, *J. Mol. Biol.* **225,** 543–549 (1992).
134. S. Hansen, E. Hough, L. A. Svensson, Y. L. Wong, and S. F. Martin, *J. Mol. Biol.* **234,** 179–187 (1993).
135. H. Ide, K. Tedzuka, H. Shimzu, Y. Kimura, A. A. Purmal, S. S. Wallace, and Y. W. Kow, *Biochemistry* **33,** 7842–7847 (1994).
136. K. Kim, S. Biade, and Y. Matsumoto, *J. Biol. Chem.* **273,** 8842–8848 (1998).
137. R. Gary, K. Kyung, H. L. Cornelius, M. S. Park, and Y. Matsumoto, *J. Biol. Chem.* **274,** 4354–4363 (1999).

138. J. J. Harrington and M. R. Lieber, *Genes Dev.* **8,** 1344–1355 (1994).
139. R. Murante, L. Huang, J. J. Turchi, and R. Bambara, *J. Biol. Chem.* **269,** 1191–1196 (1994).
140. T. Lindahl, *Eur. J. Biochem.* **18,** 407–414 (1971).
141. R. S. Murante, J. A. Rumbaugh, C. J. Barnes, J. R. Norton, and R. A. Bambara, *J. Biol. Chem.* **271,** 25888–25897 (1996).
142. J. J. Harrington and M. R. Lieber, *EMBO J.* **13,** 1235–1246 (1994).
143. E. Warbrick, D. P. Lane, D. M. Glover, and L. S. Cox, *Oncogene* **14,** 2313–2321 (1997).
144. E. Warbrick, *Bioessays* **20,** 195–199 (1998).
145. L. Li, X. Lu, C. A. Peterson, and R. J. Legerski, *Mol. Cell. Biol.* **15,** 5396–5402 (1995).
146. Z. O. Jonsson, R. Hindges, and U. Hubscher, *EMBO J.* **17,** 2412–2425 (1998).
147. M. E. Budd and J. L. Campbell, *Mol. Cell. Biol.* **17,** 2136–2142 (1997).
148. D. J. Hosfield, G. Frank, Y. Weng, J. A. Tainer, and B. Shen, *J. Biol. Chem.* **273,** 27154–27161 (1998).
149. K. Y. Hwang, K. Baek, H.-Y. Kim, and Y. Cho, *Nat. Struct. Biol.* **5,** 707–713 (1998).
150. T. C. Mueser, N. G. Nossal, and C. C. Hyde, *Cell (Cambridge, Mass.)* **85,** 1101–1112 (1996).
151. T. A. Ceska, J. R. Sayers, G. Stier, and D. Suck, *Nature (London)* **382,** 90–93 (1996).
152. Y. Kim, S. H. Eom, J. Wang, D. S. Lee, S. W. Suh, and T. A. Steitz, *Nature (London)* **376,** 612–616 (1995).
153. M. R. Shen, I. M. Jones, and H. Mohrenweiser, *Cancer Res.* **58,** 604–608 (1998).
154. C. J. Barnes, A. F. Wahl, B. Shen, M. S. Park, and R. A. Bambara, *J. Biol. Chem.* **271,** 29624–29631 (1996).
155. D. X. Tishkoff, N. Filosi, G. M. Gaida, and R. D. Kolodner, *Cell (Cambridge, Mass.)* **88,** 253–263 (1997).
156. J. K. Schweitzer and D. M. Livingston, *Human Mol. Genetics* **7,** 69–74 (1998).
157. T. Izumi and S. Mitra, *Carcinogenesis* **19,** 525–527 (1998).

Potential Double-Flipping Mechanism by E. coli MutY

PAUL G. HOUSE,*
DAVID E. VOLK,†
VARATHARASA THIVIYANATHAN,†
RAYMOND C. MANUEL,*
BRUCE A. LUXON,†
DAVID G. GORENSTEIN,† AND
R. STEPHEN LLOYD*

*Center for Molecular Science
†Sealy Center for Structural Biology
University of Texas Medical Branch
Galveston, Texas 77555-1071

I. DNA Glycosylases 350
II. MutY: An Adenine Glycosylase 350
III. Review of the Reaction Mechanism of DNA Glycosylases and
Glycosylase/AP Lyases 352
IV. Structure of the Catalytic Domain of MutY 353
V. MutY: Glycosylase with an Opportunistic Lyase Activity 355
VI. Role for the C-Terminal Domain of MutY in Substrate Specificity . 356
VII. Solution Structure of the C-Terminal Domain of MutY 357
VIII. Proposal for a Double-Flipping Mechanism by MutY 361
References 362

To understand the structural basis of the recognition and removal of specific mismatched bases in double-stranded DNAs by the DNA repair glycosylase MutY, a series of structural and functional analyses have been conducted. MutY is a 39-kDa enzyme from *Escherichia coli*, which to date has been refractory to structural determination in its native, intact conformation. However, following limited proteolytic digestion, it was revealed that the MutY protein is composed of two modules, a 26-kDa domain that retains essential catalytic function (designated p26MutY) and a 13-kDa domain that is implicated in substrate specificity and catalytic efficiency. Several structures of the 26-kDa domain have been solved by X-ray crystallographic methods to a resolution of up to 1.2 Å. The structure of a catalytically incompetent mutant of p26MutY complexed with an adenine in the substrate-binding pocket allowed us to propose a catalytic mechanism for MutY. Since reporting the structure of p26MutY, significant progress has been made in solving the solution structure of the noncatalytic C-terminal 13-kDa domain of MutY by NMR spectroscopy. The topology and secondary structure of this domain are very similar to that of MutT, a pyrophosphohydrolase. Molecular modeling techniques employed to integrate the two domains of MutY with DNA suggest that MutY can wrap around the DNA and initiate catalysis by potentially flipping adenine and 8-oxoguanine out of the DNA helix. © 2001 Academic Press.

I. DNA Glycosylases

During the last decade, outstanding progress has been made in the study of DNA glycosylases that initiate the base excision repair pathway (BER) (*1, 2*). A combination of structural and biochemical studies from several laboratories has enabled the dissection of the multiple steps involved in the recognition and binding to damaged/mismatched DNA substrates and the mechanistic steps in the removal of the base. The active sites have been well characterized and the amino acid residues that participate in the catalytic events have been identified. Within the last five years, the three-dimensional structures of several DNA glycosylases have been determined. Some of these structures are protein–DNA complexes, and these have contributed significantly to the testing of hypotheses for the reaction mechanisms of these enzymes.

II. MutY: An Adenine Glycosylase

In *Echerichia coli*, there is a complex mechanism that limits mutagenesis due to the formation or incorporation of 8-oxoguanine (8oxoG) in DNA or due to the replication of DNA containing 8oxoG lesions (*3*) (Fig. 1). During replication, cytosine or adenine can be incorporated opposite 8oxoG depending on the DNA polymerase involved (*4*). When the daughter strand contains a C opposite 8oxoG, Fapy DNA glycosylase (MutM) removes 8oxoG and allows the BER machinery to fill the gap with a G and thus restore the original sequence (*5*). However, when A is incorporated opposite 8oxoG, the mispaired adenine is removed by MutY. The complete repair of the lesion is not accomplished until C pairs with 8oxoG and creates a substrate for MutM (*6*). To augment the cell's ability to minimize the effects of oxidative stress on mutagenicity, an additional protein, MutT, functions as an 8oxodGTPase and degrades this mutagenic dNTP from the nucleotide pool (*7*). In addition to its role in the repair of oxidative lesions, MutY also removes adenine when it is mispaired with guanine. Thus, the biological role of MutY is to avoid G:C to T:A transversion mutations (*8*). *In vitro* studies have shown that MutY also recognizes and removes other purine analogs with variable efficiencies (*9–12*).

In recent years, several laboratories have investigated the reaction mechanism of MutY. This adenine glycosylase exhibits the characteristics of both a

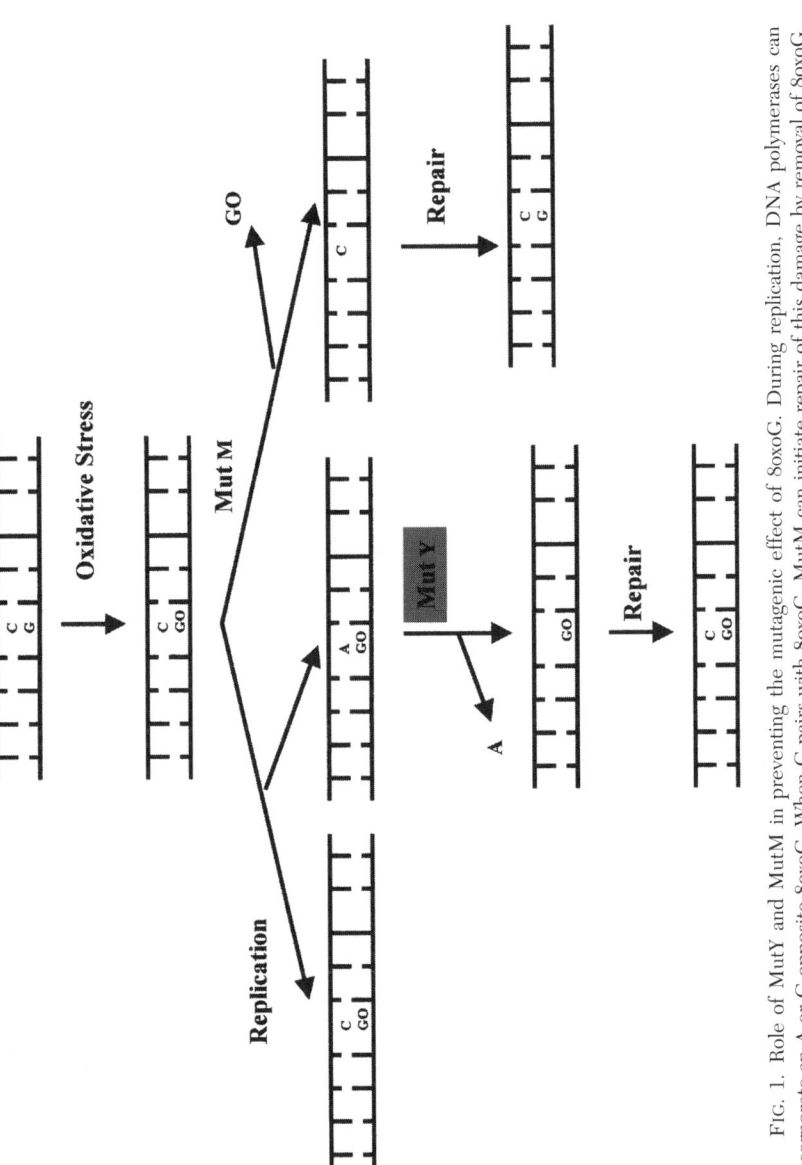

FIG. 1. Role of MutY and MutM in preventing the mutagenic effect of 8oxoG. During replication, DNA polymerases can incorporate an A or C opposite 8oxoG. When C pairs with 8oxoG, MutM can initiate repair of this damage by removal of 8oxoG. However, when A is incorporated opposite 8oxoG, MutY removes A, and thus provides an opportunity for repair polymerases to incorporate a C opposite 8oxoG. This creates a substrate for MutM. If a second round of replication takes place on the A:8oxoG mismatch without the action of MutY, the mutation will be fixed in one of the daughter strands. Reprinted from Ref. (3) with the permission of The American Society for Microbiology.

simple glycosylase and a glycosylase/AP lyase (11). Within the HhH glycosylase superfamily (13–19), Endonuclease III exhibits chemical properties that are consistent with a combined glycosylase/AP lyase (17, 18, 20). The reaction mechanism of AlkA has been thoroughly studied and is a true representative of a pure glycosylase (no lyase activity) (14, 16; see also Hollis et al., this volume). At the initiation of this work, a complete model for the reaction mechanism of MutY was lacking. However, through the past four years, by combining structural and biochemical analyses (21–24), a rection mechanism has been proposed and tested that is consistent with the chemical behavior of MutY.

III. Review of the Reaction Mechanism of DNA Glycosylases and Glycosylase/AP Lyases

Based on the identity of the nucleophile that attacks the C1' of the sugar, a unified catalytic mechanism has been proposed for DNA glycosylases (25). This model proposes that a primary or secondary amine serves as a nucleophile in DNA glycosylase/AP lyases. In DNA glycosylases that do not have AP lyase activity, either a water molecule is activated to labilize the glycosidic bond or the reaction may proceed through a direct S_N1 mechanism (16). Glycosylase/AP lyases cleave the phosphodiester bond 3' to the AP site via a β-elimination reaction. The reaction proceeds through the formation of an imino intermediate, which can be reduced in the presence of $NaBH_4$ to yield a covalent, dead-end protein–DNA complex (25, 26). In contrast, pure glycosylases do not produce covalent complexes in the presence of $NaBH_4$, since they do not proceed through an imino intermediate. Additionally, it was recognized that an acidic amino acid would likely be required to labilize the glycosyl bond through either protonation of the base or the endocyclic oxygen of the deoxyribose, although alternative proposals have been raised that negate the need for proton transfer (27).

The unified catalytic mechanism was tested by examining the reaction mechanisms of several glycosylases and glycosylase/AP lyases (28). The data presented within that study provided corroborative evidence for our hypothesis, since all of the enzymes that were known to possess the combined glycosylase/AP lyase activities were efficiently trapped as covalent intermediates on specific substrates in the presence of $NaBH_4$, while none of the pure glycosylases could be trapped. Emanating from these initial studies, it has been demonstrated that this mechanism is valid for a number of other DNA glycosylase/AP lyases: *Micrococcus luteus* pyrimidine dimer glycosylase—*Mlu*-pdgI (29, 30); Chlorella virus, PBCV1, pyrimidine dimer glycosylase—cv-pdg (31–33); *Bacillus sphaericus* pyrimidine dimer glycosylase—*Bsp*-pdg (34); and *Neisseria mucosa* pyrimidine dimer glycosylases I and II—*Nmu*-pdg I and *Nmu*-pgd II (35). $NaBH_4$ trapping experiments were also extended to DNA polymerase β, an enzyme that has

POTENTIAL DOUBLE-FLIPPING MECHANISM   353

been shown to possess AP lyase activity in addition to its polymerase function (36). Additionally, a thorough analysis of the catalytic mechanisms of all known dRPases was completed and is highly consistent with our previous hypotheses; all of the enzymes that possess authentic dRPase activities proceeded via a covalent imino intermediate (37).

IV. Structure of the Catalytic Domain of MutY

To gain insight into the structural basis of MutY's substrate recognition and catalysis, strategies were designed to identify a minimal catalytically active form of MutY. Digestion of MutY with trypsin yielded two stable domains: the 26-kDa and 13-kDa fragments, which are designated as p26 and p13, respectively. The

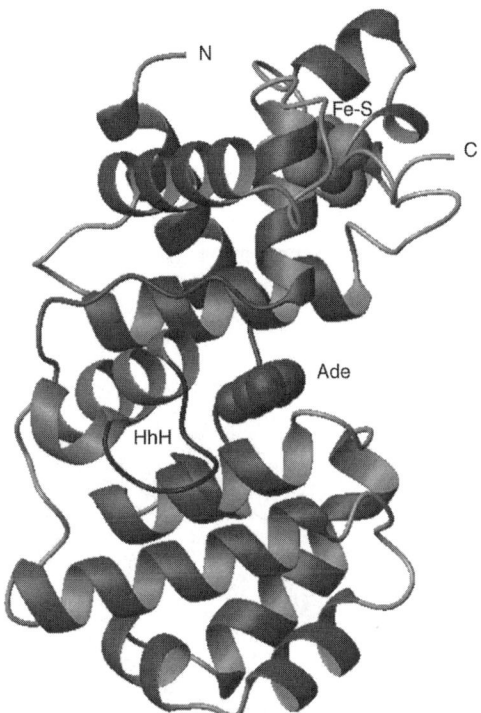

FIG. 2. Structure of p26MutY with adenine in the active site. The catalytic core of MutY (Met1–Lys225) consists of two α-helical domains; an iron–sulfur cluster domain with a [4Fe–4S] cluster (Fe–S) and a six-helix barrel domain. These two domains are linked by two short loops with adenine (Ade) bound in the interface of the domains. A helix–hairpin–helix (HhH) motif, characteristic of many proteins that interact with DNA is present near the active site.

p26 domain of MutY was shown to retain the essential catalytic properties of the intact MutY enzyme by virtue of its ability to specifically excise adenine opposite guanine or 8oxoG (*11, 21*). In all these experiments, both the p26 domain and the intact protein catalyzed the cleavage of the glycosyl and phosphodiester bonds, as evidenced by no change in the percentage of DNA products formed when secondarily treated with piperidine. The substrate specificity of the intact MutY and the p26 domain was also assessed (*11*). The coding region of the catalytic domain of MutY (Met1 to Lys225) was subcloned and expressed to approximately 20% of the total soluble cellular protein.

In collaboration with Dr. John A. Tainer (The Scripps Research Institute, La Jolla, CA), three different structures of p26MutY were determined at resolutions ranging between 1.2 Å and 1.8 Å. In addition to the wild-type p26MutY, the structure of a catalytically incompetent mutant of p26MutY (D138N) was also obtained. However, the structure that provided the greatest insight on the catalytic mechanism of MutY was that of p26MutY–D138N with an adenine buried in the active site of the enzyme (Fig. 2). The catalytic domain of MutY is an elongated molecule consisting of two α-helical domains: a 6-helix barrel domain (Leu22–Leu130), and a domain formed by 4 α-helices along with a 4Fe–4S cluster (Met1–Gly18, Phe135–Lys225) (Fig. 2). The 4Fe–4S cluster is coordinated by four cysteines (192, 199, 202, and 208) and appears to protect the integrity of this domain. These two domains are connected by two loops and form a substrate-specific pocket in the center. The overall structure of p26MutY closely resembles that of *E. coli* Endonuclease III (*17*). When the coulombic electrostatic potential of the surface of p26MutY is examined, it is apparent that

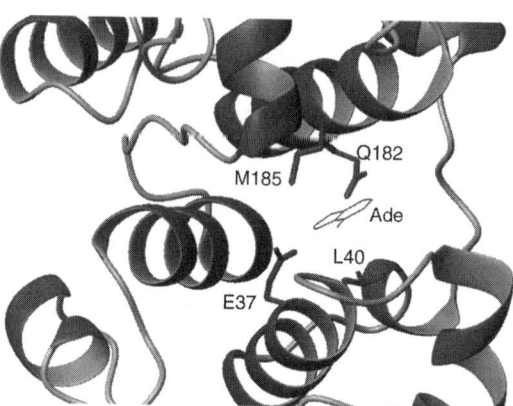

FIG. 3. Active site of MutY with adenine in the base-specificity pocket. Adenine (Ade) is packed between the hydrophobic residues Leu40 and Met185. Gln182 can potentially have two hydrogen-bonds with N1 and N6 of adenine. In addition, Glu37 can hydrogen-bond with N7 of adenine.

DNA binds to the face of the protein that contains a positively charged surface created by the side chains of nine lysines and five arginines. The structure also shows a helix–hairpin–helix (HhH) motif that is widely conserved in many other DNA interacting proteins. This structure suggests that the mismatched adenine is flipped out of the DNA helix into the substrate-specific pocket. In addition to the coordinated series of hydrogen bonds in the specific pocket formed by conserved residues at the domain interface, adenine packs between hydrophobic residues Leu40 and Met185 (Fig. 3). The structure of the p26MutY–adenine complex also revealed the hydrogen-bonding network between adenine and the side chains of amino acids in the active site (Fig. 3).

V. MutY: Glycosylase with an Opportunistic Lyase Activity

MutY displays the properties of a simple glycosylase by its inability to efficiently trap as a protein–DNA complex when $NaBH_4$ is present throughout the reaction. However, addition of $NaBH_4$ after the reaction has incubated several hours shows high levels of trapping (38). It also displays characteristics of a combined glycosylase/AP lyase in that it can cleave the phosphodiester backbone, generating a ring-opened sugar, α,β-unsaturated aldehyde (11). Inspection of the active site of MutY with adenine in the pocket of p26MutY suggested that Asp138 and Glu37 could be involved in the catalytic mechanism (Fig. 4). Cleavage of the glycosyl bond can occur through the protonation of adenine N7 by Glu37, with simultaneous attack of an activated water molecule coordinated through Asp138. To test this proposal, mutant proteins in which Asp138 and Glu37 were changed to Arg and Ser, respectively, and expressed and purified from $mutY^-$ E. coli. Biochemical analysis showed that the D138N and E37S mutants lost both glycosylase and lyase activities (22).

Further examination of the active site revealed a potential chemical mechanism to explain the lyase activity exhibited by MutY. A water-catalyzed nucleophilic displacement cleavage of the N-glycosyl bond could be followed by a β-elimination reaction, initiated by an opportunistically positioned ε-amino group of a lysine. Following the glycosyl bond cleavage, the deoxyribose sugar is available for nucleophilic attack by the ε-amino group of Lys142. This model would account for the chemical behavior of MutY in which the experimental data show both the cleavage of the glycosyl and phosphodiester bonds and the relative difficulty of trapping the covalent imino intermediate in the continuous presence of $NaBH_4$. This model has been experimentally confirmed by the identification of Lys142 as an active-site nucleophile capable of catalyzing the β-elimination reaction (38–40).

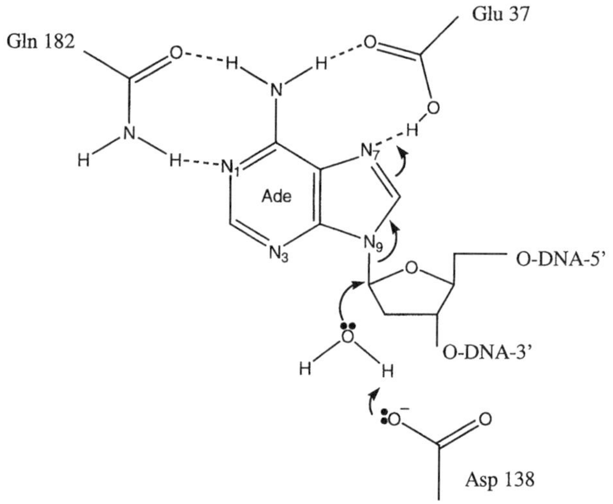

FIG. 4. Glycosylase reaction mechanism of MutY; Asp138 abstracts a proton from a water molecule, activating the water for attack on C1′ of the excised adenine. Gln182 and Glu37 stabilize the excised adenine with a series of hydrogen bonds.

VI. Role for the C-Terminal Domain of MutY in Substrate Specificity

The C-terminal domain is an important determinant of specificity for 8oxoG in the noncleaved strand of MutY substrates. This has been determined by experiments comparing the substrate and enzymatic differences between intact MutY and p26MutY. Preliminary studies in our laboratory indicated that the absence of the 13-kDa C-terminal domain caused subtle changes in the substrate specificity on A:G and A:8oxoG mismatches (11, 21).

In subsequent investigations, binding and excision of the p26 domain and MutY on 12 natural and modified substrates were analyzed (11). The mismatches were composed of adenine, and other purine analogs, nebularine, 2-aminopurine, and inosine, each paired with guanine, 8oxoG, or cytosine. Although, these data were not subjected to rigorous kinetic analyses, it was evident that the specificity of mismatch recognition and binding was altered in the absence of the C-terminal domain (11). Additional studies have also reached similar conclusions that the C-terminal domain makes major contributions to the enzyme's substrate specificity, describing the loss of specificity for A:8oxoG when the C-terminal domain was deleted (41).

Rate constants for the glycosylase reaction and enzyme turnover of the N-terminal catalytic domain and MutY on A:G and A:8oxoG substrates have been determined (42). The rate constant for adenine excision from A:8oxoG by MutY was at least 30-fold larger than the rate constants for excision of adenine from A:8oxoG by p26MutY or adenine from A:G by either MutY or p26MutY. Rate constants were also measured for the turnover of MutY and the p26 domain from DNA containing an abasic (AP) site opposite either 8oxoG or G. MutY dissociates 1500-fold slower from an 8oxoG-containing product DNA than G-containing product DNA. The catalytic domain by itself dissociates at a similar rate regardless of G or 8oxoG in DNA (42). These kinetic data compellingly suggest that the C-terminal domain has a significant effect on discriminating between A:G and A:8oxoG mismatches.

The importance of the C-terminal domain in 8oxoG specificity was further demonstrated by studies showing that intact MutY excised A from A:8oxoG at least 20-fold faster than the N-terminal catalytic domain (43). A series of binding experiments with 8oxoG paired with A, G, T, C, or inosine showed at least 18-fold lower affinity for p26MutY compared with intact MutY. In addition, deletion of the C-terminal domain caused a mutator phenotype, demonstrating the significance of this domain in biological function (43).

VII. Solution Structure of the C-Terminal Domain of MutY

While efforts were continued to obtain a crystal structure of the intact MutY, work was initiated to obtain the solution structure of the C-terminal domain. Structural analysis of the C-terminal domain with the existing structure of the catalytic domain (22) should provide insights to substrate recognition and discrimination between guanine and 8oxoG.

In order to solve the solution structure of the C-terminal domain of MutY, it was necessary to develop an efficient expression system in which large quantities of this domain would accumulate in a soluble form. Expression of residues 226–350 of MutY with an additional N-terminal methionine was not successful. It had been established earlier that small proteins in *E. coli* may accumulate if a lysine or arginine residue follows the initiating methionine (44). Thus, a lysine codon was inserted between the N-terminal methionine codon and the codon for Q226. Fortuitously, the last residue (225) in the N-terminal domain is a lysine, and insertion of a lysine resulted in the inclusion of the last residue of the N-terminal domain. *E. coli* cells transformed with the vector containing this construct expressed the C-terminal domain of MutY as 10% of total cellular protein. Surprisingly, almost all of the recombinant C-terminal domain accumulated in the periplasmic space and could be released by osmotic shock. Both ^{13}C and ^{13}C/^{15}N-labeled samples were prepared and are stable up to a concentration of 1.4 mM. NMR spectra were collected on 750 and 600 MHz Varian

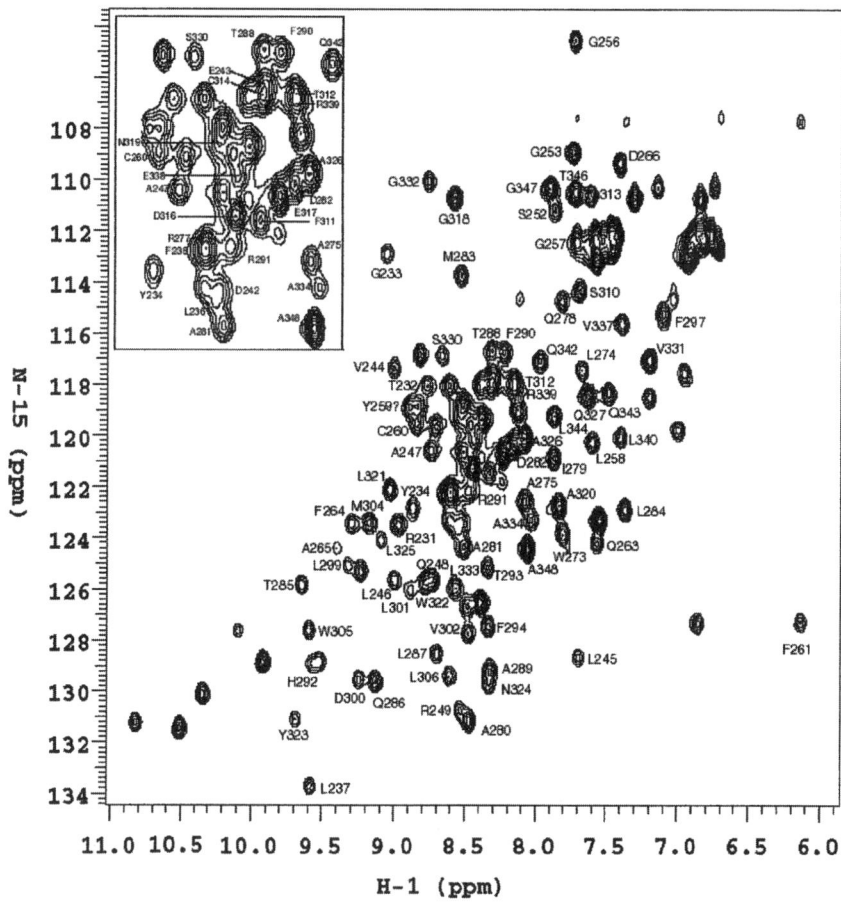

FIG. 5. ^{1}H–^{15}N-HSQC spectrum of the MutY C-terminal domain. Greater than 90% of the resonances from backbone amide protons are labeled, indicating good dispersion of signals in the nitrogen dimension.

Unity Plus instruments using triple resonance probes and pulsed field gradients (23) (Fig. 5).

Sequential backbone resonance assignments have been made by analyzing crosspeaks from the following spectra: HNCA, HNCO, HNCACB, HBCBCA (CO)NH, (HB)CBCA, CO(CA)HA. Three-dimensional HCCH–TOCSY experiments were used to determine side-chain assignments. Secondary structure predictions were made using the chemical shift index method (45) and data from chemical shifts of H_α, C_α, C_β, and C′ (23). This method predicted well-defined areas of α-helices and β-strands (Fig. 6). Verifying these predictions

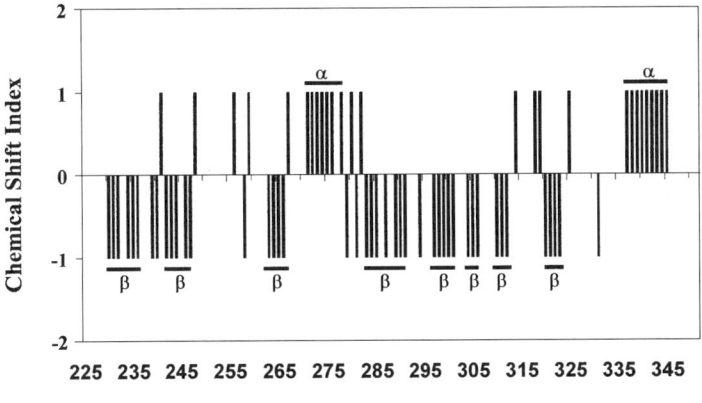

FIG. 6. Chemical shift index prediction of the MutY C-terminal domain secondary structure Consensus chemical shift index (CSI) prediction of secondary structure based on H_α, C_α, C_β, and CO chemical shifts. Indexes of +1, 0, and −1 indicate α-helical, random coil, and β-strand structure, respectively. α and β indicate secondary structure elements. Reprinted from Ref. (23) with permission from Kluwer Academic Publications.

using ^{15}N-edited NOESY data, it was concluded that the C-terminal domain consists of two α-helices and five β-strands (24). Long-range NOEs between the β-strands were used to construct two β-sheets (Fig. 7). One sheet consists of the $\beta1$, $\beta3$, and $\beta4$ strands, with $\beta4$ in the middle and parallel to $\beta1$ and antiparallel to $\beta3$. $\beta2$ and $\beta5$ compose a small sheet and are arranged in an antiparallel configuration. No long-range NOEs have been assigned between the two sheets, even though strands $\beta1$ and $\beta2$ (in different sheets) are separated by only five residues.

Comparing the secondary structure and topology of the C-terminal domain of MutY with the known NMR structure of MutT (46) strongly supports the hypothesis that the C-terminal domain of MutY and MutT are evolutionarily related (24). Figure 8 shows the similarity in the secondary structure of these two proteins. Relative to the alignment previously published (42), this alignment of MutY residues Leu270–Val308 was shifted by one residue to conserve alignment with MutT residues, Leu54, Leu67, Trp85, and Leu86. Additionally, the gap between Val308 and Ser309 was increased by one residue. These minor changes served to conserve critical hydrophobic interactions that may be important in the maintenance of the MutT structure, and conserve residues that interact with MutT substrate analogs (46). Regardless of the alignment, the amino acid sequences of the C-terminal domain of MutY (Gln226–Val350) and MutT are approximately 12% identical.

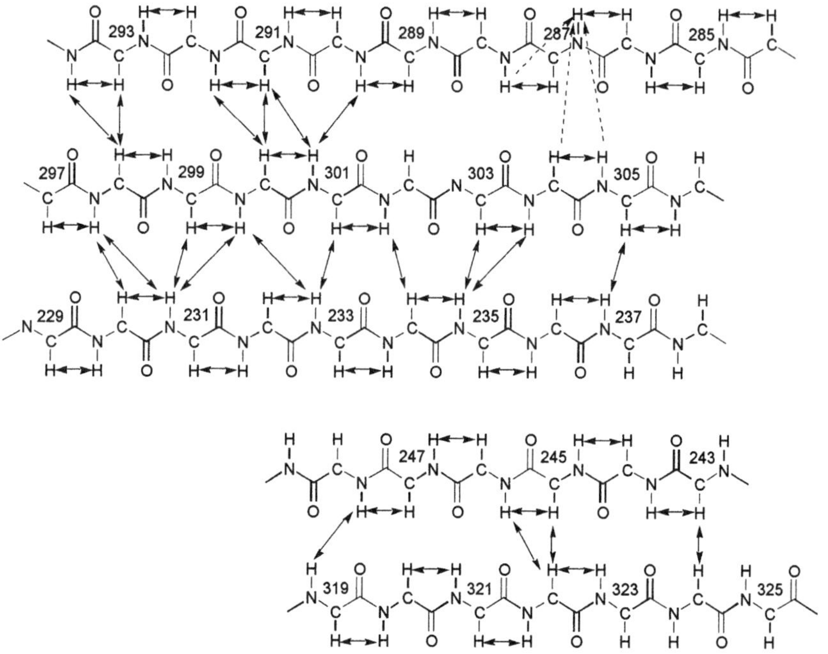

FIG. 7. Compilation of major NOE data for p13 β sheets. NOEs used to construct the β sheets and align the β strands are shown as arrows. Unusual NOEs to the amide proton of L287, which indicate some disruption of the normal β-sheet structure, are indicated by dashed arrows. Reprinted with permission from Ref. (24). Copyright 2000, American Chemical Society.

```
                                     β1             β2
MutY(E. coli)    225    KQTLPERTGYFLLLQHEDEVLLAQRPPSGLWGGLYCFPQFADEES
MutT(E. coli)      1    MKKLQIAVGIIRNENNEIFITRRAADAHMANKLEFPGGKIEMG
                                     β1             β2

                                  α1              β3            β4
MutY(E. coli)    270    ------LRQWLAQRQIAADMLTQLTAFRHTFSHFHLDIVPMWLPV
MutT(E. coli)     44    ETPEQAVVRELQEEVGITPQHFSLFEKLEYEFPDRHITLWFWLVE
                                  α1              β3            β4


                                 β5                 α2
MutY(E. coli)    309    --SSFTGCMDEGNALWYNLAQPPSVGLAAPVERLLQQLRTGAPV
MutT(E. coli)     89    RWEGEPWGKEGQPGEWMSLVGLNADDFPPANEPVIAKLKRL
                                 β5                 α2
```

FIG. 8. Alignment of MutT and p13. Sequence alignment of the *E. coli* MutY C-terminal domain and MutT proteins illustrating the common secondary structural elements present in both proteins. Conserved residues are in bold letters. The underlined residues are conserved in the alignment of Ref. (42). Reprinted with permission from Ref. (24). Copyright 2000, American Chemical Society.

VIII. Proposal for a Double-Flipping Mechanism by MutY

Given that the C-terminal domain of MutY is structurally similar to MutT and that the role of MutT is to cleanse the dNTP pools of 8oxoG-modified triphosphates, it seems reasonable to speculate that the role of the C-terminal domain of MutY is to recognize and bind 8oxoG. The X-ray crystal structure of the 26-kDa catalytic domain of MutY already has established that adenine is flipped to an extrahelical position located deep within a channel in that domain. In an attempt to visualize a model of the interaction of the C-terminal domain of MutY with 8oxoG, a structural model was created (24) using the coordinates of the p26MutY, a model structure of the C-terminal domain, and the DNA from a DNA–AlkA co-crystal complex (16). The C-terminal domain model structure was constructed from the NMR secondary constraints applied to the MutT structure using AMBER. The rationale for choosing the DNA from the AlkA co-crystal complex was the fact that the basic fold of the AlkA catalytic domain and the

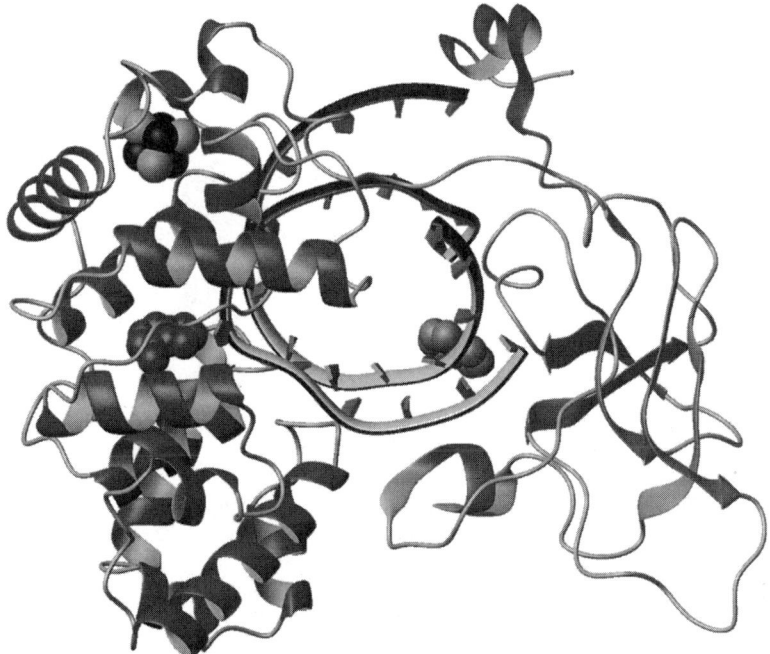

FIG. 9. Structural model of MutY complexed with DNA. The coordinates of the DNA are from duplex DNA containing 1-azaribose as an abasic site analog, co-crystallized with AlkA protein from Ref. (16). The two flipped bases are represented as adenines near the center of the figure.

catalytic domain of MutY are similar and both belong to the HhH glycosylase superfamily. This model (Fig. 9) shows MutY wrapping around the DNA, similar to the overall structure of a β-clamp polymerase processivity factor (47). This composite structure accommodates both the flipping of the adenine into the active site of the catalytic domain and positions the C-terminal domain near 8oxoG.

It is proposed that during the recognition step of MutY for an A:8oxoG mispair, that both the adenine and 8-oxoguanine are moved to extrahelical positions. The hypothesis of the 8-oxoguanine moving to an extrahelical position are supported by the biochemical differences in substrate recognition between p26 and intact MutY as well as the p13 domain structural similarity to MutT. Additionally, the 1500-fold slower turnover of MutY from DNA containing an 8oxoG opposite an abasic site compared to p26MutY turnover on the same substrate indicates that the C-terminal domain interacts with 8oxoG (42). Also, a recent study of the structure of *E. coli* Endonuclease IV complexed with DNA containing an abasic site analog provides strong evidence for a double-flipping mechanism operative in endonuclease IV (48). Based on these observations and the model generated, it seems reasonable to propose that MutY utilizes a double nucleotide-flipping mechanism. This is only a working model and awaits further confirmation.

Acknowledgments

It has been our privilege to collaborate with Dr. John A. Tainer, Scripps Research Institute, CA, in determining the X-ray crystal structure of the catalytic domain of MutY. We thank Yue Guan, Andrew Arvai, Sudip Parikh, and Clifford Mol, members of the Tainer laboratory who were instrumental in obtaining the structure of p26MutY. This work has been generously supported by grants from ES06676 (NIEHS), GM59237 (NIH), and H-1402 (Welch Foundation) to R. S. Lloyd. R. S. Lloyd holds the Mary Gibbs Jones Distinguished Chair in Environmental Toxicology from the Houston Endowment.

References

1. A. K. McCullough, M. L. Dodson, and R. S. Lloyd, *Annu. Rev. Biochem.* **68**, 255–285 (1999).
2. C. D. Mol, S. Parikh, C. Putnam, T. Lo, and J. Tainer, *Annu. Rev. Biophys. Biomol. Struct.* **28**, 101–128 (1999).
3. M. L. Michaels and J. H. Miller, *J. Bacteriol.* **174**, 6321–6325 (1992).
4. S. Shibutani, M. Takeshita, and A. P. Grollman, *Nature (London)* **349**, 431 (1991).
5. J. Tchou, H. Kasai, S. Shibutani, M. H. Chung, J. Laval, A. P. Grollman, and S. Nishimura, *Proc. Natl. Acad. Sci. U.S.A.* **88**, 4690–4694 (1991).
6. M. L. Michaels, C. Cruz, A. P. Grollman, and J. H. Miller, *Proc. Natl. Acad. Sci. U.S.A.* **89**, 7022–7025 (1992).
7. H. Maki and M. Sekiguchi, *Nature (London)* **355**, 273–275 (1992).

8. Y. Nghiem, M. Cabrera, C. G. Cupples, and J. H. Miller, *Proc. Natl. Acad. Sci. U.S.A.* **85**, 2709–2713 (1988).
9. A. L. Lu, J. J. Tsai-Wu, and J. Cillo, *J. Biol. Chem.* **270**, 23582–23588 (1995).
10. N. V. Bulychev, C. V. Varaprasad, G. Dorman, J. H. Miller, M. Eisenberg, A. P. Grollman, and F. Johnson, *Biochemistry* **35**, 13147–13156 (1996).
11. R. C. Manuel and R. S. Lloyd, *Biochemistry* **36**, 11140–11152 (1997).
12. C. L. Chepanoske, S. L. Porello, T. Fujiwara, H. Sugiyama, and S. S. David, *Nucleic Acids Res.* **27**, 3197–3204 (1999).
13. M. L. Michaels, L. Pham, Y. Nghiem, C. Cruz, and J. H. Miller, *Nucleic Acids Res.* **18**, 3841–3845 (1990).
14. J. Labahn, O. D. Schärer, A. Long, K. Ezaz-Nikpay, G. L. Verdine, and T. E. Ellenberger, *Cell (Cambridge, Mass.)* **86**, 321–329 (1996).
15. Y. Yamagata, M. Kato, K. Odawara, Y. Tokuno, Y. Nakashima, N. Matsushima, K. Yasumura, K. Tomita, K. Ihara, Y. Fujii, Y. Nakabeppu, M. Sekiguchi, and S. Fujii, *Cell (Cambridge, Mass.)* **86**, 311–319 (1996).
16. T. Hollis, Y. Ichikawa, and T. Ellenberger, *EMBO J.* **19**, 758–766 (2000).
17. C. F. Kuo, D. E. McRee, C. L. Fisher, S. F. O'Handley, R. P. Cunningham, and J. A. Tainer, *Science* **258**, 434–440 (1992).
18. M. M. Thayer, H. Ahern, D. Xing, R. P. Cunningham, and J. A. Tainer, *EMBO J.* **14**, 4108–4120 (1995).
19. H. M. Nash, S. D. Bruner, O. D. Schärer, T. Kawate, T. A. Addona, E. Spooner, W. S. Lane, and G. L. Verdine, *Curr. Biol.* **6**, 968–980 (1996).
20. Y. W. Kow and S. S. Wallace, *Biochemistry* **26**, 8200–8206 (1987).
21. R. C. Manuel, E. W. Czerwinski, and R. S. Lloyd, *J. Biol. Chem.* **271**, 16218–16226 (1996).
22. Y. Guan, R. C. Manuel, A. S. Arvai, S. S. Parikh, C. D. Mol, J. H. Miller, R. S. Lloyd, and J. A. Tainer, *Nature Struct. Biol.* **5**, 1058–1064 (1998).
23. D. E. Volk, V. Thiviyanathan, P. G. House, R. S. Lloyd, and D. G. Gorenstein, *J. Biomol. NMR* **14**, 385–386 (1999).
24. D. E. Volk, P. G. House, V. Thiviyanathan, B. A. Luxon, S. Zhang, R. S. Lloyd, and D. G. Gorenstein, *Biochemistry* **39**, 7331–7336 (2000).
25. M. L. Dodson, M. L. Michaels, and R. S. Lloyd, *J. Biol. Chem.* **269**, 32709–32712 (1994).
26. M. L. Dodson, R. D. Schrock, and R. S. Lloyd, *Biochemistry* **32**, 8284–8290 (1993).
27. M. Fuxreiter, A. Warshel, and R. Osman, *Biochemistry* **38**, 9577–9589 (1999).
28. B. Sun, K. A. Latham, M. L. Dodson, and R. S. Lloyd, *J. Biol. Chem.* **270**, 19501–19508 (1995).
29. C. E. Piersen, M. A. Prince, M. L. Augustine, M. L. Dodson, and R. S. Lloyd, *J. Biol. Chem.* **270**, 23475–23484 (1995).
30. R. S. Lloyd, *Prog. Nucleic Acid Res.* **62**, 155–175 (1998).
31. A. K. McCullough, M. T. Romberg, S. Nyaga, Y. Wei, T. G. Wood, J. S. Taylor, J. L. Van Etten, M. L. Dodson, and R. S. Lloyd, *J. Biol. Chem.* **273**, 13136–13142 (1998).
32. J. F. Garvish and R. S. Lloyd, *J. Biol. Chem.* **274**, 9786–9794 (1999).
33. J. F. Garvish and R. S. Lloyd, *J. Mol. Biol.* **295**, 479–488 (2000).
34. D. A. Vasquez, S. G. Nyaga, and R. S. Lloyd, *Mutat. Res.* **459**, 307–316 (2000).
35. S. G. Nyaga and R. S. Lloyd, *J. Biol. Chem.* **275**, 23569–23576 (2000).
36. C. E. Piersen, R. Prasad, S. H. Wilson, and R. S. Lloyd, *J. Biol. Chem.* **271**, 17811–17815 (1996).
37. C. E. Piersen, A. K. McCullough, and R. S. Lloyd, *Mutat. Res.* **459**, 43–53 (2000).
38. D. O. Zharkov and A. P. Grollman, *Biochemistry* **37**, 12384–12394 (1998).
39. S. D. Williams and S. S. David, *Biochemistry* **38**, 15417–15424 (1999).
40. P. M. Wright, J. Yu, J. Cillo, and A. L. Lu, *J. Biol. Chem.* **274**, 29011–29018 (1999).
41. A. Gogos, J. Cillo, N. D. Clarke, and A. L. Lu, *Biochemistry* **35**, 16665–16671 (1996).
42. D. M. Noll, A. Gogos, J. A. Granek, and N. D. Clarke, *Biochemistry* **38**, 6374–6379 (1999).

43. X. Li, P. M. Wright, and A. Lu, *J. Biol. Chem.* **275,** 8448–8455 (2000).
44. R. M. Belagaje, S. G. Reams, S. C. Ly, and W. F. Prouty, *Protein Sci.* **6,** 1953–1962 (1997).
45. D. S. Wishart and B. D. Sykes, *J. Biomol. NMR* **4,** 171–180 (1994).
46. D. N. Frick, D. J. Weber, C. Abeygunawardana, A. G. Gittis, M. J. Bessman, and A. S. Mildvan, *Biochemistry* **34,** 5577–5786 (1995).
47. X. P. Kong, R. Onrust, M. O'Donnell, and J. Kuriyan, *Cell (Cambridge, Mass.)* **69,** 425–437 (1992).
48. D. J. Hosfield, Y. Guan, B. J. Haas, R. P. Cunningham, and J. A. Tainer, *Cell (Cambridge, Mass.)* **98,** 397–408 (1999).

Properties and Functions of Human Uracil-DNA Glycosylase from the UNG Gene

HANS E. KROKAN, MARIT OTTERLEI, HILDE NILSEN, BODIL KAVLI, FRANK SKORPEN, SONJA ANDERSEN, CAMILLA SKJELBRED, MANSOUR AKBARI, PER ARNE AAS, AND GEIR SLUPPHAUG

Institute of Cancer Research and Molecular Biology
Norwegian University of Science and Technology
N-7489 Trondheim, Norway

I. Introduction	366
A. Discovery of Uracil-DNA Glycosylase and the Base Excision Repair Pathway	366
B. Prediction of Functions of Uracil-DNA Glycosylase and the Increasing Number of Enzymes Capable of Releasing Uracil from DNA	367
II. Uracil-DNA Glycosylase from a Conserved Gene Family	368
III. Recent Developments in Studies on Uracil-DNA Glycosylase: The Products of Human *UNG* and Murine *Ung*	370
A. The Human *UNG* Gene and Murine *Ung* Gene	370
B. Chromosomal Localization of the *UNG* Gene	371
C. The Structure of the Core Catalytic Domain (Residues 85–304 according to UNG1 Numbering)	372
D. Mitochondrial UNG1 (Human) and Ung1 (Murine)	372
E. Nuclear UNG2 (Human) and Ung2 (Murine)	374
F. Phosphorylation of UNG2	375
IV. Regulation of Expression of the *UNG* Gene	376
A. Differential Regulation of UNG1 and UNG2 at the Cellular Level	376
B. Regulation of Promoters P_A and P_B	376
V. *UNG* Mutants That Remove Normal Pyrimidines in DNA and Their Use in Studying the Biology of AP Sites	377
A. Active-Site Mutants of UNG Expressing Cytosine-DNA Glycosylase (CDG) or Thymine-DNA Glycosylase (TDG) Activities	377
B. The Use of CDG and TDG to Study Handling of AP Sites in the *E. coli* Chromosome *In Vivo*	377

Abbreviations: PCNA: proliferating cell nuclear antigen; RPA: replication protein A; MPP: mitochondrial processing peptidase; AP sites: apurinic/apyrimidinic sites.

VI. Recent Information Indicates an Essential Role for Ung2 in Removal of Misincorporated Uracils, but Is This All? 378
   A. Bacterial *Ung* Has a Verified Role in Removal of Misincorporated dUMP Residues and in Repair of Deaminated Cytosines 378
   B. A Role for Human UNG2 and Murine Ung2 in Removal of Misincorporated dUMP Residues. 379
   C. Is Murine Ung2 Involved in Removal of Uracil from Deaminated Cytosine Residues?. 380
VII. Is a Major Function of DNA Glycosylases a Long-Term Protection of the Mammalian Genome during Evolution?. 381
   A. Knockout Mice Deficient in DNA Glycosylases 381
   B. DNA Glycosylases in Long-Term Maintenance of DNA 382
   References 384

The human *UNG*-gene at position 12q24.1 encodes nuclear (UNG2) and mitochondrial (UNG1) forms of uracil-DNA glycosylase using differentially regulated promoters, P_A and P_B, and alternative splicing to produce two proteins with unique N-terminal sorting sequences. PCNA and RPA co-localize with UNG2 in replication foci and interact with N-terminal sequences in UNG2. Mitochondrial UNG1 is processed to shorter forms by mitochondrial processing peptidase (MPP) and an unidentified mitochondrial protease. The common core catalytic domain in UNG1 and UNG2 contains a conserved DNA binding groove and a tight-fitting uracil-binding pocket that binds uracil only when the uracil-containing nucleotide is flipped out. Certain single amino acid substitutions in the active site of the enzyme generate DNA glycosylases that remove either thymine or cytosine. These enzymes induce cytotoxic and mutagenic abasic (AP) sites in the *E. coli* chromosome and were used to examine biological consequences of AP sites. It has been assumed that a major role of the *UNG* gene product(s) is to repair mutagenic U:G mispairs caused by cytosine deamination. However, one major role of UNG2 is to remove misincorporated dUMP residues. Thus, knockout mice deficient in Ung activity ($Ung^{-/-}$ mice) have only small increases in GC→AT transition mutations, but $Ung^{-/-}$ cells are deficient in removal of misincorporated dUMP and accumulate approximately 2000 uracil residues per cell. We propose that BER is important both in the prevention of cancer and for preserving the integrity of germ cell DNA during evolution. © 2001 Academic Press.

I. Introduction

A. Discovery of Uracil-DNA Glycosylase and the Base Excision Repair Pathway

The base excision repair pathway (BER) represents a major mechanism for repair of spontaneous and some induced forms of damage to bases in DNA.

This pathway is initiated by DNA glycosylases that release different damaged or inappropriate bases (reviewed in Ref. *1*). The first DNA glycosylase discovered, uracil-DNA glycosylase (UDG), was found in extracts from *Escherichia coli* in a search for an enzymatic activity that would act on DNA containing deaminated cytosine residues (*2*). This clearly initiated research on an entirely novel DNA repair pathway. It is usually stated that the function of uracil-DNA glycosylase is to remove uracil resulting from deaminated cytosines, as well as uracil resulting from misincorporated dUMP residues. In accordance with this view, major UDGs encoded by genes in the conserved gene family (*ung* in *E. coli* and *UNG* in man) efficiently remove uracil both from U:A pairs (as found after misincorporation of dUMP) and U:G mispairs (as found after deamination of cytosine) (reviewed in Refs. *1* and *3*).

B. Prediction of Functions of Uracil-DNA Glycosylase and the Increasing Number of Enzymes Capable of Releasing Uracil from DNA

Biochemical and genetic evidence has demonstrated that the function of the Ung enzyme in bacteria and yeast is essentially as originally predicted. Thus, bacterial or yeast mutants deficient in uracil-DNA glycosylase display a severalfold increase in GC→AT transition mutations, indicating a deficiency in removal of uracil resulting from cytosine deamination (*4, 5*). In addition, *E. coli ung$^-$* cells accumulate incorporated dUMP, demonstrating a role for the Ung protein in removal of misincorporated uracil (*6*). However, the functions of the corresponding enzymes in mammalian cells (UNG1 and UNG2) have only been assumed by analogy, but not functionally tested until recently (*7, 8*). In fact, the roles of the mammalian enzyme appear to be somewhat different as compared with bacteria and yeast, at least quantitatively. This is apparently because different "backup" activities, which contribute a minor fraction of the total UDG activity measured *in vitro*, may play important roles in some organisms, but may have different or less important roles, or may be absent in other cell systems. These backup activities may modify the significance of UDGs encoded by the conserved family. The mammalian repertoire of uracil-removing enzymes is larger than that in bacteria, and possibly also that in yeast. Thus, mammalian cells appear to have at least six enzymes that can release uracil from DNA. These include mitochondrial UNG1 and nuclear UNG2, which are products of the *UNG* gene (*9–11*); the MUG/TDG enzyme, which removes thymine, uracil, and 3,N^4-ethenocytosine and which is essentially specific for double-stranded DNA (*12, 13*); the cyclin-like UDG2 encoded by the *UDG2* gene (*14*); and the recently discovered activities of SMUG1 (*15*) and MBD4 (*16*) (Table I). The putative amino acid sequences of these proteins are essentially unrelated, but the structures of the UNG core enzyme and MUG/TDG are clearly related (*17*).

TABLE I
Mammalian Uracil-DNA Glycosylase Activities[a]

Name	Size (aa)	Lyase activity	Cellular localization	Chromosome localization	Known substrates[b]
hUNG1	304	No	Mitochondria	12q24.1	ssU > U:G > U:A > 5-FU (very poor:
hUNG2	313	No	Nuclei	12q24.1	5-hydroxyU, isodialuric acid, alloxan)
mUng1	295	No	Mitochondria	5	U, 5-FU
mUng2	306	No	Nuclei	5	U, 5-FU
hSMUG1	270	No	—	12q13.1-q14	ssU > U:A, U:G
hTDG	410	No	Nuclei	12q24.1	U:G > εC:G > T:G
hMBD4	580	?	—	3q21	U:G,T:G, U or T in U/TpG5-meCpG
hUDG2	327	No	—	5	U:A

[a]h, human; m, mouse.
[b]In mismatches, the target base is at the left.

Whether this indicates an ancient relationship is not clear. However, a number of DNA glycosylases that have different specificities are closely or distantly related at the amino acid level; but unfortunately, the structures are known only for a few of these enzymes (1). Possibly, structures are better conserved than amino acid-specifying codons and very few amino acids may determine the structure of a certain domain (18).

II. Uracil-DNA Glycosylase from a Conserved Gene Family

Uracil-DNA glycosylases from a conserved family of genes have been extensively explored. Genes for these enzymes have been identified in a variety of organisms, including poxviruses, herpesviruses, bacteria, yeast, fish, and mammalian cells (1, 3, 10, 19). They are apparently absent in archeabacteria and possibly in some insects such as *Drosophila melanogaster*. These DNA glycosylases are often considered a prototype DNA glycosylase. However, they have some properties that are different from other DNA glycosylases, the most striking being a turnover number (in those examined) that is at least 2–4 orders of magnitude higher than those of other DNA glycosylases. Furthermore, many, but not all, other DNA glycosylases can release several different modified bases with almost equal rates. These substrates are, in some cases, not very related at the structural level. As one example, AlkA has a substrate specificity that includes both modified purines and pyrimidines and can even remove normal purines, particularly guanosine, although with low rates (20). In contrast, the uracil-DNA glycosylases from the conserved family have very narrow substrate ranges and

TABLE II
THE HUMAN UNG GENE AND ITS GENE PRODUCTS[a]

	UNG1	UNG2	C-Terminal catalytic domain
Expression			
Gene (12q24.1)	UNG	UNG	UNG
Expressed exons	(exon IB + II–VI)	(exon IA + II–VI)	—
Promoter	P_B	P_A	—
Major stimulation by	TFII-I	AP-2, UNG2	—
mRNA (nucleotides)	2062	2057	—
Structure/physical properties			
Length of polypeptide	313	304	227
Presequence (AA)	35	44	—
Localization	Mitochondria	Nuclei (repl. foci)	—
pI (measured/calculated)	ND/9.56	ND/9.37	10.5/9.04
Phosphorylation	No	Yes	No
Motifs in presequence	Amphiphilic α-helix	Essential NLS	—
	MPP cleavage site	PCNA-binding	
	RPA2-binding (1)	RPA2-binding (2)	
Activity			
Preferred substrates	ssU > U:G > UA	ssU > U:G > UA	ssU > U:G > UA
Turnover (min^{-1}, depending on conditions)	600–1000	600–1000	600–1000
Stimulated by divalent cations	Weakly	Yes	No
Active in presence of EDTA	Yes	Yes	Yes
Activity dependent on sequence context	ND	Yes	Yes

[a] MPP, mitochondrial processing peptidase; NLS, nuclear localization signal; RPA, replication protein A; PCNA, proliferating cell nuclear antigen.

essentially only release uracil at reasonable rates (Table II). Thus, bacterial Ung and human UNG2/UNG1, as the best studied examples, are very efficient and specialized DNA repair enzymes that release uracil with a turnover number of 500–1000 per min, depending on reaction conditions. In addition, they release some derivatives of cytosine, e.g., alloxan, dialuric acid, and 5-OH uracil, resulting from oxidative stress or ionizing radiation. These products are identical to uracil in positions 2, 3, and 4, all of which are essential in substrate recognition and binding in the uracil-binding pocket (21, 22). Uracil-DNA glycosylases from this family of conserved genes are monofunctional enzymes that do not contain an associated lyase activity (1, 3). All known DNA glycosylases are active in the presence of EDTA and do not require divalent cations. However, the activity of human UNG2 is stimulated severalfold by Mg^{2+}, in contrast to the core catalytic domain, which is not stimulated by Mg^{2+}. It should be noted, however, that Mg^{2+}

effects are dependent both upon substrate concentrations and concentrations of monovalent cations. The difference between UNG2 and the core enzyme is represented by a 94-amino acid long N-terminal region (using UNG2 numbering of residues), which somehow must be responsible for the Mg^{2+} effect (Slupphaug, unpublished data). This N-terminal region also has other distinct functions; it contains a nuclear localization signal, a motif for binding of proliferating cell nuclear antigen (PCNA), and two different motifs that bind replication protein A (RPA). Uracil-DNA glycosylase from bacteria, yeast, and mammalian sources are all more active on single-stranded versus double-stranded DNA and release uracil both from U:A pairs and U:G mispairs. Although the rates are usually higher with U in a mispair, uracil-DNA glycosylase (from *E. coli* and mammalian sources) has a striking sequence specificity, and in some sequence contexts uracil is released more rapidly from U:A pairs than from U:G mispairs (*19*).

III. Recent Developments in Studies on Uracil-DNA Glycosylase: The Products of Human *UNG* and Murine *Ung*

A. The Human *UNG* Gene and Murine *Ung* Gene

The human (*UNG*) and murine (*Ung*) genes are approximately 13.5 kb (*11*) and 10 kb (*8, 23, 24*), respectively. The size differences are essentially due to a smaller size of the last intron (intron 6) in the murine gene. The overall structures of the human and murine genes are very similar in organization of the promoters and exons. Thus, they both contain two TATA-less promoters (P_A and P_B) that are used to make transcripts for nuclear and mitochondrial forms of the enzyme, respectively. Many putative transcription factor-binding elements are found in the promoters in both species, but with somewhat different organization. Overall, P_A is much less conserved than P_B. Furthermore, whereas the human promoter P_A is stronger than P_B, the situation is reversed for the murine promoters. Murine P_B contains within 603 bp a duplication of 280 bp with 80% identity (*24*). This results in the duplication of several putative transcription factor-binding elements and may explain the increased relative strength of the murine P_B (*24*). Secondly, the seven exons (1A, 1B, 2–6) are organized in essentially identical ways in the two genes and have conserved exon–intron boundaries; and with the exception of the small intron 3 (385 bp), the introns are poorly conserved. Exon 6 is the largest, but only a small part of this exon is translated (*24*). Possible functions of the nontranslated part of exon 6 have not been identified. The amino acid sequence of the core catalytic domain is well conserved (approximately 90%) between human UNG and murine Ung proteins.

In contrast, the N-terminal mitochondrial targeting sequences in human UNG1 and murine Ung1 are poorly conserved. N-terminal nuclear targeting sequences in UNG2 and Ung2 are better conserved, with a 100% identity in the first 13 residues and blocks of conserved motifs downstream in the 45 (human) or 42 (murine) N-terminal residues unique to the nuclear enzymes.

B. Chromosomal Localization of the UNG Gene

The human *UNG* gene has previously been localized to chromosome 12q23–q24.1 by radiation hybrid mapping (*11*). Recently, a sequence-ready physical map of 12q24.1 covering the region 116.6–120.5cM was released by the Albert Einstein College of Medicine Human Genome Research Center (http://sequence.aecom.yu.edu/chr12/). On this map, the *UNG* gene is flanked by the markers WI-20815 (centromeric) and WI-3957 (telomeric), and is close to the D-amino acid oxidase gene (GenBank accession No. NM001917) and the acetyl-CoA carboxylase 2 gene (GenBank accession No. U89344), found centromeric and telomeric to *UNG*, respectively (Fig. 1). Yet another gene

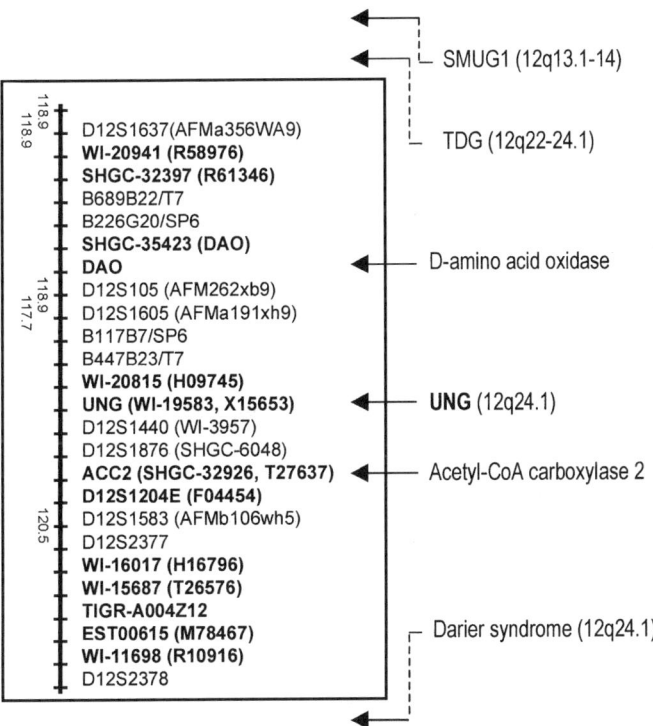

FIG. 1. Localization of the human *UNG* gene in position 12q24.1 relative to markers and identified genes (modified from http://sequence.aecom.yu.edu/chr12/).

(GenBank accession No. H09745) is found immediately 3' to *UNG*, but the product of this gene has not been characterized. The murine *Ung* gene is located on chromosome 5 (*23*). Interestingly, the human genes for two other uracil-releasing proteins, hTDG and hSMUG1, are also located on chromosome 12, centromeric to the *UNG* gene (Table I).

C. The Structure of the Core Catalytic Domain (Residues 85–304 according to UNG1 Numbering)

A fully active form of UNG was the first mammalian DNA repair enzyme for which a structure was solved (*25*) and the first eukaryotic DNA repair enzyme for which the structure of the enzyme-target DNA complex was solved (*26*). Structural and functional interactions in subsequent steps have revealed a highly interesting choreography of BER and was recently reviewed (*27*). Structural aspects of uracil-DNA glycosylase are comprehensively dealt with elsewhere in this volume (Hosfield *et al.*) and will only be briefly addressed here. The core enzyme is a single-domain protein containing eight α-helices and a central twisted β-sheet made up of four parallel β-strands. The DNA-binding groove contains several positively charged residues and is located over the C-terminal end of the β-sheet. A deep and narrow uracil-binding pocket is located near one end of the DNA binding groove. It was concluded from the structure of the uracil-binding pocket that uracil in DNA could only be accommodated in the groove if the base was flipped out (*25*). The amino acid residues forming this pocket are highly conserved and structurally identical in HSV-1 (*28*) and human enzymes (*25*). All residues implicated in catalysis by site-directed mutagenesis are located within this groove. Uracil binds by H-bonding, base stacking, and shape complementarity. The closely related pyrimidines, cytosine and thymine, are excluded by unfavorable electrostatic and structural interference ($4'$-NH_2 in cytosine verses $4'$-O in uracil) and steric hindrance (a bulky $5'$-CH_3 in thymine versus $5'$-H in uracil), respectively. Nucleotide flipping was later demonstrated in co-crystals of target DNA and UNG core enzyme (*26*). The mechanism of flipping and catalysis has later been refined (*29, 30*).

D. Mitochondrial UNG1 (Human) and Ung1 (Murine)

There is good evidence for the presence of BER, direct damage reversal, mismatch repair, and recombinational repair mechanisms in mitochondria. However, mitochondria have a complete or partial deficiency in nucleotide excision repair, depending on the type of damage (reviewed in Ref. *31*). Complete BER has been reconstituted using mitochondrial extracts (*32*). Human UNG1 has been overexpressed, purified, and studied in some detail (*33, 34*). The human *UNG* gene was in fact the first gene identified for a distinct mitochondrial DNA repair enzyme (*9*). The unique N-terminal sequences in human UNG1 (35 amino acids) (9), and murine Ung1 (31 amino acids) (24) are directing the proteins to mitochondria, but are poorly conserved between the two species. Apparently, the human mitochondrial targeting sequence is much stronger than the corresponding murine signal. Thus, when overexpressed, the murine protein translocates to both nuclei and mitochondria, whereas the human enzyme exclusively enters mitochondria. Disruption of the amphiphilic structure has been demonstrated as one important factor that decreases mitochondrial translocation (35, 36). UNG1 and Ung1 both have the potential of forming amphiphilic helices of similar sizes, but at slightly different localizations, with the Ung1 helix being closer to the N terminus. The mitochondrial targeting signal has been best studied for UNG1. The 35 N-terminal amino acids are required and sufficient for mitochondrial translocation. The N-terminal 12 amino acids are absolutely required for mitochondrial import, although only two (amino acids 11–12) of these residues reside in the putative amphiphilic helix (amino acids 11–29). Site-directed mutagenesis of positively charged amino acids (R13, K14) as well as an LSRL motif (amino acids 26–29) also reduced mitochondrial import, but no single substitution abolished sorting (34).

Import to mitochondria of human UNG1 after transfection of UNG1 cDNA into insect cells results in two processed forms in the mitochondria. We have not been able to identify an unprocessed form of UNG1 in the cytoplasm or mitochondria, indicating rapid mitochondrial import after translation. The longest processed form, UNG1Δ29 (31 kDa), lacks 29 N-terminal amino acids and is almost certainly cleaved by mitochondrial processing peptidase (MPP) for which UNG1 has a perfect recognition sequence. The second species lacks 75 or 77 amino acids, UNG1Δ75/77 (26 kDa). These two (three) forms have been purified to homogeneity and sequenced, thus defining exact sites of cleavage. Forms with identical sizes, as judged by gel electrophoresis and Western blotting, have also been identified in extracts of HeLa cell mitochondria. Furthermore, UNG1Δ29 is cleaved to the UNG1Δ75/77 form by mitochondrial extracts, but poorly by extracts from other cellular compartments, indicating that a specific mitochondrial protease is responsible for further processing of the UNG1Δ29 form (33). The shortest form essentially constitutes the core catalytic domain. Although the two forms share the same catalytic domain, they do not have identical biochemical properties. Thus, the shortest form is inhibited by AP sites in micromolar concentrations, while the UNG1Δ29 form is much more resistant. They are both stimulated in their catalytic rates by human AP endonuclease HAP1. This is apparently because the release of UNG protein from the AP site created by the enzyme is in fact a rate-limiting step in catalysis and HAP1 stimulates the dissociation (29). Presumably, HAP1 competes with UNG protein for the AP site and physically dislocates the UNG protein. It is not simply removal of an inhibitor by cleavage of the AP site by HAP1 that causes the increased catalytic rate. Interestingly, the rate of dissociation of thymine (uracil)-DNA glycosylase is also significantly increased by AP endonuclease (37), suggesting that the functional cooperation between DNA glycosylases and AP endonucleases may be more general. One attractive explanation for this functional coupling would be that the DNA glycosylase shields the reactive and mutagenic AP site until it is cleaved by AP endonuclease, as previously suggested (29). In the cells, the 31-kDa form dominates quantitatively. Two forms of UDG have also been demonstrated in rat liver mitochondria, one of which apparently is formed by proteolysis of the other (38). It remains unclear whether the two forms serve different functions. It should be noted, however, that the 31-kDa form contains an RPA-binding site that is partially deleted in the 26-kDa form (39). Although RPA is not present in mitochondria, mitochondrial single-strand DNA binding protein(s) have been demonstrated. These are related to bacterial ssDNA-binding proteins (40). Thus, it is possible that the N-terminal part of the 31-kDa may have a function in interaction with ssDNA-binding protein or other proteins required for BER in mitochondria.

E. Nuclear UNG2 (Human) and Ung2 (Murine)

Most available information on the nuclear UNG2 is derived from studies on human UNG2 (7, 9, 34). The 44 unique N-terminal amino acids in UNG2 constitute a sequence region that has several functions: (1) It is essential for nuclear sorting (9, 34), (2) it binds PCNA and trimeric RPA via an overlapping motif (7), (3) it modifies biochemical properties of the core catalytic domain and is required for stimulation of the enzyme by Mg^{2+} (Slupphaug et al., unpublished results); (4) it is phosphorylated (41; Slupphaug et al., unpublished); and (5) it may interact with AP-endonuclease HAP1 (Slupphaug et al., unpublished results). Presumably, the N-terminal sequence of murine Ung2 has similar functions. UNG2 is rapidly translocated to the nucleus after biosynthesis, and no cytoplasmic accumulation has been observed in transfection experiments. Interestingly, the complete nuclear localization signal (NLS) in UNG2 extends more than 50 amino acids downstream of the 44 N-terminal residues unique to UNG2. Within the first 100 amino acids there are several clusters of 2–4 positively charged amino acids, some of which are essential for nuclear sorting (e.g., K18 and R19), while others are involved in sorting, but of limited significance (K49, K50). In addition, some positively charged residues are apparently involved in subnuclear distribution (combination of K99 and positively charged residues downstream) (34). The 26 N-terminal amino acids are highly conserved and the 13 N-terminal residues (MIGQKTLYSFFSP) are identical in UNG2 and Ung2 (9) Interestingly, the 13 first residues of the N-terminal sequence are also highly homologous to the N-terminal residues in hMSH6/GTBP which is involved in mismatch repair (42). For both proteins, a consensus PCNA-binding motif (QxxLxxFF) is present in this region. The same sequence region in UNG2 contains a motif demonstrated to bind trimeric RPA (7).

F. Phosphorylation of UNG2

Multiple phosphorylations at serine and threonine residues in the UNG2 N-terminal region were recently demonstrated, two of which were localized to Thr31 and Ser40 (41). The functional implications of UNG2 N-terminal phosphorylations have yet to be identified, although it is tempting to speculate that this may modulate enzyme activity or protein–protein associations. We have recently demonstrated that the N-terminal domain contains phosphorylation targets for several kinases *in vitro*, having K_M values for ATP in the micromolar range (Slupphaug *et al.*, unpublished results). These kinases have been associated with various cellular functions such as cell-cycle regulation, mitogen-stimulated proliferation, and DNA repair. Furthermore, using N-terminal UNG2 deletion mutants, the various kinases have been shown to phosphorylate distinct regions along the entire N-terminal domain. None of these phosphorylations, however, was accompanied by detectable alterations in enzyme activity. We are currently investigating modulation of nuclear and subnuclear targeting by phosphorylation, by site-directed mutagenesis of N-terminal serines and threonines, and by expression of EGFP-fused mutants in HeLa cells.

The co-localization of UNG2 with PCNA and RPA in replication foci, as well as the fragmentation of newly replicated DNA in isolated nuclei after removal of uracil by UNG2, suggests that UNG2 is active at or near the replication fork (7). Furthermore, the specific binding of UNG2 to RPA and PCNA may indicate that these interactions actively recruit UNG2 to the replication fork. We have recently demonstrated that the amount of phosphorylated UNG2 peaks during the S-phase of the cell cycle (Slupphaug *et al.*, unpublished results), suggesting that phosphorylation status may play a role in the binding to RPA/PCNA or other proteins involved in replication and repair. This type of mechanism was recently demonstrated for human DNA ligase I, which is associated with long-patch PCNA-dependent BER. Lig I was shown to bind PCNA in its dephosphorylated state in G_1/S, and was released from PCNA in its fully phosphorylated state in G_2/M (43). Interestingly, HAP1, the second enzyme in the BER pathway was also shown to be phosphorylated, although this phosphorylation apparently modulates the redox capacity of HAP1 toward the transcription factor AP1, and hence its DNA-binding capability (44). We have previously demonstrated that HAP1 increases the activity of UNG at high concentrations, and have attributed this to an increased UNG off rate in the presence of HAP1 (29). Interestingly, HAP1 is also able to bind UNG2 *in vitro*, and this binding is modulated by the phosphorylation status of both enzymes (Slupphaug *et al.*, unpublished results). If this holds true *in vivo*, direct protein–protein interactions are invoked at each of the essential steps in BER. The functional significance of such interactions might be to shield the cell from potential toxic and mutagenic effects of the BER intermediates.

IV. Regulation of Expression of the *UNG* Gene

A. Differential Regulation of UNG1 and UNG2 at the Cellular Level

mRNAs for UNG1 and UNG2 are differentially expressed in human tissues. In the cell cycle, UNG2 (and to a smaller extent UNG1) is upregulated severalfold in late G_1 phase and S phase at both the mRNA and protein level (*45*). In general, mRNA for UNG1 is detectable in most tissues, with the highest levels found in skeletal muscle, heart, and testis, whereas mRNA for UNG2 is expressed at appreciable levels only in tissues containing a significant proportion of proliferating cells, such as testis, placenta, colon mucosa, and small intestine. In "resting" tissues, UNG2 mRNA is hardly detectable, or not detectable, by Northern blot analysis. The mechanism(s) that contribute to differential expression of UNG1 and UNG2 transcripts are not well understood. Previous studies have indicated that UDG activity, resulting from *UNG* expression is regulated mainly at the level of transcription. However, recent studies have indicated that post-translational modification of UNG2 may also play an important role during different stages of the cell cycle (Slupphaug *et al.*, unpublished results). Transcript levels for UNG1 and UNG2 are both increased in the S phase, rising 2.5- and 5-fold, respectively. While UNG2 mRNA rapidly disappears at the mid-S phase, UNG1 transcripts decline slowly after mid-S phase, demonstrating that both the induction and degradation of UNG mRNAs are differentially regulated.

B. Regulation of Promoters P_A and P_B

The two promoters in the *UNG* gene, P_A and P_B, are both GC-rich and lack typical TATA elements. For both promoters, elements required for full activity (>80%) are located within the first 150 bp upstream of the transcription initiation areas. An E2F element present in P_B is likely to be involved in cell-cycle-regulated expression of UNG1 mRNA. Recently, a putative element for E2F-binding has also been identified in P_A, and we are now investigating the significance of this element for cell-cycle-regulated expression of UNG2 mRNA. Site-directed mutagenesis has revealed that, in general, mutation in single transcription factor binding elements has little effect on *UNG* promoter activity; usually 0–40% stimulation or inhibition are observed. Mutation of E boxes, however, located at position −93 in P_A and −96 in P_B, reduces promoter activity to approximately 25% of normal. Members of the helix–loop–helix (HLH) family of transcriptional regulators (e.g., c-Myc, USF), as well as the multifunctional transcription factor TFII-I, originally identified as an initiator (INR)-binding factor, bind to E boxes. We have recently demonstrated that TFII-I, but not c-Myc or USF, is able to stimulate P_B activity severalfold, and this stimulation is at least in part mediated via the −96 E box (Skorpen *et al.*, unpublished results). On the other hand, deletion of consensus INR elements located in the transcription start region in P_B does not abrogate the stimulatory effect of TFII-I. Several muscle-specific proteins sharing remarkable homology to repeat domains within TFII-I have been identified and we are currently investigating whether these factors may contribute to high-level expression of UNG1 in skeletal-muscle and heart. A 50-bp region in P_A (-55 to -105) is conserved between the human and murine promoter, suggesting that this region may be important for promoter activity. This region contains the -93 E box as well as several putative elements for the binding of transcription factor AP-2. Overexpression of AP-2 selectively stimulates P_A activity 3-fold. AP-2-binding elements also appear to be occupied by AP-2 in footprint analysis of P_A with HeLa nuclear extract, suggesting that AP-2 may be one central factor in UNG2 expression.

A surprising observation was that overexpression of UNG2 protein stimulates transcription from a P_A-reporter-gene construct more than 3-fold in transient transfection experiments, whereas P_B activity is only marginally affected. Although speculative, this indicates the possibility that UNG2 expression may be subject to a novel mechanism of autoregulation. In this regard, it is interesting to note that the human T:G mismatch-specific thymine-DNA glycosylase (TDG), which is structurally and functionally related to the UNG proteins despite low sequence identity, interacts with several transcription factors and potentiates transcription in transient transfection experiments (46). This makes it tempting to speculate that UNG2 may be an additional example of a protein with dual roles in DNA repair and transcription.

V. UNG Mutants That Remove Normal Pyrimidines in DNA and Their Use in Studying the Biology of AP Sites

A. Active-Site Mutants of UNG Expressing Cytosine-DNA Glycosylase (CDG) or Thymine-DNA Glycosylase (TDG) Activities

Certain amino acid residues in the active site of UNG proteins were predicted to be important for the specificity of the enzyme toward uracil. The functional role of these residues was demonstrated experimentally by site-directed mutagenesis. Thus, resulting single amino acid substitutions altered the specificity of the enzyme so that it could bind and release either normal cytosines or thymines, as predicted from the structural information (47).

B. The Use of CDG and TDG to Study Handling of AP Sites in the E. coli Chromosome In Vivo

These remarkable novel enzymes create multiple cytotoxic and mutagenic AP sites in DNA (47) and are now being used to study the biology of AP sites *in vivo* in chromosomal DNA (*47a*). Multiple AP sites were introduced in the *E. coli* chromosome *in vivo* by expressing mutants of uracil-DNA glycosylase that remove either thymine or cytosine, and resulting substitution mutations scored as rifampicin-resistant colonies. This system detects a number of substitution mutations, but not frameshift or mutations resulting in stop codons. Increasing deficiencies in the incision step (AP endonuclease activity) of base excision repair (BER), or in recombination (RuvA), resulted in increased cytotoxicity and mutation rates. RecA deficiency resulted in rapid and massive cell death upon induction of AP sites so that too few cells survived for measurements of mutational frequencies. Mutation spectra of strains deficient in AP endonuclease or RuvA-dependent recombination demonstrated increased frequency of transversions. Differences in cytotoxicity and mutation rates between wild-type and UmuC-deficient cells after introduction of AP sites were moderate, but the mutation spectra were very different, and indicated that preferential incorporation of dAMP opposite an AP site ("A-rule") *in vivo* requires UmuC, while a dGMP-preference was observed in the absence of UmuC-dependent translesion synthesis. These data demonstrate that BER, RuvA-dependent recombination, and UmuC-dependent translesion bypass are all involved in handling of AP sites (*47a*). Similar experiments using mammalian cells have so far been of limited success, apparently because of a selection against stable transfectants (carrying the cDNA in a Tet-on vector) expressing CDG or TDG. Transfectants carrying the expressed transgene were identified, and Tet-induction resulted in cytotoxicity, but this effect was lost during propagation, apparently due to negative selection or self-inactivation of the transgene. Nevertheless, the use of these engineered enzymes may represent a new method to study the biology of AP sites in chromosomal DNA *in vivo* both in prokaryotic and eukaryotic cells, provided the expression can be completely shut off in the uninduced state.

VI. Recent Information Indicates an Essential Role for Ung2 in Removal of Misincorporated Uracils, but Is This All?

A. Bacterial *Ung* Has a Verified Role in Removal of Misincorporated dUMP Residues and in Repair of Deaminated Cytosines

The function of the proteins encoded by the *UNG* gene has been assumed to be removal of uracil from U:G mispairs resulting from deamination of cytosine and from U:A pairs resulting from misincorporation of dUMP during replication

(1, 3, 19). The basis for this view is the following: First, bacterial (4) or yeast (5) mutants deficient in UDG activity have some 3–4-fold increased overall frequencies of GC→AT transition mutations, increasing to 20–30-fold in certain sequence contexts, but little change in frequencies of other mutations. This is as expected, since deamination of cytosine to uracil would result in incorporation of dAMP opposite of the deaminated cytosine. Second, E. coli mutants deficient in dUTPase (dut mutants) accumulate short fragments of newly synthesized DNA smaller than normal Okazaki fragments while dut/ung double mutants do not accumulate such small fragments (6, 48). Together, these findings demonstrate convincingly that Ung removes uracil from deaminated cytosine residues and misincorporated dUMP residues close to the replication fork.

B. A Role for Human UNG2 and Murine Ung2 in Removal of Misincorporated dUMP Residues

Since the major mammalian uracil-DNA glycosylase is conserved and is very similar to the bacterial and yeast counterparts, the functions have been assumed to be essentially identical. A role for human UNG2 in the removal of misincorporated dUMP residues was recently indicated (7). In this work, isolated nuclei were used for *in vitro* replication in the presence or absence of dUTP. It was found that dUTP was initially rapidly incorporated, but then removed within a few minutes. This removal was prevented by UNG antibodies so that DNA synthesis in the presence of dUTP proceeded like normal synthesis in the presence of dTTP. This strongly indicated that the nuclear enzyme (UNG2) is required for removal of misincorporated dUMP, but it is not required for strand elongation during replication. Recently, knockout mice deficient in the *Ung* gene were constructed by insertion of the *neo* gene into exon 4 in the murine gene (8). The *Ung* knockout mice develop normally and have no obvious phenotype after 18 months. These $Ung^{-/-}$ mice are deficient in Ung activity, but display low uracil-DNA glycosylase activities in extracts from different organs. This activity is not inhibited by Ugi, a peptide inhibitor that efficiently inhibits the activity of bacterial Ung and mammalian UNG proteins by structure-specific binding to the DNA binding groove (49). Unlike the UNG proteins, the UDG activity in cell extracts is more active on double-stranded DNA than on single-stranded DNA. These characteristics strongly indicate that the low "backup" UDG activity is not a product of the *Ung* gene. The biochemical properties also exclude MUG/TDG as a major contributor to backup activities, but cannot be clearly distinguished from mammalian SMUG1. It is also entirely possible that the backup activity may be a previously unrecognized uracil-DNA glycosylase. Heterozygous mice ($Ung^{+/-}$) have approximately 50% of the wild-type activity, indicating that both alleles are required for expression of normal levels of Ung proteins. Embryonic fibroblasts from $Ung^{-/-}$ mice have approximately 3–4%

of the total UDG activity of wild-type mice and the biochemical characteristics of this activity are different from those of Ung protein. Isolated nuclei from $Ung^{-/-}$ fibroblasts are competent in DNA synthesis in the presence of dTTP, but display a deficiency in removal of incorporated dUMP residues. Together, these results demonstrate that one function of the human UNG2 and murine Ung2 is to remove misincorporated dUMP residues from the genome. Furthermore, comet assays demonstrate accumulation of approximately 2000 uracil residues per cell in the genome of $Ung^{-/-}$ fibroblasts. This is at least 10-fold higher than in wild-type cells in which the uracil content is below, or close to the detection limit using this assay.

C. Is Murine Ung2 Involved in Removal of Uracil from Deaminated Cytosine Residues?

If the *Ung* gene is essential for removal of uracils in U:G mispairs resulting from deamination of cytosine, one would expect that $Ung^{-/-}$ mice would display a mutator phenotype and increased GC→AT transitions. To examine this question, $Ung^{-/-}$ mice were established in *Big Blue* transgenic mice (carrying multiple copies of *lacI*). Mutation frequencies were examined in adult thymus and spleen, which have the lowest levels of uracil-DNA glycosylase backup activity (2%). Surprisingly, spontaneous mutation frequencies were increased only 1.3- and 1.4-fold in thymus and spleen, respectively, when compared with wild-type mice. The observed mutation frequencies in wild-type were consistent with previously published values (50, 51). Analysis of the mutation spectra of the *lacI* gene in plaques reflected a corresponding increase in GC→AT transitions in the spleen, but not thymus, of $Ung^{-/-}$ mice. Furthermore, 5% (spleen) and 12% (thymus) of all mutations in $Ung^{-/-}$ mice were +1 or −1 frameshifts, which were not detected at all in organs from wild-type mice. Taken together, these data indicate that the *Ung* gene is not essential for repair of U:G mispairs in mammalian organs. The most direct and likely explanation for this observation is that backup activities are sufficient for the repair of the majority of U:G mismatches. These results indicate that the major UDG activity encoded by the *Ung* gene is not solely responsible for repair of U:G mismatches. In fact, it may be a minor contributor to this process. Another possible explanation could be that spontaneous deamination of cytosines is not as frequent in mammalian cells as assumed from studies on naked DNA and bacteriophages, and that their repair consequently is of limited significance in the prevention of mutations. If this is the case, the *Ung* gene product may still be responsible for removal of most, if not all, uracils resulting from cytosine deamination. There are, however, other explanations. In mammalian cells, differing from bacteria and yeast in having relatively long G_1 and G_2 phases, the enzyme that removes deaminated cytosines need not have a very high turnover number to do the job. In fact, a slow enzyme that remains bound to the mutagenic AP site until the AP endonuclease is recruited may be advantageous. In contrast, the enzyme that operates close to replication should have a turnover number that matches the rate of polymerization. Among the glycosylases identified, only the *UNG* gene product has a turnover number that is consistent with the observed rapid removal of incorporated uracils (7, 8, 52). The biochemical properties of the *UNG* gene product makes it suitable for this specialized task. The mutagenic load that would result from deaminated cytosines (100–500 deaminations per cell per day) must in all probability be high enough to cause heritable disease and cancer if left unrepaired. Even if deficiency in the *UNG* gene itself does not predispose for cancer, it may seem likely that a collective deficiency in repair of deaminated cytosine residues would be likely to do so.

VII. Is a Major Function of DNA Glycosylases a Long-Term Protection of the Mammalian Genome during Evolution?

A. Knockout Mice Deficient in DNA Glycosylases

In the last few years, knockout mice deficient in *Aag/Mpg* (alkyladenine-DNA glycosylase), the major substrates of which are 3-methyladenine, ethenoadenine, and hypoxanthine (53); *Ogg1* (8-oxoG-DNA glycosylase) (54); and *Ung* (uracil-DNA glycosylase) (8) have been generated and analyzed. Unexpectedly, these mice develop normally, are fertile, and show no obvious phenotype. In particular, no increased cancer frequency has been observed. The spontaneous mutation frequencies are apparently slightly or moderately (1.3–3-fold) increased in $Ung^{-/-}$ (8) and $Ogg1^{-/-}$ mice (54), but not in $Agg1^{-/-}$ mice (55). However, $Aag^{-/-}$ mice are sensitive to alkylating agents, as expected (53), and accumulate mutations after exposure to alkylating agents (54). Different mutation frequencies were observed between different organs both in $Ung^{-/-}$ mice (8) and in $Ogg1^{-/-}$ mice in which no increased mutation frequency was observed in testis (54). $Ogg^{-/-}$ and $Ung^{-/-}$ cells accumulate 8-oxoG (15,000 per cell) and uracil (2000 per cell), respectively, in their genomes. In contrast to the situation for glycosylase knockouts, deficiency in DNA polymerase β that participates in the common steps of the BER pathway is lethal (56). However, this does not allow the conclusion that BER is an essential pathway because DNA polymerase β is also involved in homologous recombination during meiosis I (57). Similarly, knockouts in AP endonuclease are not viable, but the enzyme is also involved in transcription regulation through the redox function of another domain of the protein (58). A dominant negative DNA polymerase β mutant introduced into cultured mouse LN12 cells resulted in an 8-fold overall increase in spontaneous mutations (59). Thus, this gives an indication of the overall significance of BER in mutation prevention, but may underestimate the significance since long-patch repair does not have an absolute requirement for DNA polymerase β.

B. DNA Glycosylases in Long-Term Maintenance of DNA

The small effect of glycosylase deficiencies is intriguing, considering the strong evolutionary conservation of the genes for these enzymes and the almost ubiquitous presence of such glycosylase activities in different prokaryotic and eukaryotic species. As one example, the enzymes from the *E. coli ung* gene and human *UNG* gene have 55.7% identity at the amino acid level and over 70% conservation, when disregarding the human N-terminal sequences that are not part of the catalytic domain in UNG proteins and is lacking in the bacterial *ung* gene (10). The *UNG* gene in the fish *Xiphophorus* (R. B. Walter, personal communication) is strongly related to the human *UNG* gene and has consensus splice sites in exactly the same positions as those verified in the human *UNG* gene (9) and murine *Ung* gene (8, 23, 24). Since fish and mammals separated more than 450 million years ago, this indicates very strong conservation of both coding sequences and splice patterns. The *UNG* gene is apparently, along with genes for some other DNA glycosylases and mismatch repair enzymes (but not NER genes), among the most ancient DNA repair enzymes that originated before separation of prokaryotes and eukaryotes (reviewed in Ref. 60). The prokaryote *ung* gene and a number of other DNA repair genes are lost in archae. There are also several examples of loss of certain repair genes in some bacterial species. The functions are apparently conserved, however, by enzymes encoded by other genes that carry out similar functions (reviewed in Ref. 60). Therefore the evolutionary conservation of DNA glycosylases and/or corresponding functions points to an important, but so far not well understood, role of DNA glycosylases. It is possible that these enzymes are important for long-term integrity of the genome, rather than for protection at the individual level. Thus, our preoccupation with cancer and disease at the individual level, however important, may be too narrow-minded from a broader functional point of view. Assuming that most DNA glycosylases have as their only function to remove different types of base damage, they would not affect everyday DNA transactions, such as DNA replication or transcription, or specific functions during maturation of germ cells, such as homologous recombination. The absence of these enzymes may instead have long-term detrimental effects that can be observed only after a very significant number of generations. Their absence may result in a slow accumulation of germ cell mutations. Furthermore, the tools regularly used to detect aberrant phenotypes may be too limited to detect subtle, but functionally important deficiencies. One important question in mutagenesis and carcinogenesis is: Is there a threshold that must be exceeded before an increased mutation frequency makes somatic cells of an individual of a species "cancer prone"? Results from knockout mice in different DNA glycosylases indicate that a general increase of 1.4–3-fold is not sufficient. In fact, these numbers are far below the spontaneous mutation frequencies that have been reported for mismatch repair-deficient cells, and even in these individuals, cancers do not usually occur until adulthood. We have searched for mutations in the *UNG* gene in normal cells and in fibroblasts (over 100 different cell lines), but have not found any that abolishes or even reduces UNG activities significantly, not even heterozygotes. This indicates that mutations in the *UNG* gene are rare. Although many, if not most, mutations are "silent" in an individual, they may much more frequently be harmful from a population genetics point of view. Thus, a mutation that results in a small reduction in "fitness" (less than 10%) that would not be observed in an individual may have a large effect on the survival (or fertility) of successive generations. On average a mutation has been estimated to survive some 80 generations. Mutations are then eventually eliminated by "truncation selection" which, in one step, picks off several mutations in the same "victim" and thus contributes to maintaining the average number of mutations in the population below a certain level (reviewed in Ref. 61). In population genetics, mutations usually considered "recessive" are of major importance because they are most frequently partially dominant. Because heterozygotes are much more frequent than homozygotes, a small degree of selection against heterozygotes may be more important for mutant elimination than selection against homozygotes (reviewed in Ref. 61). It is likely that *UNG* mutations result in reduced fitness and truncation elimination, and probably after a shorter than average number of generations. If heterozygous mutations in *UNG* were completely neutral, one would expect a relatively high frequency of such mutations in the population because there would be no selection against them, but the frequency is apparently relatively low. We have found that $Ung^{+/-}$ have approximately 50% of the $Ung^{+/+}$ uracil-DNA glycosylase activity (H. Nilsen, unpublished results). Taken together, these results indicate that a 50% reduction in enzyme activity reduces fitness. Furthermore, the mutator phenotype or some other subtle phenotype in individuals carrying *UNG* mutations, however weak, would enhance the rate of accumulation of mutations and thus reduce the number of generations before the mutation is eliminated by truncation selection. Finally, it should be kept in mind that individuals who do not carry *UNG* mutations but have ancestors who carried such mutations would be likely to have higher than average number of mutations and thus be candidates for truncation selection.

In conclusion, the *UNG* gene encodes two uracil-DNA glycosylases, UNG1 and UNG2, that initiate BER in mitochondria and nuclei, respectively. These enzymes do not have a role in the common, everyday transactions in the cell and thus represent a self-protective gene, nobility that is not instantly missed when absent. The main function of these enzymes is probably the protection of the species from slow deterioration caused by spontaneous DNA decay over several generations.

ACKNOWLEDGMENTS

This work was supported by The Research Council of Norway, The Norwegian Cancer Society, The Cancer Research Fund at the Regional Hospital, Trondheim, and the Svanhild and Arne Must Fund for Medical Research.

REFERENCES

1. H. E. Krokan, R. Standal, and G. Slupphaug, *Biochem. J.* **325**, 1–16 (1997).
2. T. Lindahl, *Proc. Natl. Acad. Sci. U.S.A.* **71**, 3649–3653 (1974).
3. H. E. Krokan, R. Standal, S. Bharati, M. Otterlei, T. Haug, G. Slupphaug, and F. Skorpen, *in* "Base Excision Repair" (I. D. Hickson, ed.), pp. 7–30. R. G. Landes Company, 1997.
4. B. K. Duncan and B. Weiss, *J. Bacteriol.* **151**, 750–755 (1982).
5. K. J. Impellizzeri, B. Anderson, and P. M. Burgers, *J. Bacteriol.* **173**, 6807–6810 (1991).
6. B. K. Tye, P. O. Nyman, I. R. Lehman, S. Hochhauser, and B. Weiss, *Proc. Natl. Acad. Sci. U.S.A.* **74**, 154–157 (1977).
7. M. Otterlei, E. Warbrick, G. Slupphaug, T. Nagelhus, T. Haug, M. Akbari, P. A. Aas, K. Steinsbekk, O. Bakke, and H. E. Krokan, *EMBO J.* **18**, 3834–3844 (1999).
8. H. Nilsen, I. Rosewell, P. Robins, C. Skjelbred, S. Andersen, G. Slupphaug, H. E. Krokan, T. Lindahl, and D. E. Barnes, *Mol. Cell* **5**, 1059–1065 (2000).
9. H. Nilsen, M. Otterlei, T. Haug, K. Solum, T. A. Nagelhus, F. Skorpen, and H. E. Krokan, *Nucleic Acids Res.* **25**, 750–755 (1997).
10. L. C. Olsen, R. Aasland, C. U. Wittwer, H. E. Krokan, and D. E. Helland, *EMBO J.* **8**, 3121–3125 (1989).
11. T. Haug, F. Skorpen, K. Kvaløy, I. Eftedal, H. Lund, and H. E. Krokan, *Genomics* **36**, 408–416 (1996).
12. P. Neddermann and J. Jiricny, *Proc. Natl. Acad. Sci. U.S.A.* **91**, 1642–1646 (1994).
13. T. E. Barrett, O. D. Schärer, R. Savva, T. Brown, J. Jiricny, G. L. Verdine, and L. H. Pearl, *EMBO J.* **18**, 6599–6609 (1999).
14. S. J. Muller and S. Caradonna, *Biochim. Biophys. Acta* **1088**, 197–207 (1991).
15. K. A. Haushalter, M. W. Todd Stukenberg, M. W. Kirschner, and G. L. Verdine, *Curr. Biol.* **9**, 174–185 (1999).
16. B. Hendrich, U. Hardeland, H. H. Ng, J. Jiricny, and A. Bird, *Nature (London)* **401**, 301–304 (1999).
17. T. E. Barrett, R. Savva, G. Panayotou, T. Barlow, T. Brown, J. Jiricny, and L. H. Pearl, *Cell (Cambridge, Mass.)* **92**, 117–129 (1999).
18. S. J. Hamill, E. Cota, C. Chothia, and J. Clarke, *J. Mol. Biol.* **295**, 641–649 (2000).
19. H. E. Krokan, F. Skorpen, M. Otterlei, S. Bharati, K. Steinsbekk, H. Nilsen, T. A. Nagelhus-Hernes, T. Haug, P. A. Aas, M. Akbari, C. Skjelbred, B. Kavli, R. Standal, and G. Slupphaug, *in* "DNA Damage and Repair: Oxygen Radical Effects, Cellular Protection and Biological Consequences" (M. Dizdaroglu, ed.), pp. 221–236. Plenum Press, New York, 1999.
20. K. G. Berdal, R. F. Johansen, and E. Seeberg, *EMBO J.* **17**, 363–367 (1998).

21. Z. Hatahet, Y. W. Kow, A. A. Purmal, R. P. Cunningham, and S. S. Wallace, *J. Biol. Chem.* **269,** 18814–18820 (1994).
22. M. Dizdaroglu, A. Karakaya, P. Jaruga, G. Slupphaug, and H. E. Krokan, *Nucleic Acids Res.* **24,** 418–422 (1996).
23. P. C. Svendsen, H. A. Yee, R. J. Winkfein, and J. H. van de Sande, *Gene* **189,** 175–181 (1997).
24. H. Nilsen, K. Steinsbekk, M. Otterlei, G. Slupphaug, and H. E. Krokan, *Nucleic Acids Res.* **28,** 2277–2285 (2000).
25. C. D. Mol, A. S. Arvai, G. Slupphaug, B. Kavli, I. Alseth, H. E. Krokan, and J. A. Tainer, *Cell (Cambridge, Mass.)* **80,** 869–878 (1995).
26. G. Slupphaug, C. D. Mol, B. Kavli, A. S. Arvai, H. E. Krokan, and J. A. Tainer, *Nature (London)* **384,** 87–92 (1996).
27. S. S. Parikh, C. D. Mol, D. J. Hosfield, and J. A. Tainer, *Curr. Opin. Struct. Biol.* **9,** 37–47 (1999).
28. R. Savva, K. McAuley-Hecht, T. Brown, and L. Pearl, *Nature (London)* **373,** 487–493 (1995).
29. S. S. Parikh, C. D. Mol, G. Slupphaug, S. Bharati, H. E. Krokan, and J. A. Tainer, *EMBO J.* **17,** 5214–5226 (1998).
30. S. S. Parikh, G. Walcher, G. D. Jones, G. Slupphaug, H. E. Krokan, G. M. Blackburn, and J. A. Tainer, *Proc. Natl. Acad. Sci. U.S.A.* **97,** 5083–5088 (2000).
31. D. L. Croteau, R. H. Stierum, and V. A. Bohr, *Mutat. Res.* **434,** 137–148 (1999).
32. K. G. Pinz and D. F. Bogenhagen, *Mol. Cell. Biol.* **18,** 1257–1265 (1998).
33. S. Bharati, H. E. Krokan, L. Kristiansen, M. Otterlei, and G. Slupphaug, *Nucleic Acids Res.* **26,** 4953–4959 (1998).
34. M. Otterlei, T. Haug, T. Nagelhus, G. Slupphaug, T. Lindmo, and H. E. Krokan, *Nucleic. Acids Res.* **20,** 4611–4617 (1998).
35. W. Bandlow, G. Strobel, and R. Schricker, *Biochem. J.* **329,** 359–367 (1998).
36. P. K. Hammen and H. Weiner, *J. Exp. Zool.* **282,** 280–283 (1998).
37. T. R. Waters, P. Gallinari, J. Jiricny, and P. F. Swann, *J. Biol. Chem.* **274,** 67–74 (1999).
38. J. D. Domena, R. T. Timmer, S. A. Dicharry, and D. W. Mosbaugh, *Biochemistry* **27,** 6742–6751 (1988).
39. T. A. Nagelhus, T. Haug, K. K. Singh, K. F. Keshav, F. Skorpen, M. Otterlei, S. Bharati, T. Lindmo, S. Benichou, R. Benarous, and H. E. Krokan, *J. Biol. Chem.* **272,** 6561–6566 (1997).
40. U. Curth, C. Urbanke, J. Greipel, H. Gerberding, V. Tiranti, and M. Zeviani, *Eur. J. Biochem.* **221,** 435–443 (1994).
41. S. Muller-Weeks, B. Mastran, and S. Caradonna, *J. Biol. Chem.* **273,** 21909–21917 (1998).
42. N. C. Nicolaides, F. Palombo, K. W. Kinzler, B. Vogelstein, and J. Jiricny, *Genomics* **31,** 395–397 (1996).
43. R. Rossi, A. Villa, C. Negri, I. Scovassi, G. Ciarrocchi, G. Biamonti, and A. Montecucco, *EMBO J.* **18,** 5745–5754 (1999).
44. G. Fritz and B. Kaina, *Oncogene* **18,** 1033–1040 (1999).
45. T. Haug, F. Skorpen, P. A. Aas, V. Malm, C. Skjelbred, and H. E. Krokan, *Nucleic Acids Res.* **26,** 1449–1457 (1998).
46. S. Um, M. Harbers, A. Benecke, B. Pierrat, R. Losson, and P. Chambon, *J. Biol. Chem.* **273,** 20728–20736 (1998).
47. B. Kavli, G. Slupphaug, C. D. Mol, A. S. Arvai, S. B. Petersen, J. A. Tainer, and H. E. Krokan, *EMBO J.* **15,** 3442–3447 (1996).
47a. M. Otterlei, B. Kavli, R. Standal, C. Skjelbred, S. Bharati, and H. E. Krokan, *EMBO J.* **19,** 5542–5551 (2000).

48. B. M. Olivera, P. Manlapaz-Ramos, H. R. Warner, and B. K. Duncan, *J. Mol. Biol.* **128,** 265–275 (1979).
49. C. D. Mol, A. S. Arvai, R. J. Sanderson, G. Slupphaug, B. Kavli, H. E. Krokan, D. W. Mosbaugh, and J. A. Tainer, *Cell (Cambridge, Mass.)* **82,** 701–708 (1995).
50. H. Nishino, V. L. Buettner, J. Haavik, D. J. Schaid, and S. S. Sommer, *Environ. Mol. Mutat.* **28,** 299–312 (1996).
51. J. G. de Boer, S. Provost, N. Gorelick, K. Tindall, and B. W. Glickman, *Mutagenesis* **13,** 109–114 (1998).
52. E. Wist, O. Unhjem, and H. Krokan, *Biochim. Biophys. Acta* **520,** 253–270 (1978).
53. B. P. Engelward, G. Weeda, M. D. Wyatt, J. L. Broekhof, J. de Wit, I. Donker, J. M. Allan, B. Gold, J. H. Hoeijmakers, and L. D. Samson, *Proc. Natl. Acad. Sci. U.S.A.* **94,** 13087–13092 (1997).
54. A. Klungland, I. Rosewell, S. Hollenbach, E. Larsen, G. Daly, B. Epe, E. Seeberg, T. Lindahl, and D. E. Barnes, *Proc. Natl. Acad. Sci. U.S.A.* **96,** 13300–13005 (1999).
55. R. H. Elder, J. G. Jansen, R. J. Weeks, M. A. Willington, B. Deans, A. J. Watson, K. J Mynett, J. A. Bailey, D. P. Cooper, J. A. Rafferty, M. C. Heeran, S. W. Wijnhoven, A. A. van Zeeland, and G. P. Margison, *Mol. Cell. Biol.* **18,** 5828–5837 (1998).
56. H. Gu, J. D. Marth, P. C. Orban, H. Mossmann, and K. Rajewsky, *Science* **265,** 103–106 (1994).
57. A. W. Plug, C. A. Clairmont, E. Sapi, T. Ashley, and J. B. Sweasy, *Proc. Natl. Acad. Sci. U.S.A.* **94,** 1327–1331 (1997).
58. S. Xanthoudakis, R. J. Smeyne, J. D. Wallace, and T. Curran, *Proc. Natl. Acad. Sci. U.S.A.* **93,** 8919–8923 (1996).
59. C. A. Clairmont, L. Narayanan, K. W. Sun, P. M. Glazer, and J. B. Sweasy, *Proc. Natl. Acad. Sci. U.S.A.* **96,** 9580–9585 (1999).
60. J. A. Eisen and P. C. Hanawalt, *Mutat. Res.* **435,** 171–213 (1999).
61. J. F. Crow, *Proc. Natl. Acad. Sci. U.S.A.* **94,** 8380–8386 (1997).

Index

A

AA8 CHO cells
 8-oxoG, 101
  kinetics, 102
AAG, 306
AAG-DNA complex
 human, 307
Abasic (AP) site. See AP site
A549 cells
 DNA
  electropherogram, 144
Adenine, 209
 8-oxoguanine
  hMYH, 85–88
Adenocarcinoma
 pol beta, 21
African green monkey kidney cells, 236
Agarose gel electrophoresis
 M13mp2op14 DNA, 173
Aging, 274
 mtDNA, 295–296
 ROS, 96
AGT, 316–319
 O^6-MeG, 43–44
 pol beta-deficient cells, 52
 regulation, 47–49
Aldehyde
 structure, 8
Aldehyde-reactive probe-slot-blot method
 AP site, 62
AlkA
 Escherichia coli, 308–310
Alkali-labile sites
 comet assay, 62
Alkaline single-cell electrophoresis (SCGE), 18
Alkoxyamine-adducted site, 12
Alkyladenine DNA glycosylase
 human
  structure, 307–308
Alkyladenine glycosylase (AAG), 306
Alkylated bases
 recognition, 311–312

Alkylating agents, 30, 42
 chromatin-bound PCNA complex, 23
Alpha-hOgg1, 96
ALS, 278
Alzheimer's disease, 278
Amyotrophic lateral sclerosis (ALS), 278
ANPG, 14
APE, 42, 90, 129, 131, 287, 328–338
 adaptive response, 50
 AP lyase, 262
 BER, 46–47
 CKII, 49
 dRP lyase, 262
 mtDNA
  BER, 263–265
 posttranscriptional regulation, 49
 promoter
  CRE binding site, 50
 Xenopus, 264
APE1, 203, 329–335
 flipped-out AP-DNA, 330–331
 human, 191
 penetrating loop mutants, 332–333
AP endonucleases
 eukaryotic, 264
APEX. See APE1
Aphidicolin, 10, 11, 19
 uracil-DNA BER, 177–180
AP lyase
 AP endonuclease, 262
 HhH, 342
 Mmh mutant
  liver extracts, 113
 mouse liver crude extracts, 116
 mtDNA
  BER, 261
 mtDNA ligase, 268–269
 OGG1, 204
 reactions
  postulated intermediates, 225
 substrate specificity, 117
Apn1p, 31

Apoptosis
  DNA alkylation damage, 42–43
  mtDNA, 274
  O^6-MeG, 43–44
  tumor necrosis factor, 274
AP site, 1, 4, 30, 36, 107, 165, 209, 287
  aldehyde-reactive probe-slot-blot method, 62
  chemical structure, 8
  detection
    endo IV, 337
  DNA structure, 20
  methoxyamine-modified, 11
    structure, 8
  oxidized, 12–13
    structure, 14
  plasmid substrate, 125
  repair
    PCNA-dependent, 130–132
    pol beta, 22
    reduced, 13
Apurinic (AP) site. See AP site
Apurinic/apyrimidinic (AP) site. See AP site
Asp256, 65
Assays
  oxidative DNA damage, 140–143
Ataxia telangiectasia, 236

B

Bacillus cereus, 336
Base excision repair (BER)
  enzyme
    mitochondria, 292–293
    oxidative purine lesions, 212
    oxidative thymine lesions, 211–212
    structural implications, 299–304
    three-dimensional structures, 301
  factors
    gene expression, 21–23
  intermediates pathways
    pol beta-deficient cells, 51
  mutant strains
    menadione, 34
  replication-coupled
    DNA glycosylases, 134–137
  short-patch. See Short-patch base excision repair
  uracil, 166–187
  dUTPase, 320
  yeast, 29–38
Base excision repair (BER) pathway, 1, 33, 188, 208
  APE, 46–47
  cell cycle, 23–24
  cell-free extracts, 11–20
  complexities, 125–126
  coordination, 291
  damage type, 9–10
  defect strain
    hyper-rec phenotype, 36
  deficient cells
    hypermutable, 71
  DNA glycosylase-AP lyase, 16
  DNA ligase I, 161
  DNA ligase III, 161
  DNA structure, 20–21
  DNA substrates
    with AP site analogs, 11–12
    containing BER lesions, 12–16
  higher eukaryotes, 125
  mammalian cells, 5–7, 58
  mammalian DNA beta-polymerase, 57–72
    functional scheme, 70
  mammalian DNA ligases, 151–162
  mitochondrial
    vs. nuclear, 259–260
  monofunctional DNA glycosylase, 15
  MPG, 46
  mtDNA, 255–256
    enzymology, 259–270
    further study, 270
    mammalian, 275–276
  multiple
    switch mechanism, 3–25
  8-OH-G, 194
  oxidative base lesions, 287–288
  8-oxoguanine, 194
  PARP nuclear protein, 161–162
  PCNA
    molecular mechanism, 129–137
  permeabilized cell systems, 9–11
  pol beta, 47
    mutant cell lines, 16–20
  process, 4
  redundancy, 24
  regulation, 47–49
  steps, 146

INDEX

versatility, 24
viability, 42
yeast, 33–36
Bcl-2, 51
Benzo(a)pyrene diol epoxide (BPDE)
    pol beta-defective cells, 17
BER. *See* Base excision repair
Beta-hOGG1, 96
Beta-lyase 8-oxoG-DNA glycosylase (OGG1), 14
Beta-pol null cells, 57
Bipartite antimutagenic processing
    8-OH-G, 197–198
Bipyrimidine adducts, 30
Bleomycin, 11, 13
Bloom's syndrome, 236
BPDE
    pol beta-defective cells, 17
BrdU, 141
Bromodeoxyuridine (BrdU), 141
Budding yeast, 130

C

Cancer, 274
    ROS, 96
    TDG, 250–251
Capillary electrophoresis laser-induced fluorescence detection (CE-LIF), 141–143
    thymine glycol, 143–145
Casein kinase (CKII)
    APE, 49
CDC9 gene
    Saccharomyces cerevisiae, 151–152
CDG
    UNG expressing
        active-site mutants, 377–378
CE-LIF, 141–143
    thymine glycol, 143–145
Cell cycle
    BER pathways, 23–24
Cell-free extracts
    BER pathways, 11–20
Cell lines
    human
        hMMH type 1a protein, 112
        human OGG1/MMH type 1a protein, 110

Chinese hamster ovary (CHO) cell line
    AA8
        8-oxoG, 101, 102
        8-oxoguanine, 85–103
CHO cell line
    AA8
        8-oxoG, 101, 102
        8-oxoguanine, 85–103
Chronic progressive external ophthalmoplegia, 274
Cisplatin, 275–276
    pol beta-defective cells, 17
CKII
    APE, 49
Closed circular plasmids
    8-oxoG, 97–98
Cockayne syndrome B (CSB), 147
Colon adenocarcinoma
    LoVo cell extracts
        uracil-initiated BER, 171–175
Colon tumors
    TDG, 250–251
Comet assay, 140
    DNA damage, 62
    MMS, 62
CREB/ATF transcription factor-binding site, 48
CREB binding site
    APE promoter, 50
Crithidia fasciculata, 265
CSB, 147
Cystine-DNA glycosylase (CDG)
    UNG expressing
        active-site mutants, 377–378
Cytochrome C, 281–282
Cytosine, 209
    8-oxyguanine
        hOGG1, 83–85

D

DdTTP, 10, 19
Demyelinating polyneuropathy, 274
Denaturing polyacrylamide gel
    autoradiography, 13
2′-deoxyribose 5′-phosphatase (dRPase)
    pol-beta, 65, 67, 68
Deoxyribose phosphate (dRP), 1, 191
    MMS-induced cytotoxicity, 63–69

Deoxyribose phosphate (dRP) lyase, 266
  AP endonuclease, 262
Deoxyuridine triphosphatase.
  See Deoxyuridine triphosphate
  pyrophosphatase
Deoxyuridine triphosphate pyrophosphatase
  (dUTPase), 319–323
DHT, 214
  repair activity, 215
DHU, 198–199
Diabetes, 274
DideoxyTTP (ddTTP), 10, 19
Dihydrothymine (DHT), 214
  repair activity, 215
Dihydrouracil (DHU), 198–199
Dimethyl methanesulfonate
  O^6-MeG, 45–46
DMS, 275–276
DNA
  alkylation damage
    apoptotic cell death, 42–43
  bases
    nucleotide flipping, 333–334
  BER
    glycosylases, 323–328
    multiple pathways, 1–25
  damage
    comet assay, 62
    Fenton-type reactions, 208
    ionizing irradiation, 208
    oxidative, 140–143, 286–291, 293
    radiation-induced, 140–141
    reversal and avoidance, 316–323
    Saccharomyces cerevisiae, 37
    sequence context effects, 216–218
    sources, 140
  damage tolerance pathways, 31
    disruption
      oxidizing agent sensitization, 33
  damaging agents
    pol beta-deficient mouse cells, 17
  defined oxidative base lesions,
    208–210
  endonuclease
    eukaryotic, 268
  glycosylase deficient
    knockout mice, 381–382
  ligase-associated proteins, 155–158
  ligase-deficient cell lines
    phenotype, 158–160
  mitochondrial. *See* Mitochondrial DNA
  nuclear
    oxidative DNA damage, 286–291
  plasmid
    AA8 cells, 97–98
    RR, 31
  structure
    AP sites, 20
    BER pathways, 20–21
  substrates
    with AP site analogs, 11–12
    containing BER lesions, 12–16
    oxidative base lesions, 209
  synthesis, 57
  uracil, 290
DNA glycosylase, 4, 14, 24, 350
  classes, 189–190
  DNA long-term maintenance, 382–384
  duplex DNA, 189
  features, 189–190
  mammalian genome
    evolution, 381–382
  molecular mass, 190
  monofunctional
    BER, 15
  mtDNA, 261–263
  multiple enzymes, 191
  reaction mechanisms, 352–353
  replication-coupled BER, 134–137
  substrate bases, 190
  substrates, 13–16
  uracil
    conserved gene family, 368–369
DNA-glycosylase 3-methyladenine DNA
  glycosylase (ANPG), 14
DNA ligase, 129
  mammalian
    BER, 151–162
    encoding, 152
DNA ligase I, 131, 155–157, 250,
    267–268
  BER, 161
  deficient cell lines, 158–159
  DNA replication, 153–154
  one-nucleotide gap, 157
  PCNA, 132, 156
DNA ligase III, 90, 154, 157–158, 250,
    267–268
  BER, 161
  deficient cell lines, 159

INDEX   391

DNA ligase IV, 155, 158, 267–268
   deficient cell lines, 159–160
   Saccharomyces cerevisiae, 150
DNA mismatch repair
   uracil-initiated BER
      LoVo cells, 182–183
DNA-N-glycosylases
   human, 134
DNA polymerase
   inhibitors
      permeable mammalian cell systems, 10
DNA polymerase beta (pol beta), 4, 24, 42, 58, 125, 129
   adenocarcinoma, 21
   AP site repair, 22
   BER, 47
   deficient cells
      alternative repair pathway, 50–51
      BER intermediates pathways, 51
      MGMT, 52
   deficient mouse cells
      DNA-damaging agents, 17
   deleted mouse cell extracts
      AP site repair
         pol beta, 22
   dependent pathway, 11
   dRPlase, 65, 67, 68
   inhibition, 10
   long-patch BER, 289–291
   mammalian
      BER, 57–72
      mutagenesis, 69–71
      POL3 mutants, 21
   MMS, 58–59
   mtDNA
      BER, 256–266
   mutant cell lines
      BER pathways, 16–20
   mutant mouse cell lines, 16–20
   null cells
      MMS, 61
   null genotype
      mammalian cells, 59–62
   null mouse fibroblasts
      POL3 mutants, 21
   polymerase active site, 63
   quiescent cells, 23
   recombinant, 60
   short-patch BER, 289–291
   yeast, 21

DNA polymerase delta (pol delta), 129, 165
DNA polymerase epsilon (pol epsilon), 165
DNA polymerase gamma (pol gamma), 90, 265–266
DNA single-strand breaks
   factories, 25
   induction, 18
DNL4 gene
   Saccharomyces cerevisiae, 151–152
Domain-dependent repair
   nucleotide excision repair, 216
Double pathway defect mutants
   spontaneous mutator phenotypes, 33–36
Drosophila melanogaster, 248–249, 368–369
   TDG homologs, 237
DRP, 1, 191
   MMS-induced cytotoxicity, 63–69
DRPase
   pol-beta, 65, 67, 68
DRP lyase, 266
   AP endonuclease, 262
DUTP
   recognition, 322
DUTPase, 319–323
   beta hairpin, 321
   biological function, 319
   biological trimer, 322
   uracil BER, 320
DUTP nucleotidohydrolase. *See* Deoxyuridine triphosphate pyrophosphatase

E

Electrophoretic mobility shift assay (EMSA)
   TDG, 242
EMS
   pol beta-defective cells, 17
EMSA
   TDG, 242
Endo G
   mitochondrial, 295–296
Endonuclease, 12
   AP
      eukaryotic, 264
   DNA
      eukaryotic, 268
Endonuclease G (Endo G)
   mitochondrial, 295–296

Endonuclease III, 190, 214, 223
  alpha complex, 232
  beta complex, 232
  HhH enzyme family, 323–324
  homologs
    substrate specificity, 210–215
  protein, 292–293
  substrate bases, 190
  Tg, 216–218
  urea, 216–218
Endonuclease IV, 214, 335–338
  AP site detection, 337
  crystal structure, 336
  structure, 334
Endonuclease VIII, 190
  E. coli, 198–199
  repair activity, 215
  substrate specificity, 199
Enzymatic method
  vs. phosphoramidite method, 210
Escherichia coli
  AlkA, 308–310
  cell free extracts
    uracil-initiated BER, 183
  DNA repair, 9
  fpg, 223–233
    cleavage activity, 228
    enzymes, 228
    oligonucleotides, 226–227
    pre-steady-state burst kinetics, 231
    results, 228–231
    substrates, 226–227
  MuT
    C-terminal region, 87
    secondary structure, 82
  mutator mutants, 76
  MutM
    8-OH-G, 107
  MutY
    double-flipping, 361–362
  nei, 198–199
  OGG2, 198–199
  Tag, 191
Ethenoadenine
  mitochondrial extracts, 293–294
Ethenocytosine, 244, 249–250
Ethyl methanesulfonate (EMS)
  pol beta-defective cells, 17
Eukaryotic AP endonucleases, 264
Eukaryotic DNA endonucleases, 268

F

Family-A DNA polymerase, 266
FEN1. *See* Flap endonuclease 1
Fenton-type reactions
  DNA damage, 208
Flap endonuclease 1 (FEN1), 125, 129, 131, 146–147, 287
  long-patch BER, 338–342
  PCNA, 132
  structure, 340
5-fluorouracil (5-FU)
  recombinant human TDG, 245
Formamidopyrimidine N-glycosylase (fpg), 77, 232
  beta-elimination products, 228–230, 233
  Escherichia coli, 223–233
  homologs, 215–216
  K57A mutant
    oligonucleotides, 231
  K155A mutant
    oligonucleotides, 231
    transient kinetics assay, 231
  oxidative pyrimidine products, 223
  protein, 77
5-formyluracil, 208
Fpg. *See* Formamidopyrimidine N-glycosylase
5-FU
  recombinant human TDG, 245

G

Gamma-irradiated human cells
  thymine glycol, 139–148
Gamma-LIZ shuttle vector, 69
Gap-filling synthesis, 156
Gas chromatography/mass spectrometry (GC-MS), 140
GC-MS, 140
Gene expression
  BER factors, 21–23
Genetic instability phenotype, 37
Glial cells, 280
Glycosylase
  DNA. *See* DNA glycosylase
Glycosylase/AP lyases
  reaction mechanisms, 352–353

Glycosylase step
  assay, 147
Glycosylic bond cleavage
  catalysis, 312–313
GO system, 194–195
$Gpt/Mmh^{-/-}$ (O) mouse liver cells
  mutation frequency, 119
$Gpt/Mmh^{+/+}$ (W) mouse liver cells
  mutation frequency, 119
G-T mismatch correction
  TDG, 247–248
Guanine, 209

H

Handoff model base excision repair, 203–204
HAP1, 24, 243. *See also* APE1
HAPE1, 191
HeCNU
  pol beta-defective cells, 17
HeLa S3 cells
  hMMH type 1a protein
    depletion, 110–112
    depletion analysis, 112
    detection, 111
Helical clamp, 341
Helix-hairpin-helix (HhH), 353–354
  AP lyases, 342
  glycosylases, 342
Helix-loop-helix (HLH), 376–377
Helix-3turn-helix (H3TH), 339–340
Hemiacetal
  structure, 8
Hereditary nonpolyposis colon cancer (HNPCC)
  TDG, 251
HhH, 353–354
  AP lyases, 342
  glycosylases, 342
Higher eukaryotes
  BER pathways, 125
HLH, 376–377
HMMH type 1a protein, 107
  depletion
    whole HeLa S3 cell extract, 110–112
  detection
    HeLa S3 cell, 111

HeLa S3 cells
  depletion analysis, 112
  human cell lines, 112
HMTH1
  amino acid residues, 80
  C-terminal region
    sequence alignment, 87
  mRNA
    SNP, 79–80
  2-OH-dATP, 91
  8-oxoG, 91
  protein
    hypothetical representation, 90
    purine nucleoside triphosphates, 75
  secondary structure, 82
HMTH1 gene, 77–82
  mRNA, 77–79
HMYH, 88–90
  C-terminal region
    sequence alignment, 87
  2-hydroxyadenine, 85–88
  imuunological detection, 87
  8-oxoguanine
    adenine, 85–88
HMYH gene
  genome structure
    schematic representation, 89
  RT-PCR, 86–87
HMYH protein
  substrate specificity, 86
HNPCC
  TDG, 251
HOGG1
  MMH type 1a protein
    human cell lines, 110
    8-OH-G, 110–112
  multiple mRNA species, 196
  8-oxoG, 91
  8-oxyguanine
    cytosine, 83–85
  polypeptides
    MTS, 83
HOGG1-2a, 88–90
  C-terminal region, 83–85
HOGG1 gene, 96
  exons, 83
  genome structure, 84
HPLC-electrochemical detection, 140
H3TH, 339–340

Human AAG-DNA complex, 307
Human alkyladenine DNA glycosylase
  structure, 307–308
Human APE 1(hAPE1), 191
Human cell lines
  hMMH type 1a protein, 112
  human OGG1/MMH type 1a protein, 110
Human colon adenocarcinoma
  LoVo cell extracts
    uracil-initiated BER, 171–175
Human DNA-N-glycosylases, 134
Human lymphoblastoid cell lines
  mitochondrial extracts
    8-oxoG lesion, 294
  oxidative DNA damage repair
    mitochondria, 293
Human proteins
  PCNA-binding motif, 133
Human U251
  uracil-initiated BER, 174
    mutation frequency, 182
Human UNG2
  dUMP residues, 379–380
Human uracil-DNA glycosylase (UDG), 4
  UNG gene, 365–384
  HUNG gene, 370–371
HX, 14
HXTH2, 264
Hydrogen peroxide, 18, 33–34, 76
  chromatin-bound PCNA complex, 23
  DNA single-strand breaks, 19
  endonucleases, 214
  pol beta-defective cells, 48
  yeast, 31
2-hydroxyadenine
  hMYH, 85–88
5-hydroxycytosine, 210
Hydroxylethyl-N-chloroethylnitrosourea
  (HeCNU)
  pol beta-defective cells, 17
Hydroxyl radicals, 76
2-hydroxy (OH)-dATP, 88, 90
  hMTH1, 75, 91
5-hydroxyuracil, 210
Hyperrecombination phenotypes
  BER pathway defect strain, 36
  single and double pathway defect mutants,
    33–36
Hypoxanthine (HX), 14
  mitochondrial extracts, 293–294

I

Immunoassays, 140
  thymine glycols, 126
Ionizing radiation
  DNA damage, 208
  DNA lesions, 140
  mutagenic effects, 139
  pol beta-defective cells, 17, 48
IPTG, 136
Ischemic heart disease, 274
Isoguanine. *See* 2-OH-A
Isopropyl beta-D-thiogalactopyranoside
  (IPTG), 136

K

K57A mutant fpg
  oligonucleotides, 231
K155A mutant fpg
  oligonucleotides, 231
  proteins
    transient kinetics assay, 231
K155A protein, 228
Kearns-Sayre syndrome, 274
Klenow fragment, 108–109
Knockout mice
  DNA glycosylase deficient,
    381–382

L

LacZalpha DNA reversion assay
  uracil-initiated BER, 167–169
Leber's hereditary optic neuropathy, 274
Ligase
  mitochondrial, 267–269
LIG1 gene, 126, 151–154
LIG3 gene, 126, 151, 154–155
LIG4 gene, 126, 151, 155, 159–160
Long-patch base excision repair, 6–9, 12, 15,
  23, 146, 160–161, 167, 291
  FEN-1, 338–342
  mtDNA, 287–289
  pol beta, 289–291
  uracil-DNA glycolyses, 289
  vs. single-nucleotide BER, 58
LoVo cell extracts

INDEX

human colon adenocarcinoma
  uracil-initiated BER, 171–175
patch size
  uracil-DNA BER, 175–181
  uracil-initiated BER
    mutation frequency, 182
LoVo cells
  DNA mismatch repair
    uracil-initiated BER, 182–183
Lymphoblastoid cell lines
  mitochondrial extracts
    8-oxoG lesion, 294
  oxidative DNA damage repair
    mitochondria, 293
Lysine 57, 226
Lysine 155, 226, 231

M

Mafosfamide
  pol beta-defective cells, 17
MAG
  substrate bases, 190
Mammalian cells
  BER pathway, 5–7, 58
  DNA ligase genes, 126
  error-avoidance mechanisms, 77
  mtDNA, 291–294
  8-OH-G, 109
  pol-beta null genotype, 59–62
Mammalian DNA beta-polymerase
  BER, 57–72
  mutagenesis, 69–71
Mammalian DNA ligases
  BER, 151–162
  encoding, 152
Mammalian MMH
  8-OH-G, 119–120
Mammalian $Ogg1/Mmh$ gene
  8-OH-G, 107–121
Mammalian pol beta
  POL3 mutants, 21
MED1/MBD4, 134
Melphalan
  pol beta-defective cells, 17
Menadione, 31, 33–34, 280–281
Methanococcus jannaschii, 339
Methoxyamine-modified AP site,
  11–12

repair, 13
structure, 8
3-methyladenine DNA glycosylase, 37
Methylmethanesulfonate (MMS), 18,
  42, 57
comet assay, 62
cytotoxicity, 58–59
5′-deoxyribose phosphate, 63–69
DNA ligase, 160
DNA single-strand breaks, 19
fibroblast cell lines
  cytotoxic sensitivity, 64
  mutations, 69–70
O^6-MeG, 45–46
pol beta, 58–59
  defective cells, 16–17, 48
  null cells, 61
Methylnitrosourea (MNU), 10, 11, 275–276,
  280
  pol beta-defective cells, 17
Methyl-N′-nitro′N-nitrosoguanidine (MNNG),
  10
O^6-MeG, 44
  pol beta-defective cells, 17
Methylpurine-DNA glycosylase (MPG), 42,
  134, 307–308
  BER, 46
  substrate bases, 190
MGMT, 41. See O^6-alkylguanine-DNA
  alkyltransferase
Mismatched uracil
  TDG, 248–249
Mismatch repair (MMR), 41, 96
  O^6-MeG, 43–45
  regulation, 47–49
Mismatch-specific uracil DNA glycosylase
  (Mug), 237
Mitochondria
  BER enzymes, 292–293
  oxidative DNA damage repair
    human lymphoblastoid cell lines, 293
  oxidative phophorylation, 76
  Xenopus laevis, 292–293
Mitochondrial DNA (mtDNA)
  aging, 295–296
  apoptosis, 274
  BER, 255–256
    AP endonuclease, 263–265
    DNA glycosylases, 261–263
    enzymology, 259–270

Mitochondrial DNA (*Cont.*)
 further study, 270
 mammalian, 275–276
 damage, 258–259
  incidence, 258–259
  lifetime, 258
 ligase
  AP lyase, 268–269
  long-patch BER, 287–289
  mammalian cells, 291–294
  NO-induced damage, 276–279
  nucleotides, 277–279
  oxidative damage, 276
  point mutations, 258
  progressive cell death, 274
 repair
  cell-specific differences, 279–282
  enzymes, 261–270
  mechanisms, 279
Mitochondrial endonuclease G, 295–296
Mitochondrial extracts
 ethenoadenine, 293–294
 hypoxanthine, 293–294
Mitochondrial ligase, 267–269
 identification, 267–268
Mitochondrial localization signal (MLS), 257
Mitochondrial short-patch base excision repair
 common pathway, 260
Mitochondrial targeting signal (MTS), 80
 hOGG1 polypeptides, 83
Mitochondrial UNG1, 372–374
Mitochondrial uracil DNA glycosylase (mtUDG), 295–296
Mitomycin C
 pol beta-defective cells, 17
MLS, 257
MMH
 mammalian
  8-OH-G, 119–120
Mmh deficient mice
 8-OH-G, 118
Mmh homozygous mutations, 121
Mmh knockout mice
 generation, 113
 8-OH-G, 113–119
  Mmh mutant, 113
Mmh mutant
 AP lyase
  liver extracts, 113

 mutation frequency, 115–119
 8-OH-G, 115
 targeted disruption, 115
M13mp2*lacZ alpha*
 DNA-based reversion assay, 168, 181–184
M13mp2op14 DNA
 agarose gel electrophoresis, 173
MMR, 41, 96
 O^6-MeG, 43–45
 regulation, 47–49
MMS. *See* Methylmethanesulfonate
MNHT1
 Tg, 216–218
 urea, 216–218
MNNG, 10
 O^6-MeG, 44
 pol beta-defective cells, 17
MNU, 10, 11, 275–276, 280
 pol beta-defective cells, 17
MOGG1, 215–216
Monofunctional DNA glycosylase
 BER, 15
MPG, 42, 134, 307–308
 BER, 46
 substrate bases, 190
MRNA
 hMTH1
  SNP, 79–80
 hMTH1 gene, 77–79
MtDNA. *See* Mitochondrial DNA
MTH, 134
MTH1a protein
 mitochondrial targeting sequences, 81
MTH1d protein
 mitochondrial targeting sequences, 81
 submitochondrial localization, 79
MTH1 gene
 SNP, 75
MTS, 80
 hOGG1 polypeptides, 83
MtUDG, 295–296
Mug, 237, 249–250
 base flipping, 240
 crystal structure, 240
 UDG, 240
Multiple base excision repair pathways
 switch mechanism, 3–25
 control, 9–24

Multiple DNA repair pathways
  disruption
    oxidizing agent sensitization, 33
Multiple mRNA species
  hOGG1, 196
Murine UNG2
  dUMP residues, 379–380
  uracil, 380–381
Murine *UNG* gene, 370–371
MuT
  amino acid residues, 80
  Escherichia coli
    C-terminal region, 87
    secondary structure, 82
Mutagenesis
  mammalian DNA beta-polymerase, 69–71
Mutation
  frequency
    Mmh mutant mice, 115–119
    Mmh homozygous, 121
    MMS, 69–70
  rates
    spontaneous recombination, 35
    uracil-initiated BER, 181–186
Mutator mutants
  E. coli, 76
Mutator phenotype, 36–37
MutM, 194–197. *See also* OGG1
  Escherichia coli
    8-OH-G, 107
  Fapy-DNA glycosylase, 190
  substrate bases, 190
  substrate specificity, 196, 199
  yeast, 195
MutT
  homolog proteins
    primary structure, 82
MutY, 350–352
  active site, 355
  catalytic domain
    structure, 353–354
  c-terminal domain
    solution structure, 357–360
    substrate specificity, 356–357
  Escherichia coli
    double-flipping, 361–362
  glycosylase reaction mechanism, 356
  OGG, 200–201
  opportunistic lyase activity, 354–355
  8-oxoG, 350–352

p13, 360
structural model, 361
substrate bases, 190
MutY gene, 109
MYH, 134
  replication-coupled BER, 136
  UDG2
    PCNA, 135
MYH protein, 129
  hypothetical representation, 90
  mitochondrial targeting sequences, 81
  submitochondrial localization, 79

N

Nei. *See* Endonuclease VIII
NER. *See* Nucleotide excision repair
Nicking assay
  TDG, 242
Nitric oxide (NO)
  induced damage
    mtDNA, 276–279
Nitrogen mustard, 275–276
NO
  induced damage
    mtDNA, 276–279
Ntg1, 293
Ntg2, 293
Ntg1p, 30, 31
Ntg2p, 30, 31
Nth. *See* Endonuclease III
Nuclear DNA
  oxidative DNA damage, 286–291
Nuclear short-patch base excision repair
  common pathway, 260
Nuclear UNG1, 374
Nucleotide
  flipping
    DNA bases, 333–334
    mtDNA, 277–279
Nucleotide excision repair (NER), 33, 96
  domain-dependent repair, 216
  mutant strains
    menadione, 34
  oxidative base lesions, 287–288
  sequence context-dependent repair, 216
Nucleotide excision repair (NER) pathway, 30–31
  yeast, 33–36

O

O^6-alkylguanine-DNA alkyltransferase (AGT), 316–319
  O^6-MeG, 43–44
  pol beta-deficient cells, 52
  regulation, 47–49
OGG, 193, 194
  mtDNA
    BER, 261
    MutY, 200–201
    substrate bases, 190
OGG1, 14, 134, 193. *See also* MutM
  AP lyase, 204
  apurinic sites
    assay, 98
    incised apurinic sites
      AA8 CHO cells, 102–103
    8-OH-G, 203–204
  Saccharomyces cerevisiae, 107
OGG2, 193–194
  E. coli, 198–199
  FPLC, 196
  8-OH-G, 199–200
  substrate specificity, 196
  transcription-coupled repair, 202
  yeast, 196
OGG1-2a protein
  mitochondrial targeting sequences, 81
  submitochondrial localization, 79
OGG1 gene
  Saccharomyces cerevisiae, 109
OGG1 knockout mutant mice, 202
OGG1/Mmh gene
  mammalian
    8-OH-G, 107–121
OGG1 protein
  hypothetical representation, 90
8-OH-G. *See* 8-oxoguanine
8-OH-G:FPG
  repair, 109
Oligonucleotides
  K57A mutant fpg, 231
  K155A mutant fpg, 231
O^6-MeG. *See* O^6-methylguanine
O^6-methylguanine-DNA methyltransferase.
  See O^6-alkylguanine-DNA alkyltransferase
O^6-methylguanine-DNA methyl transferase (MGMT), 41
O^6-methylguanine (O^6-MeG), 43–44

apoptosis, 43–44
dimethyl methanesulfonate, 45–46
genotoxicity, 45–46
methyl methanesulfonate, 45–46
MGMT, 43–44
MMR, 43–45
MNNG, 44
Oxidative base lesions
  BER, 287–288
  DNA substrates, 209
  NER, 287–288
Oxidative DNA damage
  assays, 140–143
  nuclear DNA, 286–291
Oxidative DNA damage repair
  mitochondria
    human lymphoblastoid cell lines, 293
Oxidative purine lesions
  BER enzyme, 212
Oxidative pyrimidine products
  fpg, 223
Oxidative thymine lesions
  BER enzymes, 211–212
Oxidized AP sites, 12–13
  structure, 14
Oxidizing agents, 30
  wild-type sensitivity
    yeast BER mutants, 31–32
8-oxo-G. *See* 8-oxoguanine
8-oxoguanine-DNA glycosylase (OGG), 193, 194
  mtDNA
    BER, 261
    MutY, 200–201
    substrate bases, 190
8-oxoguanine (8-OH-G), 14, 55, 88, 96, 107, 193, 287–288
  AA8 CHO cells, 98–103, 101
    kinetics, 102
    transcription-coupled repair, 100–102
  adenine
    hMYH, 85–88
  antimutagenic processing, 194–195
  BER, 98–100
  BER pathway, 194
  bipartite antimutagenic processing, 197–198
  cell line and culture conditions, 97
  CHO cell line, 85–103
  closed circular plasmids, 97–98
  Escherichia coli MutM, 107

hMTH1, 91
  dATP, 75
  dGTP, 75
hMYH, 91
hOGG1, 91
  cytosine, 83–85
  HPLC, 108
  human OGG1/MMH type 1a protein, 110–112
  mammalian cells, 109
  mammalian MMH, 119–120
  Mmh deficient mice, 118
  Mmh knockout, 113–119
  Mmh mutant mice, 115
  mutagenesis, 96
  MutY, 350–352
  OGG1, 203–204
  OGG2, 199–200
  plasmid DNA
    AA8 cells, 97–98
    potassium bromate, 108
  ROS, 194
  short-patch BER, 99

P

P13
  MutY, 360
Parkinson's disease, 274, 278
PARP nuclear protein
  BER, 161–162
Patch size
  uracil-DNA BER
    LoVo cell extracts, 175–181
PCNA. *See* Proliferating cell nuclear antigen
PCR-based assays, 140
Pearson's syndrome, 274
Penetrating loop mutants
  APE1, 332–333
Penicillium citrinum, 336
Permeable mammalian cell systems
  DNA repair
    DNA polymerase inhibitors, 10
Phosphodiester bond
  formation, 152
  hydrolysis, 337–338
Phospholipase C, 336
Phosphoramidite method, 208
  vs. enzymatic method, 210

Pinch-push-pull model
  UDG, 325
Plasmid DNA
  AA8 cells
    8-oxoG, 97–98
Plasmid substrate
  Ap site, 125
P26MutY
  structure, 354
Point mutations
  mtDNA, 258
Pol beta. *See* DNA polymerase beta
Pol delta, 129, 165
Pol epsilon, 165
Pol gamma, 90, 265–266
POL3 mutants
  mammalian pol beta, 21
Potassium bromate
  8-OH-G, 108
^{32}P-postlabeling, 140
P21 protein
  PCNA, 132–134
Progressive cell death
  mtDNA, 274
Proliferating cell nuclear antigen (PCNA), 8, 25, 58, 125, 129, 147, 287, 339
  BER
    molecular mechanism, 129–137
  binding motif
    human proteins, 133
  dependent AP site repair, 130–132
  dependent BER, 8–9, 11, 129
  DNA glycosylases
    replication-coupled BER, 134–137
  DNA ligase I, 132, 156
  FEN1, 132
  MYH
    UDG2, 135
  p21 protein, 132–134
  proteins, 132–134
Proteins
  human
    PCNA-binding motif, 133
Pyr, 307–308
Pyrimidine bases
  TDG, 249–250
Pyrimidines
  UNG mutants, 377–378
Pyrococcus furiosus, 339
Pyrrolidine abasic nucleotide (pyr), 307–308

Q

Quiescent cells
  pol beta, 23

R

Radiation-induced DNA damage
  measurement
    problems, 140–141
Reactive nitrogen species (RNS), 278
Reactive oxygen species (ROS), 12–13, 30, 75, 76, 208
  aging, 96
  carcinogenesis, 96
  8-OH-G, 194
Recombinant human thymine-DNA glycosylase
  5-FU, 245
Recombinant pol beta, 60
Recombinational repair (RR)
  disruption
    oxidizing agent sensitization, 33
    DNA, 31
  mutant strains
    menadione, 34
Recombinational repair (RR) pathways
  yeast, 33–36
Reduced AP site
  repair, 13
  structure, 8
Ref-1. *See* APE
Repair patch size, 132
Replication-coupled base excision repair
  DNA glycosylases
    PCNA, 134–137
    MYH, 136
    UDG2, 136
Replication factor C (RF-C), 8, 125, 129, 131
Replication protein A (RPA), 58, 125, 370
RF-C, 8, 125, 129, 131
RNS, 278
ROS, 12–13, 30, 75, 76, 208
  aging, 96
  carcinogenesis, 96
  8-OH-G, 194
RPA, 58, 125, 370
RR
  disruption
    oxidizing agent sensitization, 33
    DNA, 31
  mutant strains
    menadione, 34

S

Saccharomyces cerevisiae, 30, 214
  CDC9 gene, 151–152
  DNA damage
    processing, 37
  DNA ligase, 153
  DNA ligase IV, 150
  DNL4 gene, 151–152
  homologs, 30
  hydrogen peroxide, 32
  menadione, 32
  OGG1, 107
  OGG1 gene, 96, 109
ScAPN2, 264
SCGE, 18
Schizosaccharomyces pombe, 214
  DNA ligase, 153
  TDG homologs, 237
Sequence context-dependent repair
  nucleotide excision repair, 216
Short-patch base excision repair, 5, 15, 147, 160–161, 167
  nuclear
    common pathway, 260
  8-oxoG, 99
  pol beta, 265, 289–291
  uracil, 294
Short-patch mismatch
  thymine DNA glycosylase, 236–237
Single-nucleotide base excision repair
  vs. long-patch BER, 58
Single-nucleotide gap, 58
Single-nucleotide patch repair, 291
Single-nucleotide polymorphism (SNP)
  hMTH1 mRNA, 79–80
  MTH1 gene, 75
Single pathway defect mutants
  spontaneous mutator phenotypes, 33–36
Single-strand breaks (ssb)
  factories, 25
  genomic
    comet assay, 62
    induction, 18

INDEX 401

SNP
   hMTH1 mRNA, 79–80
   MTH1 gene, 75
Spontaneous mutator phenotypes
   single and double pathway defect mutants, 33–36
Spontaneous recombination
   mutation rates, 35
Ssb
   factories, 25
   genomic
      comet assay, 62
   induction, 18
Strand break-based assays, 140
Superoxide, 76
Superoxide dismutase, 76
SV40
   DNA
      transfection, 236
Switch mechanism
   multiple BER pathways, 3–25

T

Tag
   Escherichia coli, 191
TCR, 95
   OGG2, 202
   8-oxoG
      AA8 CHO cells, 100–102
TDG. *See* Thymine-DNA glycosylase
Tetrahydrofuran, 11
   structure, 8
Thermus aquaticus, 339
Thymine, 209
Thymine-DNA glycosylase (TDG), 134, 191, 235–251, 377–378
   biochemistry, 241–246
   biology, 246–251
   damaged pyrimidine bases, 249–250
   EMSA, 242
   enzymatic activities, 241–243
   evolutionary conservation, 238–239
   G-T mismatch correction, 247–248
   HNPCC, 251
   human cancer, 250–251
   minimal catalytic domain, 245
   mismatched uracil, 248–249
   nicking assay, 242

   physiological role, 251
   protein
      structure, 237–238
   recombinant human
      5-FU, 245
   short-patch mismatch, 236–237
   substrate specificity, 243–246
   three-dimensional structure, 239–247
   transcription-associated DNA repair, 250
Thymine glycol (Tg), 214, 287–288
   A549 cells
      kinetics, 146
   BER, 4
   CE-LIF, 143–145
   DNA
      human cells, 147
   endo III, 216–218
   gamma-irradiated human cells, 139–148
   immunoassay, 126
   measurement, 141, 144
   mNHT1, 216–218
   proliferative status, 148
   repair activity, 215
   structure, 140
TIM, 335
TLS mutant strains
   menadione, 34
TLS pathways
   yeast, 33–36
Transcription-associated DNA repair
   TDG, 250
Transcription-coupled repair (TCR), 95
   OGG2, 202
   8-oxoG
      AA8 CHO cells, 100–102
Triose phosphate isomerase (TIM), 335
Tumor necrosis factor
   apoptosis, 274

U

U251
   human
      uracil-initiated BER, 174, 182
UDG. *See* Uracil DNA glycosylase
UDG2, 23, 129, 134
   MYH
      PCNA, 135
      replication-coupled BER, 136

Ugi
  uracil-DNA BER, 179–181
Ugi-resistant uracil-DNA glycosylase-initiated DNA repair, 185–186
  human colon adenocarcinoma LoVo cell extracts, 172–175
Ultraviolet radiation, 275–276
  pol beta-defective cells, 17
UNG
  CDG
    active-site mutants, 377–378
UNG1
  nuclear, 374
UNG2
  human
    dUMP residues, 379–380
  phosphorylation, 375
UNG gene
  chromosomal localization, 371–372
  expression, 376–377
  human uracil-DNA glycosylase, 365–384
  promoters, 376–377
UNG mutants
  pyrimidines, 377–378
Uracil, 126, 135
  BER
    dUTPase, 320
  DNA, 290
  mismatched
    TDG, 248–249
  murine UNG2, 380–381
  short-patch BER, 294
Uracil DNA glycosylase (UDG), 136, 189, 323–324, 324–328
  conserved gene family, 368–369
  discovery, 366–367
  DNA repair
    human colon adenocarcinoma LoVo cell extracts, 171–172
  enzyme-induced distortions, 326
  functions, 367–368
  human, 4
    UNG gene, 365–384
  long-patch BER, 289
  mammalian activities, 368
  mtDNA
    BER, 261
  Mug, 240
  pinch-push-pull model, 325
  substrate bases, 190

UNG gene, 365–384
Uracil glycol, 210
Uracil-initiated base excision repair, 166–187
  aphidicolin, 177–180
  DNA mismatch repair
    LoVo cells, 182–183
  E. coli BH156, 183
  error frequency, 186
  Escherichia coli cell free extracts
    mutation frequency, 183
  human colon adenocarcinoma LoVo cell extracts, 171–175
  human U251, 174
  lacZalpha DNA reversion assay, 167–169
  measurement, 170
  mutational spectrum, 184–186
  mutation frequency
    human U251 cell extracts, 182
    LoVo cell extracts, 182
  mutations, 181–186
  nucleotide incorporation, 168, 169–171
  patch size
    LoVo cell extracts, 175–181
  Ugi, 179–181
Urea
  endo III, 216–218
  mNHT1, 216–218

W

Wild-type sensitivity
  yeast BER mutants
    oxidizing agents, 31–32

X

Xenopus
  APE, 264
  mitochondrial enzyme, 263
Xenopus laevis, 11, 130
  mitochondria, 292–293
Xeroderma pigmentosum (XP), 236, 279
XP complementation group A (XP-A), 279
XP complementation group D (XP-D), 279
XPG protein, 147
X-rays
  endonucleases, 214

XRCC1 gene, 157–158
XRCC4 gene, 159

Y

Yeast
  hydrogen peroxide, 31
  MutM, 195
  OGG2, 196
  pol beta, 21
  recombination, 30–31
Yeast base excision repair, 29–38
  characteristics, 36
  network, 36–38
  oxidizing agents, 31–32
    sensitize, 33
  phenotypes, 33–36
  wild-type sensitivity
    oxidizing agents, 31–32
YOGG1, 215–216

ISBN 0-12-540068-3